HANDBOOK OF
MAGNETIC
MEASUREMENTS

Series in Sensors

Series Editors: Barry Jones and Haiying Huang

Other recent books in the series:

HANDBOOK OF MAGNETIC MEASUREMENTS

S Tumanski

Warsaw University of Technology
Poland

CRC Press
Taylor & Francis Group
Boca Raton London New York

CRC Press is an imprint of the
Taylor & Francis Group, an **informa** business
A TAYLOR & FRANCIS BOOK

CRC Press
Taylor & Francis Group
6000 Broken Sound Parkway NW, Suite 300
Boca Raton, FL 33487-2742

First issued in paperback 2019

© 2011 by Taylor & Francis Group, LLC
CRC Press is an imprint of Taylor & Francis Group, an Informa business

No claim to original U.S. Government works

ISBN-13: 978-1-4398-2951-6 (hbk)
ISBN-13: 978-0-367-86495-8 (pbk)

Visit the Taylor & Francis Web site at
http://www.taylorandfrancis.com

and the CRC Press Web site at
http://www.crcpress.com

Contents

Preface

This book is a continuation of my previous book entitled *Principles of Electrical Measurement*. However, the challenge this time was much more difficult. Electrical measurements are quite well recognized—there is a lot of good literature on this subject. In the case of magnetic measurements, the availability of up-to-date books is limited, and in some parts I had to compile the state-of-the-art knowledge from dispersed specialized papers.

Of course, there are several excellent books and review papers concerning magnetic measurements. For many years, we had our "magnetic" bibles, such as *Ferromagnetism* by Bozorth (first published in 1951), *Introduction to Magnetic Materials* by Cullity (1972*), and *Experimental Methods in Magnetism* by Zijlstra (1967). Today, however, these books are reprinted mainly as historical headstones (maybe also due to the lack of more up-to-date alternatives).

Two modern, excellent books on the subject can be recommended as additional readings: *Measurement and Characterization of Magnetic Materials* by Fiorillo and *Magnetic Sensors* edited by Ripka. Both these books, however, present rather a relatively narrow part of the wide spectrum of problems related to magnetic measurements.

In my university, the lecture on magnetic measurements is not obligatory. And, indeed, it is very difficult to collect a sufficient number of students ready to voluntarily study this subject in more detail. Most of them are prejudiced against magnetic measurements after the previous (compulsory) lecture on the fundamentals of the theory of electricity. In their opinion, magnetic measurements are exceptionally difficult and full of extremely complex mathematics. Indeed, the Maxwell rules; complicated terms such as gradient, divergence, and curl; and the theory of electromagnetic field all require exceptional intellectual effort. For this reason, many students fail to realize that once these difficulties are overcome, they would be able to recognize the new fascinating world of science and technology. As a result, the general knowledge of magnetism is rather poor among students and, as a consequence, practicing engineers.

Recently I reviewed a doctorate thesis on power electronics in which the magnetic part played quite an important practical role. The PhD student had concluded this subject as follows: "In a basic form, the magnetic circuit consists of a magnetic core, usually of ferromagnetic materials, so-called soft magnetic material." And

this was the only sentence on the subject of magnetism in the whole thesis! I was taken aback. My astonishment can be compared to the reaction of a French person who has just learned that all that is to be known about wines is that "they are divided into red and white ones." It is obvious that similar to wines, magnetic materials are represented by hundreds of types, which are further divided into hundreds of grades—each with its own specific properties.

Magnetic devices are commonly used in a wide range of applications. This book is written not for a specialist but rather for a greater audience of engineers and students with perhaps very little knowledge of "magnetic measurements," as described in the anecdote above. It should help them navigate in the jumble of sometimes impenetrable terms, but most of all toward the design of electromagnetic devices in a more effective way. Therefore, this book begins with the easy-to-follow and relatively broad topic of "fundamentals of magnetic measurements." This way, the readers can familiarize themselves with the essential topics before moving on to the other parts.

Specialists active in magnetic measurements discuss and meet each other at many international conferences, such as INTERMAG, Soft Magnetic Materials, Magnetic Sensors, Magnetic Measurements, and Workshop on 2D Magnetic Measurements, to mention a few. A large number of "magnetic" papers are published every year. But for the same reason, this circle of advanced readers can often be somewhat adrift and lost in the jungle of information. I hope that this book can be helpful as an organizer, guide, and reference for these readers too.

Magnetic measurements, especially methods of testing of magnetic materials, are quite complex, because many factors, such as the shape of the sample or magnetizing conditions, influence the results. Therefore, most of the testing methods are precisely described by appropriate standards. Unfortunately, despite globalization, different countries still use different standards. In this book, I refer to International Electrotechnical Commission (IEC) standards assuming that differences between the main American, European, and Japanese standards are not very significant, although, of course, they cannot be neglected.

Also terms and units are sometimes different in different countries. Although the SI system (the International System of Units) is obligatory, in some environments it is not fully adopted. For instance, physicists still quite commonly use the previous system—Gauss as a unit of flux density, Oersted as a unit of magnetic field strength,

---

* Second edition partially updated by Graham in 2009.

or Angstrom as a unit of small dimensions. In this book, I consistently use only SI units. The "old" units can be easily converted to the "new" with the following simple and easy-to-remember relation:

$$1\,\text{Oe} \rightarrow 1\,\text{Gs} \rightarrow 100\,\mu\text{T} \rightarrow 0.1\,\text{mT} \rightarrow 0.796\,\text{A/cm} \rightarrow 79.6\,\text{A/m}$$

In "magnetic" literature, some terms are often used in a misleading way. Two main fundamental terms often have varying names; for example, *magnetic flux density B* is also denoted as *magnetic induction B* (indeed, the first is longer to write). Similarly, we have *magnetic field strength H* as well as *magnetic field intensity H*. Moreover, in everyday communication, the *magnetic field* is commonly understood as the *magnetic field strength* (e.g., in the expression "magnetic field of 100 A/m"). To avoid such ambiguity, I always use the *magnetic field* to describe the physical phenomenon (regardless of its units) in contrary to the *magnetic field strength* or the *magnetic flux density* as values with units.

The lack of a comprehensive, essential book on magnetic measurements arises from a very wide spectrum of information from sometimes distant disciplines, such as physics, medicine, material science, geophysics, electrical engineering, and informatics. It is simply impossible to be an expert in so many subjects. I was fully aware of the risk when preparing for the publication of this book. On the other hand, a book written by a single author has the advantage of being consistent in its presentation of the main idea in comparison with a publication created jointly by many separate specialists. Fortunately, there are many excellent books on the market devoted to more specialized subjects, such as magnetic materials, magnetic resonance, magnetic imaging, physics of magnetism, SQUID, etc. Most of the directly relevant are included in the list of references, which, though numbering in the hundreds, is not exhaustive by any means.

I would like to express my gratitude to Professor Tony Moses from Cardiff University for his invaluable help. I would especially like to thank Dr. Stan Zurek from Megger Ltd., who was the first reader of this book and who provided significant inputs in improving its quality.

Slawomir Tumanski

For MATLAB® and Simulink® product information, please contact:

The MathWorks, Inc.
3 Apple Hill Drive
Natick, MA, 01760-2098 USA
Tel: 508-647-7000
Fax: 508-647-7001
E-mail: info@mathworks.com
Web: www.mathworks.com

Author

Slawomir Tumanski was a professor at Warsaw University of Technology for many years and was actively involved in the study of magnetic sensors and the testing of magnetic materials. He has published two books, *Thin Film Magnetoresistive Sensors* and *Principles of Electrical Measurements*. He also serves as the editor-in-chief of the scientific journal *Przeglad Elektrotechniczny* (*Electrical Review*).

Units, Symbols, and Constants Used in This Book

a	dimensions (m)		M	mutual inductance (H)
A	area (cross section) (m^2)		m	magnetic moment (A m^2)
$\mathbf{A}$	magnetic potential vector (T m)		m	mass (kg)
B	magnetic flux density (magnetic induction) (T)		m_e	electron mass, $m_e = 9.109 \times 10^{-28}$ g
$b(t)$	instantaneous flux density (T)		m_l	magnetic quantum number
$B_j(x)$	Brillouin function		m_s	spin quantum number
B_r	remanent flux density (T)		n	number of turns
B_s	saturation flux density		n	principal quantum number
BH	energy product (J/m^3)		N_A	Avogadro's number, $N_A = 6.022 \times 10^{26}$ atoms kg/mol
c	velocity of the light, $c = 2.998 \times 10^8$ m/s		N_d	demagnetization factor
e	electronic charge, $e = -1.602 \times 10^{-19}$ C		P	power (W)
C	capacity (F)		P_A	axial power loss
C_j	Josephson constant, $C_j = 483.597891$ GHz/V		P_r	rotational power loss
d	diameter (m)		P_w	specific power loss (W/kg)
e	electromotive force (V)		P_H	hysteresis power loss
E	electric field strength (V/m)		P_{EC}	eddy current power loss
E	energy density (J/m^3)		P_{Ex}	excess power loss
f	frequency (Hz)		q	electric charge (C)
f_0	resonance frequency		R	resistance (Ω)
F	force (N)		R	reluctance
g'	Lande g-factor		R_H	Hall coefficient
H	magnetic field strength (magnetic field intensity) (A/m)		r	radius (m)
$h(t)$	instantaneous magnetic field strength (A/m)		s	spin angular moment
H_c	coercive field (A/m)		S	apparent power loss
H_{cr}	critical field (in superconductors) (A/m)		S	spin total moment
H_d	demagnetizing magnetic field (A/m)		S	shielding factor
H_k	anisotropy field (A/m)		S	spin number
H_t	tangential component of magnetic field strength (A/m)		t	time (s)
			t	thickness (m)
$_BH_c$	coercivity for $B(H)$		T	temperature (K)
$_JH_c$	coercivity for $J(H)$		T	time constant
I	current (A)		T_c	Curie temperature
I_{cr}	critical current (in superconductors)		V	voltage (V)
J	electronic angular moment		V	volume (m^3)
j	current density (A/m^2)		v	velocity (m/s)
J	magnetic polarization (T)		w	width (m)
J_s	saturation polarization (T)		x	distance (m)
k_B	Boltzmann's constant, $k_B = 1.381 \times 10^{-23}$ J/K		Z	atomic number
K_0, K_1, K_2	anisotropy constants (J/m^3)		Z	impedance (Ω)
K_u	uniaxial anisotropy constant (J/m^3)		Z_0	free space impedance, $Z_0 = 377\,\Omega$
K_K	von Klizing constant, $K_K = 25812 \times 10^{-7}\,\Omega$		α	direction angle
			β	rotation of the light in magneto-optical effects
l	length (m)			
l	orbital quantum number		β	McCumber factor
L	angular moment		γ	density (kg/m^3)
L	magnetic inductance (H)		γ	gyromagnetic ratio
L	anisotropy axis		γ_p	proton gyromagnetic ratio, $\gamma_p = 42.5760812$ MHz/T
M	magnetization			
M_s	saturation magnetization		γ_e	electron gyromagnetic ratio, $\gamma_e = 28.02468$ GHz/T
M	magnetic moment (A m^2)			

δ	skin effect depth (m)	μ_r	relative permeability
σ	mechanical stress	μ_{rev}	reversible permeability
ε	permittivity (C/V m)	μ_{in}	initial permeability
ε_0	permittivity of free space, $\varepsilon_0 = 8.854 \times 10^{-12}$ C/V m	μ_{rec}	recoil permeability
		μ_Δ	differential permeability
ε	angle of direction of magnetic strip	ρ	resistivity (Ω m)
Φ	magnetic flux density (Wb)	ϑ	angle between direction of the current and magnetization
Φ_0	flux quantum, $\Phi_0 = 2.067833667 \times 10^{-15}$ Wb		
λ_s	saturation magnetostriction	φ	angle between magnetization diction and anisotropy axis
$\hbar$	Planck's constant, $\hbar = 6.626 \times 10^{-34}$ J s		
μ_m	magnetic moment	ψ	scalar magnetic potential
μ_l	orbital magnetic moment	ω	angular frequency (rad/s)
μ_s	spin magnetic moment	ω_0	precession frequency
μ_B	Bohr magnetron, $\mu_B = 1.165 \times 10^{-29}$ J m/A $= 9.274 \times 10^{-24}$ A m^2	χ	susceptibility
		ν	Verdet constant
μ_N	nuclear magnetron, $\mu_N = 5.050583 \times 10^{-27}$ A m^2	ν	reluctance
μ	permeability	θ	tilt angle
μ_0	permeability of the free space, $\mu_0 = 4\pi \times 10^{-7}$ Wb/A m $= 4\pi \times 10^{-6}$ H/m	τ	torque (N m)

1

Introduction to Magnetic Measurements

In this book, magnetic measurements are discussed much wider than just obvious testing of magnetic materials or the measurement of the magnetic field. Direct and indirect measurements of magnetic parameters like magnetic field strength H, flux density B, permeability μ, susceptibility χ, or magnetostriction λ are commonly used in other areas of science and technology, such as paleomagnetism, magnetoarcheology, mines detection, displacement or proximity testing, current measurement, nondestructive testing of materials, and medicine diagnostics, to name just a few. The practical applications of magnetic measurements are almost unlimited. For example, the magnetostriction sensors can be used for testing exotic quantities such as acidity pH, blood coagulation, concentration of castor oil (ricin), or even presence of *Salmonella* bacteria.

Although magnetic and electric measurements require similar "tools" (both are based on the determination of voltages or currents), usually the magnetic measurements are more complex or even more ambiguous. For many years, the assumption remained that the magnetization of magnetic materials is reliably represented by two main values: magnetic field strength H and flux density B. However, several years ago, it was concluded (and international standards followed) that the less commonly known term polarization J ($J = B - \mu_0 H$) better describes state of magnetization than flux density. Moreover, nowadays there are specialists suggesting that even better would be to use the magnetization M. There are two schools of measurement of magnetic field strength in the magnetic materials: one preferring indirect measurement by applying Ampére's law (by measurement of the magnetizing current) and second preferring direct measurement by search coils (thus by applying Faraday's law). In the case of certain sensors of magnetic field, it is still under discussion what they really measure: magnetic field strength H or magnetic flux density B. Thus, in comparison with electrical measurements where main terms are better established, many fundamental problems of magnetic measurements are still discussed.

In electrical measurements, there exists relatively simple relation between current and voltage known as Ohm's law. Upon difference in electric potentials represented by the supplying voltage V, there is a current I flowing in the tested sample (Figure 1.1). The value of this current depends on the material parameter known as resistivity ρ, which gives rise to resistance $R = \rho l / A$ (where l is the length of the sample, A is the cross section). According to Ohm's law this current is

$$I = \frac{V}{R} \tag{1.1}$$

More precisely the response of material with resistivity ρ under the electric field E is represented by the current density J.

Figure 1.1 also presents the response of the magnetic material represented by the permeability μ under influence of the magnetic field strength H, which is generated by the current I flowing in the coil. This field gives rise to the magnetic flux Φ. In other words, the response of the material with permeability μ subjected to magnetic field strength H is represented by the flux density $B = \Phi/A$. This flux density is usually detected as electric voltage induced in the secondary coil with n turns—the value of this voltage depends on the time derivative of B. Between exciting magnetic field strength H and the response B, there is relatively simple relation

$$B = \mu H \tag{1.2}$$

But tests of magnetic materials are usually much more complex than it is the case of electrical measurements, because

- Usually in a typical sample of current conducting material (e.g., cuboid or cylinder), the current distribution is uniform with some exceptions (e.g., due to skin effect for high-frequency currents). In the case of magnetic materials, this condition is almost completely opposite—due to demagnetizing fields the sample under test is usually magnetized nonuniformly with few exceptions (e.g., ellipsoidal shape of the sample).

- In typical conducting materials, the relation between current and voltage is highly linear with some exceptions. Thus, the materials can be described by a single scalar value—its resistance R. In contrary, in most typical magnetic materials, the relation between magnetic field

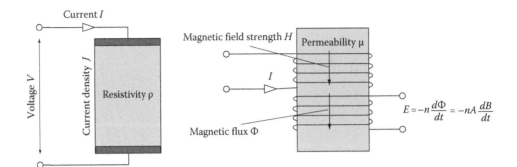

FIGURE 1.1
Analogy between electrical and magnetic measurements.

strength and the flux density is highly nonlinear. Therefore, generally it is necessary to use the whole relation $B = f(H)$ known as the magnetization curve.

- Most of the current-conducting materials are homogeneous and hence distribution of current is also uniform. In the case of most of magnetic materials, the samples are magnetized nonuniformly due to the grain and domain structure.

- Most of current-conducting materials are isotropic with a few exceptions where the tensor of resistivity is really required. Most of the magnetic materials exhibit anisotropy of their properties.

Thus, let us assume that we would like to test the sample of a typical magnetic material. Because distribution of flux density strongly depends on the shape of the sample, we cannot simply discuss properties of the material. We can only determine the properties of certain preselected sample shapes (e.g., ring core, Epstein frame, or sheet/strip). Generally, closed magnetic circuit is recommended and therefore open sample needs to be closed by the external yoke.

Even if we prepare the recommended sample, there is another problem to overcome. Because materials are nonlinear, both magnetic field strength and flux density can be non-sinusoidal (if the AC testing method is used). The results of testing are quite different when we force either the sinusoidal flux density or magnetic field strength to be sinusoidal. Therefore, it has been widely accepted in standards that sinusoidal flux density is required. It is necessary to use special digitally controlled magnetizing devices, which can be very sophisticated for high value of flux density.

However, even if we ensure correct shape and correct magnetizing waveform, there are further other problems to be overcome. Because materials are usually anisotropic, we should precisely determine the direction of magnetization. This problem is of course much greater in two- or three-dimensional tests. The results of the test depend strongly on the frequency of magnetization. In the case of DC (static) magnetization, additional problems appear because, according to Faraday's law, to induce voltage the magnetic flux should be changed. Due to a large number of difficulties accompany testing of magnetic materials, most of them are performed only in professional laboratories.

The *hysteresis loop* is commonly recognized as a symbol of magnetism. Indeed, it is one of the most frequently tested characteristics of magnetic materials.

Figure 1.2 presents a typical method of testing of the hysteresis loop. Because flux density is represented by its time derivative, often the integrating amplifier is necessary although modern oscilloscopes are capable of performing this operation. The exciting magnetic field strength is usually determined from the value of the magnetizing current (voltage drop V_H in Figure 1.2) according to Ampére's law $H = I_1 n_1/l$ (where l is the length of magnetic path—in the case of the ring sample of appropriate dimensions close to the mean circumference, n_1 is the number of turns of the magnetizing winding). The flux density is usually derived from Faraday's law $dB/dt = -V_2/n_2 A$.

Also the *permanent horseshoe magnet* (Figure 1.3) is commonly used as a symbol of magnetism. Modern rare earth magnets can exhibit energy product BH above $400\,kJ/m^3$, which is close to the theoretical limit assumed as $485\,kJ/m^3$. Such extremely large energy related to coercivity of above $1200\,kA/m$ stimulates development of new methods of testing hard magnetic materials. One of the solutions is to use the pulsed field techniques.

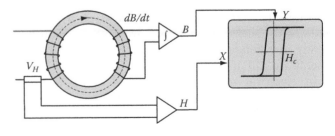

FIGURE 1.2
Typical approach for determination of the hysteresis loop.

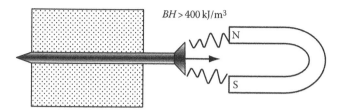

FIGURE 1.3
Modern rare-earth magnets with extremely large energy product *BH*.

The magnetic field can influence many other physical properties, such as resistivity, mechanical stress, optical properties, etc. Hence, these phenomena are commonly used for design of magnetic field sensors.

Magnetic field sensors such as Hall effect sensors, magnetoresistive sensors, and inductive sensors are frequently used not only for detection of magnetic field but also for testing magnetic materials in nondestructive material evaluation and for measurement of other values, especially mechanical and electrical. Of course, they can be used for the design of commercially available measuring instruments. On the market are present different measuring instruments starting from SQUID devices with ability to measure the magnetic field as low as several fT (10^{-15} T), which is comparable with magnetic field of the human brain activity. For measurements of small magnetic fields, the fluxgate-sensors-based instruments are commonly used. For measuring of the large magnetic field, the instruments based on the Hall effect sensors are very popular (Figure 1.4).

Earth's magnetic field can pose a great problem. It is possible to measure the magnetic field as small as several fT but these measurements must be performed in the presence of Earth's magnetic field, which is over million times stronger (around 50 μT). Moreover, this field is accompanied by industrial magnetic interferences of

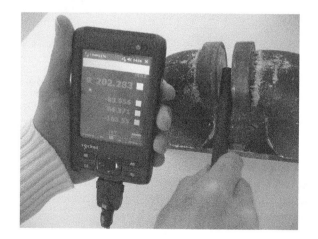

FIGURE 1.4
Measurement of magnetic field by three-component Hall sensor magnetometer. (Courtesy of Metrolab Instruments SA, Geneva, Switzerland.)

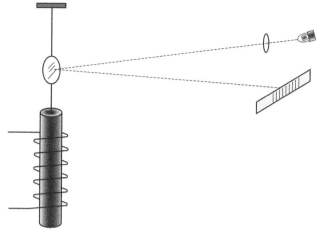

FIGURE 1.5
Einstein–de Haas experiment.

similar and larger value. The special shielded rooms as well as active field cancelation techniques for such measurements are indispensable.

Although the basis of the theory of magnetism was developed in nineteenth century, the research is still in progress because new findings in physics help in the understanding of magnetic phenomena. And conversely, discoveries in magnetic measurement techniques, for example, the Zeeman effect or the Mossbauer resonance, have crucial importance in physics. No wonder that almost all famous physicians such as Einstein or Curie contributed to magnetic measurements. Figure 1.5 presents the experiment performed by Einstein (known as Einstein–de Haas experiment).

Einstein and de Haas were looking for a relation between angular mechanic moment and magnetic moment. In the experiment presented in Figure 1.5, after change in direction of magnetization of the iron rod (by reversing the current in the coil surrounding the specimen) suspended on a torsion fiber, a small angular deflection was detected. This change of angular moment, ΔG, caused by the change of magnetization, ΔM, can be described as follows:

$$\frac{\Delta M}{\Delta G} = g' \frac{e}{2m_e c} \tag{1.3}$$

where g' is the magnetomechanical factor or simply g'-factor, e is electron charge, m_e is a mass of electron and c is the velocity of the light.

From the experiment on ferromagnetic rod, the g'-factor was determined as approximately 2, which proved that the magnetization of iron is essentially due to electron spin.

Many other famous physicians investigated the magnetic phenomena and many of them were later honored by the Nobel prize (Table 1.1).

TABLE 1.1

Some Nobel Prize Laureates for the Works Related to Magnetic Phenomena

Hendrik Lorentz Pieter Zeeman	1902	For research on the influence of magnetism upon radiation phenomena
Felix Bloch Edward Purcell	1952	For development of new methods for nuclear magnetic precession measurements (nuclear magnetic resonance)
Alfred Castler	1966	For research on the relation between magnetic and optical resonance (optical pumping)
Louis Neel	1970	For discoveries concerning antiferromagnetism and ferrimagnetism
Brian Josephson	1973	For tunneling effect in superconductors (Josephson effect)
Nevill Mott Philip Anderson	1977	For work on the electronic structure of magnetic and disordered systems
Klaus von Klitzing	1985	For quantum Hall effect
Paul Lauterbur Peter Mansfield	2003	For their discoveries concerning magnetic resonance imaging
Albert Fert Peter Grünberg	2007	For discovery of giant magnetoresistance

FIGURE 1.6
Participants of the Solvay Congress, which was devoted to magnetism (front row: Th. De Donder, P. Zeeman, P. Weiss, A. Sommerfeld, M. Curie, P. Langevin, A. Einstein, O. Richardson, B. Cabrera, N. Bohr, W.H. De Haas; second row: E. Herzen, E. Henriot, J. Verschaffelt, C. Manneback, A. Cotton, J. Errera, O. Stern, A. Pickard, W. Gerlach, C. Darwin, P.A.M. Dirac, E. Bauer, P. Kapitsa, L. Brillouin, H.A. Kramers, P. Debye, W. Pauli, J. Dorfman, J.H. Van Vleck, E. Fermi, W. Heisenberg). (Photo of Benjamin Couprie. From the Collection of Institute International de Physique Solvay.)

Starting from 1911, the International Solvay Institute of Physics and Chemistry organized famous conferences devoted to outstanding problems of physics and chemistry collecting most prominent scientists. In 1930, the conference devoted to magnetism was organized. Figure 1.6 presents the photo of participants of this conference. Practically, most of the notable physicists of that time met at this Congress.

2

Fundamentals of Magnetic Measurements

2.1 Historical Background

Magnetism and magnetic fields were one of the oldest physical phenomena investigated and subjected to measurements. The effect of attraction (or repulsion) between pieces of some lodestones has been known as long as iron smelting.* Moreover, it was observed that small pieces of this rock spontaneously rotated along north–south direction. Thus, no wonder that such an effect was applied to pointing north direction in the compass—written records show that ancient Chinese used the compass (*Si Nan*) between 300 and 200 BC[†] (Campbell 2001).

Thales of Miletus (640–546 BC) investigated the effect of attraction between lodestone and iron, although this information comes from other written sources. Socrates (470–399 BC) wrote "that lodestone not only attracts iron rings, but imparts to them a similar power" (Keithley 1999).

A more detailed study of magnetism and compass was performed by Pierre de Maricourt (also known as Peter Peregrinus). In 1269, he published the work "Epistola de magnete." In this work, the polarity of magnets was described for the first time (also evidence of two *magnetic poles*—after division of a magnet into two smaller parts, the polarity was retained). Probably, he also was the first to propose to use the magnet in the construction of a machine—to obtain the *perpetual motion*.

The work of Peregrinus was continued by Wiliam Gilbert (1544–1603). For the first time in the work "De magnete" (1600 AD), he described the movement of a magnetic needle in a compass being caused by the Earth's magnetic field—he concluded that Earth acts as huge spherical magnet.

Charles-Augustin de Coulomb (1736–1806) (and independently John Michell 1750) constructed a torsion balance and investigated electrostatic and magnetic forces of attraction. He formulated *Coulomb's law* describing mathematically the force F between hypothetical

magnetic poles m_1, m_2 distanced by r (magnetic dipole—Figure 2.1a) as

$$F = \frac{1}{4\pi\mu_0}\frac{m_1 m_2}{r^2} \tag{2.1}$$

where μ_0 is permeability of free space ($\mu_0 = 4\pi \times 10^{-7}$ Wb/Am).

The magnetic field tries to align the magnetic dipole parallel to direction of this field with torque τ:

$$\tau = M \times B \tag{2.2}$$

where M is the magnetic moment (Am^2).

In 1820, Hans Christian Oersted discovered that a magnetic needle moved near a wire with electric current—thus that electric current generates a magnetic field that encircles the wire. This revolutionary discovery inspired Andre Marie Ampère (1775–1836) to formulate the hypothesis that an electric current is a source of every magnetic field (also generated by a magnet)[‡] (Figure 2.1b). In 1820, he demonstrated that two parallel wires conducting current attract each other.

In the same year, Jean-Baptiste Biot and Felix Savart proposed the rule (known as *the Biot–Savart law*) enabling the determination of the magnetic field strength around a current conducting wire. The magnetic field strength δH generated by current i in infinitesimal length δl of a conductor at a radial distance a is

$$\delta H = \frac{1}{4\pi a^2}i\delta l \times u \tag{2.3}$$

where u is a unit vector along the radial direction. Thus, the magnetic field strength H generated by a current flowing in circuit C is

$$H = \frac{1}{4\pi}I\int_C \frac{dl \times u}{a^2} \tag{2.4}$$

---

* Magnetic lodestones were found near Magnesia in ancient Macedonia. According to Lucretius (98–55 BC), the term magnet is derived from Magnesia (Keithley 1999).
[†] It has been claimed that ancient Chinese used compass before 2500 BC (Morrish 2001).

[‡] He assumed that magnet comprises rotating "electrodynamic molecules"—it was about 80 years before the discovery and identification of the electron and about 100 years before the Rutherford–Bohr model of the atom with orbiting electrons.

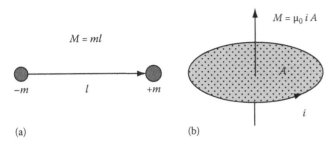

FIGURE 2.1
Two models of elementary entity in magnetism: dipole model proposed by Coulomb (a) and circular loop conducting current proposed by Ampère (b) (M, magnetic moment).

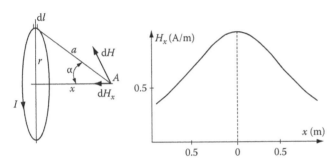

FIGURE 2.2
A circular wire and axial component of magnetic field $H(x)$ generated at the distance x (calculated for $I = 2$ A, $r = 1$ m).

Let us calculate the magnetic field H in the point A (Figure 2.2) generated by the current in a circular wire of radius r. The axial component of this field is

$$dH = \frac{I}{4\pi} \frac{dl}{a^2} \sin\alpha \tag{2.5}$$

while $a = \sqrt{r^2 + x^2}$ and $\sin\alpha = r/\sqrt{r^2 + x^2}$.

The magnetic field H at the distance x from the axis is

$$H = \oint dH = \frac{I}{2\pi a^2} \sin\alpha \oint dl = \frac{1}{2} I \frac{r^2}{\left(\sqrt{r^2 + x^2}\right)^3} \tag{2.6}$$

because $\oint dl = 2\pi r$.

Thus, the magnetic field strength at the center of the loop (for $x = 0$) is $H = I/2r$.

In 1826, Ampère formulated one of the fundamental laws of magnetism—Ampere's circuital law:

$$\oint \mathbf{H} \cdot dl = nI \tag{2.7}$$

where n is the number of conductors each carrying a current I.

As it will be demonstrated later (Section 2.9.2), by using Ampère's law it is also possible to calculate the magnetic field around a current-conducting wire (similarly as using Biot–Savart law).

In 1831, Michael Faraday discovered another fundamental law of magnetism. He stated that if a magnetic flux Φ linking an electrical circuit changes, it induces in this circuit electromotive force (voltage) V proportional to the rate of change of the flux:

$$V = -\frac{d\Phi}{dt} \tag{2.8}$$

This effect is called *electromagnetic (EM) induction*. The minus sign in Equation 2.8 shows that the voltage is induced in a direction opposing the flux change, which produces it (*Lenz's law*).

We will finish our overview of the fundamental discoveries of the magnetism with a work published in 1873 by James Clerk Maxwell. In "A Treatise on Electricity and Magnetism," he proposed a set of 20 equations (later summarized to a set of just four: two Gauss's, Ampère's, and Faraday's), which are fundamental to all analyses of magnetic and electric fields. Maxwell equations are described in Section 2.9.6.

2.2 Main Terms

2.2.1 Magnetic Field Strength *H*, Magnetic Flux Φ, and Magnetic Flux Density *B*

Magnetic field strength H and magnetic flux density B are most commonly used magnetic parameters. Other parameters such as permeability (B/H), losses ($H \cdot dB/dt$), polarization ($B - \mu_0 H$), magnetization [$(B/\mu_0) - H$], and magnetizing curve ($B = f(H)$) depend on these two values.*

Lorentz showed in 1982 that an EM field acts with a force on a charge q moving with velocity v:

$$F = q(E + v \times \mu_0 H) \tag{2.9}$$

Usually, this force can be divided into two parts, the first one is caused by the electric field E:

$$F' = qE \tag{2.10}$$

---

* International standards recommend using the polarization J instead of the flux density B to describe the properties of magnetic materials.

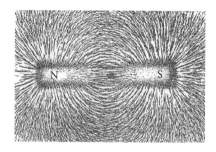
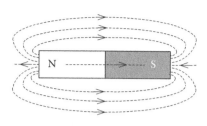

FIGURE 2.3
The detection of the presence of magnetic field and direction of magnetic flux by using iron filings. Right side, calculated lines of magnetic flux for comparison. (From *Practical Physics*, Macmillan and Company, 1914.)

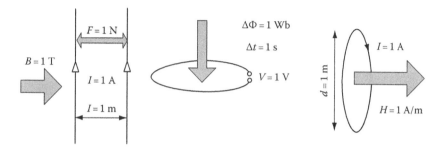

FIGURE 2.4
Various definitions of magnetic field units: tesla (a), weber (b), and A/m (c)

and the second one is caused by the magnetic field H:

$$F'' = q\mathbf{v} \times \mu_0 H \qquad (2.11)$$

Thus in an EM field, the electric field results in force acting on any charges (stationary and moving), while the magnetic field acts only on movable charges.*

The unit of a magnetic field strength H (sometimes called magnetic field intensity) is 1 A/m (ampere per meter).

The presence of a magnetic field H in an area A results in a magnetic flux Φ depending on the magnetic properties of the medium described by permeability μ and magnetization M of a material. In free space, there is no magnetization and permeability is equal to μ_0, so the magnetic flux Φ caused by a magnetic field H is

$$\Phi = \mu_0 A H \qquad (2.12)$$

The unit of magnetic flux is 1 Wb (weber) or Vs.

The magnetic flux density B (sometimes also called as a magnetic induction) is a more commonly used quantity and is equal to

$$B = \frac{\Phi}{A} \qquad (2.13)$$

From (2.12) and (2.13), we can see that in a free space the relationship between magnetic field strength H and flux density B is

$$B = \mu_0 H \qquad (2.14)$$

The unit of magnetic flux density is 1 T (tesla).

The presence and direction of magnetic flux can be easily detected with iron filings (Figure 2.3).

According to Equation 2.14, in a free space, the relationship between the magnetic field strength H and the flux density B is linear (constant factor μ_0). For this reason, it does not matter which quantity is used as the reference source. Independently of what is the cause and what is the result, most often, the flux density standard is used as a reference for the magnetic field. The standard is defined by the relationship between the magnetic field generated by an electric current and mechanical force. *A magnetic flux density B of 1 T generates a force of 1 N (perpendicular to the direction of the magnetic flux) for each 1 m of a conductor carrying a current of 1 A* (Figure 2.4a).

We can also determine the flux density according to Faraday's law: *The weber is the magnetic flux which, linking a circuit of one turn, would produce in it an electromotive force of 1 volt if it were reduced to zero at a uniform rate within 1 second* (Figure 2.4b).

Also the unit of the magnetic field strength can be described according to the Biot–Savart law. We have shown above that a current of 1 A generates in a circle of

---

* According to Einstein's theory of relativity, for the stationary observer, this statement is obvious, but, if the observer is moving with the charge, the effects of magnetic and electric field are reversed. So we can say that the magnetic field is a relativistic correction to the electric field (see Jiles 1998).

TABLE 2.1

Conversion Factors for Common Magnetic Units

	Tesla (T)	(A/m)	Gauss (G)	Oersted (Oe)
A/m	1.256×10^{-6}	1	12.56×10^{-3}	12.56×10^{-3}
Oe	10^{-4}	79.6	1	1
T	1	7.96×10^5	10^4	10^4
γ	10^{-9}	7.96×10^{-5}	10^{-5}	10^{-5}
G	10^{-4}	79.6	1	1

radius 1 m a magnetic field equal to $I/2r$ (Equation 2.6). So a magnetic field strength of 1 A/m is generated at the center of a single circular turn with a diameter of 1 m of a conductor that carries a current of 1 A.

All three definitions described above are illustrated in Figure 2.4. It is worth noting that we also have other good-quality sources of a standard magnetic field—Helmholtz coil (described in Section 2.12.1) and magnetic field source controlled by a high-accuracy magnetic resonance meters (described in Sections 2.7 and 4.7).

In many countries, the *SI* units (*SI*—"systeme internationale") are required and the use of the old units gauss ($1 G = 10^{-4}$ T) and oersted ($1 Oe = 10^3/4\pi$ A/m) cannot be legally used. Nevertheless, the old units (gauss, oersted, and $\gamma = 10^{-5}$ Oe) are still in use in several countries, especially in the United States. In fact, the old units are convenient because, in the free space, the field strength of 1 Oe corresponds with the flux density of 1 G. Table 2.1 summarizes the conversion factors between the Gaussian and the SI units.

2.2.2 Magnetization *M*

The relationship (2.14) for any material other than free space is

$$B = \mu_0(H + M) \tag{2.15}$$

where *M* is the vector of magnetization.

In this relationship, the component $\mu_0 H$ represents the contribution of the external source, while $\mu_0 M$ represents the internal contribution of a magnetized material. We see that, even if the external magnetic field strength is equal to zero, the material can exhibit some flux density because it is magnetized (spontaneously or as result of previous magnetizing).

We can assume that every magnetized material comprises a large number of elementary dipoles. Such dipoles result from orbiting of an electrons around the nuclei or their own rotation—spin. These dipoles are represented by magnetic moments *m* (for elementary current loop of area *A*—see Figure 2.1b—this moment is equal to $m = I \cdot A$). In the state of complete demagnetization of a material, the average magnetic moment is balanced out

and the resultant magnetization is zero. If the material is magnetized, its magnetization *M* is equal to

$$M = \frac{\sum m_i}{V} \tag{2.16}$$

The magnetization is defined as the resultant magnetic moment per unit volume (the unit of magnetization is the same as that of magnetic field strength A/m).

After magnetizing the material with an external magnetic field in diamagnetic materials, the magnetization is not in the same direction as the applied field, so the resultant flux density is weakened; in paramagnetic materials, the magnetization is in the same direction as the applied field, so the flux density is reinforced. Yet in ferromagnetic and ferrimagnetic materials, the magnetization can be significantly larger and in the same direction as the applied field.

2.2.3 Magnetic Polarization *J*

In the early literature, the state of magnetic materials was described by magnetic induction *B*. Recently, many standards have recommended substituting the flux density *B* by magnetic polarization *J* (intrinsic flux density or intensity of magnetization):

$$J = B - \mu_0 H \tag{2.17}$$

So, polarization is equal to $\mu_0 M$. Because in typical applications of soft magnetic materials the value of magnetic field strength is usually not larger than 1 kA/m and μ_0 is $4\pi \times 10^{-7}$ Wb/Am, the difference between the flux density *B* and the polarization *J* is negligibly small. In the case of hard magnetic materials, this difference can be significant and often both relationships $B = f(H)$ and $J = f(H)$ are presented.

2.2.4 Permeability μ

The relationship between the flux density *B* and the magnetic field strength *H* of a magnetized material is

$$B = \mu H \tag{2.18}$$

This way of describing properties of a material is not convenient in practice and usually the permeability of material is described in relation to the permeability of free space, that is, *relative permeability* $\mu_r = \mu/\mu_0$. The dependence (2.18) can be, therefore, written as

$$B = \mu_r \mu_0 H \tag{2.19}$$

Theoretically, permeability μ could be the best factor for describing the properties of magnetic materials because it informs directly about the relationship between two main material parameters: the flux density B and the magnetic field strength H. But in practice, the situation is much more complex, because:

- The relationship between B and H is almost always nonlinear, and therefore permeability depends on the working point (value of magnetic field strength). Figure 2.5 presents a curve $\mu = f(H)$ determined for a typical electrical steel. We can see that the maximum value of relative amplitude permeability reaches about 40,000, but, at higher flux density, it is much lower (for deep saturation it is very small—the material practically is not ferromagnetic). Similarly, the initial permeability (for very small magnetic field) is also significantly smaller. Therefore, a fixed value of permeability can give information only about a fixed operating point.

- The magnetization of the material is shape dependent—permeability of a magnetized body can be completely different than permeability of a raw material (this problem will be discussed later in Chapter 5). Generally, the magnetization is nonuniform in the whole body and we can only determine the mean value.

- Most magnetic materials are polycrystalline and therefore they have different properties (also permeability) for various directions of magnetization (we say that the material is anisotropic). Thus, permeability should be described in the form of a tensor:

$$\begin{bmatrix} H_x \\ H_y \\ H_z \end{bmatrix} = \begin{bmatrix} \mu_{xx} & \mu_{xy} & \mu_{xz} \\ \mu_{yx} & \mu_{yy} & \mu_{yz} \\ \mu_{zx} & \mu_{zy} & \mu_{zz} \end{bmatrix} \begin{bmatrix} B_x \\ B_y \\ B_z \end{bmatrix} \tag{2.20}$$

Usually we limit this problem to two-dimensional (2D) case but even 2D magnetization is very complex.

- Permeability depends on many other factors: frequency, the presence of harmonics (deviation from sinusoidal shape of flux density), etc. For higher frequency, we should take into account real and imaginary parts of permeability (*complex permeability*) and we have $\mu = \mu' + j\mu''$.

Therefore, although permeability is a very useful factor from a physical point of view, in technical applications (e.g., in the design process), it is more useful to apply the magnetization curve as a description of magnetizing process. Nevertheless, in some applications, the permeability can be the most important factor—for instance, for magnetic shielding design (the higher the value of permeability, the more effective shielding) or a magnetic concentrator design. In such devices, it is necessary to use a magnetic material with permeability of the highest possible value. Currently, it is possible to obtain ferromagnetic materials with relative permeability as high as 1,000,000 (1 million). Table 2.2 presents typical values of maximum relative amplitude permeability of some commercial ferromagnetic materials.

Instead of *total amplitude permeability* (the slope of the line which connect the origin 0 and any point on the magnetization curve—Figure 2.5), we can also determine *differential permeability* as the slope of the part of the magnetization curve:

$$\mu_\Delta = \frac{\Delta B}{\Delta H} \tag{2.21}$$

In many applications, the slope of the minor hysteresis loop is important—for example, when we bias a magnetic material using a direct magnetic field in the presence of a smaller alternating field. For such minor loops,

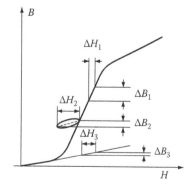

FIGURE 2.5

Dependence of relative permeability on magnetic field strength determined for a typical electrical steel and method of defining of other permeabilities: total amplitude permeability $\mu_r = (B/H)/\mu_0$, differential permeability $\mu_\Delta = (\Delta B_1/\Delta H_1)/\mu_0$, reversal permeability $\mu_{rev} = (\Delta B_2/\Delta H_2)/\mu_0$, and initial permeability $\mu_{in} = (\Delta B_3/\Delta H_3)/\mu_0$.

TABLE 2.2

Typical Values of Maximum Relative Amplitude
Permeability of Some Ferromagnetic Materials

Material	μ_{rmax}
Iron	6,000
Pure 99.9 iron	350,000
Silicon iron (nonoriented)	8,000
Silicon iron (oriented)	40,000
Silicon iron (cubic texture)	100,000
Permalloy 78Ni-22Fe	100,000
Supermalloy 79Ni-16Fe-5Mo	1,000,000
Permivar 43Ni-34Fe-23Co	400,000
Amorphous Metglas Fe40-Ni38-Mo4-B18	800,000
Amorphous Metglas Co66-Fe4-B14-Si15-Ni1	1,000,000
Nanocrystalline Nanoperm Fe86-Zr7-Cu1-B6	50,000 at 1 kHz

we can determine the *reversal permeability* (Figure 2.5) (Bozorth 1951, Morrish 2001).

2.2.5 Susceptibility χ

In some applications, especially in physical analysis of magnetization processes, susceptibility χ is used more often than permeability as the quantity linking magnetization M and the magnetic field strength H:

$$M = \chi H \qquad (2.22)$$

There is a simple relationship between permeability and susceptibility:

$$\mu_r = \chi + 1 \qquad (2.23)$$

Measurement of magnetic susceptibility is commonly used in paleomagnetism to investigate the remanent magnetization of the rocks (see Section 6.1.3). Magnetic susceptometry is often used as a method of diagnosis of iron overload of the liver (see Section 6.3.3).

2.2.6 Reluctance R

In the analysis of a magnetic circuit, it is useful to employ the concept of a reluctance R (magnetic resistance), which is related to the so-called magnetomotive force or ampere-turns nI and the magnetic flux Φ. In this way, we obtain a relation similar to Ohm's law for electricity. The reluctance of a uniform magnetic circuit with length l and cross section A is

$$R = \frac{l}{\mu_0 \mu_r A} \qquad (2.24)$$

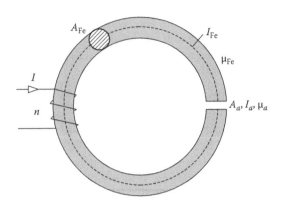

FIGURE 2.6
Toroidal magnetic circuit with an air gap.

Let us consider a ring magnetic core with an air gap (Figure 2.6). According to Ampere's law (Equation 2.7),

$$nI = H_{Fe} l_{Fe} + H_a l_a \qquad (2.25)$$

Thus, we have

$$nI = \left(\frac{B}{\mu_0} - M \right) l_{Fe} + \frac{B l_a}{\mu_a} = B \left(\frac{l_{Fe}}{\mu_{Fe}} + \frac{l_a}{\mu_a} \right)$$

$$= \Phi \left(\frac{l_{Fe}}{\mu_{Fe} A_{Fe}} + \frac{l_a}{\mu_a A_a} \right) \qquad (2.26)$$

We obtained a simple relationship similar to Ohm's law—magnetic flux Φ depends on reluctance R and magnetomotive force nI as $\Phi = nI/R$ (compare with $I = V/R$).

2.3 The Magnetization Process of Ferromagnetic Materials

2.3.1 Domain Structure

In ferromagnetic materials, there are spontaneously magnetized micro-areas, which constitute the micro-dipoles. Moreover, a typical material is polycrystalline as it is composed of many crystallites (grains). Each crystallite exhibits the privileged direction of magnetization—a so-called *easy axis of magnetization*. These grains can be randomly oriented and therefore the bulk material is composed of a large number of areas magnetized in random directions. But even if we technologically order all crystallites in one direction (we can say that we assure a good texture), the whole body can be demagnetized due to the presence of the domain structure.

The state of local magnetization depends on many factors such as—grain structure, grain size, presence of impurities, local stress, and, more importantly, the balance of local energy. Local energy can be composed of many components, for instance (Carr 1969, Hubert and Schäfer 1998):

- *Magnetostatic energy*: energy associated with demagnetizing fields
- *Magnetocrystalline energy*: energy associated with anisotropy of crystals
- *Exchange energy*: exchange interaction between neighboring magnetic moments
- *Magnetoelastic energy*: energy associated with the magnetostriction effect
- *Domain wall energy*: energy associated with interaction between neighboring electron spins

Magnetic material spontaneously creates small regions of the same direction of magnetization—*magnetic domains* in order to ensure the minimum free energy. Figure 2.7 presents the process of formation of a domain structure—each subsequent state exhibits lower energy, and the last one (of minimum magnetostatic energy) is practically without flux leakage—magnetization averages zero and the whole magnetic energy is contained within the material. In real magnetic materials, domain structure can be very complex due to the influence of grain boundaries, impurities, mechanical defects, etc.

Domain walls (DWs) separate any domains of opposite direction of magnetization (Figure 2.8). Such walls are relatively thin (thinner than 10 μm) and, within such a small volume, the elementary magnetic dipoles reverse their alignment (as shown in Figure 2.8).

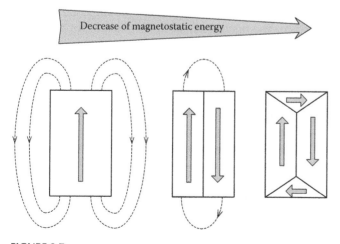

FIGURE 2.7
Various domain structures—each subsequent state exhibits lower total energy.

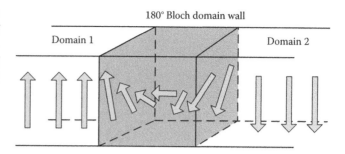

FIGURE 2.8
Magnetic structure of the Bloch domain wall.

The presence of magnetic domains and domains' walls significantly influence the process of magnetization represented by the hysteresis loops and the magnetization curves. The domain structure is discussed in Section 2.10.7.

2.3.2 Magnetization Curve

The magnetization curve represents a relationship between the polarization J (or the flux density B) and the magnetic field strength H. It contains fundamental information about a given magnetic material and it is usually presented in material catalogues (and used by the designers).

Figure 2.9 presents a typical magnetization curve. The process of magnetization can be divided into several characteristic parts. Let us start from a fully demagnetized state. When a small magnetic field is applied, the domains whose direction of spontaneous magnetization is closest to the direction of the applied field start to grow at the expense of other domains. For a small magnetic field, this process is reversible—if we remove the magnetic field, the material returns to its previous state without hysteresis.

The next part of the magnetization curve is characterized by the maximum permeability. In this part, the DWs movements are irreversible. If we remove the magnetic field, the material remains partially magnetized due to the new positions of DWs—the hysteresis effect appears.

The individual DW movements are detectable because the displacements of walls are discontinuous as they "jump" from one pinning site to another. Such irregular movements change the magnetization and can generate voltage pulses in a coil wrapped around the magnetized material. This phenomenon is called a Barkhausen effect (Barkhausen 1919). The Barkhausen emission effect is presented in Figure 2.10—note the discontinuous changes of the flux related to the wall movements—this part of the curve in Figure 2.9 is not smooth in magnification.

The Barkausen noise depends strongly on the microstructure and the mechanical stress;

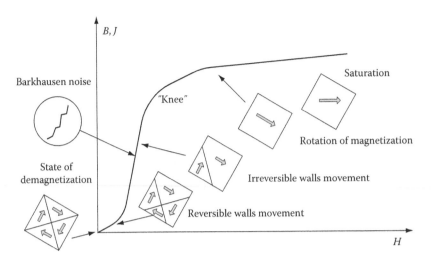

FIGURE 2.9
Typical magnetization curve. (After Brailsford, F.: *Magnetic Materials*. 1948. Copyright Wiley-VCH Verlag GmbH & Co. KGaA.)

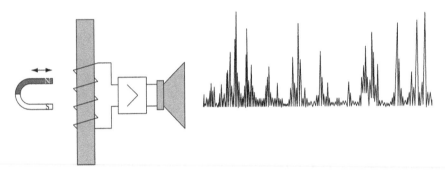

FIGURE 2.10
The Barkhausen effect.

therefore, it is commonly used for material evaluation and nondestructive testing (Bray and McBride 1992).

When we further increase the magnetic field (above the knee point), the process of DWs movements vanishes and the magnetization is realized by rotation of the direction of magnetization—the magnetizing field tries to force the directions of magnetizations along the direction of this field. The change of the value of polarization is now much smaller with the increase of the magnetic field strength—we reach the state near the saturation* (saturation polarization J_s).

The primary magnetization curve can be obtained by measuring the changes of the flux density caused by changes of the direct magnetic field (starting from the state of demagnetization†). More common and actually simpler is magnetizing the sample with an alternating magnetic field—the magnetization curve is then a result of connecting the end points of hysteresis loops (see the next section). When we determine the magnetization curve by using alternating excitation, the magnetic field strength or the flux density or both can be non-sinusoidal. Therefore, the relationship $B = f(H)$ is usually determined for the magnitude of B,H signals or for their *rms* values (or average values).

2.3.3 Hysteresis Loop

The hysteresis is a behavior characteristic practically for all ferromagnetic materials (Bertotti 1998, Della Tore 1999)—it is often a synonymous with the very symbol of magnetism. Typical hysteresis loop is presented in Figure 2.11.

When we start from the demagnetized state, the first part of the path between 0 and 1 is similar to the primary magnetization curve. But, if we stop increasing the magnetic field strength and start decreasing it, the return path 1–2 is different than the ascending magnetizing curve due to irreversible change of DW locations.

---

* According to the relationship (2.17) for saturated polarization J_s, the flux density B still increases with H (see Figure 2.13).
† A material can be demagnetized by applying the large alternating magnetic field and then by gradual, slow decrease of the amplitude of this field to zero. This way we realize smaller and smaller hysteresis loops. To obtain real fully demagnetized state, the material should be heated above its Curie point and slowly reduced to the room temperature (but such a method can sometimes irreversibly change the material properties).

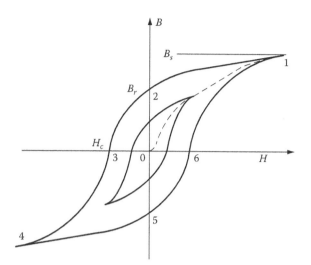

FIGURE 2.11
Typical hysteresis loop and its characteristic points B_r and H_c

Thus, for zero magnetic field strength (point 2), the material remains magnetized and this magnetization is called the residual or remanent flux density B (or remanent magnetization).

To obtain again zero value of the flux density (point 3), it is necessary to apply the magnetic field in the opposite direction—this field is called a coercivity H_c.* The coercivity is a very important parameter of soft magnetic materials because the magnetic loss depends on the area of the hysteresis loop. It is obvious then that the smaller the value of the coercivity, the smaller the power loss. Table 2.3 collects some typical values derived from a hysteresis loop for various soft magnetic materials.

The other hysteresis parameters are required from the hard magnetic materials—the permanent magnets.

TABLE 2.3

Main Parameters of Hysteresis Loop of Various Soft Magnetic Material

Material	H_c (A/m)	B_s (T)
Iron	70	2.16
Pure 99.9 iron	0.8	2.16
Silicon-iron (nonoriented)	40	1.95
Silicon-iron (oriented)	12	2.01
Silicon iron (cubic)	6	2.01
Permalloy 78Ni-22Fe	4	1.05
Supermalloy 79Ni-16Fe-5Mo	0.15	0.79
Permendur 50Fe-50Co	160	2.46
Amorphous Metglas Fe40-Ni38-Mo4-B18	8	0.88
Amorphous Metglas Co66-Fe4-B14-Si15-Ni1	0.24	0.55
Nanocrystalline Nanoperm Fe86-Zr7-Cu1-B6	3	1.52

TABLE 2.4

Typical Parameters of Hysteresis Loop for Various Hard Magnetic Materials

Material	$_BH_c$ (kA/m)	B_r (T)	$(BH)_{max}$ (kJ/m³)
Ferrite $BaFe_{12}O_{19}$	144	0.35	26
Alnico Fe—Co-Ni-Al	52	1.30	44
$SmCo_5$	690	0.92	200
Sm(Co-Fe-Cu-Zr)	560	1.12	240
Nd-Fe-B	780	1.35	320

Source: McCurrie, R.A., *Ferromagnetic Materials—Structure and Properties*, Academic Press, San Diego, CA, 1994.

In this case, the value of the remanent flux density and the value of the coercivity should be as high as possible because the $(B \cdot H)_{max}$ value is responsible for stored magnetic energy (and the same the attraction force of the permanent magnet) (see Figure 2.24). Table 2.4 collects the typical parameters of hysteresis loop for various hard magnetic materials.

After approaching the point 3 on the hysteresis loop, we can continue increasing the magnetic field (in negative direction) and arrive at the opposite tip—close to negative saturation B_s (point 4). Next, if we continue the change of the magnetic field (from negative to positive value), we will not return to the starting point 0 but we will close the loop at the point 1. If we magnetize the material by an alternating sinusoidal magnetic field, we will keep following the loop (every cycle of alternating field will correspond to a full run round the loop). The hysteresis loops are different for a different peak value of the magnetizing field—by changing this value, we can obtain a family of hysteresis loops (Figure 2.12a). By connecting the tips of these loops, the curve very close to the primary magnetization curve is drawn.

If at some arbitrary point the change of the magnetizing field is reversed, then the loop does not follow the same contour but branches out to form a *minor hysteresis loop* (Figure 2.12b).

In the case of soft magnetic materials, the hysteresis loops $B(H)$ and $J(H)$ are practically the same. This is not the case with hard magnetic materials (Figure 2.13), for which significant differences appear due to much higher values of the magnetic field strength (compare the values of coercivity in Tables 2.3 and 2.4).

2.4 Anisotropy and Texture

2.4.1 Magnetocrystalline Anisotropy

The material exhibits *anisotropy* if its properties change with the direction, for example, in the case of magnetic materials if magnetic properties depend on direction of

---

* If we determine the value of coercivity from $B(H)$ loop, we denote it usually as $_BH_c$—as opposed to the value of the coercivity determined from $J(H)$ loop and indicated as $_JH_c$ (Figure 2.13).

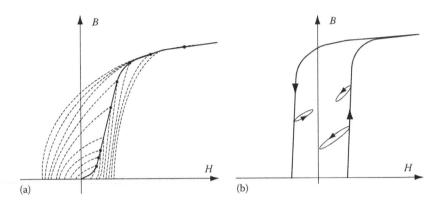

FIGURE 2.12
The family of hysteresis loops obtained for various amplitudes of the magnetizing fields and the magnetization curve obtained by connecting the tips of these loops (a); major and minor hysteresis loops (b).

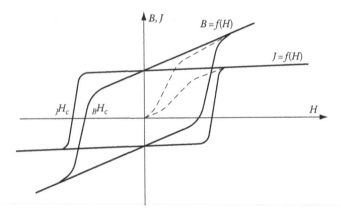

FIGURE 2.13
Hysteresis loops determined for a hard magnetic material.

magnetization. If there are no directional variations (the same properties in all directions), the material is isotropic. The anisotropy can be an intrinsic phenomenon (for instance, due to the crystal structure) or can be induced, for example, by plastic deformation (rolling), annealing, irradiation, etc. (Cullity and Graham 2009). Anisotropy can also be created during the manufacturing process—it is possible to induce the uniaxial anisotropy in thin magnetic film by deposition in the presence of a magnetic field. Even if the material is basically isotropic, the sample acquires anisotropy due to *shape anisotropy*—demagnetization field (see the next section).

Most of ferromagnetic materials consist of multiple crystals—they are polycrystalline. It is assumed that the strong magnetic anisotropy of crystal is caused by spin–orbit–lattice interaction. Figure 2.14 presents the magnetization curves of iron crystal, which has a body-centered cubic (BCC) structure. It can be seen that iron crystal exhibits excellent magnetic properties along [100]* axis (we say that it is an easy axis of magnetiza-

---

* Usually, we use Miller indexes for description of crystallographic axes and planes. Figure 2.15 presents the examples of Miller index for cubic lattice.

tion), medium properties along the [110] axis, and poor magnetic properties along the [111] axis (a hard axis of magnetization) (Figure 2.15).

Typical polycrystalline materials comprise randomly distributed crystals and therefore the magnetization curve is an average of the presented in Figure 2.14. However, the following conclusions can be drawn from Figure 2.14:

- If we know the direction of magnetic flux in the designed product (e.g., in a wound core or in a limb of the transformer), it would be desirable to have the excellent properties of the [100] direction by ordering the crystals along this axis. Indeed, there are magnetic materials with ordered direction of crystals (or simply *the texture*)—for example, grain-oriented electrical steel. The preferred direction is normally synonymous with the *rolling direction* because

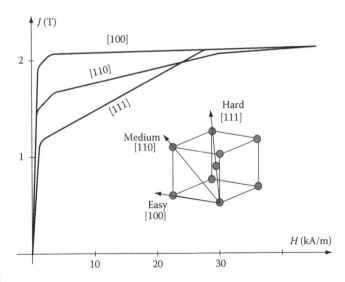

FIGURE 2.14
Magnetization curves of single crystal of iron (monocrystal).

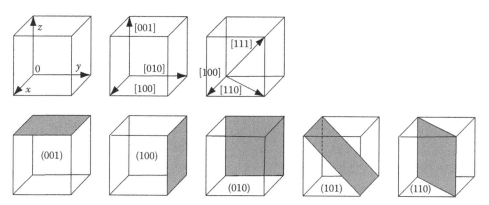

FIGURE 2.15
Illustration of usage of Miller index.

the texture is often induced by special kind of thickness reduction by hot or cold rolling.

- If the magnetic flux changes its direction (for instance, in rotating electrical machines), it is important to eliminate anisotropy because it generates additional power loss. There are several groups of isotropic materials: nonoriented electrical steel, amorphous, or nanocrystalline. However, in reality too, these materials exhibit some degree of anisotropy due to manufacturing process: rolling, casting, or annealing.

As it was described earlier, the domain structure forms in such a way as to remain on a state of minimum energy—the preferable orientation of domain magnetization tries to follow an easy axis of local crystallites. The anisotropy energy E_k of cubic structure is described by the following relationship (Akulov 1929):

$$E_k = K_0 + K_1(\alpha_1^2\alpha_2^2 + \alpha_2^2\alpha_3^2 + \alpha_3^2\alpha_1^2) + K_2(\alpha_1\alpha_2\alpha_3)^2 \quad (2.27)$$

where
   $\alpha_1, \alpha_2, \alpha_3$ are the directional cosines of the magnetization with respect to the cube axes
   K_1, K_2 are the anisotropy constants (K_0 is the isotropy energy contribution)

For iron, the anisotropy constants are $K_1 = 4.8 \times 10^4$ J/m³ and $K_2 = -1.0 \times 10^4$ J/m³.

Anisotropic properties of a given material are often presented in the form of polar figures. Two examples of such plots—energy and magnetic field strength versus the direction of magnetization are presented in Figure 2.16. It is evident that the worst performances are for the angle 54.7°, which corresponds with the [111] hard axis of magnetization.

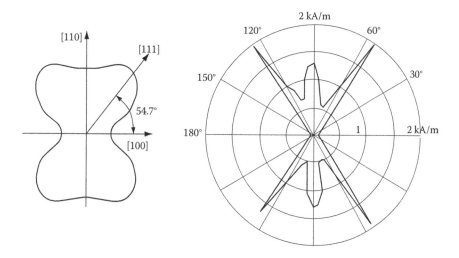

FIGURE 2.16
Polar plot of crystal anisotropy energy for iron crystal and polar plot of experimentally determined magnetic field strength for grain-oriented steel.

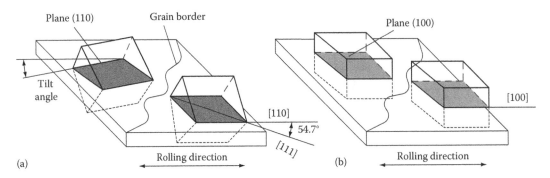

FIGURE 2.17
Two textures with preferred easy axis of magnetization: cube on edge (Goss) (a) and face-centered cubic (b).

2.4.2 Texture

In 1934, Goss patented an invention of producing textured silicon-iron electrical steel (Goss 1934, 1935). By using special metallurgical processes (combination of rolling and annealing), he developed a technology resulting in a cube-on-edge or (110)[100] texture (Figure 2.17a). In this texture, the crystals are positioned with one easy axis close to the rolling direction. This technology (with many improvements) is still commonly used in manufacturing of grain-oriented electrical steel worldwide.

Another texture with an easy axis in an induced direction, face-centered cubic [100](100), is presented in Figure 2.17b. Such texture exists in Permalloy alloys (Graham 1969) and in laboratory scale in silicon-iron steel (the cube-on-face or double-oriented steel).

For the Goss texture, we can substitute the directional cosines in Equation 2.27 by

$$\alpha_1 = \sin\theta\sin 45°; \quad \alpha_2 = \sin\theta\cos 45°; \quad \alpha_3 = \cos\theta \quad (2.28)$$

where θ is the angle of magnetization. After simple calculation, the energy is (Chikazumi 2009)

$$E_k = K_0 + K_1(0.25\sin^4\theta + \sin^2\theta\cos^2\theta) + 0.25K_2\sin^4\theta\cos^2\theta \quad (2.29)$$

Figure 2.18 presents dependence of the energy on the angle of magnetization computed from this relationship as well as results calculated for cubic texture (Chikazumi 2009)

$$E_k = K_0 + K_1\cos^2\theta\sin^2\theta \quad (2.30)$$

Figure 2.19 presents the measured magnetic field strength versus the direction of magnetization. Although proportions are different than those in the case of theoretical results for energy (Figure 2.18), all three axes 0°, 55°, and 90° are clearly detectable.

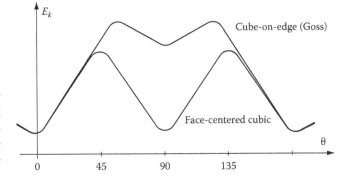

FIGURE 2.18
Angular dependence of magnetocrystalline energy calculated for the Goss and cubic texture.

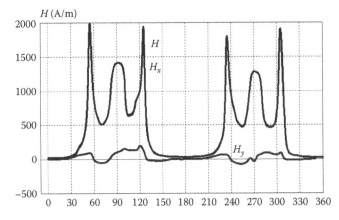

FIGURE 2.19
Angular dependence of magnetic field strength determined experimentally for the grain-oriented steel. (From Tumanski, Bakon 2001.)

The most reliable method of texture analysis is to use x-ray diffraction, which was developed by Laue (Nobel Prize 1914). From picture of the x-ray diffraction on elements of crystal lattice are obtained pole figures—Figure 2.20 presents pole figures for Goss and cubic textures. It is also possible to use other diffraction methods—for instance, the electron backscattered diffraction (EBSD)

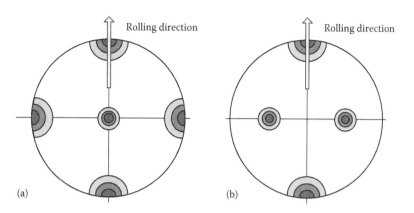

FIGURE 2.20
Pole figures representing the cubic (a) and the Goss (b) texture. (From Graham, C.D., Textured magnetic materials, in *Magnetism and Metallurgy*, Berkowitz, A.E. and Kneller, E. (Eds), Academic Press, San Diego, CA, Chapter XV, 1969.)

method. The EBSD data can be analyzed in MATLAB® toolbox MTEX, which was developed especially for the texture analysis.

There is no direct unit of the texture. Usually, texture analysis comprises statistical analysis of diffraction data—what kind of texture is dominating and what is the angular dispersion from rolling direction (tilt and yaw angles) (Bunge 1982, Blandford and Szpunar 1989, Wenk and Van Houtte 2004). The orientation distribution function (ODF) parameter is often given as a result of analysis. It informs about volume fraction of grain orientation along certain direction.

2.4.3 Shape Anisotropy

When a piece of ferromagnetic material is magnetized, it creates its own magnetic field (de facto this magnetized sample operates as a magnet due to poles induced at its ends). This demagnetizing fields H_d depends on the shape of the sample and can be expressed by demagnetization factor N_d:

$$H_d = N_d M \tag{2.31}$$

The demagnetizing field H_d is in the opposite direction to the magnetizing field H and decreases the internal field of the sample H_{in}:

$$H_{in} = H - N_d M \tag{2.32}$$

Because the demagnetizing field influences the internal field, in practice, we can never investigate the properties of ferromagnetic material but only the properties of the actual sample of the material—the magnetic performances are shape dependent. The main problem is that the demagnetizing field is usually nonuniform and difficult to calculate—thus sample is magnetized nonuniformly and we do not know exactly the value of internal field H_{in}. There is one exception—it is the shape of ellipsoid (which also include sphere). An ellipsoid is magnetized uniformly in an external applied field (Figure 2.21) and, for a given ellipsoid, we can determine relatively easy the demagnetizing factor N_d.

For an elongated ellipsoid with length l and diameter d (at the center), the demagnetizing factor can be described by the following simplified relationship (Osborn 1945):

$$N_d \approx \frac{1}{k^2}(\ln 2k - 1) \tag{2.33}$$

where k is the shape factor of ellipsoid (length/diameter ratio). Figure 2.22 presents the dependence of demagnetizing factor on the "slimness" of ellipsoid. For $k=1$ (sphere) $N_d=1/3$. For $k=10$ (long ellipsoid) N_d is 0.02.

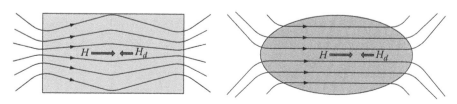

FIGURE 2.21
The sample is magnetized nonuniformly in a uniform external field (a) with exception of ellipsoid shape (b).

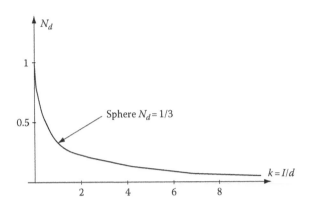

FIGURE 2.22
Dependence of the demagnetizing factor of ellipsoid on its shape factor l/d (length to diameter ratio).

The demagnetizing field significantly deteriorates the apparent permeability of a ferromagnetic sample. The resultant permeability μ_c is

$$\mu_c = \frac{\mu_r}{1 + N_d(\mu_r - 1)} \tag{2.34}$$

For example, for a magnetic core with dimensions $l = 300\,mm$ and $D = 10\,mm$ made from material of permeability $\mu_r = 100,000$, the resultant permeability is only about 300. However, the influence of the shape anisotropy can also be advantageous. If μ_r is very large, the relationship (2.34) can be approximated to

$$\mu_c \approx \frac{1}{N_d} \tag{2.35}$$

This means that for a high-permeability material, the apparent permeability of the core practically does not depend on the material features (e.g., changes with temperature) but only on the core dimensions.

Both, the magnetization curve and the hysteresis loop can be significantly affected by the shape anisotropy. Figure 2.23 presents differences between those obtained for the sample and for the material (i.e., under conditions where shape of the sample can be neglected). The difference is represented by the demagnetization line with slope $B/H\mu_0 = -1/N_d$.

Taking into account the shape anisotropy, it is clear that we should prefer closed ring-shaped cores because in such case demagnetization factor is near zero.

For a sample with an air gap (as for instance in Figure 2.6), the relationship (2.25) can be written as

$$nI = H_{Fe}(l - l_p) + H_a l_a \tag{2.36}$$

Without the gap in the sample, the magnetic field strength would be $H' = nI/l$, and thus

$$H' - H_{Fe} = M \frac{l_a}{l} \tag{2.37}$$

Due to continuity of magnetic flux $B_{Fe} = B_a$ and $\mu_0(H_{Fe} + M) = \mu_0 H_a$ and

$$H_a = M + H_{Fe} \tag{2.38}$$

So from (2.37) and (2.38), the demagnetizing factor of the ring core with an air gap is

$$N_d \cong \frac{l_a}{l} \tag{2.39}$$

In the case of permanent magnet, the air gap always exists. Due to demagnetizing field, the remanent induction of the material B_r is decreased to the value B_m as it is depicted in Figure 2.24. The right side of the graph in Figure 2.24 shows the dependence of magnetic energy (BH) on the induction, whereas the left side shows a

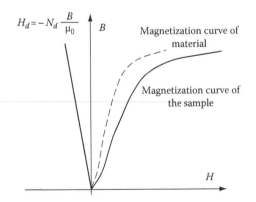

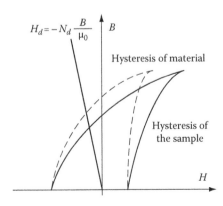

FIGURE 2.23
Influence of shape anisotropy on magnetization curve and hysteresis loop.

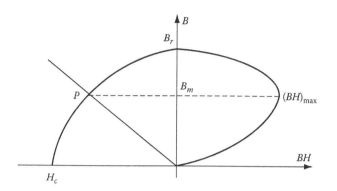

FIGURE 2.24
Influence of shape anisotropy on remanent induction in permanent magnet.

part of the hysteresis loop in the third quadrant. Correct choice of a shape of the permanent magnet should ensure that the operating point P corresponds with the maximum stored magnetic energy (this problem is discussed in Section 3.6.1).

2.5 Electromagnetic Loss

2.5.1 Axial Magnetization Power Loss

Losses are an important parameter of electrical steel. The grades of electrical steel (and of course its price) strongly depend on loss. The loss is mainly dissipated as heat, thus it is wasted energy. Although efficiency of modern power transformers is as high as 99%, it is estimated that annual losses of energy just in the United Kingdom are equivalent to about 7×10^6 of barrels of oil, what is equivalent to about 35,000 ton of SO_2 and 4×10^6 ton of CO_2 (Moses 1990). It is estimated that annual magnetic core losses in the United States amount to nearly 45 billion kWh costing about 3 billion dollars (Moses 1990, Moses and Leicht 2004). No wonder that all steel manufacturers try to improve electrical steel quality because even small improvement translates into significant economical and environmental benefits.

Figure 2.25 presents historical improvement of iron loss of electrical steel. Over the last 100 years, the losses were reduced close to the theoretical value $P_{1.5/50} = 0.4\,\text{W}/$kg* starting from about 15 W/kg. During this history, four revolutionary improvements occurred: addition of Si in 1900, Goss texture in 1934, development of HiBi steel in 1970, and development of amorphous material in 1980.

Power loss can be estimated by two different approaches—physical and engineering concept (Sievert 2005) (Figure 2.26). The first one utilizes the time t and area a integral of the Poynting vector[†] (over one cycle of the magnetization process)

$$P = \frac{1}{T} \int_0^T \oiint_a (E \times H)\, da\, dt \qquad (2.40)$$

Equation 2.40 can be transformed[‡] into known dependence for magnetic loss in volume V (Bertotti 1998):

$$P = \frac{V}{T} \int_0^T H \frac{dB}{dt} dt \qquad (2.41)[§]$$

Taking into account the engineering concept, we simply measure the power delivered to a magnetized circuit with the primary winding n_1 and secondary winding n_2 (Fiorillo 2004):

$$P = \frac{1}{T} \frac{n_1}{n_2} \int_0^T i_1(t) u_2(t) dt \qquad (2.42)$$

Because the current I_1 in the primary winding is proportional to the magnetic field strength H (according to Ampere's law $H\,l = I_1 n$) and the emf induced in the secondary winding is proportional to dB/dt (according to Faraday's law), both concepts realize the same idea.

By convention, power loss is determined as specific power loss P_w related to the mass m of the sample

$$P_w = \frac{P}{m} = \frac{V}{mT} \int_0^T H \frac{dB}{dt} dt \qquad (2.43)$$

or taking into consideration that specific density γ is $\gamma = m/V$

$$P_w = \frac{1}{\gamma T} \int_0^T H \frac{dB}{dt} dt \qquad (2.44)$$

[†] Poynting vector $S = E \times H$ (E, electrical field H, magnetic field) represents energy flux vector of electromagnetic field delivered to magnetized body (therefore, vector is directed toward the magnetized surface). The Poynting vector theory was developed by Umov; therefore, it is also referred to as a Pointing–Umov vector.
[‡] The derivation of losses from the Poynting's vector is described later (Section 2.9.5).
[§] Dependence (2.41) is sometimes presented with substitution of polarization instead of induction.

---

* $P_{1.5/50}$ means loss determined for induction 1.5 T and for frequency 50 Hz.

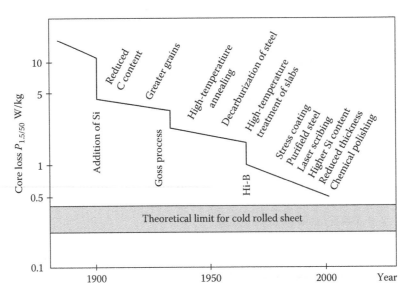

FIGURE 2.25
Historical improvement in iron loss of silicon steel. (From Stodolny, J., *Metall. Foundry Eng.*, 21, 307, 1995.)

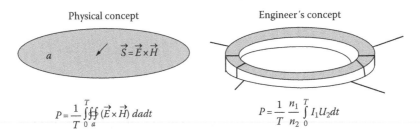

FIGURE 2.26
Two concepts of power loss determination. (From Sievert, J., *Przegl. Elektrotech.*, 5, 1, 2005.)

TABLE 2.5

Specific Power Loss of Common Soft Magnetic Materials (at 50 Hz)

Material	$P_{1.3}$ (W/kg)	$P_{1.5}$ (W/kg)	$P_{1.7}$ (W/kg)
Silicon steel oriented	0.6	0.8–1.1	1.2–1.6
Silicon steel nonoriented		2.3–13	
Silicon steel (laboratory)	0.13	0.17	0.21
Silicon steel HiB			0.6–1
Amorphous Metglas 2605	0.11	0.27	

TABLE 2.6

Specific Power Loss of Common Soft Magnetic Materials (at 1 T)

Material	400 Hz	1 kHz	5 kHz
SiFe oriented, 0.05 mm	8	22	200
SiFe nonoriented, 0.1 mm	11	33	350
Nanocrystalline	0.2	0.9	30
Amorphous Metglas 2605	1.6	4	23
Permalloy	3	10	150

Source: Waeckerle, T. and Alves, F., Aliages magnetiques amorphes, in *Materiaux magnetiques en genie electrique* 2, Kedous-Lebous A. (Ed.), Lavoisier, Chapter 1, 2006.

The relationship (2.44) is useful if we do not know the mass of the sample, for example, when we determine localized specific loss.

Tables 2.5 and 2.6 present a specific power loss of some common soft magnetic materials. The loss depends approximately on the square of flux density as it is presented in Figure 2.27. The power loss depends also on frequency f as it is presented in Figure 2.28.

The power loss depends on the flux density B_m, thickness of the sample t, frequency f, and resistivity of the material ρ, according to the approximate relationship:

$$P = C_0 B_m^2 f + \frac{\pi^2 t^2}{6\rho}(B_m f)^2 + C_1 B_m f^{3/2} = P_H + P_{EC} + P_{EX}$$

(2.45)

The first part of the dependence (2.45) is related to static hysteresis loss because it depends on the energy necessary to magnetize the material and is proportional to

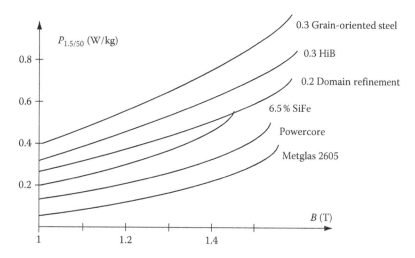

FIGURE 2.27
Variation of specific loss versus the flux density for various materials. (From Moses, A.J., *IEE Proc.*, 137, 233, 1990.)

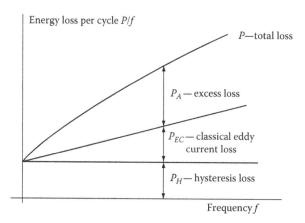

FIGURE 2.28
Method of loss separation.

the area of static hysteresis loop. The second part of Equation 2.45 is related to the loss caused by eddy currents and can be reduced by decrease of thickness of the sheet and increase of the resistivity of the material (it is one of the reasons for addition of the silicon to the iron). The last part is called excess loss (or sometimes anomalous loss). This third component of losses is thought to be caused by DW movement. It was named as "excess" because the total value of measured loss was exceeding the values determined by a classical theory based only on hysteresis and eddy current (although the excess loss is also related to eddy currents) (Pry and Bean 1958).

Because each component of total loss depends on frequency in its own way, it is relatively easy to separate these components by preparation of $P/f = f(f)$ plot (Figure 2.28). In nonoriented steel, hysteresis losses are about 60% of the total loss, but in amorphous materials, the excess loss on its own can account up 90% of the total loss.

2.5.2 Power Loss under Rotational Magnetization

Beside uniaxial magnetization, materials can also be magnetized by a field rotating in a plane or, more general, in a space. Such kind of magnetization is present obviously in rotating electrical machines but also parts of transformer cores can be subjected to a rotating magnetic field (Moses 1992). Rotational loss can be in many cases higher than the "normal" axial loss. The rotational loss can also be derived from the Poynting vector as (Pfützner 1994)

$$P_r = \frac{1}{\gamma T} \int_0^T \left(H_x \frac{dB_x}{dt} + H_y \frac{dB_y}{dt} \right) dt \qquad (2.46)$$

To determine the rotational loss, it is necessary to determine both components of magnetic field strength and magnetic flux density (directed along x and y axis). Moreover, the rotational loss depends on the relation between both components of magnetizing field—loss for circular magnetization ($B_x = B_y$) can be almost two times higher in comparison with elliptical magnetization ($B_x = k B_y$) (Anuszczyk and Pluta 2009). This loss can also depend on the direction of rotation—clockwise and counterclockwise (Zurek and Meydan 2006, Todaka et al. 2009).

Similarly, as in the case of axial magnetization between the magnetic field strength and the flux density (or polarization), there is a phase shift, which causes hysteresis—this time this phase shift is a spatial angle φ_r (Figure 2.29). The rotational loss can be described as

$$P_r = \frac{1}{T} \int_0^T H_r B_r \sin \varphi_r dt \qquad (2.47)$$

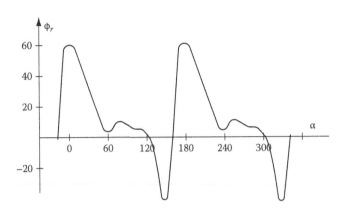

FIGURE 2.29
Phase angle between *H* and *B* under rotational magnetization. (From Moses, A.J. and Leicht, J., *Przegl. Elektrotech.*, 12, 1181, 2004.)

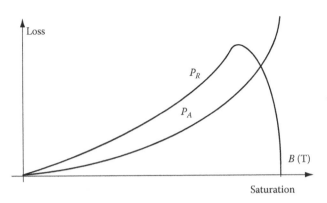

FIGURE 2.31
Comparison of rotational and alternating losses as a function of flux density.

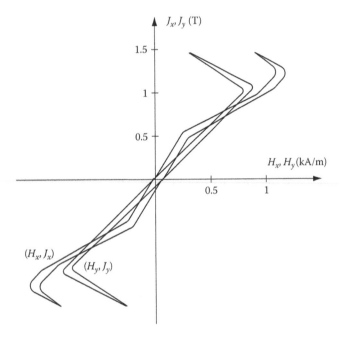

FIGURE 2.30
Rotational hysteresis loops determined under controlled circular polarization (NO SiFe 3%, *f* = 10 Hz). (*Measurement and Characterization of Magnetic Materials*, Fiorillo, F., Copyright (2004), from Elsevier.)

Figure 2.30 presents rotational hysteresis loops determined under controlled circular polarization.

Under rotational magnetization, the anisotropy of material becomes an important factor because in the case of the Goss texture, the material must also be magnetized in its hard axis (54.7°) four times per cycle of rotation. Although the rotational loss is caused by effects similar as under axial magnetization (hysteresis, eddy currents, wall motion), the mechanism is different and it is not possible to

predict these losses by, for example, averaging the longitudinal and transversal axial losses. Figure 2.31 presents comparison of rotational and axial losses versus the flux density value—the axial loss increases with induction while rotational loss is zero at saturation (because at saturation phase shift does not exist between *B* and *H**).

In contrary to the axial losses, the rotational losses are not internationally standardized. Therefore, steel manufacturers seldom inform about level of these losses. Figure 2.32 presents the variation of rotational losses with flux density for various materials. Table 2.7 presents the comparison of axial and rotational losses of various materials of different anisotropy.

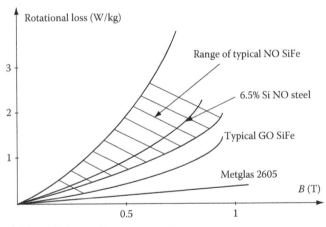

FIGURE 2.32
Variation of rotational losses with flux density for various materials. (From Moses, A.J., *IEE Proc.*, 137, 233, 1990.)

---

* In practice, the rotational loss exists also in saturation due to eddy currents (Zurek 2009).

TABLE 2.7

Comparison of an Axial P_A and a Rotational P_R Losses ($B = 1\,\text{T}, f = 50\,\text{Hz}$)

Material	P_A (W/kg)	P_R (W/kg)
SiFe nonoriented 2.7% Si	1.4	3.5
SiFe nonoriented 1.2% Si	1.23	4
SiFe oriented 3.2% Si	0.46	1.84

Source: Moses, A.J., *J. Mater. Eng. Perform.*, 1, 235, 1992.

2.6 Influence of the Magnetic Field on Physical Properties of a Material

2.6.1 Magnetostriction and Other Magnetoelastic Effects

Changes of magnetic field can cause changes of dimensions of magnetized material. This effect (discovered by Joule in 1842) is called the *magnetostriction* or *Joule effect*.

Magnetostriction can be used in actuators—for example, as generators (vibrators) of ultrasound. But in most devices, it is an unwanted parasitic effect. It causes extra energy losses well known as acoustic noise emitted by transformers (Weiser et al. 2000). Therefore, if possible, it is recommended to use the material with low magnetostriction.

Magnetostriction effect is described by longitudinal changes in length λ,

$$\lambda = \frac{\Delta l}{l} \qquad (2.48)$$

As a material parameter, the saturation magnetostriction λ_s is used—determined when the material is magnetized from zero to saturation. Table 2.8 presents values of saturation magnetostriction of various materials.

The simplified picture of physical origin of the magnetostriction is illustrated in Figure 2.33. The rotation of magnetization or the DW displacement changes the strain inside a crystal causing its change of dimensions. Therefore, the 180° DWs do not contribute to the magnetostriction but the movements of 90° walls is related to more significant changes of dimensions (generally changes of dimensions are rather small—on the level of 10^{-6}, see Table 2.8).

The mechanism of magnetostriction is complex. It depends on the crystal symmetry and in atomic scale on spin–orbit interaction (Du Tremolet de Lacheisserie 1993, Chikazumi 2009, Cullity and Graham 2009). It depends also on texture and material composition and it is possible to obtain material with negligible magnetostriction (see Table 2.8). And vice versa, it is possible to obtain

TABLE 2.8

Typical Values of Saturation Magnetostriction of Various Materials at Room Temperature

Material	λ_s (×10⁻⁶)
Iron	−7
SiFe 3.2% Si	+9
SiFe 6.5% Si	0
Nickel	−33
Permalloy 45% Ni–55% Fe	+27
Permalloy 82% Ni–18% Fe	0
Amorphous Metglas 2605	30
Amorphous Co72Fe3B6Al3	0
SmFe₂	−1560
Terfenol D Tb0.3Dy0.7Fe1.93	+2000

Source: McCurrie, R.A., *Ferromagnetic Materials—Structure and Properties*, Academic Press, San Diego, CA, 1994.

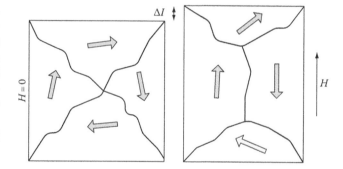

FIGURE 2.33
Simplified mechanism of magnetostriction caused by domain wall movement.

materials with large values of the magnetostriction—the most known is so-called giant magnetostrictive material Terfenol D, with saturation magnetostriction of about 2000×10^{-6}.

Because magnetostriction is strongly related to crystal symmetry, it is therefore an anisotropic effect and depends on direction of stress. For cubic crystal symmetry, such as in iron, the magnetostriction along the direction of magnetization can be described as

$$\lambda_c = \lambda_{100} + 3(\lambda_{111} - \lambda_{100})(\alpha_1^2\alpha_2^2 + \alpha_2^2\alpha_3^2 + \alpha_3^2\alpha_1^2) \qquad (2.49)$$

where
$\lambda_{100}, \lambda_{111}$ are the saturation magnetization along the directions < 100 > and < 111 >
α are the direction cosines of magnetization relative to the crystal axis

Figure 2.34 presents the magnetostriction on the magnetization determined for single crystal of iron.

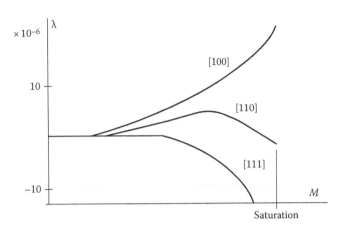

FIGURE 2.34
Magnetostriction of the single crystal of iron determined along the principal axes. (From Cullity, B.D. and Graham, C.D., *Introduction to Magnetic Materials* © (2009) IEEE; Webster, W.I., *Proc. R. Soc.*, A109, 570, 1925.)

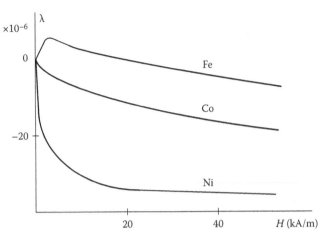

FIGURE 2.35
Magnetostriction of the polycrystalline samples of the main ferromagnetic metals. (From Cullity, B.D. and Graham, C.D., *Introduction to Magnetic Materials* © (2009) IEEE; Lee, E.W., *Rep. Prog. Phys.*, 18, 184, 1955.)

For an isotropic material ($\lambda_{100} = \lambda_{111} = \lambda_i$), the magnetostriction depends on the angle of magnetization θ:

$$\lambda = \frac{3}{2}\lambda_i\left(\cos^2\theta - \frac{1}{3}\right) \qquad (2.50)$$

and for polycrystalline sample (Cullity and Graham 2009),

$$\lambda = \frac{1}{2\pi}\int_0^{\theta=\pi/2}\int_0^{\phi=\pi/2}\lambda\sin\phi\,d\phi\,d\theta = \frac{1}{5}(2\lambda_{100} + 3\lambda_{111}) \qquad (2.51)$$

where θ, ϕ angles of direction of magnetization for x and z axes, respectively. Figure 2.35 presents the magnetostriction characteristics of polycrystalline samples of the main ferromagnetic metals.

Magnetostriction exhibits the hysteresis as shown in Figure 2.36. This hysteresis means that under AC magnetization, the sample vibrates at twice the frequency of magnetization—therefore, humming sound of transformers is two times of the power frequency.

There is an effect inverse to magnetostrictive one—the Villari effect in which the mechanical stress causes changes of magnetic properties of material. The Villari effect can be advantageous in sensor applications—for example, it can be used in the pressure sensors. Figure 2.37 presents simplified demonstration of the inverse magnetostriction effect as the equivalence to the situation presented in Figure 2.33. Under initial conditions, the material is in demagnetized state. After application of the stress σ, the domains with directions coinciding with the direction of the stress change their dimensions.

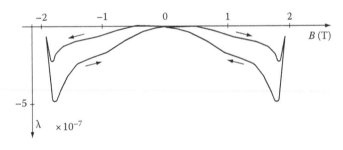

FIGURE 2.36
Hysteresis of the magnetostriction. (From Yabumoto, M., *Przegl. Elektrotech.*, 1, 1, 2009.)

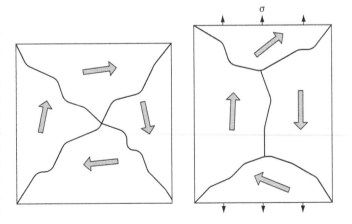

FIGURE 2.37
Simplified illustration of the inverse magnetostriction effect (see also Figure 2.33).

The magnetoelastic energy created by stress directed at an angle γ can be described as (Cullity and Graham 2009)

$$E_{el} = -\frac{3}{2}\lambda_{100}\sigma(\alpha_1^2\gamma_1^2 + \alpha_2^2\gamma_2^2 + \alpha_3^2\gamma_3^2)$$
$$-3\lambda_{111}\sigma(\alpha_1\alpha_2\gamma_1\gamma_2 + \alpha_2\alpha_3\gamma_2\gamma_3 + \alpha_3\alpha_1\gamma_3\gamma_1) \quad (2.52)$$

and for an isotropic sample ($\lambda_{100} = \lambda_{111} = \lambda_i$), this energy is

$$E_{el} = \frac{3}{2}\lambda_i\sigma\sin^2\theta \quad (2.53)$$

In some materials, stress can improve its parameters, which is the reason why the electrical steel is often covered by special insulation coating that creates tensile stress in the material.

Due to Villarri effect, the stressed material, primary demagnetized can exhibit the magnetization detectable by magnetic field sensor—such phenomenon was applied to investigate fatigue effects of ferromagnetic materials (Figure 2.38) (Kaleta et al. 1996).

The inverse magnetization effect causes that due to the stresses introduced during various technologic processes (cutting, punching, welding, riveting, etc.), the performances of the final product can be significantly worse than the performances of the virgin material (Wilczynski et al. 2004).

Beside the magnetostriction, there are also other magnetoelastic effects. For example, material can twist when a helical magnetic field is applied—this phenomenon is called the *Wiedemann effect*. If we push through a magnetostrictive wire pulses of current that generate helical magnetic field and add an external axial magnetic field, then the mechanical twists create the ultrasonic waves.

The inverse Wiedemann effect called also as *Matteucci effect* is often used for torque measurement. When a material is subjected to a torque, its magnetization changes, for example, if it is a magnetostrictive wire.

2.6.2 Magnetoresistance

Practically, all materials change their resistivity in applied magnetic field but for mostly substances to observe such effect, the magnetic field should be very large and temperature should be low. There are several materials that exhibit useful large changes of resistivity in the room temperature under not too large magnetic field—semiconductors as InSb, bismuth, and some ferromagnetic materials* (Tumanski 2001). Currently, most often, the ferromagnetic thin-film elements are used as magnetoresistive sensors.

The magnetoresistive effect in ferromagnetic materials was discovered in 1857 by Thomson (Lord Kelvin) (Thomson 1857) and was called *anisotropic magnetoresistance* (*AMR*) because the changes of resistance were different in longitudinal and transverse directions with respect to the direction of magnetizing field.

It was necessary to wait more than 100 years until practical application of this disclosure was possible. In bulk material, this effect is hard to detect and the magnetization of this material by external magnetic field is complex. The era of magnetoresistive sensors started when technology of thin ferromagnetic film was available. Thin ferromagnetic films, made most often from non-magnetostrictive Permalloy 80Ni/20Fe, exhibit special features—they have uniaxial anisotropy in the film plane and are magnetized similarly to a single domain—by rotation of magnetization. The anisotropy is usually induced during the deposition process—it is performed in the presence of magnetic field. Due to such anisotropy, a thin film is magnetized initially along the easy axis of anisotropy. If we apply external magnetic field perpendicular to the anisotropy axis, the direction of magnetization starts to rotate by some angle φ.

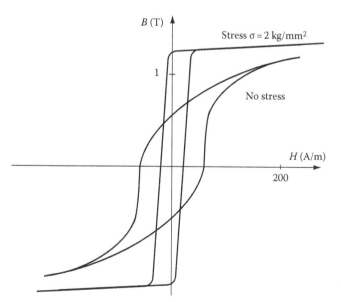

FIGURE 2.38
The change of magnetic material (Permalloy 68) performances after application of the stress. (From Bozorth, R.M., *Ferromagnetism*, Van Nostrand, New York, 1951.)

---

* Bismuth was used as magnetic field sensor many years ago—today it is only the technical curiosity because, to obtain this effect, a large magnetic field is necessary (about 2 T). Semiconductor magnetoresistors are today substituted by ferromagnetic because they also need relatively high magnetic field (about 0.1 T) and it is difficult to eliminate the influence of temperature.

The direction of magnetization can be determined from the balance of free energy. This energy E is described as

$$E = -HM + K_u \sin^2 \varphi \qquad (2.54)$$

where K_u is anisotropy constant.

The first term represents the magnetostatic energy and the second one the anisotropy energy. We can rewrite Equation 2.54 as

$$E = -\mu_0 M_s H_x \sin \varphi - \mu_0 M_s H_y \cos \varphi + \frac{1}{2}\mu_0 M_s H_k \sin^2 \varphi \qquad (2.55)$$

where $H_k = 2K_u/M_s$ is the anisotropy field (parameter of the material).

The simplest ferromagnetic magnetoresistive sensor is created in form of the current conducting thin-film narrow strip. Measured is the change of its resistance. To enlarge the effect, often this strip is created in form of a meander.

The minimum of energy with respect to the magnetization direction φ is when $\partial E / \partial \varphi = 0$. From this condition, we obtain information about the direction of magnetization in the following form:

$$\sin \varphi = \frac{H_x}{H_k + H_y} \qquad (2.56)$$

The relative change of resistance of thin-film strip depends on the direction of magnetization with respect to the direction ϑ of the current (Figure 2.39):

$$\frac{\Delta R_x}{R_x} = -\frac{\Delta \rho}{\rho} \sin^2 \vartheta \qquad (2.57)$$

where $\Delta\rho/\rho$ is magnetoresitivity coefficient (maximum change of resistivity of the material)—for Pemalloy $\Delta\rho/\rho$ is about 2%.

If thin-film strip is directed along anisotropy axis ($\varepsilon = 0$ and $\varphi = \vartheta$), the response of the magnetoresistor to external magnetic field H_x is

$$\frac{\Delta R_x}{R_x} = \frac{\Delta \rho}{\rho} \frac{1}{(H_k + H_y)^2} H_x^2 \qquad (2.58)$$

This dependence is nonlinear (parabolic—see Figure 2.40). We can obtain linear dependence $\Delta R_x / R_x = f(H_x)$ in two ways—by biasing the thin film with an additional field H_{xo} field (i.e., by shifting the operation point into linear part of the characteristic) or by changing the initial direction of current with respect to the anisotropy axis (by changing the angle ε). For the angle $\varepsilon = 45°$, we obtain an almost linear effect and more importantly, differential with respect to the case when $\varepsilon = -45°$ (see Figure 2.40). For $\varepsilon = \pm 45°$, the dependence $\Delta R_x / R_x = f(H_x)$ is

$$\frac{\Delta R_x}{R_x} \cong \frac{\Delta \rho}{\rho}\left(\frac{1}{2} + \frac{1}{H_k + H_y} H_x \right) \qquad (2.59)$$

In 1988, other magnetoresistive effect in thin ferromagnetic films was discovered—the so-called *giant magnetoresistance (GMR)* (for this work, Albert Fert and Peter Grünberg obtained a Nobel Prize in 2007) (Fert 2008).

Discovery of the GMR effect was made in two steps. The first one was the observation that if the thickness of a conductive film (spacer) between two parallel

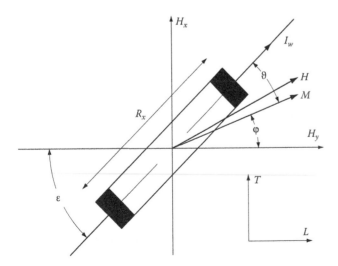

FIGURE 2.39
The thin-film ferromagnetic device—the simplest ferromagnetic magnetorersistive sensor (L, anisotropy axis).

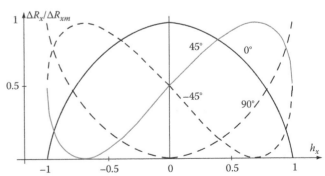

FIGURE 2.40
The change of the resistance of a magnetoresistor with the strip inclined of the angle ε with respect to the anisotropy axis ($H_y = 0$ and $H_x = H_x/H_k$).

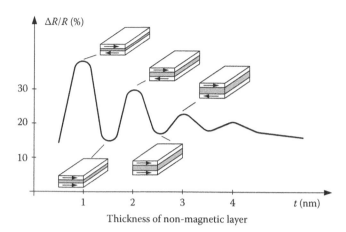

FIGURE 2.41
The evidence of oscillatory exchange coupling in thin-film multilayers.

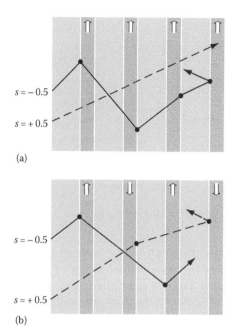

(a)

(b)

FIGURE 2.43
The spin-dependent scattering of electrons in the multilayer system with magnetization aligned parallel (a) and antiparallel (b). (After White, R.L., *IEEE Trans. Magn.*, 28, 2482, 1992.)

ferromagnetic films is very small, then due to the atomic coupling between these films, direction of magnetization became antiparallel. Moreover, the direction of these magnetizations changes periodically (between parallel and antiparallel) with thickness (Figure 2.41).

The second discovery was announced in 1988 by Baibich (Fert team). They stated that transition between antiparallel and parallel magnetization causes large (giant) change of resistance—even about 200% (Figure 2.42).

The GMR effect is explained as a result of spin–orbit interaction and because the change of resistance is spin dependent, the whole technique is called as *spintronics* (Daughton 2003, Parkin et al. 2003, Wolf et al. 2006, Bandyopadhyay and Cahay 2008). In simplified form, the spin-dependent scattering of electrons is presented in Figure 2.43.

In the illustration of scattering of electrons in a multilayer system (Figure 2.43), there are two classes of electrons—with spin 1/2 and spin –1/2. Scattering

of electrons on the interface between layers depends on relationship between the spin and the direction of magnetization. In the case of antiparallel order of layers, both classes of electron are scattered on every second interface. In the case of parallel order, one of the electrons class is privileged—that with the spin coinciding with direction of magnetization and travelling almost without scattering. Thus, transition between antiparallel and parallel order means decrease of resistance.

AMR and later GMR sensors became great scientific, technical, and commercial success because they are used in reading heads of disk memory devices. The original multilayer GMR effect is recently used rather seldom because it requires large external magnetic field to overcome atomic coupling between the layers (to force transition between antiparallel and parallel state). Instead of this, different mechanism called *spin valve* (*SV*) is used (Dieny 1991). Great progress is also observed in a special kind of the GMR effect called *tunnel magnetoresistance* (*TMR*). In the tunnel magnetoresistor, a spacer is made not from conductive material but from an insulator. Theory of the tunneling magnetoresistance was described many years ago (Slonczewski 1989) but only today it is possible technologically to obtain very thin (several atoms) layer of insulator. Lately, a TMR experiment based on MgO insulator layer with magnetoresistivity coefficient as large as 400% was reported (Parkin et al. 2004, Yuasa et al. 2004).

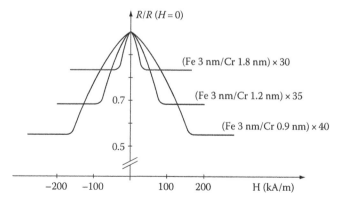

FIGURE 2.42
The first announcement of the GMR effects—results obtained by Baibich. (From Baibich, M.N., *Phys. Rev. Lett.*, 61, 2472, 1988.)

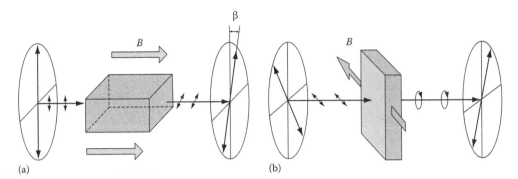

FIGURE 2.44
The principle of Faraday (a) and Voigt (b) effects.

Beside the thin film, the magnetoresistive effect also appears in some crystal materials—for example, extraordinary large effect called *colossal magnetoresistance* was discovered in manganese-based perovskite oxides (Jin 1994, Tumanski 2001). In colossal magnetoresistors, it is possible to switch between insulator and conductor state (magnetoresistance as high as 10^8% (Chen 1996). Unfortunately, colossal effect requires low temperature (below 100 K) and large magnetic field (above 2000 kA/m).

In thin-film ferromagnetic structures and in ferromagnetic nanowires supplied by high-frequency current, large (even 400% for a relatively small magnetic field not exceeding 1 kA/m) change of impedance, *giant magnetoimpedance* (GMI) effect is detectable. The magnetoimpedance effect arises from the change of permeability with external magnetic field and consequently, the change of the depth of the skin effect (Velazquez 1994, Panina 1995).

Thin-film magnetoresistive sensors are described later in this book in Section 4.4.

2.6.3 Magneto-Optical Phenomena

There are several effects where light (or generally EM wave) can be altered by the magnetic field (Pershan 1967, Freiser 1968). In 1845, Faraday discovered that in optically transparent materials, the linearly polarized light changes the angle of polarization depending on the magnetic field B (*magneto-optical Faraday effect*). This rotation angle β in the sample of the length l is described by the following equation:

$$\beta = \nu Bl \qquad (2.60)$$

where ν is the Verdet constant. There are some materials, for example, terbium gallium garnet (TGG) with extremely large Verdet constant, as −134 rad/Tm for light wavelength 633 nm (Verdet constant strongly depends on the light wavelength) (Figure 2.44).

Similar effect appears when the magnetic field is perpendicular to the direction of the light. This effect is known as *Voigt effect* (analogous effect in liquids is known as *Cotton–Mouton effect*). In such cases, this phenomenon is more complex due to double refraction (*birefringence*)—as a result, transmitted light is polarized elliptically. Generally, both effects (Faraday and Voigt) can exist simultaneously, which complicates the analysis because a rotation of polarization in the Voigt effect is quadratic function of B.

As the explanation of the Faraday effect, it is assumed that propagation of circularly polarized light is different for clockwise and anticlockwise state and in the output light, the phase difference between both components is different than in the input beam.*

The Faraday effect is limited only to optically transparent materials. For the analysis of opaque or metallic surfaces, another effect is used in which the reflected polarized light changes its angle of polarization depending on the magnetization of material—*magneto-optical Kerr effect* (MOKE) (Kerr 1877). The reflected polarized light changes its angle θ of polarization and ellipticity a/b as a function of magnetization M (Figure 2.45). This effect can be used for testing of the local value of magnetization (or flux density) of a magnetized body and is commonly used to investigate the domain structures (Hubert 1998).

The analysis of the Kerr effect is complex because it consists of various effects depending on the direction of magnetization—polar effect when magnetization is perpendicular to the plane of the material, longitudinal effect when magnetization is in the plane of the material and parallel to the plane of the incidence, transverse effect when magnetization is in the plane of the material and perpendicular to the plane of the incidence. Moreover, this first-order effect is often accompanied by the second-order quadratic effects.

---

* We can assume that the linearly polarized light beam is composed of two oppositely polarized circular light beams.

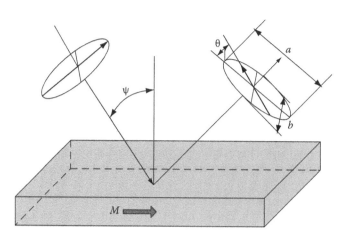

FIGURE 2.45
The Kerr effect.

The total signal amplitude relative to the incident amplitude A_{tot} depends on the polarization ψ, reflection coefficients R_p, R_s, R_k^{pol}, R_k^{lon}, R_k^{tra}, and components of magnetization m_{pol}, m_{lon}, and m_{tra} (Hubert 1998):

$$A_{tot} = -R_p \cos\psi\sin\theta + R_s \sin\psi\cos\theta + R_k^{pol}\cos(\theta-\psi)m_{pol}$$
$$+ R_k^{lon}\cos(\theta+\psi)m_{lon} - R_k^{tra}\cos\psi\sin\theta m_{tra} \qquad (2.61)$$

where θ is the angle of rotation of the direction of polarization (see Figure 2.45).

By applying the numerical analysis of the Kerr effect for various sample configurations, it is possible to separate these effects and determine the value of magnetization (Rave et al. 1993, Defoug et al. 1996, Hubert 1998). However, due to difficulties and poor resolution of the method, it is used mainly for qualitative rather than quantitative studies (for instance, domain structure observation).

Figure 2.46 presents the Kerr testing system designed for measurements of local magnetization and hysteresis loop in the thin-film samples (Wrona 2002). The Glan–Thompson prism transmits only linearly polarized light that is then reflected from a polished ferromagnetic sample. This reflected light is split into two linearly polarized beams and detected by two photodetectors. This way, it is possible to determine positions of two main axes of ellipsoid.

Dutch physicist Zeeman has observed that spectral lines of sodium vapor change their width when the magnetic field was applied (Zeeman 1897).* He discovered that the energy levels are split in the presence of magnetic field (*Zeeman effect*). The spectral energy levels are split into several sublevels with different wave-

---

* Zeeman was awarded a Nobel Prize in 1902 for this discovery.

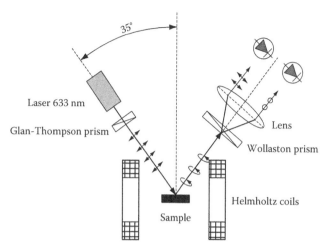

FIGURE 2.46
An example of the hysteresis loop tracer based on the Kerr effect. (From Wrona, J., Magnetometry of the thin film structures, PhD thesis, AGH University of Science and Technology, Cracow, Poland, 2002.)

lengths and the distance between these sublevels ΔE depends on the magnetic field (Figure 2.47):

$$\Delta E = \hbar\omega_0 = \frac{e\hbar}{m_e}H = \hbar\gamma H \qquad (2.62)$$

where
$\hbar$ is the Planck constant
e, m_e are the charge and mass of electron, respectively
γ is the gyromagnetic ratio

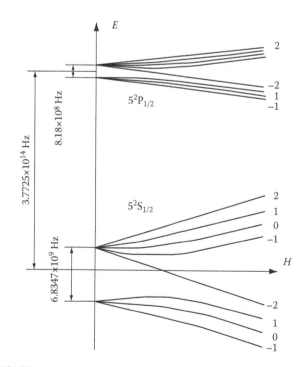

FIGURE 2.47
The splitting of energy levels due to Zeeman effect—an example of energy level splitting for Rb87. (After Hartmann, 1972.)

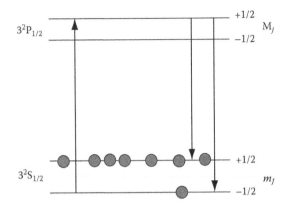

$3^2P_{1/2}$ $+1/2$ M_j $-1/2$

$3^2S_{1/2}$ $+1/2$ m_j $-1/2$

FIGURE 2.48
The principle of optical pumping.

The splitting of the energy levels can be observed as the emission (or absorption—in this case, it is the *inverse Zeeman effect*) of the light of the wavelength (or frequency) corresponding with transitions of the electrons between sublevels of the energy. The absorption or dispersion of the light exhibits resonance feature versus magnetic field or frequency (resonance frequency is ω_0). These effects are used for measurements of very small magnetic fields. Alkaline vapors are used commonly as a specimen for such measurements. Figure 2.47 presents the energy levels splitting of the rubidium Rb87.

To observe the energy levels splitting, the optical pumping technique* is commonly used. If we deliver circularly polarized light beams, then an atom of the energy level m_j absorbs a photon and is excited to the level M_j (Figure 2.48). According to the *law of conservation of angular moment* (selection rule), when the polarized light is absorbed, it must be related to a change of spin moment.

The principle of optical pumping is explained in Figure 2.48 on an example of sodium vapor. If we deliver circularly polarized light of appropriate wavelength (tuned laser or filtered light), the electrons are excited from state 3^2S to 3^2P. But only electrons with a spin $-1/2$ can be excited for selected polarization of light beam (from $m_j = -1/2$ to $M_j = +1/2$). Thus, after certain period of time, the two low sublevels are not occupied symmetrically—sublevel $+1/2$ is occupied because transition from this state is forbidden (for selected light polarization).

In a "fully pumped" state, the sample is practically transparent for the light beam because all electrons that can absorb light are excited. Optically pumped sample tries to return to the state of equilibrium and we can determine the external magnetic field value knowing the Zeeman sublevel splitting. This resonance method utilizing *change of opacity* of gaseous cell as

dependence of external magnetic field (optical pumping magnetometer) is described in Section 4.7.4.

2.6.4 Magnetocaloric Effect

Magnetic field can cause changes of temperature of materials. It is expected that this *magnetocaloric effect* (MCE) can be used in magnetic refrigeration. Some materials heat up when they are magnetized and cool down when they are removed from magnetic field (Tishin and Spichkin 2003). This effect was discovered by Warburg in 1881 for pure iron and explained by Debye (1926) and Giaugue (1927). Debye and Giaugue proposed to use this effect for cooling by an adiabatic demagnetization.

Recently, materials with relatively high MCE, mainly gadolinium and its alloys, have been found. Such materials exhibit MCE as high as 3–4 K for 1 T. Unfortunately, gadolinium is rather expensive; therefore, other materials based on manganite compounds are proposed (Phan 2007, Szymczak and Szymczak 2008).

By applying MCE, it was possible to obtain temperature below 0.3 K (Giaugue and MacDougal 1933).

The principle of adiabatic demagnetization is presented in Figure 2.49. The specimen is magnetized and in this process the temperature increases. Next in the state of full magnetization, the heat is removed using, for example, liquid helium. Next after demagnetization, we obtain cooling of the specimen.

2.7 Magnetic Resonance

2.7.1 Gyromagnetic Ratio and Larmor Precession

Electrons, nuclei, or atoms are rotating, which is expressed by their *angular moment J* quantized by the spin number *I*:

$$J = \hbar I \tag{2.63}$$

Because of this rotation, the particles having electric charge exhibit the *magnetic moment* μ depending on their mass *m* and charge *q*:

$$\mu = \frac{q\hbar}{2m} = \gamma\hbar \tag{2.64}$$

where the γ coefficient is called *gyromagnetic ratio*.

For example, for free electron with mass m_e and charge *e*, this magnetic moment is

$$\mu_B = \frac{e\hbar}{2m_e} \tag{2.65}$$

---

* This technique was developed by 1966 Nobel prize winner Alfred Kastler.

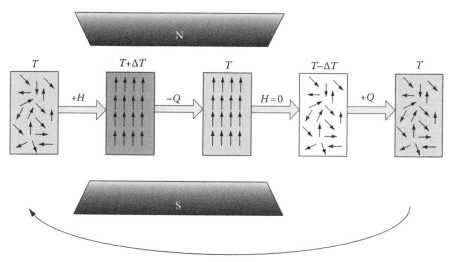

FIGURE 2.49
The principle of magnetic refrigeration.

This elementary magnetic moment is called the *Bohr magneton.** The magnetic moment is a vector value and depends on which spin we take into account. For example, for electron with angular spin this moment $\boldsymbol{\mu}_s$ is

$$\boldsymbol{\mu}_S = -g_s \mu_B \left(\frac{\mathbf{S}}{\hbar} \right) \qquad (2.66)$$

where

g_s is a coefficient called the electron g-factor (equal to -2.002319)

S is spin number[†]

If we put such particle rotating with magnetic moment $\boldsymbol{\mu}$ into external magnetic field, the magnetic field exerts torque τ:

$$\tau = -\boldsymbol{\mu} \times \boldsymbol{B} \qquad (2.67)$$

This torque causes that the particle acts as a gyroscope rotating with precession around the direction of the external magnetic field (Figure 2.50). The torque from angular moment should be equal to the torque caused by the external field

$$\hbar \frac{d\mathbf{I}}{dt} = \boldsymbol{\mu} \times \boldsymbol{B} \qquad (2.68)$$

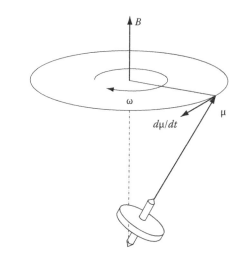

FIGURE 2.50
Free precession of a particle with magnetic moment μ in external magnetic field.

and relationship describing this gyroscope is

$$\frac{d\boldsymbol{\mu}}{dt} = \gamma \boldsymbol{\mu} \times \boldsymbol{B} \qquad (2.69)$$

For a simplified case of constant magnetic field B directed along one of the coordinate axes, the solution of Equation 2.69 is relatively simple (Kittel 2004) and we obtain that

$$\omega_0 = \gamma B \qquad (2.70)$$

Equation 2.69 describes precession effect of elementary particle. For the whole magnetized sample, we can

---

[*] Bohr magneton is equal to 9.274009×10^{-24} JT^{-1}. Similarly we can determine the nuclear magneton $\mu_N = e\hbar/m_p$ (m_p, mass of the proton) and $\mu_N = 5.050583 \times 10^{-27}$ JT^{-1} (which is, therefore, three order of magnitude lower than the Borh magneton).

[†] Similarly we can determine: electron orbital spin magnetic momentum as $\boldsymbol{\mu}_L = g_l \mu_B(\mathbf{L}/\hbar)$ (g_l is equal to 1, $\mathbf{L}$ is spin number); proton spin magnetic momentum as $\boldsymbol{\mu}_P = g_p \mu_B(\mathbf{I}/\hbar)$ (g_p is equal to $5.585694\ldots$).

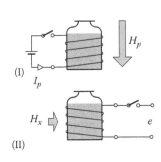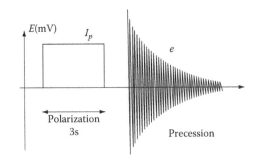

FIGURE 2.51
Free proton precession detected in polarized cell with water.

extend this relation assuming that the magnetization M is $M = \sum \mu_i$ and

$$\frac{dM}{dt} = \gamma M \times B \qquad (2.71)$$

This means that in external magnetic field, a free particle is rotating with frequency depending on the value of this field called *Larmor frequency*. The gyromagnetic ratio γ for free electron is expressed as

$$\gamma_e = \frac{eg_s}{2m_e} = \frac{g_s \mu_B}{\hbar} = 1.76085977 \times 10^{11} \text{ rad/sT}$$

and

$$\gamma_e' = \frac{\gamma_e}{2\pi} = 28.1481 \text{ GHz/T}$$

For a proton, the gyromagnetic ratio is

$$\gamma_p = \frac{eg_p}{2m_p} = 2.67515255 \times 10^8 \text{ rad/sT}$$

and

$$\gamma_p' = \frac{\gamma_p}{2\pi} = 42.576375 \text{ MHz/T}$$

Thus, the frequency of rotation strictly depends on the value of external magnetic field and can be determined with high precision. Moreover, we can see that the precession of electrons enable obtaining much higher sensitivity to external magnetic field than the proton precession.

The evidence of the proton precession can be confirmed in relatively simple experiment presented in Figure 2.51. We can use a container with proton-rich medium (e.g., water). In the first step, we polarize this water by connecting large DC current I_p to the coil generating the magnetic field. Then, the magnetizing source is disconnected and to the same coil is connected an oscilloscope. We can observe decaying oscillations of induced voltage e of the frequency proportional to external magnetic field.

2.7.2 Nuclear Magnetic Resonance

In the nuclear magnetic resonance as the precession of the nuclei is investigated. Table 2.9 presents properties of typical nuclei used in the nuclear magnetic resonance (NMR) analysis. We can see that if we use the NMR for magnetic field measurement, then the recommended material is that rich of hydrogen ^1H because it exhibits large gyromagnetic ratio (and therefore it enables measurements with highest sensitivity). Therefore, typically a water or benzene is used as a medium. Moreover, we can see that components of organic materials also have large gyromagnetic ratio, which predestines this method for investigation in medicine (*NMR tomography*).

The sample needs to be initially polarized (ordered) because otherwise spins are oriented chaotically and NMR is not detectable. The most frequently used

TABLE 2.9

Properties of Various Nuclei Used in NMR Analysis

Nuclei	Gyromagnetic Ratio, $\gamma/2\pi$ (MHz/T)	Spin
^1H	42.576	1/2
^2H	6.536	1
^{13}C	10.71	1/2
^{23}Na	11.26	3/2
^{14}N	3.08	1
^{39}K	1.99	3/2
^{17}O	30	5/2

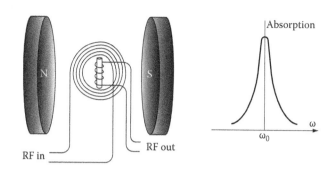

FIGURE 2.52
The nuclear magnetic resonance observed due to the presence of the additional perpendicular AC field H_1 and the resonance curve of absorption.

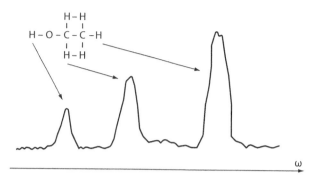

FIGURE 2.53
An example of the NMR spectrogram.

method of polarization is to magnetize the sample with static, large magnetic field. If we remove polarization, we can observe free precession of protons as it is demonstrated in Figure 2.51. The analysis of the free precession is a typical application of the NMR phenomenon, for example, in magnetic field measurements and in the NMR tomography.

Beside passive observation of free precession, the NMR analysis is also carried out under the resonance conditions. To obtain such conditions, the additionally high-frequency magnetic field $H_1 \sin \omega_1 t$ is used (directed perpendicularly to the static magnetic field H_0). Next, this ω_1 frequency is tuned and the precession signal is detected. For the frequency equal exactly to Larmor frequency, the delivered energy is absorbed in the most efficient way, which is observed as a resonance curve (Figure 2.52).

The delivered energy of a microwave beam $\hbar \omega_1$ can be absorbed if it causes change of energy level, thus the condition of resonance is

$$\hbar \omega_1 = \hbar \gamma B \qquad (2.72)$$

As a consequence, it can be expected that the resonance frequency ω_0 is equal to Larmor frequency

$$\omega_1 = \omega_0 = \gamma B \qquad (2.73)$$

In the case of multielement sample, several resonance peaks are obtained as shown in Figure 2.53. This way, we can analyze the chemical composition of the investigated sample and such method is called *NMR spectroscopy* (Rouessac and Rouessac 2000).

In the NMR analysis, the dynamic properties of the medium are very important. If we introduce the polarizing magnetic field, it is necessary to wait certain amount of time to obtain equilibrium conditions. This time is defined by longitudinal relaxation time T_1 (called also *spin lattice relaxation time*) and

$$M_z = M_0(1 - e^{t/T_1}) \qquad (2.74)$$

Moreover, precession of individual spins start with different Larmor frequencies and phases and only after certain time is obtained the equilibrium condition. This relaxation transverse time T_2 (called also *spin–spin relaxation time*) causes that the magnetization in xy plane is established after certain period of time:

$$M_{xy} = M_{xy0} e^{-t/T_2} \qquad (2.75)$$

During observation of the free precession, we try to decrease the relaxation time because it prolongs the time of sampling. In medicine applications (NMR tomography), the investigated patient is placed in a nonuniform static magnetic field and by changing the frequency of an alternating field, we can localize a given part of a body. In the tomography method, the body is excited by a pulsed magnetic field, and for tests of tissue, the analysis of relaxation times is used because they are very sensitive to the state of the tissue. Table 2.10 presents example of relaxation times for various human tissue.

TABLE 2.10

Example of Relaxation Times of Various Human Tissues

Tissue Type	T_1 (ms)	T_2 (ms)
Bone	0.001–1	0.001–1
Muscle	540–930	120–240
Fat	180–300	47–63
Body fluid	1000–2000	150–480
Brain gray matter	340–610	90–130
Brain white matter	220–350	80–120

Source: Guy, C. and Ffyche, D., *The Principles of Medical Imaging*, Imperial College Press, London, U.K., 2005.

In the analysis of the precession, the relaxation times can be taken into account by using the *Bloch equations*,

$$\frac{dM_{xy}}{dt} = \gamma(M \times B)_{xy} - \frac{M_{xy}}{T_2}$$

$$\frac{dM_z}{dt} = \gamma(M \times B)_z + \frac{M_0 - M_z}{T_1} \tag{2.76}$$

For the case presented in Figure 2.52, when the sample is magnetized by the magnetic field **B**,

$$B = B_0 z + B_1 \left(x \cos \omega t - y \sin \omega t \right) \tag{2.77}$$

the solution of Bloch equations is (Kittel 2004)

$$M = \frac{\gamma M_z T_2 B_1}{[1 + (\omega_0 - \omega)^2 T_2^2]^{1/2}} \tag{2.78}$$

It is the relationship describing the resonance curve (Figure 2.52) with resonance for $\omega = \omega_0$.

2.7.3 Electron Spin Resonance

Due to smaller mass of electron, the gyromagnetic ratio of *electron spin resonance (ESR)* (also called electron paramagnetic resonance [EPR]) is much larger than in the case of NMR, which enables investigation of much lower magnetic fields (e.g., in the outer space).* Due to large difference of Larmor frequencies for both phenomenons, it is possible easy separation of both resonances—NMR and ESR.

Generally, the ESR is used less frequently than the NMR because

- It is limited to certain group of chemical compounds having unpaired electrons
- It needs high-frequency microwave testing devices in GHz range, which is technically more demanding
- The picture of spectral lines can be complicated and difficult to interpret due to the presence of two electron resonances (from electron rotation and orbital movement) and moreover it can be disturbed by nuclear resonance (hyperfine coupling)

Nevertheless, ESR is used in two main important applications:

- Measurement of very small magnetic fields (optical pumping magnetometers)
- Spectroscopy analysis, most of all for diagnosis and detection of free radicals, important in the cancer risk analysis

The ESR is based on the Zeeman energy levels splitting under the influence of external magnetic field. Because the Zeeman energy separation is $\Delta E = g\mu_B B$, the condition of resonance is

$$\hbar\omega_0 = g\mu_B B = \gamma B \tag{2.79}$$

Methods of the analysis of the resonance are very similar to those used in the NMR with the difference being the use of microwave techniques—in the presence of polarizing, static magnetic field B_0 is used the additional AC magnetic field B_1, which is perpendicular to B_0. The ESR effect can be observed by changing the value of magnetic field for fixed value of frequency—the maximum absorption of delivered microwave appears. It is possible to reverse this operation—observation of the absorption of delivered microwave by changing its frequency for fixed B_0 field (Figure 2.54).

Instead of polarization by a static magnetic field B_0, the optical pumping technique (see Section 2.7.3) is commonly used for magnetic field measurement. An example of such magnetometer is presented in Figure 2.55. The gaseous sample in a glass vessel (cell) is placed between a light source and a light detector. Between the light source and the cell, there are lens, filter, and circular polarizer. The filter ensures that an appropriate wavelength of light is delivered—for example, in the case of ^{4}He it is $\lambda = 1.083\,\mu m$ and in the case of ^{87}Rb it is $\lambda = 0.795\,\mu m$. Often, in the light bulb, the same substance as in the cell is added. In the fully pumped state, the transparency of the cell is highest

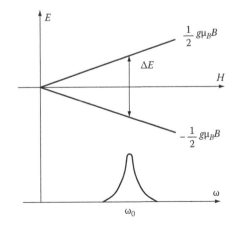

FIGURE 2.54
The principle of ESR phenomenon.

---

* The resolution of the NMR magnetic field measurements is not high—the Larmor frequency for Earth's magnetic field is only about 2 kHz while for the ESR it is around 1.5 MHz.

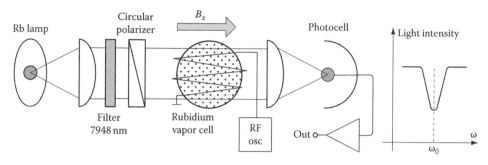

FIGURE 2.55
The example of an ESR magnetometer utilizing optical pumping technique.

due to a minimum of the light absorption. The optical pumping plays the same role as the polarization by magnetic field in the NMR.

In the circuit presented in Figure 2.55, an additional RF magnetic field with a frequency close to Larmor frequency is used to initiate precession. In the state of resonance, the absorption of the light is highest. To detect the resonance condition, an additional modulating AC magnetic field is usually added.

Table 2.11 presents properties of typical substances used in optical pumping magnetometers. To obtain alkali metals in gaseous state, it is necessary to heat them—the temperature of evaporation of rubidium is about 40°C, while cesium needs 22°C. In the case of helium isotope, heating is not necessary because it is gaseous.

2.7.4 Ferromagnetic Resonance

Ferromagnetic materials exhibit spontaneous magnetization (even in the absence of external magnetic field). Thus, if they have magnetic moments, we can expect that also in such materials the resonance phenomenon occurs. Indeed in most of the ferromagnetic materials, *ferromagnetic resonance* is detectable. We can assume that all local magnetic moments are represented by a resultant magnetization vector M of the whole sample. Similarly as in other resonances, this effect can be analyzed by polarization of the sample with large static magnetic field B_0 and the change of perpendicular to this field additional magnetic field B_1 of high frequency.

TABLE 2.11

Typical Media Used in Optical Pumping Magnetometers

Substance	Number of Energy Sublevels	Gyromagnetic Ratio, γ (GHz/T)	λ (μm)
^{85}Rb	7	4.667	0.795
^{87}Rb	5	6.999	0.795
^{133}Cs	9	3.499	0.894
^{4}He	3	28	1.083

This resonance frequency depends on the gyromagnetic ratio γ_{Fe} and magnetic field value:

$$\hbar\omega = \gamma_{Fe} B_0 \qquad (2.80)$$

In comparison with earlier discussed resonances, we should take into account several differences. Firstly, there is a significant influence of the shape of the sample represented by the demagnetization factor N_d. The resultant magnetic field in the sample with anisotropy field H_k is

$$H = H_0 + H_1(t) - N_d M + H_k \qquad (2.81)$$

The resonance can be significantly influenced also by the domain structure. Therefore, usually the magnetic field B_0 is high enough to obtain magnetic saturation and remove the DWs. High magnetic field B_0 results in high resonance frequency, even in GHz range. But high frequency of the field B_1 complicates the resonance effect due to increased damping (caused, e.g., by eddy current and skin effect). To describe the ferromagnetic resonance effect, the *Landau and Lifshitz equation* (Landau and Lifshitz 1935) of motion can be used:

$$\frac{dM}{dt} = \gamma(M \times B) - \frac{\alpha\gamma}{M}[M \times (M \times B)] \qquad (2.82)$$

where α is $\alpha = \lambda/M\gamma$ and λ is damping factor.

The ferromagnetic resonance is very useful for analysis of the ferromagnetic thin films. In this case, the influence of demagnetization field is negligible (the magnetization vector is in the film plane and sample is often circular) and eddy currents and skin effects are also small. Moreover, because in thin ferromagnetic films the domain structure is close to one-domain state, the polarizing field (and resonance frequency) can be relatively lower.

Figure 2.56a presents an example of resonance curve of a thin-film sample with uniaxial anisotropy obtained

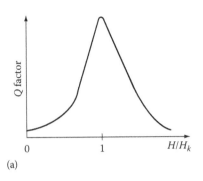
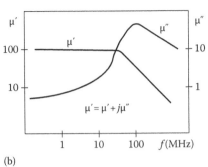

FIGURE 2.56
The method of testing of the anisotropy field of a thin film (a) (after Hasty, T., *J. Appl. Phys.*, 34, 1079, 1963) and the complex permeability determined for nickel-zinc ferrite (b) (after Smit, J. and Wijn, H.P.J.: *Ferrites.* 1959. Copyright Wiley-VCH Verlag GmbH & Co. KGaA).

in RF range (Hasty 1963). From this relationship, the anisotropy field and direction of the anisotropy axis can be detected.

When ferromagnetic sample is placed inside the coil, we can analyze the complex permeability by the measurement of the impedance of a coil. Figure 2.56b presents the results of tests of the complex permeability of ferrite. The imaginary part of this permeability exhibits resonant properties.

2.7.5 Overhauser Resonance

Overhauser predicted that it should be possible to polarize the sample in NMR not by a conventional large static magnetic field B_0 but by the ESR resonance spin coupling (Overhauser 1953). Such method is de facto a double resonance system NMR/ESR. Polarization of the sample is transferred from ESR to the NMR via cross-relaxation. This technique (called *dynamic nuclear polarization [DNP]*) (Barker 1962) is recently used with success in commercially available magnetometers (Kernevez and Glenat 1991, 1992, Duret et al. 1995).

The cross effect requires unpaired electrons to initiate the ESR and, at the same time, a proton-rich sample for the NMR resonance. For this goal, the free radicals can be used (Beljers et al. 1954). In *Overhauser magnetometers*, a nitroxide-free radical (Tempone® nitroxide-free radical) is commonly used as a medium (Figure 2.57a) with unpaired free electron diluted in a proton-rich solvent.

Figure 2.57b presents the energy levels of tempone ^{15}N specimen. Without a magnetic field, the energy levels are split due to hyperfine coupling A corresponding to a frequency about 60 MHz. In the presence of external magnetic field, the tripled split occurs due to the Zeeman effect. Initial polarization can be obtained by inserting the sample inside a 60 MHz cavity resonator. It is possible to obtain a dynamic nuclear polarization very high, and in this way the sensitivity of the NMR resonance

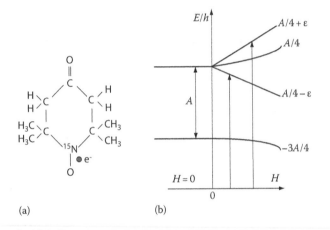

FIGURE 2.57
Nitroxide free radical (unpaired electron near the nitrogen atom is indicated by the dot) (a) and the energy levels of this substance (b).

in the Overhauser magnetometer is comparable with optically pumped helium magnetometers—10 pT/Hz$^{1/2}$ (Kernevez et al. 1992).

The Overhauser magnetometers have following advantages in comparison to the conventional NMR effect:

- Short-time polarization (also while measurement process is performed) that enables close to continuous measurements—while in conventional NMR we have to reserve time break for polarization and relaxation.

- The conventional NMR requires high static magnetic field for polarization while the Overhauser effect needs only small RF field (with frequency 60 MHz for nitroxide). It means that the power consumption is decreased radically, which is especially important in the space investigation (Duret et al. 1995).

- Polarization via DNP is much stronger than by magnetic field; therefore, the signal-to-noise ratio and the sensitivity are better.

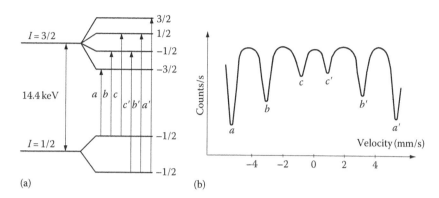

FIGURE 2.58
The Zeeman splitting of energy levels of ^{57}Fe (a) and the Mössbauer effect spectrogram of this sample (b).

2.7.6 Mössbauer Resonance

The resonance effect occurs when we deliver energy corresponding with the energy necessary to change the energy level (usually transition to a higher level). In the previous sections was described as example of the resonance methods of detection of a Zeeman splitting by absorption or emission of the light. Figure 2.58a presents the Zeeman splitting of nuclear energy levels of ^{57}Fe. The transition from the $I = 1/2$ ground state to $I = 3/2$ level requires the absorption and emission of the γ-ray corresponding to energy gap $\Delta E = 14.4$ keV. Unfortunately, attempts to observe the resonance effect for γ-ray were, for a long time, practically impossible.

The main reason of difficulties in observation of γ-ray resonance is the energy lost for recoil. This energy E_R depends on the γ-ray energy E_γ and is related to the mass m_n of nuclei $E_R = E_\gamma^2/2m_nc^2$. The emitted photon has energy decreased by recoil energy while absorbed photon should have energy increased by the recoil energy if we take into account loss of the energy. Thus, between these two lines (absorption and emission) exist mismatch equal to $2E_R$ and therefore we do not observe the resonance.

Mössbauer was first to prove that it is possible to obtain emission and absorption of x-ray in solid iridium without the recoil (Mössbauer 1958). If the investigated nucleus is not free but bound to the lattice (e.g., in crystal), the recoil energy is significantly smaller because it is related to much larger mass. Moreover, another trick was employed to decreasing of the mismatch between the absorption and emission lines. If the source of emission (or absorbing sample) is moved, then due to the Doppler effect the energy line is also shifted. It was sufficient to move the source with relatively low speed (several cm/s) to detect spectral lines corresponding to the split of energy levels. The shift of energy of the source moving with the speed v due to the Doppler effect is

$$\Delta E = \frac{Ev}{c} \tag{2.83}$$

For 30 keV γ-ray and velocity 1 cm/s, this energy shift is about 10^{-6} eV (Seiden 1969). Figure 2.58b presents the Mössbauer effect spectral lines of ^{57}Fe. This result was not possible to obtain by other methods. Figure 2.59 presents the principle of the Mössbauer spectrophotometer. By applying the Mössbauer resonance, it was possible to investigate fundamental properties of materials, for example, magnetic moment. But it is also possible to use this effect in technical applications, for example, for non-destructive testing of materials (Bray and McBride 1992).

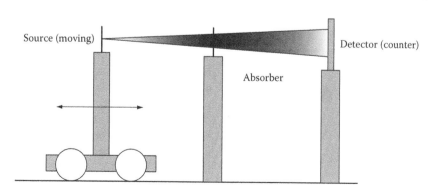

FIGURE 2.59
The principle of Mössbauer spectrophotometer.

2.8 Superconductivity Used in Magnetic Measurements

2.8.1 Superconductivity and Magnetism— The Meissner Effect

It is commonly known that superconductivity means sudden vanishing of electrical resistance in certain materials in low temperature (Figure 2.60a). The material in magnetic field at low temperature becomes not only perfect conductor but also exhibits a perfect diamagnetism (Figure 2.60b). At low temperature, the flux density inside a superconducting material is trapped. In other words, surface of magnetized body is a perfect shield for magnetic field. This effect is known as *Meissner effect*.

Due to Meissner effect, magnetic flux is expelled from the sample (Figure 2.61a). Therefore, in superconducting state, the magnetic flux is trapped inside the sample and current flowing only near the surface acts as shield for magnetic field. Both current and magnetic flux penetrate only near the surface on the depth λ (several μm)

$$B(x) = \mu_0 H_{ex} e^{-x/\lambda} \qquad (2.84)$$

The spectacular experiment confirms this effect (Figure 2.61b). A ferromagnetic material is attracted by a

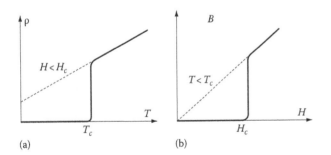

FIGURE 2.60
The material becomes a perfect conductor (a) and perfect diamagnetic (b) at low temperature.

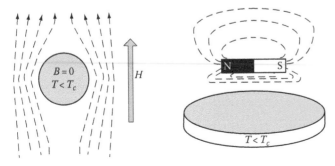

FIGURE 2.61
Two experiments illustrating the Meissner effect.

magnet but the diamagnetic material is repelled. If we put the magnet above the superconducting material and decrease the temperature (to induce the superconductivity), the magnet starts levitation several millimeters above the superconductor (because the force of repelling and force of gravitation are in equilibrium).

The superconducting effect occurs below certain *critical temperature T_c*. This temperature depends also on the level of magnetic field and if this field exceeds a certain value of *critical magnetic field H_c*, then the superconducting effect collapses (Figure 2.62). There are two types of superconducting materials. In *superconductors type I*, the superconducting effect collapses suddenly after the critical field H_c. In *superconductors type II*, there are two critical values of magnetic field (Figure 2.62). If the magnetic field does not exceed the first critical value H_{c1}, the material behaves as a "regular" (Meissner type) superconductor. Above the H_{c1} but below H_{c2}, there is so-called mixed state where the transition to a non-superconducting begins. In the mixed state, the magnetic flux progressively penetrates into the sample through the vortices. Above the second critical magnetic field value H_{c2}, the superconducting effect is completely suppressed.

There are 38 elements and hundreds of alloys and compounds exhibiting superconductivity—the typical examples are presented in Table 2.12. We can divide these materials into two practical groups. The *low-temperature superconductors* (LTS) require a liquid helium (temperature 4.2 K) to remain in superconducting mode. The *high-temperature superconductors* (HTS) can operate at liquid nitrogen (temperature 77 K). Although HTS materials seems to be attractive for measuring applications, they have several disadvantages, most of all electrical noise. Due to relatively high critical temperature, niobium is most often used as a superconducting material.

2.8.2 Josephson Effect

The resistance of almost all conductors decreases with decrease of the temperature due to reduced scattering of free electrons. But the superconductivity is related to other mechanism. Conducting occurs due to pairs of electrons with opposite moments and spins—*Copper pairs*. The Cooper pair comprises two mutually bound electrons existing only at low temperature, and such pair represents an elementary particle with electric charge $q = 2e$ and mass $m = 2m_e$. The phenomenon of conduction at low temperature by a Cooper pair results in various extraordinary quantum effects: conduction through a thin insulator barrier (*tunneling effect* or *Josephson effect*), magnetic flux quantization, quantum Hall effect, etc. These effects are very important also in magnetic measurements.

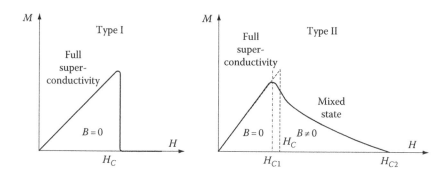

FIGURE 2.62
Magnetization as a function of magnetic field for two types of superconductors.

TABLE 2.12

Typical Superconducting Materials

LTS Materials		HTS Materials	
Material	T_c (K)	Material	T_c (K)
Al	1.2	$YBa_2Cu_3O_7$	92
Hg	4.2	$Bi_2Sr_2Ca_2Cu_3O_{10}$	110
Pb	7.2	$Tl_2Bi_2Ca_2Cu_3O_{10}$	125
Nb	9.3	$HgBa_2Ca_2Cu_3O_8$	135
Nb_3Al	16.0		

Josephson predicted theoretically[*] in 1962 that in two superconductors separated by a very thin (thinner than coherence distance between electron in the pair: 0.1–2 nm) insulator barrier (*weakly coupled superconductors or Josephson junction*), there can be a flow of extraordinary current due to existence of Cooper pairs. This junction can be realized by a tip, thin-film layer, or bridge connection (Figure 2.63). The current in the junction is described by two following equations:

$$i_s = I_c \sin\theta \qquad (2.85)$$

$$f = \frac{d\theta}{dt} = \frac{2e}{h}V \qquad (2.86)$$

The first Josephson Equation 2.85 describes the DC current i_s in the junction. The current I_c is a parameter of the junction and it is the *critical current*—if current exceeds this value, the superconducting effect vanishes. The angle θ (*an order parameter*) is the phase difference between wave functions of electrons in pair at both sides of the junction.[†]

The second Josephson Equation 2.86 describes the alternating current when we supply the junction by DC

voltage V. This current oscillates with frequency $2ev/h$. Thus, the supercurrent in the junction changes with frequency

$$f = C_j v \qquad (2.87)$$

where $C_j = 2e/h$ is Josephson constant.

The Josephson constant depends only on physical fundamental constants and therefore is known with high accuracy and

$$C_j = \frac{2e}{h} = 483.597891\,\text{GHz/V} \qquad (2.88)$$

If we put the Josephson junction in a high-frequency magnetic field, we obtain the quantum standard of voltage. In this standard, voltage versus current changes stepwise (is quantized) with the step value (Figure 2.64):

$$V(n) = nf\frac{h}{2e} = nf\frac{1}{C_j} \qquad (2.89)$$

Note that the voltage $V(n)$ (n is a number of the step) depends on the very well-defined values: h, Planck's constant; e, electron charge; and f, microwave frequency that we are able to measure with high accuracy.

Figure 2.65 presents the dependence of the current versus voltage across the junction. In superconducting state, the voltage across the junction is equal to zero until the current reaches the critical value I_c. When the current exceeds the critical value, beside the Cooper pairs also normal electrons participate in conduction, which causes that the superconductor acts as a normal conductor.

The $V = f(I)$ characteristics of a Josephson junction can have two shapes: with or without hysteresis (Figure 2.65). This behavior depends on the R, L, C parameters of the junction (*McCumber factor* β):

---

[*] Confirmed experimentally next year by Anderson and Rowell.
[†] The phase difference of wave functions of electrons in pairs is called order parameter according to Ginzburg–Landau theory.

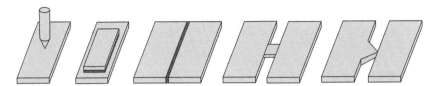

FIGURE 2.63
Various forms of a Josephson junction.

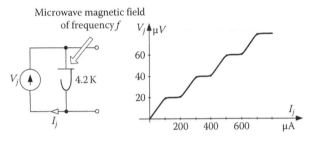

FIGURE 2.64
The Josephson junction as a voltage standard.

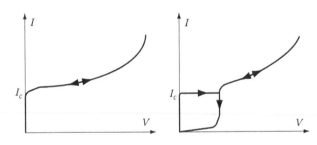

FIGURE 2.65
Current versus voltage characteristic of Josephson tunnel junction.

$$\beta = \frac{2\pi R^2 C I_c}{\Phi_0} = \frac{2\pi L I_c}{\Phi_0} \qquad (2.90)$$

If $\beta < 1$, then $V = f(I)$ characteristic is without the hysteresis.

2.8.3 The SQUID Devices

SQUID is the acronym of *Superconducting Quantum Interference Device*. It is a superconducting ring sample with one or two Josephson tunneling junctions. The flux in the ring with Josephson junction Φ_{in} is quantized and is described by the following relationship:

$$\theta + 2\pi \frac{\Phi_{in}}{\Phi_0} = 2\pi n \qquad (2.91)$$

where Φ_0 is the magnetic flux quantum. The magnetic flux quantum (fluxon) depends only on physical constants,

$$\Phi_0 = \frac{h}{2e} = 2.067833667 \times 10^{-15} \text{ Wb} \qquad (2.92)$$

For a ring with a cross section of $1\,\text{cm}^2$, the flux quantum corresponds to the flux density of 2.067×10^{-11} T and therefore it is commonly used for measurements of extremely small magnetic fields.

Taking into account Equation 2.85, we can determine the current in the ring as

$$i_s = I_c \sin \frac{2\pi \Phi_{in}}{\Phi_0} \qquad (2.93)$$

If the ring of the inductance L is placed in external magnetic field Φ_{ex}, the internal flux is $\Phi_{in} = \Phi_{ex} - L i_s$ and

$$\Phi_{in} = \Phi_{ex} - L I_c \sin \frac{2\pi \Phi_i}{\Phi_0} \qquad (2.94)$$

The relationship (2.94) describes the RF SQUID device—a ring of superconducting material with one Josephson junction. This relationship is graphically presented in Figure 2.67. Again it can be with or without hysteresis—this depends on the inductance L and critical current I_c of the junction (β parameter—see Equation 2.90). Both cases can be used as RF SQUID. When the flux increases, the jumps occur with the intervals equal to quantum flux Φ_0.

The RF SQUID device consists of superconducting ring coupled by a mutual inductance M with resonance

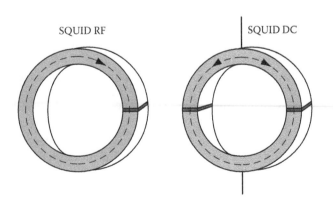

FIGURE 2.66
Two types of SQUID devices.

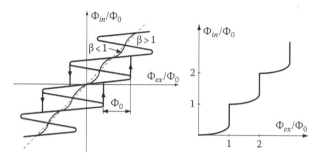

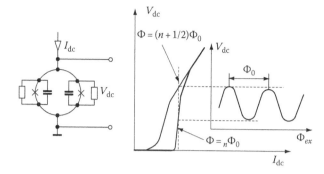

FIGURE 2.67
The magnetic flux inside the RF SQUID versus the external magnetic flux. (From Koch, H., SQUID sensors, in *Magnetic Sensors*, W. Göpel (Ed.), VCH Verlageselschaft, Weinheim, Germany, Chapter 10, 1989.)

FIGURE 2.69
The DC SQUID supplied directly by the current I_{dc}. (After Clark, J., *Proc IEEE.*, 77, 1208, 1989.)

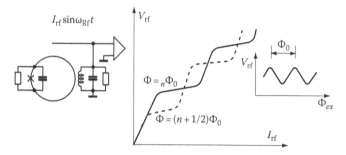

FIGURE 2.68
The RF SQUID coupled with resonance tank circuit. (After Clark, J., *Proc IEEE.*, 77, 1208, 1989.)

2.8.4 Quantum Hall Effect

The normal Hall effect, discovered in 1879, is the basis of the most popular sensor of magnetic field. In the Hall effect, in the plate conducting the current I with magnetic field B perpendicular to the plate, the electric field transverse to electric current appears (due to Lorentz force acting on the moving charges). Therefore, voltage V_H induced on the electrodes on opposite side versus current electrodes (see Figure 2.79) can be used to measure the magnetic field and

circuit LC excited by an RF oscillator (Figure 2.68). The current in the ring is the sum of induced current and superconducting current i_s. The reverse jumps of flux in the ring influences the voltage of the resonance circuit. Figure 2.66 presents typical dependence of voltage on the resonance circuit versus the current and versus the external magnetic flux. The output is a triangle wave with the cycle equal to the magnetic flux quantum.

$$V_H = \frac{1}{tne} IB = K_H B \qquad (2.95)$$

where
t is the thickness of the plate
n is a charge carrier density

It is possible to determine the external magnetic field by simply counting the number of cycles, but more commonly the SQUID device acts as a zero detector due to negative feedback. Thus, we apply only one slope of the output signal.

Some semiconductor materials as InSb, InAs, or GaAs exhibit exceptionally large K_H coefficient.

At very low temperature (below 2 K) and high magnetic field (several Tesla), in 1980, Von Klitzig discovered extraordinary *quantum Hall effect*. This effect exists in 2D electron gas structures. Such structures can be obtained in MOSFET transistors (at the border between Si and SiO_2 insulating cover) on in GaAs/AlGaAs heterostructures.

The SQUID device can be supplied directly by a DC voltage. Such device called DC SQUID has two Josephson junctions shunted by resistors—it consists of two half-ring devices (Figure 2.69). In the first half ring is the sum of current $I/2 + i_s$, in the second one there is a difference of these currents. The resultant characteristic of the voltage versus current for two external magnetic fluxes is presented in Figure 2.69. For fixed value of the I_{dc} current, we obtain periodical dependence of voltage on external magnetic field with the cycle equal to quantum flux Φ_0. Similarly as in the case of RF SQUID, commonly, the DC SQUID device works with negative feedback—as the detector of zero magnetic field.

The quantum Hall effect is most of all used as the standard of resistance because the Hall resistance defined as $R_H = V_H / I$ is

$$R_H(n) = \frac{h}{2e^2} \frac{1}{n} = \frac{1}{n} K_K \qquad (2.96)$$

where n is the number of the step on the characteristic $R_H = f(B)$ (Figure 2.70). The K_K (*von Klitzing constant*) depends only on fundamental physical constants and is known with high accuracy— $K_K = h/2e^2 = 25,812.807 \times 10^{-7}\ \Omega$.

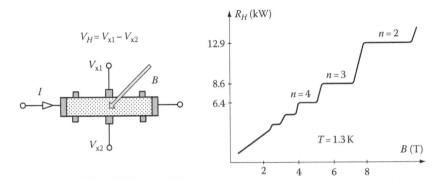

FIGURE 2.70
The quantum Hall standard of resistance. (After Nawrocki, W., *Introduction to Quantum Metrology* (in Polish), Poznan University of Technology, Poznan, Poland, 2007.)

2.9 Main Rules of Magnetics

2.9.1 The Biot–Savart Law

The Biot–Savart law helps in calculation of the magnetic field strength H in selected point A distanced by R from the elementary part dl of a loop C with current I and by r from the cross point of the axes (Figure 2.71):

$$H(r) = \int_C \frac{Idl \times u}{R^2} \qquad (2.97)$$

where u is a unit vector pointing from dl to point A

$$u = \frac{r - r'}{|r - r'|}$$

In section 2.1.1 was presented an example of application of the Biot–Savart law to calculation of the magnetic field generated by a ring coil (one turn coil). Let

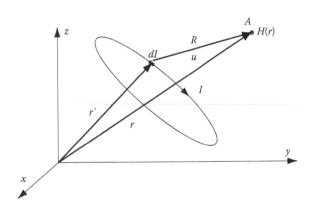

FIGURE 2.71
Definition of the parameters in Biot–Savart Equation 2.97.

us consider another typical case—the magnetic field generated by a long straight wire (Figure 2.72).

If we have infinitely long conductor carrying a current I, the magnetic field δH at the point A resulting from the current in dl (Figure 2.72) is

$$dH = \frac{1}{4\pi R^2} Idl \sin(90 - \alpha) \qquad (2.98)$$

as $dl = Rd\alpha/\cos\alpha = xd\alpha/\cos2\alpha$, we obtain

$$dH = \frac{I\cos\alpha}{4\pi\alpha} dl \qquad (2.99)$$

Thus, the whole magnetic field in point A is

$$H = \int_{-\pi/2}^{\pi/2} \frac{I}{4\pi x} \cos\alpha d\alpha = \frac{I}{2\pi x} \qquad (2.100)$$

Let us consider the magnetic field inside a long solenoid (Figure 2.73). If we assume that the magnetic field H is the superposition of elementary fields generated by a ring coil dl carrying the current $(nI/2L)dl$, this elementary field can be calculated as

$$dH = \frac{nI}{2L} \frac{a^2}{R^3} dl \qquad (2.101)$$

where n is the number of turns of the coil of the length $2L$ and radius a.

Because $R = a/\sin\alpha$, $l = x - actg\alpha$, and $dl = a\, d\alpha/\sin^2\alpha$, we can write

$$dH = \frac{nI}{4L} \sin\alpha d\alpha \qquad (2.102)$$

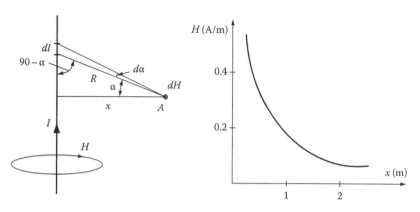

FIGURE 2.72
Magnetic field around long conductor carrying current I (calculated using Equation 2.100 for $I=1\,A$.

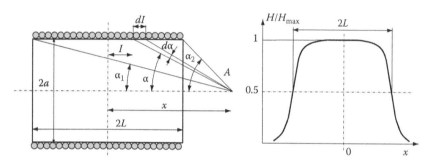

FIGURE 2.73
Magnetic field inside the long solenoid.

and finally

$$H = \int_{\alpha 1}^{\alpha 2} dH = \frac{nI}{4L}(\cos\alpha_1 - \cos\alpha_2)$$

$$= \frac{nI}{4L}\left[\frac{L+x}{\sqrt{a^2+(L+x)^2}} + \frac{L-x}{\sqrt{a^2+(L-x)^2}}\right] \quad (2.103)$$

At the geometric center of the coil ($x=0$), the magnetic field is

$$H = \frac{nI}{2\sqrt{a^2+L^2}} \quad (2.104)$$

The distribution of magnetic field inside the solenoid is presented in Figure 2.73.

2.9.2 Ampère's Circuital Law

Also, Ampère's law helps in calculation of magnetic field generated around a closed loop conducting a current. This relation is as follows:

$$\oint_C Hdl = \iint_S JdS \quad (2.105)$$

where
   C is an arbitrary closed curve
   S is the surface enclosed by the curve C
   J is the density of the current I flowing in the loop

This relationship can also be written as

$$\oint_C Hdl = I \quad (2.106)$$

We can see that magnetic field generated by the current in a closed loop C is equal to the current in a surface enclosed by this curve independently of the shape of this curve. Hence, in this aspect, Ampère's law is equivalent to Biot–Savart law. For example, if we consider discussed in the previous section the magnetic field around the current conducting wire (Equation 2.100), we can use relationship (2.106). If we assume a closed loop in a form of a circle with a radius x and we perform integration along this path, we obtain

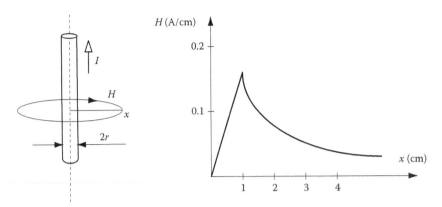

FIGURE 2.74
Magnetic field around long conductor carrying current I (calculated using Equation 2.109 for $I=1\,\mathrm{A}$, $r=1\,\mathrm{cm}$).

$$\int_0^{2\pi} Hxd\phi = H \cdot 2\pi x = I \quad \text{and} \quad H = \frac{I}{2\pi x} \qquad (2.107)$$

and as result, we have the same relationship as Equation 2.100.

This relation can be extended taking into account the magnetic field inside the wire of the radius r (Figure 2.74). The current density is $I=J\pi r^2$, and

$$\int_0^{2\pi} Hxd\phi = H2\pi x = \int_0^{2\pi}\int_0^{x} Jxdxd\phi = \frac{2\pi Ix^2}{2\pi r^2} \qquad (2.108)$$

Therefore,

$$H = \frac{I}{2\pi r^2}x \qquad (2.109)$$

If we have several closed loops of current (*ampere-turns* nI) wound around a magnetic circuit (yoke) with path of magnetic flux l, we can use Ampère's *circuital law* in form

$$Hl = nI \qquad (2.110)$$

The relationship (2.110) is commonly used to determine the magnetic field strength in a magnetic circuit. For example, if we have magnetic yoke with air gap of length l_a, thus we can write

$$H_{Fe}l_{Fe} + H_a l_a = nI \qquad (2.111)$$

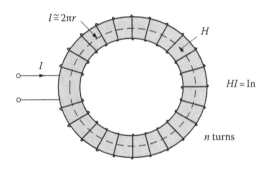

FIGURE 2.75
The circuital law used to determine the magnetic field strength in a toroidal magnetic circle.

If the air gap is negligibly small, the magnetic field strength H in the magnetic circuit presented in Figure 2.75 is

$$H = \frac{nI}{l} \qquad (2.112)$$

The length of magnetic path l can be determined with sufficient accuracy* only for a ring-shaped sample of the mean radius r (toroidal shape) as $l=2\pi r$.

2.9.3 Faraday's Law of Induction

Faraday's law of induction is a fundamental law of magnetism determining the electromotive force (voltage) $V^{\dagger}$ induced in closed circuit in magnetic field B (magnetic flux Φ):

---

* According to standards IEC 60404, the ratio between external and internal radius should be smaller than 1.1.
† Or electric field **E**.

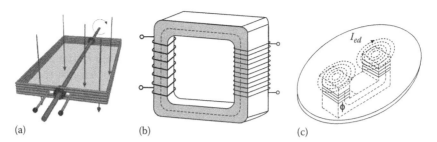

FIGURE 2.76
The examples of application of Faraday's law of induction: generator/induction sensor (a), transformer (b), and eddy currents (c).

$$V = \int_C \boldsymbol{E} d\boldsymbol{l} = -\frac{d}{dt} \int_S \boldsymbol{B} d\boldsymbol{S} = -\frac{d\Phi}{dt} \qquad (2.113)$$

The voltage induced in a closed loop is proportional to the rate of change of magnetic flux linking the loop. According to the Lenz's law, the induced voltage is in such direction that opposes the flux changes.

Faraday's law is the basis of many EM devices or effects (Figure 2.76). If a closed conducting loop is moving in the magnetic field (thus, there is $d\Phi/dx$ and consequently $d\Phi/dt$), the voltage is induced in this loop (Figure 2.76a)—it is a principle of a *generator*. If non-movable loop placed is in varying magnetic field (thus there is $d\Phi/dt$) (Figure 2.76b), we obtain induced voltage—it is a principle of a *transformer*. Both principles can be used to determine the magnetic field in an *inductive sensor*. If the conducting material (of the resistivity ρ and thickness d) is moving in magnetic field or is placed into time-varying magnetic field, the *eddy currents* J_{ed} are induced in this material (Figure 2.76c):

$$J_{ed} = -\frac{d}{2\rho} \frac{dB}{dt} \qquad (2.114)$$

The eddy currents generate magnetic field that tends to oppose the external magnetic field—this phenomenon is used in an inductive watt-hour meters (Ferraris engine). Eddy currents dissipate energy as heating—it is one of the sources of energy loss in magnetic materials.

If we test the magnetic circuit, we can determine the magnetic flux density B in the sample according to Faraday's law because induced voltage V depends on the changes of magnetic flux $d\Phi/dt = A$, $dB/dt = -V$, and

$$B = -\frac{1}{nA} \int_0^T V dt \qquad (2.115)$$

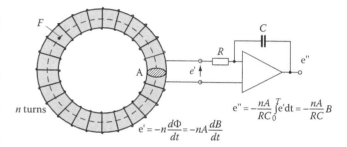

FIGURE 2.77
The principle of measurements of the flux density B in a magnetic circuit.

Therefore, in order to determine the magnetic flux density from a voltage induced in n turns of the coil, it is necessary to use an integrating circuit (Figure 2.77). For pure sinusoidal voltage $V = V_m \sin\omega t$, we can determine the flux density as

$$B(t) = \frac{1}{2\pi f n A} V_m \cos\omega t \qquad (2.116)$$

2.9.4 Lorentz Force

If an electric particle q is moving with velocity v in the EM field E and B, then a force F (*Lorentz force*) acts on this particle, given as

$$F = q(E + \mathbf{v} \times \mathbf{B}) \qquad (2.117)$$

If we consider only magnetic field B, this force is called as *magnetic force* or *Laplace force*, and

$$F_m = q(\mathbf{v} \times \mathbf{B}) \qquad (2.118)$$

The direction of the Lorentz force can be determined using left-hand rule as it is presented in Figure 2.78.

An example of a practical application of the Lorentz force is Hall effect sensor. If electric charges (electrons

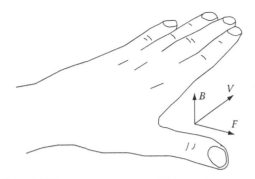

FIGURE 2.78
The left-hand rule for determination of magnetic force.

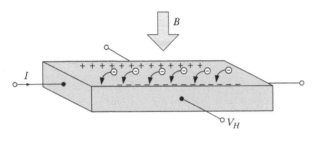

FIGURE 2.79
The principle of operation of the Hall effect sensor.

or holes) are moving in a semiconductor plate (under influence of longitudinal electric field E_L), the magnetic force

$$F = q\mu_n(E_L \times B) \qquad (2.119)$$

is causing the change of their moving trajectory (Figure 2.79). As a result, an additional transverse component of electric field E_T appears:

$$E_T = -R_H(J \times B) \qquad (2.120)$$

where
R_H is the Hall coefficient
J is the current density

For the plate with the thickness t, output voltage of the Hall sensor is

$$V_H = \frac{R_H}{t}IB \qquad (2.121)$$

Another important application of the Lorentz force is the mechanical force acting on a current conducting wire (e.g., in electrical machines). In this case, the Lorentz force can be described as

$$F = I(L \times B) \qquad (2.122)$$

where L is the vector of the length of the wire. Thus, the force acting on the wire of the length l inclined by the angle α to the direction of the magnetic field B is

$$F = IlB \sin\alpha \qquad (2.123)$$

2.9.5 Poynting's Vector

According to Poynting's theorem, the energy density u of EM field in a volume V is described by the following relationship:

$$u = \int_V \frac{1}{2}(\varepsilon E^2 + \mu H^2)\,dv \qquad (2.124)$$

The instantaneous power related to this energy is

$$p = -\frac{\partial u}{\partial t} = -\int_V \left(\varepsilon E\frac{\partial E}{\partial t} + \mu H\frac{\partial H}{\partial t}\right)dV \qquad (2.125)$$

After transformation of this equation, we obtain

$$p = \int_V \sigma E^2 dV + \oint_A (E \times H)\,dA \qquad (2.126)$$

where σ is the conductance of the material.

The first part of Equation 2.126 is related to dissipated electric power and the second part is an energy flux vector of EM field streaming into the area A. This vector S is usually described by

$$S = E \times H \qquad (2.127)$$

and is often called as *Poynting's vector* (or *Umov–Pointing vector*).

Poynting's vector is a basis for the analysis of losses in ferromagnetic materials. In the general case of 2D magnetization of a thin ferromagnetic sheet, the instantaneous power streaming into the sheet through the two outer sheet surfaces (of area $2A$) is (Pfützner 1994)

$$p(t) = 2(E \times H) \cdot A = \begin{bmatrix} E_x & H_x & 0 \\ E_y & H_y & 0 \\ 0 & 0 & 2A \end{bmatrix} = 2A(E_xH_y - E_yH_x) \qquad (2.128)$$

According to Faraday's law, the electric field E is related to $-dB/dt$, so we can describe the total losses of ferromagnetic material as Equation 2.46

$$P_r = \frac{1}{\gamma T} \int_0^T \left(H_x \frac{dB_x}{dt} + H_y \frac{dB_y}{dt} \right) dt \qquad (2.129)$$

2.9.6 Maxwell's Equations

Maxwell formulated four equations (*Maxwell equations*) describing phenomena in EM field:

$$\nabla \times H = J + \frac{\partial}{\partial t} \varepsilon_0 E \quad \text{or} \quad \oint_C H \cdot dl = I + \frac{\partial}{\partial t} \int_S \varepsilon_0 E \cdot dS \qquad (2.130)$$

$$\nabla \times E = -\mu_0 \frac{\partial H}{\partial t} \quad \text{or} \quad \oint_C E \cdot dl = -\frac{\partial}{\partial t} \int_S \mu_0 H \cdot dS \qquad (2.131)$$

$$\nabla \cdot \mu_0 H = 0 \quad \text{or} \quad \oint_S \mu_0 H \cdot dS = 0 \qquad (2.132)$$

$$\nabla \cdot \varepsilon_0 E = \rho \quad \text{or} \quad \oint_S \varepsilon_0 E \cdot dS = q \qquad (2.133)*$$

where *H*, *E*, and *J* are the vectors of the magnetic field strength, the electric field, and the current density,

---

* $\nabla \cdot A$ means the *divergence operation* (*div*) and in Cartesian coordinates it is defined as

$$\nabla \cdot A = \frac{\partial A_x}{\partial x} + \frac{\partial A_y}{\partial y} + \frac{\nabla A_z}{\partial z}$$

$\nabla \times A$ is the *curl operation* and

$$\nabla \times A = \left(\frac{\partial A_z}{\partial y} - \frac{\partial A_y}{\partial z} \right) i + \left(\frac{\partial A_x}{\partial z} - \frac{\partial A_z}{\partial x} \right) j + \left(\frac{\partial A_y}{\partial x} - \frac{\partial A_x}{\partial y} \right) k$$

or in the matrix form as

$$\nabla \times A = \begin{bmatrix} i & j & k \\ \frac{\partial}{\partial x} & \frac{\partial}{\partial y} & \frac{\partial}{\partial z} \\ A_x & A_y & A_z \end{bmatrix}$$

respectively.[†] The symbol ρ means the charge density and *q* is an electric charge.

Maxwell's equations can be presented in either form—differential or integral. Both forms are equivalent according to the Gauss and Ostrogradsky theorem.

Maxwell's equations are not completely new—we can recognize that Equation 2.130 is a slightly modified Ampère's law (magnetic field generated by the current or by the change of an electric field) and Equation 2.131 is Faraday's law (relationship between induced electric field and changes of magnetic field).

Also, Equations 2.132 and 2.133 represent Gauss's laws for electric and magnetic fields. *Gauss's law* (Equation 2.133) states that electric field is produced by electric charge and the flux of electric field passing through closed surface depends on the charge contained within this surface. The total electric flux does not depend on the shape and size of that surface.

Gauss's law for magnetism (Equation 2.132) states that the total magnetic flux passing through a closed surface is zero. In other words, it means that this field is "sourceless" (or divergence-free)—we say that it is a solenoidal vector field. Such field is produced by magnetic dipole (magnetic monopoles do not exist). The magnetic field lines form closed paths—they have neither a beginning nor an end.

The importance of Maxwell's equations lies in coherent (and elegant) treatment of practically all phenomena of EM field (earlier, often electric and magnetic fields were analyzed separately). Together with Lorentz's law, these five relations are the fundamentals of classical electrodynamics and are the basis of magnetic field computation.

In the analysis of EM field, useful is to introduce the term of *vector magnetic potential* **A**, defined as

$$\mathbf{B} = \nabla \times \mathbf{A} \qquad (2.134)$$

and Maxwell's Equations 2.131 and 2.132 are

$$\nabla \cdot \mathbf{B} = \nabla \cdot (\nabla \times \mathbf{A}) = 0 \qquad (2.135)$$

$$\nabla \times \mathbf{E} = -\frac{\partial}{\partial t}(\nabla \times \mathbf{A}) = -\frac{\partial \mathbf{B}}{\partial t} \qquad (2.136)$$

Combining Equations 2.130 and 2.134, we obtain

$$\nabla \cdot \nabla \mathbf{A} = -\mu_0 \mathbf{J} \quad \text{or} \quad \nabla^2 \mathbf{A} = -\mu_0 \mathbf{J} \qquad (2.137)$$

known as *Poisson's equation*.

[†] In the Maxwell equations, often instead of εE an electric flux density *D* is used, as well instead of μH a magnetic flux density *B* is used. Also a current density *J* can be represented by γE (γ is a conductivity).

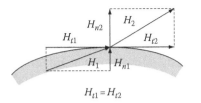
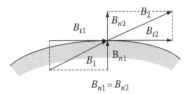

FIGURE 2.80
Boundary conditions of magnetic field.

We can also introduce the term of *scalar magnetic potential* ψ as*

$$\mathbf{B} = -\mu_0 \nabla \psi \qquad (2.138)^\dagger$$

and we get

$$\mathbf{J} = \frac{1}{\mu_0} \nabla \times \mathbf{B} = -\nabla \times \nabla \psi = 0 \qquad (2.139)$$

Combining Equations 2.137 and 2.132, we obtain

$$\nabla^2 \psi = 0 \qquad (2.140)$$

known as *Laplace's equation*.

The magnetic scalar potential ψ and magnetic vector potential $\mathbf{A}$ are very useful for computation of static magnetic field. This class of problems is called as *magnetostatics*. But often, we have to analyze also the varying magnetic field. This field induces the eddy current in the conducting material. Therefore, in such cases, it is necessary to analyze the EM field rather than just magnetic field. In the EM computation, it is useful to introduce also electric scalar potential φ and electric vector potential $\mathbf{T}$.‡

Directly from Maxwell's equations, we can derive the *boundary conditions* of magnetic field (Figure 2.80). From Equation 2.132, we obtain that the magnetic flux on the closed area is continuous, and

---

* Defined only for regions of space free of currents.
† ∇f means a *gradient* (*grad*) operation and $\nabla f(x,y,z) = (\partial f / \partial x)\mathbf{i} + (\partial f / \partial y)\mathbf{j} + (\partial f / \partial z)\mathbf{k}$.
‡ The *electric scalar potential* and *electric vector potential* are defined as

$$E = -\nabla \varphi \quad J = \gamma E = \nabla \times T \qquad (2.141)$$

with

$$H = T - \nabla \psi \qquad (2.142)$$

The *Poisson's equation* for electric potential is

$$\nabla^2 \varphi = -\frac{\rho}{\varepsilon} \qquad (2.143)$$

$$B_{n1} = B_{n2} \quad \text{or} \quad \mu_1 H_{n1} = \mu_2 H_{n2} \qquad (2.144)$$

According to Ampère's law,

$$\mathbf{H}_{t1} - \mathbf{H}_{t2} = j_S \times \mathbf{n} \qquad (2.145)$$

If there is no surface current in the investigated area, we can assume that

$$H_{t1} = H_{t2} \qquad (2.146)^\S$$

This conclusion is important for analysis of magnetic field strength in electrical steel samples because it means that we can use external tangential field sensor for direct investigation of magnetic field strength inside of the sample.

2.9.7 Computation of a Magnetic Field

In previous sections, it was demonstrated that it is possible to calculate the magnetic field using the Biot–Savart law. But these examples concerned only very simple cases: a magnetic field from a current conducting straight wire or a circle. If we analyze the magnetic field inside the magnetized body, we can be only limited to the simple shapes as ellipsoid because the demagnetizing field is otherwise generally nonuniform. The values of the demagnetizing factors N_d were calculated for more complex shapes but the results were also complex and approximated. Recently, it is much simpler to compute the distribution of magnetic field in the magnetized body. When magnetic field is static, we can limit our problems to magnetostatics. For alternating magnetic field, we should analyze the EM field, which is naturally more complicated.

Maxwell's equations are the basis of such computations. Fortunately, it is not necessary to start always from these equations because there are many open source and professional software for magnetic or EM field 2D and three-dimensional 3D computation. But it should be noted that the commercial software are very useful for rather typical problems. Generally, the problem of the magnetic field computation is not a trivial

---

§ For the electric field, the boundary conditions are $E_{t1} = E_{t2}$ and $J_{n1} = J_{n2}$.

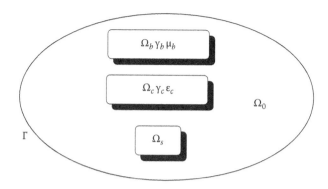

FIGURE 2.81
Definition of the object for computation.

one. For example, most of the programs are not capable of taking into account anisotropy of the magnetic materials not to mention of magnetic hysteresis. No wonder that yearly thousands of scientific papers are published on this subject and several large international conferences, as COMPUMAG or ISEM are organized.

We can compute the magnetic field knowing the sources, boundary conditions, and the material parameters. Sometimes comes across an *inverse problem*—we know the magnetic field distribution and we would like to know the sources. The classical inverse problem is a problem of computational tomography when the target is to recover 3D distribution of the results from usually 2D measurements.

The first step in computation is the definition of the object—its geometry, parameters, boundary conditions, etc. For example, in the object presented in Figure 2.81, we can separate the following parts: the area Ω_b of a ferromagnetic element with a defined $B = f(H)$ characteristics and with conductivity γ; Ω_c—the area of nonmagnetic parts with defined γ and ε; Ω_s—the area of the source of magnetic field; Ω_o—free space (air) and Γ—boundary conditions.

Next step is a division of our model to small elementary elements—*mesh discretization*. The most popular is discretization into elementary triangles with three nodes (in a 2D problems) or into tetrahedrals (triangular pyramids) (in a 3D problems). The more elements of a mesh there are, the more precise can be the calculation, but also there are more equations to solve.* The mesh does not have to be regular—it can be finer in the area of interest (e.g., in the air gap between two magnetized parts).

The division into mesh elements helps in solving the equations because the problem is simplified to determine the results only at the nodes. But because the neighboring elements have common nodes, the result

of computation influences the results of neighbors (up to border elements with known fixed boundary conditions) (Figure 2.82). For linear problems (where there is no saturation), it is possible to calculate the final solution within one step. However, in the case of nonlinear problems, because the influence of the field in neighboring element cannot be a simple superposition, it is necessary to repeat computation many times to arrive at the final result by iterations.

For every element, we formulate an equation or equations describing the physical model. Usually, it is much simpler to solve equations with magnetic vector potential A (B or H are then represented in this equation indirectly).[†] For example, by combining Maxwell's Equations 2.130 and 2.134, we obtain the following relationship for solving a magnetostatic problem:

$$\nabla \times \frac{1}{\mu_0 \mu} \nabla \times A = J \qquad (2.147)$$

In the case of alternating magnetic field, the equations are more complex. For example, one of the commercial softwares carries out a solution of the following two equations:

$$\nabla \times \frac{1}{\mu} \nabla \times A - \nabla \frac{1}{\mu} \nabla \times A - \gamma \left(\frac{\partial A}{\partial t} + \nabla \varphi \right) = 0$$

$$\nabla \cdot \gamma \nabla \varphi + \nabla \cdot \gamma \frac{\partial A}{\partial t} = 0 \qquad (2.148)$$

Theoretically, it is possible to solve such equations for every node in the mesh but it would be very complex to solve so many partial differential equations—in a moderately complex problem, there could be hundreds of thousands of mesh elements. Therefore, many numerical methods were developed for solving this problem by applying simpler approximate algorithms.

One of the simplest is the *finite difference method*. In this method, the derivatives in the differential equations are substituted by finite differences (assuming that the elementary cell of the grid is sufficiently small). Unfortunately, this method is effective only for relatively simple shapes.

Recently, the most frequently used is the *finite element method* (FEM) (also commonly referred to as *finite element analysis* [FEA]). In the FEM, the partially differential equations are substituted by the test function (functional) φ. This way the problem is reduced to the analysis of small finite elements of the mesh and then to solving a matrix of linear equations. In this method,

---

* However, the precision should not be confused with accuracy, which actually deteriorates with the number of calculations due to computational errors resulting from limited precision of number representation.

[†] Flux density B is represented by magnetic vector potential A because $B = \nabla \times A = i(\partial A_z / \partial y - \partial A_y / \partial z) + j(\partial A_x / \partial z - \partial A_z / \partial x) + k(\partial A_y / \partial x - \partial A_x / \partial y)$.

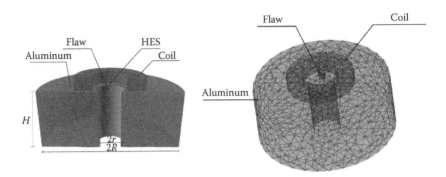

FIGURE 2.82
Creation of a mesh representing the computed model. (From Polik, Z. and Kuczmann, M., *Przegl. Eleketrotech.*, 85, 137, 2009. With permission.)

appropriate choice of the test function (functional) is crucial—this choice is mainly based on the previous experience. To check if a chosen test function is acceptable, we test the weighted residuals. In this case, it is helpful to employ the weighted residual method proposed by Galerkin (*Galerkin method*). Exact description of the FEM is outside of scope of this book—there are many books on this subject (Jin 2002, Zienkiewicz et al. 2005).

Slightly similar to the FEM method is the *boundary element method* (BEM). In this method, we determine the values on the nodes in the boundary by applying integral equation. Often for this problem, Green's function is used. By limiting the solving of the problem to the boundary, this method can be significantly simpler than in the case of FEM. And next it is possible to perform postprocessing in order to determine also the internal

values of the analyzed sample. However, as currently implemented in commercial packages, the FEM seems to be faster and more efficient computationally. Also, in this case, there are many books on the subject (Banerjee 1994, Wrobel and Aliabadi 2002).

Figure 2.83 presents results of the analysis of the magnetic field distribution in and around the sample of the measuring device used for testing of electrical steels—a single strip tester. Such analysis helped in the correct design of this apparatus. Recently, many software packages for magnetic field analysis also include CAD modeling tools.

We should remember that every computation of magnetic field operates on simplified mathematical model of a physical configuration. Figure 2.84 presents comparison of numerical computation and real

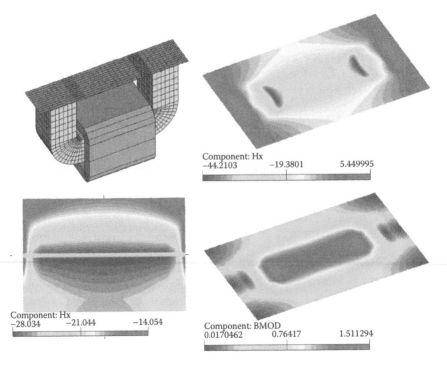

FIGURE 2.83
The results of analysis of a single strip tester apparatus. (From Polakowski et al., 1999.)

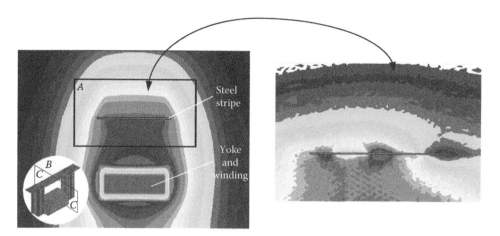

FIGURE 2.84
Numerical and experimental analysis of magnetic field in the single strip tester device.

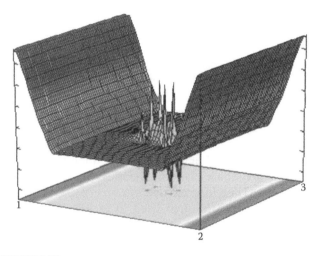

FIGURE 2.85
Computation of magnetic field with simulated four micro-holes.

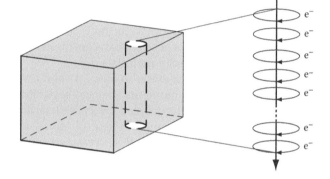

FIGURE 2.86
Imaginary representation of magnetized material by ordered atomic currents. (After O'Handley, R.C., *Modern Magnetic Materials*, John Wiley & Sons, Inc., 2000.)

measure (scanned) results. Numerical model quite well corresponds with the scanned results but better for non-oriented steel. In the case of a grain-oriented steel, the distribution of magnetic field is strongly influenced by the grain structure, which was not included into mathematical model. Therefore, if it is possible, the numerical analysis should be supported by verifying measurements. Figure 2.85 presents results of computation of the model where four micro-holes have been simulated.

2.10 Physical Principles of Magnetism

2.10.1 Magnetic Field and the Magnetic Moment

When, in 1826, Ampère discovered the relationship between current and magnetic field, he prophetically suggested (many years before Bohr) that small loops of elementary electric current are responsible for magnetism. Indeed, today we know that an electron with its rotation (spin), an atom with orbital circulation of electrons, and nucleus with its rotation act as miniature magnets (dipoles) and play a fundamental role in the magnetic effects.

But such classical picture of tiny ordered dipoles does not explain many of the magnetic phenomena in magnetic materials. O'Handley (2000) presented an impressive vision in his book. He proposed that we could analyze a line of ordered atomic current loops represented by a number of atoms aligned in a given direction (Figure 2.86). Such ordering of atoms can be analyzed as a solenoid with n turns, each of a known area proportional to the orbit of an electron. The magnetic moment μ_m of a hydrogen atom with area A and elementary current I can be calculated as $\mu_m = IA$ and

$$\mu_m = IA \approx e \frac{\omega}{2\pi} \pi r_0^2 \approx e \frac{1}{r_0} \sqrt{\frac{2E}{m}} \pi r_0^2 \approx \sim 9.27 \times 10^{-24} \ \text{Am}^2$$

where

ω = v/r_0 and v is the velocity of electron
e is the electron charge
r_0 is the radius of the loop
E is the energy of electron in the 1s shell of hydrogen
m is the mass of electron

Assuming that all atoms are ordered (e.g., in a crystal structure) and the number of atoms is equal to $N_A \approx 10^{29}$/m³ (atoms per unit volume), we obtain a resultant magnetization of about 10^6 A/m or the flux density of about 1 T. We can therefore see that even such an idealized picture of all atoms ordered in one direction is not sufficient to explain saturation induction of ferromagnetic materials exceeding 2 T (on the other hand, it should be noted that hydrogen is a diamagnetic material and therefore cannot be magnetized to such a high value of the flux density). Hence, relying just on the ordering of elementary magnetic moments of atoms is not sufficient to explain ferromagnetism although there is no doubt that the magnetic moments of atoms can contribute substantially to the magnetism of materials.

Because ferromagnetism exists mainly in solids, we cannot consider magnetic moments of atoms separately because they are mutually coupled by crystal lattice. Figure 2.87 presents the example when all atoms are coupled in the same direction in a BCC crystal structure

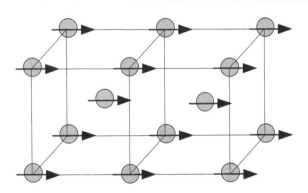

FIGURE 2.87
Ordering of atoms in a BCC iron crystal structure.

of iron. The magnetic ordering is clearly visible but we cannot assume that the crystal structure on its own is responsible for ferromagnetism because there are amorphous materials with excellent magnetic properties but without crystal ordering.

Considering the crystal structure, we can also take into account that in many cases the atoms are packed very closely—Figure 2.88 presents crystal structure with the appropriate geometrical proportions for the main ferromagnetic metals. Thus, the neighboring atoms interact spatially—they can have common electrons, which can be influenced by an electrostatic field of neighbors in the lattice. Symmetry and ordering of crystal structure results in the anisotropy and also influences other magnetic parameters (Birrs 1964, Cracknell 1975). Apart from the crystal structure, also the domain structure strongly influences the magnetic properties.

The theory of magnetism is extremely complex and many problems are still not completely explained or understood. The fundamentals of the theory of magnetism were proposed by Weiss, Curie, Langevin, and others at the turn of the nineteenth and twentieth centuries—well before quantum physics was invented. The quantum physics enabled to extend and expand classical theories—certain phenomena, for example, ferromagnetism can be explained only on the ground of quantum physics. But generally, the quantum theory does not invalidate classical theories, but rather supports them. Therefore, currently a mix of classical and quantum theories is often used to explain physics of magnetism. Such theories are out of scope of this book and they are described in details elsewhere (Jiles 1998, O'Handley 2000, Morrish 2001, Chikazumi 2009, Cullity and Graham 2009). Thus, hereafter in this section, instead of detailed explanation, rather the most important conclusions will be presented.

Considering the elementary parts as small dipoles (magnets), we should take into account that magnetic moment depends on the mass involved (Equation 2.64):

$$\mu = \frac{q\hbar}{2m} = \gamma\hbar \qquad (2.149)$$

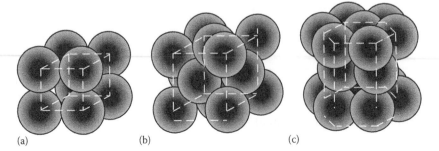

(a)　　　　　　　(b)　　　　　　　(c)

FIGURE 2.88
The crystal structure of iron (a), nickel (b), and cobalt (c).

For this reason, the magnetic moment of a nuclei with the mass almost 2000 heavier than that of an electron can be in practice neglected in the whole balance of magnetic moment of the atom. Moreover, electrons in lower energy states usually fill completely the lower shells and they are paired magnetically such that their net contribution to magnetism is zero. Thus, they also do not take part in the total balance of magnetic moment. We can see that only electrons from outer shells (or near the Fermi energy level) contribute significantly to the resultant magnetic moment. But also, in this case, the electron in the outer shell can be magnetically paired—in such situation, the net magnetic moment is zero and the atom is practically nonmagnetic. It was proved experimentally that in many materials, the value of magnetic moment of the orbital movement was negligible in comparison with magnetic moment of the spin (effect of *quenching of orbital angular moment*).

Intuitively, we feel that the elementary magnetic dipoles in presence of magnetic field should attempt to be aligned in direction of this field. But the elementary angular moments (and therefore also magnetic moments) can change their direction in quantum way—it means that they can take only such direction as is dictated by the four quantum numbers: n, principal quantum number, l, orbital quantum number, m_l, magnetic quantum number, and m_s, spin quantum number. These numbers can have only the following values:

$$
\begin{aligned}
&n = 1,2,3,4,\ldots && (K,L.M.N,\ldots) \\
&l = 0,1,2,3,\ldots,n-1 && (s,p,d,f,g,\ldots) \\
&m_l = -l,-l+1,\ldots,l-1,l \\
&m_s = -1/2,+1/2
\end{aligned}
\tag{2.150}
$$

According to the Pauli exclusion principle, in a given atom, two electrons cannot have the same four quantum numbers. Therefore, if we consider situation for n principal quantum number, we can have only n angular moments l, n^2 orbits m, and $2n^2$ electrons (see Figure 2.89).

Moreover, in transition elements, the orbital states are not filled regularly (higher orbit can be occupied before the lower orbit is full—see Table 2.13). This has significant implications for magnetic properties of such elements. Table 2.13 collects data for main 3d transition elements.

Therefore, only in certain cases, the angular moments have the same direction as the magnetic field—they are quantized in such a way that their projections onto the H direction have to be multiples of $\hbar$. In Figure 2.90a, possible states of angular moments for $l = 2$ ($m = -2, -1, 0, 1, 2$) are presented. The directions of angular moments make precession around the direction H. The value of orbital magnetic moment is

$$
|\mu_l| = \sqrt{l(l+1)} \cdot \mu_B \tag{2.151}
$$

while its projection onto H direction (we assume, in this case, to be synonymous with the z axis) is

$$
\mu_l^{(z)} = m_l \mu_B \tag{2.152}
$$

The magnetic field causes change of frequency of precession:

$$
\omega_0 = \frac{2\mu_B}{h} B = \gamma_e B \tag{2.153}
$$

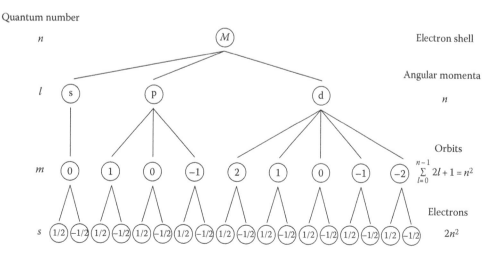

FIGURE 2.89
Various possible states of electrons for principal number $n = 3$. (After Chikazumi, 2009.)

TABLE 2.13

Electronic Configuration of 3d Transition Elements

Z			K 2 1s 2	L 8 2s 2	2p 6	M 18 3s 2	3p 6	3d 10	N 32 4s 2	L	S	J	Ground Terms
21	Sc	$3d^14s^2$	2	2	6	2	6	1	2	2	1/2	3/2	$^2D_{3/2}$
22	Ti	$3d^24s^2$	2	2	6	2	6	2	2	3	1	2	$^3F_2$
23	V	$3d^34s^2$	2	2	6	2	6	3	2	3	3/2	3/2	$^4F_{3/2}$
24	Cr	$3d^54s^1$	2	2	6	2	6	5	1	0	6/2	0	$^7S_3$
25	Mn	$3d^54s^2$	2	2	6	2	6	5	2	0	5/2	0	$^6S_{5/2}$
26	Fe	$3d^64s^2$	2	2	6	2	6	6	2	2	4/2	4	$^5D_4$
27	Co	$3d^74s^2$	2	2	6	2	6	7	2	3	3/2	9/2	$^4F_{9/2}$
28	Ni	$3d^84s^2$	2	2	6	2	6	8	2	3	2/2	4	$^3F_4$
29	Cu	$3d^{10}4s^1$	2	2	6	2	6	10	1	0	1/2	0	$^2S_0$
30	Zn	$3d^{10}4s^2$	2	2	6	2	6	10	2	0	0	0	$^1S_0$

Source: After Chikazumi, 2009.

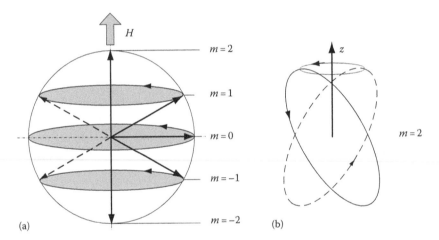

FIGURE 2.90
Spatial quantization of orbital angular moments determined for $m_l = 2$ and example of the shapes of the orbits. (After Chikazumi, 2009.)

Figure 2.90b presents the example of the precession of electron determined for $m_l = 2$.

There is a similar situation when we consider the direction of the spin moment. In this case also, only two selected quantized states are possible—for spin-up and spin-down with respect to the direction of magnetic field (Figure 2.91). The value of the spin moment is

$$|\mu_s| = 2\sqrt{s(s+1)} \cdot \mu_B \qquad (2.154)$$

while its projection onto the z direction is

$$\mu_s^{(z)} = 2m_s\mu_B \qquad (2.155)$$

where s is the spin angular moment quantum number $s = \pm 1/2$.

In an atom, the angular and magnetic moments are represented by atomic quantum numbers

$$L = \sum_i l_i \quad \text{and} \quad S = \sum_i s_i \qquad (2.156)$$

or as a sum of orbital angular moment vectors l_i and spin angular moment vector s_i

$$L = \sum_i l_i \quad \text{and} \quad |L| = \sqrt{L(L+1)} \cdot \hbar \qquad (2.157)$$

$$S = \sum_i s_i \quad \text{and} \quad |S| = \sqrt{S(S+1)} \cdot \hbar \qquad (2.158)$$

Calculation of total angular moment vector J is more complex and can be expressed as

$$J = L + S \quad \text{and} \quad |J| = \sqrt{J(1+J)} \cdot \hbar \qquad (2.159)$$

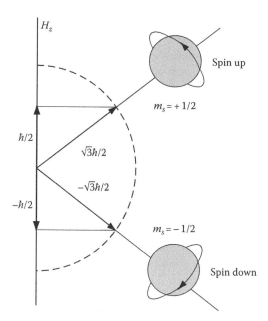

FIGURE 2.91
Spatial quantization of spin angular momenta.

where

$$J = \sum_i j_i$$

j is a total angular moment vector

The total angular moment is also quantized and its projection onto the direction z is multiple of $\hbar$. Figure 2.92 presents the vector addition of total angular and spin vectors—note that magnetic moment is proportional to $L + 2S$.

$$J = \sum l_i + \sum s \quad \text{or} \quad J = \sum j_i = \sum (l_i + s_i) \qquad (2.160)$$

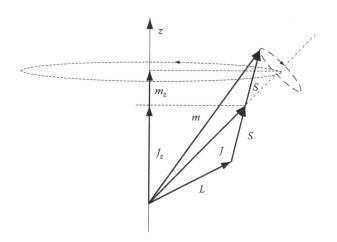

FIGURE 2.92
The vector addition of total angular and spin vectors. (After Comstock, R.L.: *Introduction to Magnetism and Magnetic Recording*. 1999. Copyright Wiley-VCH Verlag GmbH & Co. KGaA.)

In the first method (*Russel–Saunders coupling*), first we determine the total L and S vectors and then the sum of them. In the second method (*j–j coupling*), we determine values of j_i vectors and then we are summing them. The results of both methods can be different because we cannot assume that the vectors are not correlated. This correlation is the *spin–orbit coupling* and plays important role in many magnetic phenomena. Most often, the first method is used (recommended for relatively small spin–orbit coupling).

Table 2.13 presents values of the L, S, J quantum numbers determined for selected 3d transition elements. In the same table, there are *ground terms* that summarize information about L, S, J quantum numbers.*

From Figure 2.92, we can see that the total angular vector and total magnetic vector can be not collinear. The total magnetic moment can be determined as

$$|\mathbf{\mu}_J| = -g\mu_B\sqrt{J(J+1)} \qquad (2.161)$$

where g is the g-factor proposed by Landé (*Landé factor*) in a form

$$g = 1 + \frac{J(J+1) + S(S+1) - L(L+1)}{2J(J+1)} \qquad (2.162)$$

The Landé factor for $S=0$ is equal to $g=1$ while for $L=0$, it is $g=2$.

2.10.2 Magnetic Field and Band Structure Density of States

In the previous section, we derived that materials in a magnetic field can exhibit a magnetic moment. But, we still do not know why only several elements from 3d transitions metals (iron, cobalt, and nickel) can be easily magnetized and exhibit large permeability. Partially, this question is possible to explain taking into account band structure of these metals. Such a band theory of ferromagnetism was introduced by Stoner (1933) and Slater (1936).

In metals, strong coupling between atoms in the crystal lattice exists due to the close proximity from one to another. For example, 1 mg of iron contains around 10^{19} atoms. Such close-packed atoms can have common electrons and between them there are free electrons. As a result of this coupling, the line representing energy levels of electrons can be split into a large number of possible energies. It is because the Pauli exclusion principle

---

* Ground numbers are prepared according to the following system: $^{2S+1}L_J$, where L is presented in spectroscopic notation: $L=0,1,2,3,\ldots$ corresponds with letters: S, P, D, G, $\ldots$

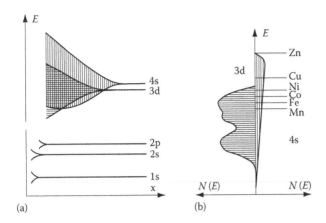

FIGURE 2.93
The splitting of energy bands for small interatomic distance *x* (a) and corresponding density of states curves (b). (After Cullity, B.D. and Graham, C.D., *Introduction to Magnetic Materials* © (2009) IEEE.)

requires additional energy levels to avoid situation that the same electrons have the energy represented by the same quantum numbers. In Figure 2.93a, such a situation is illustrated as a function of decrease of the distance between atoms.

We can see that for small distances (e.g., for crystals) now 3d and 4s states overlap energy bands. This case is illustrated in form of density-of-states curves (Figure 2.93b).* The lower levels (below 3d level) can be neglected because they are fully filled and therefore their resultant magnetic moment is zero. Thus, the magnetic properties of 3d transition metals are decided by the situation in the two last bands, 3d and 4s. The 3d

band has much larger density because it represents 10 electrons while 4s only two electrons.

The density of states presented in Figure 2.93b was confirmed numerically and experimentally (e.g., Figure 2.94).[†] Real shape of the density-of-states curve is more complex although we can assume that picture 2.93b faithfully represents 3d transition metals. Therefore, for further analysis, we can assume that the shape of density-of-states curves is the same for all transition metals. Such approximation is known as the *rigid-band-model*.

Usually, it is convenient to draw the density-of-states curves separately for spin-up and spin-down states. The highest possible level represents Fermi energy level E_f. An example of such curves is presented in Figure 2.95.

Looking at Figure 2.95, we can see that ferromagnetic features appear at those elements that have unfilled 3d band (see also Figure 2.101). For example, copper (and zinc) have completely filled 3d band and therefore they are nonmagnetic (diamagnetic). On the other hand, manganese and further lighter elements have energy less than half band and are also nonmagnetic (paramagnetic)—although their alloy FeMn (iron manganese) is commonly used as antiferromagnetic material.

In the presence of magnetic field, the electrons with privileged spin (parallel to the magnetic field) obtain different energy than those ordered antiparallel. Electrons can easily migrate to the privileged band by changing their spin direction. This leads to imbalance in energy states and magnetic moment. But what is more important—in ferromagnetic materials, due to their internal magnetic field, there is a permanent imbalance of energy states for both spins as it is

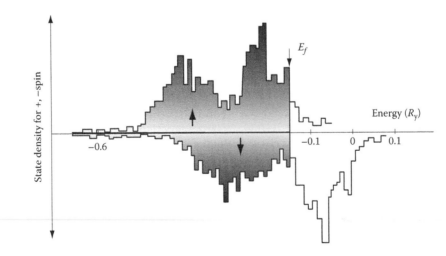

FIGURE 2.94
An example of the density-of-states curve determined for iron. (After Wakoh, S. and Yamashita, J., *J. Phys. Soc. Jpn.*, 21, 1712, 1966.)

---

* Density-of-states curve represents relationship between energy level *E* and density of states Z(*E*)—number of states per unit volume of the sample.

† In Figure 2.94, energy is expressed by *Ry* units. *Ry* (from Swedish physicist Johannes Rydberg) is an energy related to the ground state energy of the hydrogen atom 1 Ry = 13.605 6923 eV.

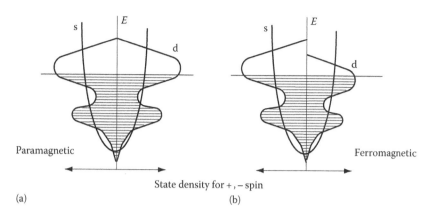

FIGURE 2.95
The density-of-states curves of a paramagnetic (a) and ferromagnetic (b) material.

presented in Figure 2.95. This imbalance called *spontaneous magnetization* causes that ferromagnetic materials can remain magnetized even in absence of magnetic field. According to Pauli exclusion principle, we can assume that at first five electrons have the same spin. Thus, if n is number of 3d and 4s electrons per atom, x is number of 4s electrons per atom, $n - x$ is number of 3d electrons per atom, we can write the following equation describing the magnetic moment: $\mu = [5 - (n - x - 5)]\mu_B = [10 - (n-x)]\mu_B$. This leads to simplified rule (Cullity and Graham 2009)

$$\mu = (10.6 - n)\mu_B \qquad (2.163)$$

Indeed, such an empirical rule quite corresponds quite well with experimental results. From the relationship (2.163), we obtain for Fe, Co, Ni the saturation magnetization values 2.6 – 1.6 – 0.6 (μ_B/atom), while measurements give the following results: 2.22 – 1.72 – 0.6, respectively.

Theoretical analysis of the band structure enables to explain why ferromagnetism is limited for only few exclusive metals. We should remember that these considerations operated on the very simplified rigid-band model. Figure 2.96 presents examples of the density-of-states representations for various crystal structures or various alloy composition.

2.10.3 Weak Magnetism—Diamagnetism and Paramagnetism

Most of substances exhibit "weak" magnetism. It means that to magnetize such materials, it is necessary to apply very large magnetic field. Such materials are classified as *diamagnetic* or *paramagnetic*. Because magnetization of these materials is very small, in order to describe their performances, it is most convenient to use susceptibility $\chi = M/H$ (in contrary to ferromagnetic materials where we use permeability $\mu = B/H$). Table 2.14 collects susceptibility values of typical diamagnetic and paramagnetic materials.

Practically all materials exhibit diamagnetism, but in non-diamagnetic materials, this effect is masked by other stronger effects. *Diamagnetic materials* respond to an external magnetic field with opposing internal field (therefore susceptibility is negative). As it was described earlier, superconductors are ideal diamagnetic materials due to the Meissner effect.

Diamagnetism is demonstrated by materials with a zero net magnetic moment (e.g., with all electrons

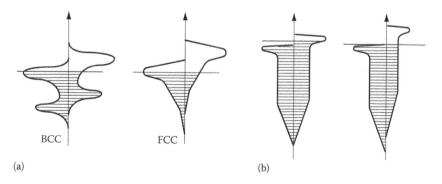

FIGURE 2.96
The density states representation of various crystal structure (a) (after O'Handley, R.C., *Modern Magnetic Materials*, John Wiley & Sons, Inc., 2000) or various alloy composition (after Chikazumi, S., *Physics of Ferromagnetism*, and 2009 of the Oxford Scientific Publication, Oxford, U.K.).

TABLE 2.14

Susceptibility of Typical Diamagnetic and
Paramagnetic Materials

Diamagnetic Materials		Paramagnetic Materials	
Material	$\chi \times 10^{-5}$	Material	$\chi \times 10^{-5}$
Ethyl alcohol	−0.72	Oxygen gas	0.19
Benzene	−0.77	Sodium	0.72
Water	−0.91	Magnesium	1.2
Copper	−1.0	Aluminum	2.2
NaCl	−1.4	Tungsten	6.8
Silver	−2.6	Platinum	26
Gold	−3.5	Uranium	40
Bismuth	−16.6	Neodymium	300
Superconductor	-10^5	Holmium	729

paired). Therefore, their response to external magnetic field does not come from ordering of elementary dipoles (as is in other materials). The diamagnetic effect was explained by Langevin (1905). In diamagnetic materials, an electromotive force is induced (according to Faraday law) in the elementary current loops. This results in some circulating current on the expense of the electron velocity. Although electromotive force is induced when magnetic flux is changed, the new value of current persists because there is no resistance in the elementary current loop. The effect of opposing the magnetic field is in accordance with Lenz's law.

We can analyze the diamagnetic effect by summing the elementary changes of magnetic moments. The susceptibility of diamagnetic materials according to Langevin's theory depends on

$$\chi = -\frac{N_A \mu_0 e^2 Z <r^2>}{6m_e} \qquad (2.164)$$

where
N_A is Avogadro's number (number of atoms per mole)
Z is the number of electrons in the atom
$<r>$ is the average radius of the current loop

In *paramagnetic materials*, there is a net magnetic moment; therefore, the external magnetic field tries to align all elementary dipoles to the direction of this field. As it was discussed earlier, as a result, the total angular moment J and the magnetic moment $|\mu_J| = -g\mu_B \sqrt{J(J+1)}$ change.

The ordering magnetic energy is disturbed by thermal energy, which causes disordering. In example presented in Figure 2.86, we obtain results of ordering equal to magnetic moment $\mu_m \approx 10^{-23}$ Am², which corresponds with potential energy—$\mu_m B \approx 10^{-23}$ J (for $B = 1$ T). The thermal energy at room temperature is

$k_B T \approx 400 \ 10^{-23}$ J (k_B, Boltzmann constant). Therefore, we can see that the thermal energy is much higher than the magnetic energy even for extremely high flux density. This explains why susceptibility of paramagnetic materials is very small and why this effect is so weak. Moreover, it also explains why the paramagnetic effect in some materials is so dependent on temperature. It should be noted that all ferromagnetic materials metamorphose to paramagnetic when the thermal energy is high enough—above certain material-specific temperature (*Curie temperature*).

The analysis of ordering/disordering effect is commonly performed by taking into account Boltzmann statistics where the probability P of an ion with energy E_i is given by

$$P_i = Ae^{-E_i/k_B T} \qquad (2.165)$$

where A is a constant.

Such analysis leads to the following relationship describing magnetization M of paramagnetic materials

$$M = N_A g \mu_B J B_j(\alpha) \qquad (2.166)$$

with $\alpha = \mu_m B/k_B T$.

In Equation 2.166, B_j means the Brillouin function given in Figure 2.97 and expressed as

$$B_j(\alpha) = \left[\frac{2J+1}{2J} \coth\left(\frac{2J+1}{2J}\alpha \right) - \frac{1}{2J}\coth\left(\frac{\alpha}{2J} \right) \right]$$

$$= \frac{J+1}{3J}\alpha - \frac{[(J+1)^2 + J^2](J+1)}{90J^3}\alpha^3 + \cdots \qquad (2.167)$$

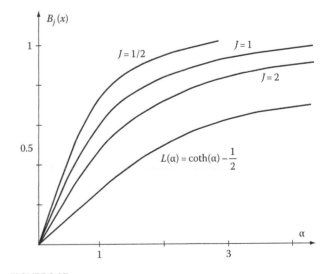

FIGURE 2.97

The Brillouin function (and the Langevin function $L(\alpha)$).

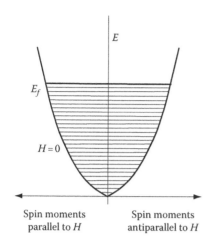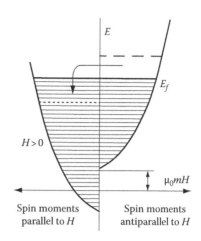

FIGURE 2.98
The Pauli paramagnetism principle.

Expanding Equation 2.167 into a Taylor series and taking into account only first term, we obtain following relation describing susceptibility of paramagnetic materials:

$$\chi \cong \frac{N_A \mu_0 g^2 \mu_B J(J+1)}{3 k_B T} \qquad (2.168)$$

Classical theory of paramagnetism introduced by Langevin expresses the magnetization M as $M = N_A \mu_m L(\alpha)$ where $L(\alpha)$ is the Langevin function and $L(\alpha) = \coth \alpha - (1/\alpha) = (\alpha/3) - (\alpha^3/45) + \cdots$ Taking into account only first term $\alpha/3$, the classical expression of susceptibility is

$$\chi \cong \frac{N_A \mu_0 \mu_m^2}{3 k_B T} \qquad (2.169)$$

The Langevin function is presented in Figure 2.97 and it is clear that classical relation is a special case of more general quantum relation.

From the relationship (2.169), we obtain classical dependence of *Curie's law*:

$$\chi = \frac{C}{T} \qquad (2.170)$$

where C is the Curie constant $C = N_A \mu_0 \mu_m^2 / 3 k_B$.*

Curie observed temperature dependence of paramagnetic materials and in 1895 proposed the relationship (2.170) known as Curie's law (Curie 1895). This law was later extended by Weiss (1906). Weiss assumed that beside magnetic field in the material, there is additional molecular field $H_w = \gamma M$ where γ is the molecular field constant. Including the relation $H_{tot} = H + H_w$ into

the dependence $\chi = M/H$, we obtain $M = CH/(T - C\gamma)$ or $M = CH/T - \theta$, and this leads to the *Curie–Weiss law*

$$\chi = \frac{C}{T - \theta} \qquad (2.171)$$

The temperature θ is a boundary of transition from a ferromagnetic to a paramagnetic state and in the case of ferromagnetic material, it is called the Curie temperature T_C.

The Curie–Weiss theory explains the magnetic phenomena in some materials, including nickel and rare-earth metals. But, in many metals, we do not observe dependence of susceptibility on temperature. The explanation is that in such materials, there are free electrons and the mechanism of the magnetization process is slightly different. This process can be explained by the *Pauli model of paramagnetism* (Figure 2.98) (Pauli 1926).

In Pauli paramagnetism, the density-of-states curves are the same in absence of external magnetic field—the material does not exhibit spontaneous magnetization. In magnetic field, electrons with antiparallel spins change their spin direction and migrate to the second part of the density-of-states curve. This leads to imbalance of energy levels and because the Fermi level should be the same for both spins, the material obtains the magnetic moment equal to $2\mu_0\mu_B H$ (see Figure 2.98). The susceptibility in this case is described by the following relationship (Jiles 1998):

$$\chi = \frac{3 N_A \mu_0 \mu_m^2}{2 k_B T_f} \qquad (2.172)$$

---

* If we assume $\mu_m = g\mu_B \sqrt{J(J+1)}$, quantum dependence corresponds with classical law.

where T_f is the Fermi temperature and $T_f = E_f / k_B$.

2.10.4 Strong Magnetism—Ferromagnetism

The ferromagnetic materials can be easily magnetized by relatively low magnetic field. In comparison with paramagnetic materials (with large temperature disorder effect), it means that in ferromagnetic materials exists internal "force" that overcomes the temperature energy. This is a basis of a fundamental theory of ferromagnetism proposed by Weiss (1906). Weiss assumed that in ferromagnetic materials exists an additional molecular field $H_w = \gamma M$ where γ is a molecular field constant. This field causes that ferromagnetic material is initially magnetized even if in absence of external magnetic field. This phenomenon is known as *spontaneous magnetization*.

The Weiss assumption leads to the question: what happens with the molecular field when the material is demagnetized? Weiss solved also this problem. He assumed that ferromagnetic material consist of small regions, called domains each of which is magnetized to saturation. In demagnetized state, these domains are ordered in such way that the resulting internal magnetic field is balanced out. But the material is "ready for magnetization process." It means that it is not necessary to overcome the thermal disorder, but it is sufficient to reorganize the domain structure (by a movement of the DWs and by a change of direction of magnetization of the domain).

Although recently, more extended theories of ferromagnetic effect were introduced (including quantum and band theories) and we know that ordering is caused not by internal magnetic field, the Weiss explanation surprisingly still helps well in understanding of the ferromagnetic phenomena. It should be stated that even today, there is no single comprehensive theory of ferromagnetism. The classical Weiss theory supported by the quantum physics quite well explains why at the Curie temperature T_c, ferromagnets are transformed into paramagnets (although the theory does not hold well for all materials, but it is especially faithful in explanation of this phenomenon in nickel). The band theory explains well why only some elements exhibit ferromagnetism. The Heisenberg model of ferromagnetism helps in explanation of the ordering principle (and origin of the "molecular field"). Spin wave theory helps in the analysis of the phenomena near the Curie temperature. Domain theory explains the magnetization process, etc.

We can estimate an approximate value of the molecular field H_w. At the Curie temperature, this field is compensated by the thermal energy and $\mu_m \cdot H_w = k_B \cdot T_c$. Assuming Curie temperature 1024°C for iron and μ_m equal to 2.2 Bohr magneton, we obtain that this field is around 5×10^8 A/m. It is difficult to imagine the generation of such a high value of magnetic field and therefore

it is currently assumed that the spontaneous magnetization results from interatomic exchange interaction (more electrostatic than magnetic).

We can determine the change of spontaneous magnetization M_s with temperature according to the Weiss theory,

$$M_s = M_{max} B_j \left(\frac{\mu_m H_w}{k_B T} \right) = M_{max} B_j \left(\frac{\mu_m \gamma M_s}{k_B T} \right) \quad (2.173)$$

where B_j is the Brillouin function.

Equation 2.173 can be rewritten in the following form:

$$M_s \frac{\mu_m \gamma}{k_B T} \cdot \frac{k_B T}{\mu_m \gamma} \cdot \frac{1}{M_{max}} = B_j \left(\frac{\mu_m \gamma}{k_B T} M_s \right)$$

$$\text{or} \quad \frac{k_B T}{\mu_m \gamma} \cdot \frac{1}{M_{max}} \cdot x = B_j(x) \quad (2.174)$$

where $x = \mu_m \gamma M_s / k_B T$.

Because in Equation 2.174 M_s appears on both sides, we commonly solve such relationship graphically by drawing two functions $B_j(x)$ and $(k_B T / \mu_m \gamma M_s)(x)$ for fixed temperature T. The intersection of both curves determines M_s value (Figure 2.99). The dashed line in Figure 2.99 is drawn for $T = T_C$. Above the temperature T_C (on the left of the dashed line), there is no intersection of both curves and $M_s = 0$.

The calculated dependence of magnetization on the temperature is presented in Figure 2.100. From results presented in Figure 2.100, arise several important conclusions:

- Results of experiment agree very well with theoretical model (curve for $J = 1/2$)—we observe difference only for low temperature.

- Quantum model represented by the Brillouin function agrees better with the experiment than the Langevin function ($J = \infty$).

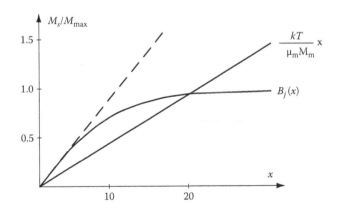

FIGURE 2.99
Graphical solution of Equation 2.174.

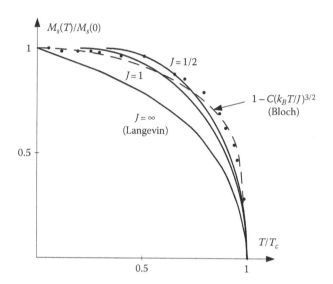

FIGURE 2.100
The dependence of magnetization ratio on the temperature ratio—points indicated experimental results. (After Springer Science+Business Media: [*Ferromagnetismus*, 1939, Becker, R. and Döring, W].)

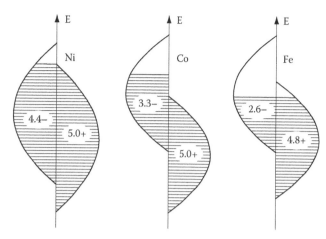

FIGURE 2.101
The band structure of the main ferromagnetic elements.

- The Brillouin function for $J = 1/2$ also agrees better with experiment than the curve for $J = 1$. This means that the angular moment makes only small contribution to the magnetic moment and the spin moment plays main role in ferromagnetism.

If magnetic material has a regular crystal lattice, the spins can precess in correlated way creating a *spin wave*. The length of this wave depends on the lattice constant. If quantized, this collective excitation is represented by quasiparticle *magnon*. When spin waves are excited by thermal energy, they reduce a spontaneous magnetization. Analyzing the spin waves, Bloch determined dependence of magnetization on the temperature as (*Bloch's law*) (Bloch 1930)

$$M_s = M_{max}\left[1 - C\left(\frac{k_B T}{J}\right)^{3/2}\right] \qquad (2.175)$$

where coefficient C depends on the crystal structure ($C = 0.1174$ for a simple cubic, $C = 0.0587$ for a BCC, and $C = 0.0294$ for a face-centered cubic lattices), J is an exchange integral. The relationship (2.175) agrees very well (better than the Brillouin function) with the experimental data for low temperatures (Figure 2.100).

As it was discussed earlier, asymmetry of electronic band of atomic structure supports creation of magnetic moment. Figure 2.101 presents the band structure of the main ferromagnetic elements.

The transition ferromagnetic metals have partially filled 3d band and in this band, there is place for 10 electrons. According to Hund's rules,* the electrons will occupy states with all spins parallel in the shell as far as it is possible (without violating the Pauli exclusion principle). Therefore, ferromagnetic metals exhibit large asymmetry of states (e.g., in contrary to Cu and Zn) as is illustrated in Figures 2.101 and 2.102.

Table 2.15 presents typical ferromagnetic elements. Cullity formulated following three criteria for the existence of ferromagnetism (Cullity and Graham 2009):

- The electrons must lie in partially filled bands such that there is a vacant energy level available for electrons with unpaired spins to move into.

- The density of levels in the band must be high so that the increase in energy caused by spin alignment will be small.

- The atoms must be appropriate distance apart so that the exchange force can cause the d-electron spins in one atom to align the spins in a neighboring atom.

Looking for the origin of the Weiss "molecular field," just the interatomic distance should be considered as the potential source of exchange force. Indeed, it was proved that the exchange force between neighboring atoms and electrons can cause parallel spin orientation.

---

* Hund's rules governed organization of electron states: (1) Spins are arranged to maximize resultant spin S. (2) For given spin arrangement, total atomic orbital momentum L is maximized. (3) When the shell is less than half full $J = L - S$, when the shell is more than half full $J = L + S$.

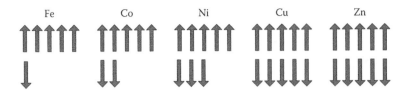

FIGURE 2.102
The spin arrangement in 3d transition metals according to Hund's first rule.

TABLE 2.15

Characteristics of Main Ferromagnetic Elements

Substance	B_s (T) (290 K)	B_s (T) (0 K)	μ_m (μ_B)	μ_{max} (μ_0)	T_c (K)
Fe	2.14	2.18	2.22	5000	1043
Co	1.81	1.82	1.72	250	1388
Ni	0.61	0.64	0.61	600	627
Gd		2.59	7.63		292
Dy		3.67	10.2		88

This exchange energy E_{ex} is described by *Heisenberg Hamiltonian* (Heisenberg 1928):

$$E_{ex} = -2J_{ex} \sum_{i=1}^{N} \sum_{j-1}^{z} \mathbf{S}_i \mathbf{S}_j \qquad (2.176)$$

where
  J is the exchange integral
  N is the number of spins in crystal
  z is the number of neighbors
  i, j are the indexes of lattice points

The dependence of the exchange integral on interatomic distance is known as *Bethe–Slater curve* (Figure 2.103). Indeed, it was proved that under some circumstances, the exchange energy between neighboring atoms can be $J>0$, so parallel order of spins is privileged (thus, in such cases, the energy of the system is smaller than for antiparallel order). For $J<0$, the antiparallel order can is privileged.

2.10.5 Mixed Magnetism—Antiferromagnetism and Ferrimagnetism

In paramagnetic materials, magnetic moments are distributed chaotically, while in ferromagnetic materials, these moments are parallel ordered (Figure 2.104). There are also materials with more complex ordering—antiferromagnetic materials in which magnetic moments are antiparallel (Figure 2.104) and ferrimagnetic materials where magnetic moments are also antiparallel, but the values do not compensate completely (Figure 2.104).

The Bethe–Slater curve suggests that the aniparallel order of spins should exist in certain elements—for example, in manganese and chromium. Indeed, although both these elements are normally paramagnetic but below certain temperature called Néel temperature T_N (100 and 308 K, respectively), they exhibit some kind of antiferromagnetism. The antiferromagnetism is attributed to oxides of transition metals, and in technical applications, Mn-based alloys with antiferromagnetic effect in a room temperature are used. In nature, some minerals as hematite Fe_2O_3 exhibit antiferromagnetism. Table 2.16 collects data of a few main antiferromagnetic materials.

The antiferromagnetic materials have features similar to paramagnetic materials—they have low susceptibility

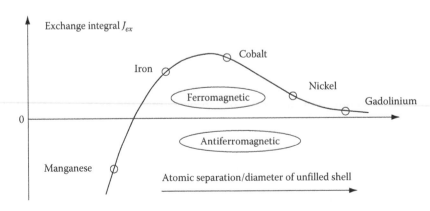

FIGURE 2.103
The Bethe–Slater curve.

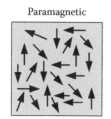
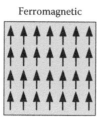
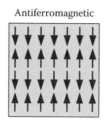
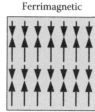

Paramagnetic Ferromagnetic Antiferromagnetic Ferrimagnetic

FIGURE 2.104
Various kinds of magnetic orders.

TABLE 2.16

Performances of Main Antiferromagnetic Materials

Substance	T_N (K)	θ (K)
MnO	122	610
FeO	198	570
NiO	523	3000
FeMn	423	
NiMn	720	
Fe_2O_3	950	2000

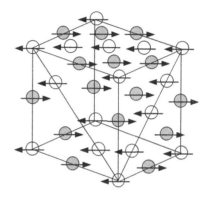

FIGURE 2.105
Ordering of ions in MnO crystal structure.

(changing with temperature) and in modest external magnetic field, exhibit only very small magnetization. But using neutron diffraction method that is sensitive to magnetic moment, it was discovered that these materials have highly ordered magnetic structure. Thus, a small magnetization results not in disorder of magnetic moments but just the opposite—there is spontaneous magnetization but antiparallel moments compensate each other.

The antiparallel order of spins can be explained as a larger distance between ferromagnetic atoms in the crystal structure because in oxides transition, metal cations are separated by oxygen anions. Thus, interatomic exchange coupling is too small to force parallel order. Moreover, there is a special kind of coupling via oxygen atoms called *superexchange* that forces antiparallel order. In understanding phenomena of antiferomagnetism, the heuristic model with an assumption that antiferromagnetic material is composed of two ferromagnetic lattices of opposite spins can be helpful. In fact, in the crystal structure of MnO (Figure 2.105), we can distinguish adjacent planes with parallel magnetized ions.

Figure 2.106 presents typical temperature dependence of susceptibility of antiferromagnetic materials. The susceptibility increases with temperature but only to the T_N temperature—above the T_N temperature, the substance becomes paramagnetic. If we look at the plot of the temperature dependence of inverse susceptibility, we easily discover that antiferromagnetic material can be described by Curie–Weiss model if

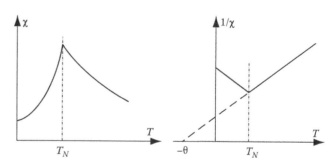

FIGURE 2.106
The dependence of susceptibility χ (and inverse susceptibility $1/\chi$) on the temperature in a typical antiferromagnetic material.

$$\chi = \frac{C}{T-(-\theta)} = \frac{C}{T+\theta} \qquad (2.177)$$

If we assume that above the T_N temperature, there are molecular fields in both sublattices $H_w^A = -\gamma M_B$ and $H_w^B = -\gamma M_A$ (and C' as Curie constant), the magnetizations of both sublattices are $M_A = C'(H - \gamma M_B)/T$ and $M_B = C'(H - \gamma M_A)/T$. Because the total magnetization is $M = M_A + M_B$, we obtain

$$M = \frac{2C'H}{1+C'\gamma} \qquad (2.178)$$

and

$$\chi = \frac{M}{H} = \frac{2C'}{T + C'\gamma} = \frac{C}{T + \theta} \qquad (2.179)$$

To determine the magnetization below the T_N temperature, the best way is to use the Brillouin function and solve the equation graphically (as it is presented in Figure 2.99). After realization of such algorithm, we obtain temperature dependence of magnetization as presented in Figure 2.107. We can see that although both sublattices exhibit spontaneous magnetization, the net magnetization is zero.

The theory of antiferromagnetic effect was proposed by Néel (1932) and confirmed experimentally by neutron diffraction by Shull and Smart (1949). This phenomenon is still treated as a scientific curiosity without significant technical application. In 1957, Meiklejohn and Bean proposed to use antiferromagnetic material for the exchange bias (Meiklejohn and Bean 1957, Nogués and Schuller 1999, Kiwi 2001). When in 1994, Dieny proposed a new GMR structure known as *SV* (Dieny 1994), just this kind of exchange was the basis of new generation of GMR sensors. Recently, antiferromagnetic thin films, such as FeMn or NiO are commonly used to bias thin-film magnetoresistive sensors. Figure 2.108 presents the hysteresis loop of permalloy shifted after biasing it by FeMn antiferromagnetic layer.

It should be noted that the antiferromagnetic state of thin-film GMR sensors was forced artificially by coupling two ferromagnetic films separated by very thin conducting layer—spacer (Figure 2.109). This kind of

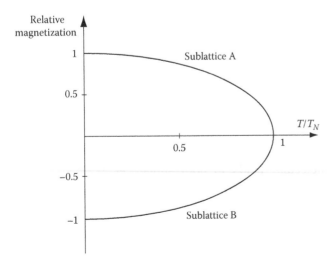

FIGURE 2.107
The spontaneous magnetization of both sublattices below the temperature T_N.

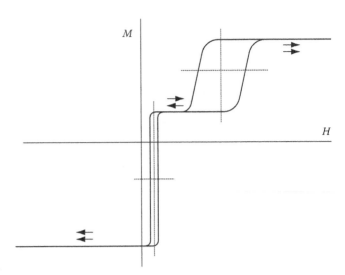

FIGURE 2.108
The antiferromagnetic state obtained in GMR sensors by applying of an antiferromagnetic FeMn layer. (From Tumanski, 2000.)

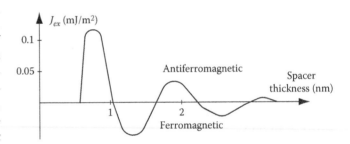

FIGURE 2.109
The antiferromagnetic state obtained in GMR sensors by atomic coupling. (From Tumanski, 2000.)

atomic coupling is known as RKKY interaction—from initials of Ruderman, Kittel, Kasuya, and Yosida.

The *ferrimagnetic phenomenon* can be considered as a special case of antiferromagnetism—when the two sublattices have different spontaneous magnetic fields and the net magnetic moment is nonzero (due to only partial mutual compensation). The ferrimagnetism is often identified with *ferrites* because most of the ferrites are ferrimagnetic materials. But the ferrimagnetism is also exhibited by other materials like rare-earth transition metal alloys. In nature, mineral magnetite Fe_3O_4, known also as lodestone, is ferrimagnetic.

The ferrites have wide applications in many branches of technology. They exist in two main structures—cubic (with general formula MO Fe_2O_3 where MO is a divalent metal ion like Mn, Ni, Fe, Co, Mg) and hexagonal (barium and strontium ferrites used as permanent magnets). The important advantage of ferrites is that they are electrically close to insulators and therefore can be used in high-frequency application as immune to eddy currents (although their magnetic parameters are generally poorer in comparison with ferromagnetic materials). The γ-Fe_2O_3 gamma iron oxide was commonly used as a magnetic recording medium.

The ferrimagnetic materials have rather complex crystal structure. The cubic ferrites are sometimes known as *spinels* (or ferrospinels) because their crystal structure resembles existing in nature mineral spinel—MgO Al_2O_3. The unit cell of a spinel contains 58 ions. An example of a spinel crystal is presented in Figure 2.110.

Ferrimagnetic materials have characteristics intermediate between those of antiferromagnetic and ferromagnetic materials. Figure 2.111 presents a comparison of a typical temperature dependence of ferrimagnets and ferromagnets.

In the ferrimagnetic materials, the permeability or Curie temperature depends on performances of both sublattices (e.g., Curie constants C_A and C_B) and on interlattice coupling constant α (Jiles 1998):

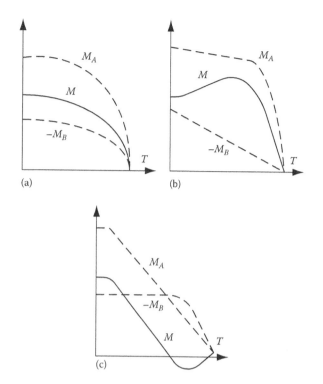

FIGURE 2.112
Three examples of dependence of the saturation magnetization versus temperature (below the T_C temperature).

$$\chi = \frac{(C_A + C_B)T - 2\alpha C_A C_B}{T^2 - \alpha^2 C_A C_B} \quad \text{and} \quad T_C = \alpha\sqrt{C_A C_B} \quad (2.180)$$

Also, the temperature dependence of saturation magnetization can be very unusual as the result of incomplete compensation of two different component curves. Three examples are presented in Figure 2.112 and as can be seen from Figure 2.112c, it is possible to compensate both components for certain temperature.

Other ferrimagnetic materials—garnets, based on rare-earth compounds found wide application. As an example, we can point yttrium iron garnet $Y_3Fe_2(FeO_4)_3$ known also as YIG. Due to high resistance (immune to eddy currents), they are used in microwave frequency range. They are also used in magnetooptical devices as lasers, Faraday effect rotators, magnetooptical imaging tools, etc. Table 2.17 presents performances of main ferrimagnetic materials.

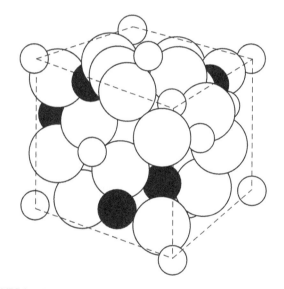

FIGURE 2.110
An example of the spinel crystal.

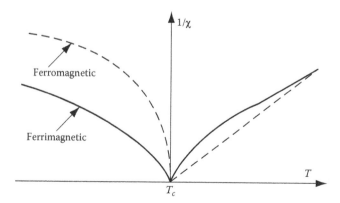

FIGURE 2.111
The temperature dependence of ferrimagnetic materials (and for comparison the same dependence of ferromagnetic material).

2.10.6 Amorphous and Nanocrystalline Materials

The *amorphous* and *nanocrystalline* materials contradict the assumption that the crystal structure is the crucial factor influencing magnetic properties of material. These materials are rapidly cooled from molten state (*rapid solidification technology*) such that there is no time

TABLE 2.17

Performances of Main Ferrimagnetic Elements

Substance	B_s (T)	T_c (K)	μ_p (μ_0)
$MnFe_2O_4$	0.5	573	250
$FeFe_2O_4$	0.6	858	70
$NiFe_2O_4$	0.34	858	10
$BaOFe_2O_3$	0.38	723	
$Y_3Fe_5O_{12}$ (garnet)	0.18	523	

to start the crystal forming process. Nevertheless, they can have excellent magnetic performances—both as soft magnetic materials as well as hard magnetic materials. Amorphous magnetic ribbons are also known as *metallic glasses*.

Typical composition of amorphous material is $(Fe,Co,Ni)_{70-85}$ $(Si,B)_{15-30}$—for example, Metglas Fe40-Ni38-Mo4-B18. In the metallic glass, it is necessary to add metalloid atoms (boron, silicon) in order to help in glass structure formation and ensure a stable glassy state. In amorphous materials, there is no *long-range order* typical for crystal structure in which atoms are distributed regularly (the crystal structure is "repeated") over distances greater than 100–200 nm. But in amorphous state instead, there is a *short-range order* called also *dense random packing*, which guarantees interatomic coupling according to Heisenberg model of ferromagnetism.

Figure 2.113 presents an example of amorphous structure. The ferromagnetic atoms are distributed randomly but they are "frozen" in close positions to each other. The physics of amorphous materials is rather complex (see O'Handley 1987), but one surprising conclusion seems to be viable—crystal structure helps in

ferromagnetic properties but crystal ordering is not a necessary condition. The appropriate interatomic distance and exchange coupling of neighboring atoms are more important.

Another surprising effect appeared during development of amorphous structures into nanocrystalline structure. Previously, it was assumed that the grain size should be as large as possible in order to decrease the coercive force—$1/D$ *rule* (Mager 1952). That is why in grain-oriented electrical steel the grains are even larger than 10 mm. Also, it was assumed that annealing and crystallization destroy amorphous material. However, controlled annealing of amorphous material (containing Cu for prevention of excessive grain growth) leads to improvement of material performances. Moreover, coercive force in small grain range depends on D^6 (Figure 2.114) Provided that the grains are smaller than around 100 nm (thus smaller that the dimensions of a long range order) and embedded in amorphous matrix, very small coercivity is attained.

In comparison with amorphous, the nanocrystalline materials are more stable with temperature, contain less boron (amorphous 25%–30%,* nanocrystalline 3%–10%), are based on less expensive materials like Fe/Si, exhibit lower magnetostriction. The concept of the technology of nanocrystalline material is presented in Figure 2.115.

Both amorphous and nanocrystalline materials have similar performaces. They have large resistivity ($\rho \approx 100$–$200\,\mu\Omega cm$ while for SiFe electrical steel $\rho \approx 40$–$50\,\mu\Omega cm$), which reduces eddy currents, low coercivity, and possibility of application in high-frequency range. They do not have microstructural discontinuities at grain boundaries or precipitates. They can have extremely high permeability 50,000–150,000 while SiFe electrical steel maximum permeability is around 40,000. Due to small coercivity also losses are significantly smaller (0.2 W/kg while SiFe 0.8 W/kg). There are two main disadvantages: low saturation induction (0.5–1.5 T compared to 2.03 T for SiFe electrical steel) and higher price.

After introduction of amorphous and nanocrystalline materials, it was expected that they would quickly conquer market, but competitive materials were also improved. Recently, amorphous and nanocrystalline materials are used for special purposes although amorphous ribbons are also used in power transformers (especially in the United States).

2.10.7 Magnetic Domains and Domain Walls

The existence of ferromagnetic domains was predicted in 1907 by Weiss as an explanation why the ferromagnetic materials can be demagnetized despite the large

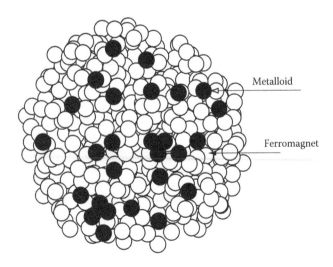

Metalloid

Ferromagnet

FIGURE 2.113
The example of a structure of amorphous material.

---

* Boron is necessary for glass structure formation, but it is expensive and decreases saturation magnetization.

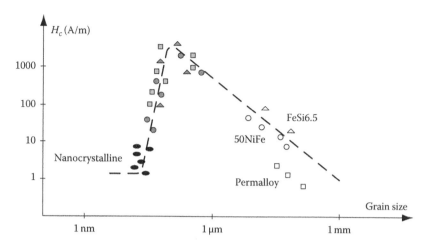

FIGURE 2.114
Dependence of coercivity on the grain size. (From Herzer G., *Nanocrystalline soft magnetic alloys* in Handbook of Magnetic Materials, Vol. 10, Elsevier, 1997.)

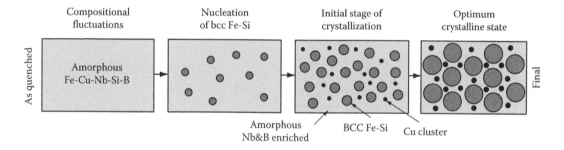

FIGURE 2.115
Formation of nanocrystalline structure during controlled annealing. (From Herzer G., *Nanocrystalline soft magnetic alloys* in Handbook of Magnetic Materials, Vol. 10, Elsevier, 1997.)

internal "molecular field." He assumed that material is divided into many small areas, called domains, each of which is magnetized to saturation. In the demagnetized state, these domains are distributed in such way that the net magnetization is zero.

The domain model of Weiss is still valid. In 1935, Landau and Lifshitz mathematically proved that material is divided into domains because such configuration ensures the lowest energy (Landau and Lifshitz 1935).

They obtained the closure model composed of several domains where net magnetization is zero (Figure 2.116b). Other, open structure (Figure 2.116a) was considered by Kittel (1949). He also proved that division of large domain into narrower strips results in minimization of free energy. The domain structures presented in Figure 2.116 are detectable experimentally in some materials (Shiling and Houze 1974, Hubert and Schäfer 1998).

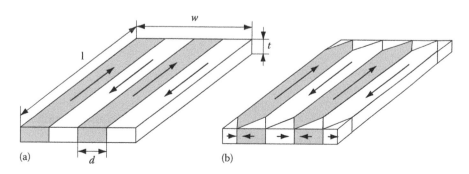

FIGURE 2.116
Two examples of domain structure.

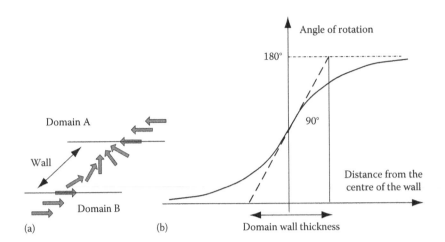

FIGURE 2.117
The change of the direction of magnetization in 180° Bloch wall.

Let us consider the total free energy of a ferromagnetic sample. We should take into account the following components: exchange energy E_{ex}—exchange interaction between magnetic moments, magnetocrystalline energy E_k—energy associated with anisotropy of crystallites, magnetostatic energy E_{ms}—energy associated with demagnetizing field, magnetoelestic energy E_λ—energy associated with the magnetostriction effect, potential energy E_H (Zeeman energy)—energy of the domains in the presence of an applied field and the DW energy E_W. Thus,

$$E = E_{ex} + E_k + E_{ms} + E_\lambda + E_H + E_W \qquad (2.181)$$

The exchange energy E_{ex} is described by the Heisenberg model of ferromagnetism as

$$E_{ex} = -2JS^2 \cos\varphi = -2A\cos\left(\frac{\partial\varphi}{dx}\right) \qquad (2.182)$$

where
  φ is the direction of magnetization
  A is the *exchange stiffness* $(A = nJS^2/a)$
  n is the number of the atoms per unit cell
  a is the lattice parameter

The magnetocrystalline energy depends on anisotropy constants K_1, K_2, and for uniaxial anisotropy, it is

$$E_{k1} = K_1 \sin^2\varphi + \cdots \qquad (2.183)$$

for cubic anisotropy with magnetization in (100) plane, it is

$$E_{kc} = K_1 \sin^2\varphi\cos^2\varphi + \cdots \qquad (2.184)$$

The magnetostatic energy in the sample in its own demagnetizing field is

$$E_{ms} = \frac{\mu_0}{2} N_d M^2 \qquad (2.185)$$

The magnetoelastic energy depends on lattice strains e_{ij}, magnetostriction coefficient λ (or constants B_1, B_2), and for isotropic material under stress σ, it is

$$E_{\lambda 1} \approx\sim B_1 e_{33} \sin^2\theta = \frac{3}{2}\lambda_s\sigma\cos^2\theta \qquad (2.186)^*$$

The potential energy is

$$E_H = -MB \qquad (2.187)$$

The direction of local magnetization at the border between two domains is changing gradually over certain distance called *the DW*. An example of a typical DW (Bloch wall) is presented in Figure 2.117a (see also Figure 2.8).

Equation 2.182 can be simplified by taking into account that $\cos\varphi = 1 - \varphi^2/2$ and the exchange anisotropy is

$$E_{ex} = A\left(\frac{\partial\varphi}{\partial x}\right)^2 \qquad (2.188)$$

The wall energy can be determined by taking into account the exchange and magnetocrystalline energy:

---

* For cubic anisotropy, this energy is
$$E_{\lambda 2} = B_1[e_{11}(\alpha_1^2 - 1/3) + e_{22}(\alpha_2^2 - 1/3) + e_{33}(\alpha_3^2 - 1/3)$$
$$+ B_2(e_{12}\alpha_1\alpha_2 + e_{23}\alpha_2\alpha_3 + e_{31}\alpha_3\alpha_1) + \cdots$$

$$E_W = E_{ex} + E_k = K_1 \sin^2 \varphi + A \left(\frac{\partial \varphi}{\partial x} \right)^2 \qquad (2.189)$$

From the condition of minimum wall energy, we obtain the relationship (Cullity and Graham 2009)

$$x = \sqrt{\frac{A}{K_1}} \ln \left(\tan \frac{\varphi}{2} \right) \qquad (2.190)$$

This relationship is presented in Figure 2.117b. The maximum slope of the dependence $\varphi = f(x)$ is for $\varphi = 90°$ and is at the center of the wall. The thickness of a 180° Bloch wall is

$$t_{dw} = \pi \sqrt{\frac{A}{K_1}} \qquad (2.191)$$

and wall energy density $e = E/V$ (V, volume) is

$$e_W = 4\sqrt{AK_1} \qquad (2.192)$$

Similar analysis of a 90° wall (the wall between domains of magnetization direction rotated by 90°—see Figure 2.116b) gives result $t_{dw}^{90} = 1/2 t_{dw}$ (Cullity and Graham 2009). The thickness of a Bloch DW depends on the anisotropy constant and on the exchange energy. Assuming typical values for iron K_1—5×10^3 J/m^3 and $A = 10^{-11}$ J/m, we obtain that the thickness of the wall is around 140 nm. For higher values of a crystalline anisotropy, for example, for hard magnetic materials, this thickness can be even as small as 10 nm.

Beside the Bloch walls, there are also other kinds of walls—for example, in the thin films below certain critical thickness, a *Néel walls* appear in which the magnetization vector rotate in the film plane.

Let us consider the open domain structure presented in Figure 2.116. The magnetostatic energy of such structure was estimated as (Chikazumi 2009)

$$f_{ms} = 1.08 \cdot 10^5 M_s^2 d \qquad (2.193)$$

and the wall energy as

$$f_W = \frac{e_w l}{d} \qquad (2.194)$$

Both energies have been presented in Figure 2.118a. We can see that there is an optimum width of the domain that we can estimate by finding a minimum of a total energy $\partial(f_{ms} + f_w)/\partial d = 0$. This width is (Chikazumi 2009)

$$d = 3 \cdot 10^{-3} \frac{\sqrt{e_w l}}{M_s} \qquad (2.195)$$

For typical parameters of iron: $e_w = 1.6$ mJ/m^2, $M_s = 1$ T, and $l = 5$ mm, we obtain $d = 8.5$ μm.

The energy of this structure is (Chikazumi 2009)

$$f = 650 M_s \sqrt{e_w l} \qquad (2.196)$$

For the same parameters as above, we obtain $f = 1.8$ J/m^2. If it is a single domain, then the energy is described by $f_{sd} = M_s^2 l/2\mu_0$ and, for the same parameters, we obtain $f_{sd} = 0.2 \times 10^4$ J/m^2. Thus, by division of the structure into smaller domains, we get more than 1000 times lower total energy.

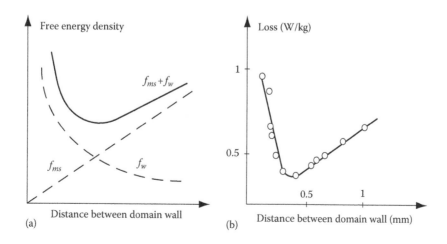

(a) Distance between domain wall

(b) Distance between domain wall (mm)

FIGURE 2.118
The domain wall energy (a) and specific power losses (b) versus the width d of domain. (After O'Handley, R.C., *Modern Magnetic Materials*, John Wiley & Sons, Inc., 2000; Nozawa et al., 1996.)

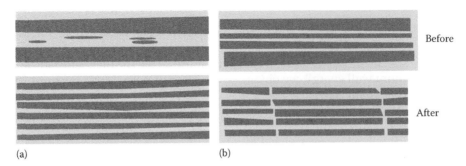

FIGURE 2.119
The domain refinement of electrical steel by additional tensile stress (a) or laser scribing (b).

We can also see that DW spacing depends on the length of domain. It is known that losses in silicon steel depend strongly on the domain width (Figure 2.118b) due to eddy current loss (Nozawa et al. 1996, O'Handley 2000). Therefore, many technologies were developed, which improve quality of electrical steel quality by domain refinement (Nozawa et al. 1996). For example, additional tensile stress is often introduced in electrical steel (by special coating layer) to divide domain into narrower ones (Figure 2.119a). In another technology used in the case of grain-oriented steel, the surface is prepared by *laser scribing* (Figure 2.119b).

The theoretical considerations presented above concerned idealized case of the monocrystal. If we have polycrystalline material, the domain structure depends strongly on the grain structure. Every deviation of the local anisotropy axis results in significant change of domain picture. If two grains have the same crystal orientation, the domains (and DWs) can pass through the grain border. But usually, the grain borders, defects, and precipitations act as nuclei of new domains. Figure 2.120 presents typical picture of domain structure determined for a sample of SiFe steel.

As it was described in Section 2.4.2, the process of magnetization is realized mainly by changes in the domain structure. Initially, we have movements of the walls such that the domains with coinciding direction of magnetization increase at the expense of their neighbor. For very small changes in the magnetizing field, this process is reversible. For higher values of the magnetizing field, movement becomes nonreversible, and hence the domain structure (number of DWs and their configuration) is a crucial factor influencing coercivity. The DW movements are the main reason for the magnetostriction.

Although the domain creation process is random due to presence of impurities, there are methods of numerical calculations of magnetization distribution is small areas. One of such methods was introduced by Brown

FIGURE 2.120
The example of the domain structure (visible dark and light regions inside the grain, and a complex domain structure outside of the grain). (Courtesy of Zurek, Wolfson Centre for Magnetics, Cardiff, U.K.)

(1963). Brown's equations describing the stability conditions are formulated as

$$m \times H_{eff} = 0 \qquad (2.197)$$

where H_{eff} is determined from H_{ex}, external magnetic field; H_d, demagnetizing field; f_{an}, density of anisotropy energy, as (Bertotti 1998)

$$H_{eff} = H_{ex} + H_d + \frac{2A}{\mu_0 M_s} \nabla^2 m - \frac{1}{\mu_0 M_s} \frac{\partial f_{AN}}{\partial m} \qquad (2.198)$$

Figure 2.121 presents an example of results of micromagnetic computation determined for thin-film MR sensor layer (Zheng and Zhu 1996).

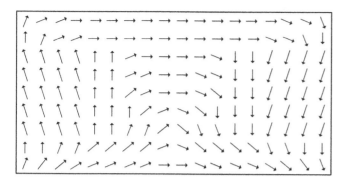

FIGURE 2.121
An example of calculated distribution of magnetization in thin-film plate. (After Zheng, Y. and Zhu, J.G., *IEEE Trans. Magn.*, 32, 4237, 1996.)

2.11 Magnetic Hysteresis

2.11.1 Magnetization Process

Let us start with the state of full demagnetization. The most common way of demagnetization is a magnetization of the sample with AC magnetic field (as high as possible to obtain saturation) and next slow gradual decrease of magnitude of this field to zero (Figure 2.122). It is only technical demagnetization—to obtain full physical demagnetization, it would be necessary to heat the material above the Curie temperature.

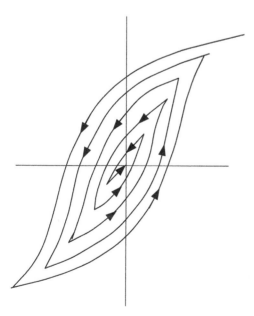

FIGURE 2.122
Demagnetization of a sample by gradual decrease of the magnetizing AC field.

What does it mean that the sample is demagnetized? It simply means that the net magnetization in a chosen direction is zero. But we know that inside a domain the magnetization can be close to saturation. The domains are organized in such a way as to minimize the total energy, which is the same as net magnetization equal to zero. Figure 2.123 presents a few cases of domain configuration with net zero magnetization in the sample. In the cubic texture, the closed configuration of domains is predominant. Even if we have a one-domain state (possible in some cases in thin film samples), we also can have zero magnetization if we measure it perpendicularly to the easy axis of anisotropy.

Looking at microstructure of magnetic materials, we should remember that there might also be residual local magnetizations resulting from the various distribution of orientation of individual grains and from local mechanical stresses.

Figure 2.124 presents possibility of different directions of magnetization caused by imperfect texture. It is assumed that dispersion of anisotropy is not more than 7° in normal grain-oriented steel. In electrical steel with improved texture (e.g., HiB steel), this dispersion can be smaller than 4°.

If we start with magnetization of the initially demagnetized sample, we obtain so-called *virgin curve of magnetization*. There are two main mechanisms of magnetization—by DWs displacement (e.g., growth of the domain parallel to the field at the expense of neighbors) or by rotation of magnetization. It is assumed that wall motion is dominating for small magnetic fields (below the "knee" of the magnetization curve) while above the knee, the rotation of magnetization is predominant. Both mechanisms can be irreversible. When the wall is displaced, it usually does not return to its previous position. Hence, if the increase of magnetic field is stopped, it is obvious that after nonreversible change of magnetization, the magnetization curve does not return to the initial position. This memory effect is a cause of hysteresis.

The virgin curve of magnetization can be obtained after demagnetization of the sample. The dependence $B = f(H)$ is then measured by changes of the magnetic field strength by some value ΔH (DC point by point method—*ballistic method*). But this method is rather lengthy; therefore, often it is substituted by AC magnetization curve obtained by connecting the tips of hysteresis loops—see Figure 2.12a.

It is also possible to obtain the so-called *anhysteretic curve of magnetization* (Figure 2.125) by applying a saturating magnetic field and then by a gradual decrease of the amplitude of the AC magnetizing field in a presence of biasing DC field of the strength H_1. This curve

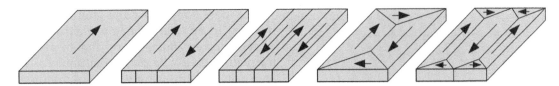

FIGURE 2.123
Various states of the demagnetization.

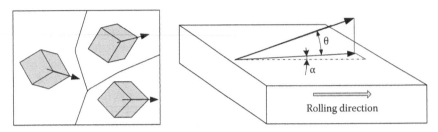

FIGURE 2.124
Various direction of magnetization caused by imperfect texture (α, yaw angle, θ, tilt angle).

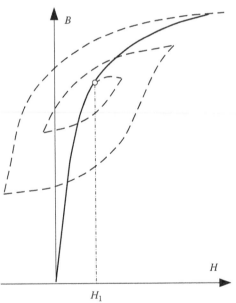

FIGURE 2.125
The anhysteretic magnetization curve—method of determination.

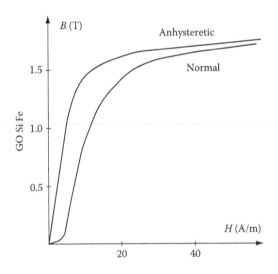

FIGURE 2.126
The comparison of anhysteretic and normal magnetization curves.

is sometimes called as *ideal curve* because it represents the magnetization process without irreversible changes. Both, the anhysteric curve and normal virgin curve are different (Figure 2.126). We see that anhysteretic curve is without the point of defection at low values of magnetic field strength, with large and almost constant permeability in initial part of magnetization.

The classical investigations of the magnetization process by DW motion were described in 1949 by Williams and Shockley. They prepared a frame (19 × 13 mm) cut from an SiFe monocrystal with the all sides parallel to the cubic [100] axis. The perfect hysteresis loop was obtained (Figure 2.127). They also observed the domain structure. In the initial state, the frame was composed of eight domains in a closed configuration. When the magnetizing field was increased, the central domains walls were moved toward the one edge of the frame. In the state of saturation, only four domains remained. A perfect correlation between the position of the wall and resulting magnetization was detected

As a result of the irreversible changes of magnetization, the hysteresis loop appears. If the magnetization is decreased from the saturation, the characteristic is different than for increase of the field, and for a zero magnetic field we return to certain state of magnetization

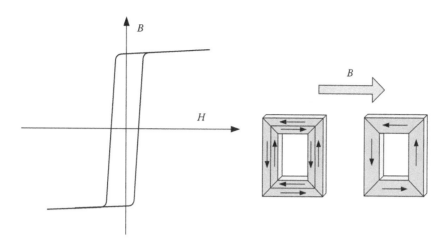

FIGURE 2.127
Experiment on a monocristal frame described by Willimas and Shockley: hysteresis and domain structures.

known as *remanence flux density* B_r (or *residual flux density*). To return to a zero magnetization, it is necessary to apply additional magnetic field in opposing direction known as *coercivity* H_c (or *coercive force* or *coercive field*) (see Figure 2.11).

One of the possible mechanisms of nonreversible changes of magnetization was presented by Rossignol et al. (2005) (Figure 2.128). If a DW is displaced by a half of the interatomic distance, it is in the state of maximum energy. Thus, from the point A (Figure 2.128b), reversible return of the wall is still possible, but if the wall is forced to the point B, it falls down to the new position of a local minimum energy. The same effect can be explained as a result of local residual stresses in magnetic material. These stresses also change in a periodic way (Becker and Kersten 1930, Becker 1930, Hoselitz 1952).

It is assumed (and confirmed by experiments) that a similar mechanism of nonreversible changes of magnetization is caused by imperfect structure of magnetic materials. The impurities, defects, grain boundaries, etc. can pin the wall (so-called *pinning effect of inclusions*). This mechanism was firstly analyzed by Kersten (1943) and Néel (1944a,b,c, 1946). If the wall bisects a nonmagnetic inclusion, the magnetic poles are redistributed, as it is shown in Figure 2.129a. With new ordering of magnetic poles, the magnetostatic energy is much lower, which causes pinning of the wall to the inclusion. Moreover, supplementary spike domains can be created causing additional decrease of the energy (Figure 2.129b).

Figure 2.130 presents a Néel model of reversible and nonreversible changes of magnetization. The wall is moved due to a force dE/dx (E, energy). When it arrives at the position B, it jumps to the position C corresponding

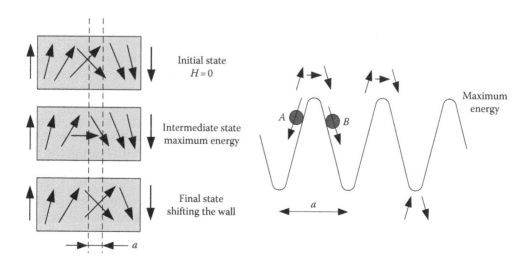

FIGURE 2.128
The energy barriers for displacement of the domain wall. (After Springer Science + Business Media: [*Magnetism—Fundamentals*, Ferromagnetism of an ideal system, Chapter 5, 2005, Rossignol, M. et al.])

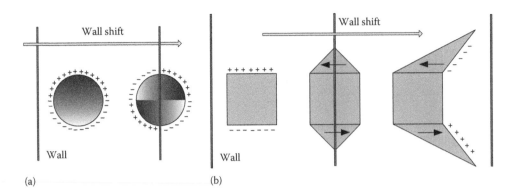

FIGURE 2.129
The effects generated when the wall bisects the inclusion.

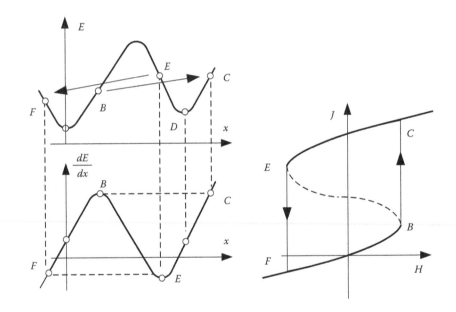

FIGURE 2.130
The Néel model of reversible and nonreversible motion of the domain wall.

with dE/dx. This jump is irreversible because if the magnetic field is removed, the wall returned to new stable position D. Similar effect appears if we decrease the magnetic field— there would be a jump from the position E to F. Such sudden jumps manifest themselves as a *Barkhausen noise*. The described motion of the wall corresponds with the elementary hysteresis curve presented in Figure 2.130c.

We can see that the DWs motion depends strongly on the impurities. The wall pinned by two locations can change its length similarly to an elastic membrane by bowing (and still being pinned) instead of purely linear motion. The inclusions impede the wall motion, so the coercive force H_c depends on the local stresses and inclusions. Hilzinger and Kronmüller (1977) determined this dependence for weak defects interaction as $H_c \sim \rho^{1/2}$ and for strong defect interaction as $H_c \sim \rho^{2/3}$ (ρ, density of defects). Thus, a small number of defects and

impurities results in smaller coercivity H_c and larger initial susceptibility χ_{in} because $\chi_{in}H_c = $ const (Jiles 1998).

Because the magnetic material near saturation is close to a one-domain state, the decrease of magnetic field (from saturation) causes nucleation of new domains (and DWs). Of course in the *nucleation process*, the presence of impurities and inclusions plays significant role.

If we magnetize the material, we deliver the energy that is dissipated as heat. During the magnetization of a demagnetized sample (as we obtain the magnetizing curve), the energy delivered W is represented by the area above the magnetizing curve (Figure 2.131a) and is described by Warburg equation (Warburg 1881):

$$W = \int_0^B HdB \qquad (2.199)$$

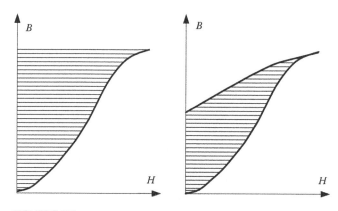

FIGURE 2.131
Graphical representation of the energy delivered to magnetized sample.

The same relationship holds for the whole hysteresis loop and therefore we can assume that the area of the hysteresis represents the power loss for one cycle of magnetization. There are various empirical relations describing the hysteresis loss P_H (e.g., proposed by Richter (1910) $P_H = aB_s + bB_s^2$ or proposed by Anderson and Lance (1922) $P_H = aB_sH_c$) but the most commonly accepted is the relationship proposed by Steinmetz in form $P_H = aB_s^{1.6}$ (Steinmetz 1891, 1892). This equation for the sample magnetized by an AC magnetic field of frequency f is usually presented as

$$P_H = C_0 f B_s^2 \qquad (2.200)$$

Figure 2.132 presents three hysteresis loops determined at various frequencies. We can observe significant differences between static and dynamic hysteresis loops determined for higher frequency. The area of the loop is much larger, and hence the power loss is also increased. Figure 2.133 presents the hysteresis loops for grain-oriented 3% SiFe electrical steel.

The additional power loss is caused by eddy currents. These eddy current losses can be calculated as (see Equation 2.45)

$$P_{EC} = \frac{\pi^2 t^2}{6\rho}(B_m f)^2 \qquad (2.201)$$

We can easily separate both types of losses because hysteresis losses depend on $C_H f$ while eddy current losses depend on $C_E f^2$ (see Figure 2.28). The hysteresis loss can be decreased by elimination of imperfections, while the eddy current loss can be suppressed by decrease of thickness t and increase of resistivity ρ.

The experiments indicate much larger losses than $P_H + P_{EC}$, as it is presented in Figure 2.28. These extra losses are known as *excess losses* P_A (previously related to ambiguous *anomalous losses*). It is assumed that these additional losses are caused by the dynamic effects of DWs motion. The movements of DWs results in micro eddy currents around the walls. Williams et al. (1950) investigated the velocity of magnetic DWs and calculated loss P_{EC} for one domain (Figure 2.134a) and the loss P_{A1} for the same domain with a single rigid DW (Figure 2.134b) (the WSK model). The difference between these losses was determined as

$$P_{A1} = P_{EC}\frac{w}{2\pi^3 t} \qquad (2.202)$$

where w is a distance between walls (the width of a domain).

The analysis of WSK was extended by Pry and Bean (1958) for multiple DWs (Figure 2.134c). The excess loss in this model was estimated as

$$P_A = 1.628 P_{EC}\frac{2w}{t} \qquad (2.203)$$

Further extensions of this analysis were presented by Bishop (1980, 1985) and Bertotti (1998). From the relationship (2.203), it is clear why the division of a wide domain into two narrower ones (e.g., by domain refinement technology, see Figure 2.119) improves quality of the electrical steel.

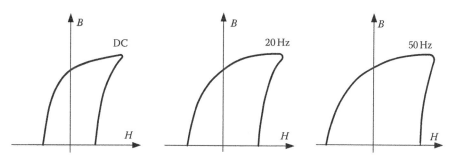

FIGURE 2.132
The hysteresis loop determined for various frequencies.

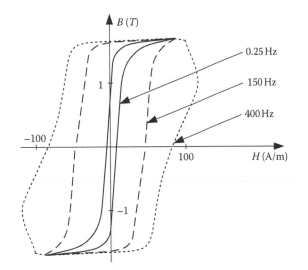

FIGURE 2.133
The hysteresis loop of GO SiFe 0.3 mm electrical steel. (*Measurement and Characterization of Magnetic Materials*, Fiorillo, F., Copyright (2004), from Elsevier.)

Slightly different method is used to describe losses in soft ferrites. In these materials, eddy currents are relatively small even for high frequency because ferrites are electrical insulators. Therefore, the hysteresis loop is relatively narrow. The losses in high-frequency materials are described by the *complex permeability* $\mu(\omega)$ related to the phase shift δ between H and B:

$$\mu(\omega) = \frac{B_m \exp[j(\omega t + \delta)]}{H_m \exp(j\omega t)} = \mu' - j\mu'' \qquad (2.204)$$

where
μ' is the real (in-phase) part of permeability
μ'' is the imaginary (out-of-phase part)

The losses depends on the loss factor* (or loss tangent) $tg\delta = \mu''/\mu'$ and are described by the following equation (Fiorillo 2004):

$$P = \pi J_p H_p \frac{\tan\delta}{\sqrt{1 + \tan^2\delta}} \qquad (2.205)$$

where J_p and H_p are initial polarization and magnetic field strength, respectively.

Because practically all parameters of ferromagnetic materials depend strongly on the operating frequency, the presence of harmonics in the flux density or magnetic field strength waveform is crucial for magnetic measurements. Unfortunately, both—the magnetization

curve and the hysteresis—are nonlinear and the shapes of B and H waveforms are non-sinusoidal. Figure 2.135 presents the shape of the flux density waveforms calculated from the hysteresis under assumption that the magnetic field strength is pure sinusoidal and the shape of the magnetic field strength waveforms calculated from the hysteresis under assumption that the flux density is pure sinusoidal.

If the magnetic field strength is sinusoidal due to saturation of the magnetizing curve, the flux density waveform is distorted. Because results of magnetization depend on the presence of harmonics, it would be impossible to compare results of investigations of various magnetic materials measured with various excitation waveforms. That is why the international standards recommend the investigations for pure sinusoidal shape of the flux density waveform. This requirement means that the magnetic field strength should be appropriately distorted as it is illustrated in Figure 2.135b.

2.11.2 Rayleigh Model of Hysteresis

Various models of the hysteresis loop were proposed (Mayergoyz 1991, Ivanyi 1997, Della Torre 1999). One of the oldest attempts to modeling of the hysteresis and the magnetizing curve was the model proposed by Lord Rayleigh in 1887 (Figure 2.136). He proposed to describe the magnetizing curve using the initial permeability μ_{in} as the parameter:

$$\mu(H) = \mu_{in} + aH \quad \text{and} \quad B(H) = \mu_{in}H + \frac{1}{2}aH^2 \qquad (2.206)$$

where a is the Rayleigh constant determined experimentally.

The first term in Equation 2.206 represents a reversible magnetization while the second term a nonreversible change. Assuming that the magnetic field strength is changed between $\pm H_1$, we obtain the following relationships describing both branches of a hysteresis loop:

$$B = -B_1 + \mu_{in}(H + H_1) + \frac{a}{2}(H + H_1)^2$$

$$ \qquad (2.207)$$

$$B = B_1 + \mu_{in}(H - H_1) + \frac{a}{2}(H - H_1)^2$$

Thus, the magnetizing curve and the hysteresis loop are described by the following equations:

$$B = \left(\mu_{in} + \frac{a}{2}H\right)H \qquad (2.208)$$

---

* Sometimes the normalized loss factor $\tan\delta/\mu = \mu''/\mu'$ is used as the quality factor.

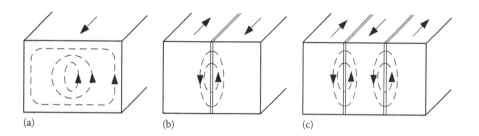

FIGURE 2.134

The domain models used for calculation of eddy current loss—classical P_{EC} (a) and anomalous P_A—model of WSK (b) and model of Pry and Bean (c).

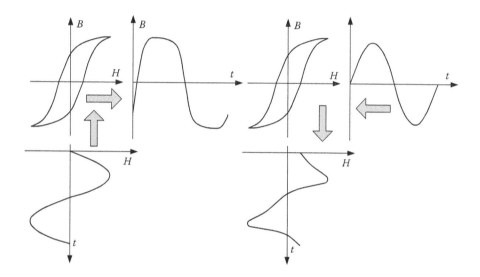

FIGURE 2.135

Waveform shapes for sinusoidal magnetic field strength and flux density calculated from the hysteresis loop.

$$B = (\mu_{in} + aH_1)H \pm \frac{a}{2}(H^2 - H_1^2) \qquad (2.209)$$

The Rayleigh constant can be determined as $a = 2B_r/H_1^2$ and both parameters of the hysteresis loop as

$$B_r = \frac{a}{2}H_1^2 \quad \text{and} \quad H_c = \sqrt{\frac{\mu_{in}^2}{a} + H_1^2} - \left(\frac{\mu_{in}}{a}\right) \qquad (2.210)$$

The hysteresis losses can be described as

$$P_H = 2\int_{-B_1}^{B_1} H dB = 2\int_{-B_1}^{B_1} H[(\mu_{in} + aH_1) + aH]dH = \frac{4}{3}aH_1^3 \quad (2.211)$$

If we assume that the magnetic field strength is varying sinusoidally as $H(t) = H_1 \sin\omega t$, we obtain the following equation describing the flux density response:

$$B = B_m \sin\omega t + 2B_r \sin\omega t \pm B_r \cos^2\omega t \qquad (2.212)$$

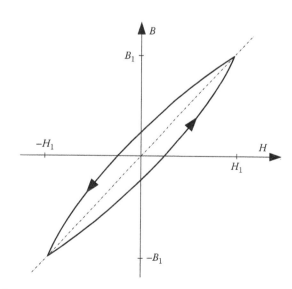

FIGURE 2.136

The Rayleigh hysteresis loop.

where $B_m = \mu_{in} H_1$. By expanding $\cos^2 \omega t$ into series, we obtain

$$B = B_\omega \sin(\omega t + \varphi) + B_{3\omega} \cos^3 \omega t \qquad (2.213)$$

where

$$B_\omega = \sqrt{(B_m + 2B_r)^2 + (8B_r/3\pi)^2}$$

$$B_{3\omega} = 8B_r/15\pi$$

$$\tan \varphi = 4aH_1/3\pi(\mu_{in} + aH_1)$$

2.11.3 Stoner–Wohlfart Model of Hysteresis

The model proposed by Stoner and Wohlfarth in 1948 (Stoner and Wohlfarth 1948, 1991) assumes that we consider a one-domain state. This way we can neglect the effects related to the domain movements and we can assume that the magnetization is caused by a coherent rotation of the magnetization. Although this model seems to be extremely simplified, it was proved to work well in the case of thin ferromagnetic films.

The ferromagnetic films exhibit usually the uniaxial anisotropy induced by the deposition process (e.g., by the deposition in a presence of magnetic field). Thus, initially the thin film is magnetized along the anisotropy axis called as an *easy axis of anisotropy*.[*] If we apply the magnetic field perpendicular to the easy axis, then the magnetization rotates by an angle φ. Even if the real thin-film sample comprises several domains due to small anisotropy dispersion, we can assume that all domains are magnetized in a similar way. If we apply the magnetic field along the easy anisotropy axis, the direction of the magnetization vector can change if we overcome the anisotropy field H_k:

$$H_k = \frac{2K_u}{M_s} \qquad (2.214)$$

where K_u is the anisotropy constant.

The induced anisotropy field can be amplified (or weakened) by the shape anisotropy according to the relationship (Fluitman 1973):

$$H_k = \sqrt{H_{k0}^2 + (N_d M_s)^2 + 2H_{k0}N_d M_s \cos^2 \varepsilon} \qquad (2.215)$$

where
H_{k0} is the anisotropy of the material (for 80/20 permalloy H_{k0} is about 250 A/m)

N_d is the demagnetizing factor
ε is the angle between the anisotropy axis and the geometrical axis of the strip (see Figure 2.39)

For typical dimensions of a ferromagnetic strip (thickness 20–150 nm, width 5–30 μm), the anisotropy field is $H_k = 1$–20 kA/m).

If we consider the energy E as (see also Equation 2.54),

$$E = -\boldsymbol{HM} + K_u \sin^2 \varphi \qquad (2.216)$$

we can determine the position of the magnetization vector from the condition of minimum energy:

$$\frac{\partial E}{\partial \varphi} = 0 \quad \text{and} \quad \frac{\partial^2 E}{\partial \varphi^2} = 0 \qquad (2.217)$$

The solution of the conditions (2.217) is a hypocycloid described by the following equation:

$$\left(\frac{H_x}{H_k}\right)^{3/2} + \left(\frac{H_y}{H_k}\right)^{3/2} = 1 \qquad (2.218)$$

known also as a normalized Stoner–Wohlfarth asteroid (Figure 2.137),

$$h_x^{3/2} + h_y^{3/2} = 1 \qquad (2.219)$$

where
$h_x = H_x/H_k$
$h_y = H_y/H_k$

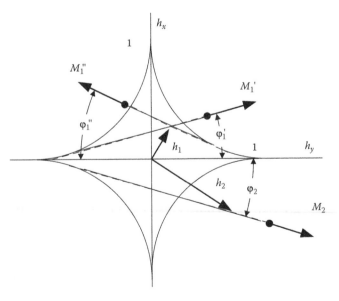

FIGURE 2.137
The Stoner–Wohlfarth asteroid and the method of determination of the position of a magnetization vector.

---

[*] The axis perpendicular to the easy axis is sometimes termed as a *hard anisotropy axis*.

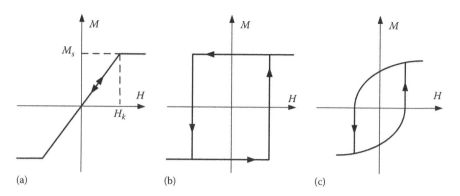

FIGURE 2.138
The hysteresis loops calculated from the Stoner–Wohlfarth model: a magnetizing field perpendicular to the easy anisotropy axis (a), a magnetizing field along the easy anisotropy axis (b), and a magnetizing field directed of about 45° from the anisotropy axis (c).

From this equation, it is possible to obtain the value φ representing the position of the magnetization vector. Equation 2.219 can be solved only by computation (Kwiatkowski et al. 1983), but it can be easily solved graphically as it is demonstrated in Figure 2.137.

Figure 2.137 illustrates how to determine the position of the magnetization vector for a given value of the magnetic field strength vector $h = h_x + h_y$. The direction of the magnetization vector M is possible to be found by drawing a line tangent to the asteroid from the tip of the applied magnetic field vector h. When the field vector h_1 is inside the asteroid, it is possible to draw two tangent lines declined by the angles φ_1' or φ_1'' with respect to the anisotropy axis. The stable position of the magnetization vector (M_1' or M_1'') depends on the "magnetic history" of the film. When the vector h_2 is outside the asteroid, only one position of the magnetization vector M_2 is possible.

Figure 2.138 presents the examples of the hysteresis loop determined from Equation 2.219. If we magnetize the sample perpendicularly to the easy axis of anisotropy, we obtain linear characteristic without hysteresis. If we magnetize the sample along the anisotropy axis, we obtain rectangular hysteresis loop. These results were confirmed in many experiments.

Even if we have the multidomain thin-film system, it is possible to use asteroid for its analysis if the dispersion of anisotropy is not too large. The result is then obtained as a average of several characteristics (Figure 2.139). Figure 2.140 presents calculation of the dependence $\sin\varphi$ (h_x) of the sample magnetized with the field inclined from the anisotropy axis (for the one-domain system) and the experimental result of investigation of the thin-film sample with some dispersion of anisotropy.

By magnetizing the multidomain sample, it is possible to technically demagnetize it. The possible mechanism is presented in Figure 2.141. Let us start with the initial state when all domains are magnetized parallel to the easy

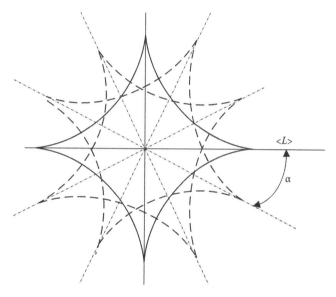

FIGURE 2.139
The asteroids representing multidomain system with dispersion of anisotropy α.

anisotropy axis (with small angular dispersion of anisotropy) (Figure 2.141a). After applying transverse field, the magnetization rotates toward the direction of the field. When magnetic field reaches the critical value (explained below), some directions of magnetizations jump in conformity with *Stoner–Wohlfarth* model (Figure 2.141b and d). When magnetic field is removed, then the localized directions of magnetization might be aligned antiparallel as it is shown in Figure 1.141c. Thus, this magnetic film may be partially demagnetized after removing of the magnetic overload. To recover the previous state of magnetization, it is necessary to apply sufficiently large magnetic field along the easy anisotropy axis. It should be remembered that even in the case of a single-domain state, the hysteresis might appear. The difference is that in the case of a single-domain state, this hysteresis is reversible and the film cannot be demagnetized.

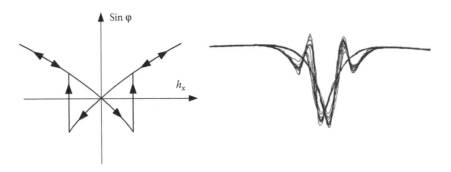

FIGURE 2.140
Comparison of the calculated $\sin\varphi$ (h_x) dependence and experimental results for the thin-film sample.

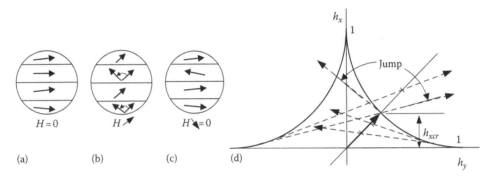

FIGURE 2.141
The mechanism of the demagnetization of multidomain sample.

Figure 2.141d explains the critical value of magnetic field. If we start the magnetization of the sample and the end of the H_x vector is inside the asteroid, two positions of magnetization vector are possible. When the end of the H_x vector arrives to the critical point H_{xcr}, it is not possible to draw the tangent line. Thus, the direction of magnetization jumps to the right. One of the methods enabling elimination of the possibility of such demagnetization of the sample is applying the additional bias field H_b (along the easy anisotropy axis). The working point is therefore moved outside the asteroid and the critical field (causing the jump of magnetization direction) is much higher (Figure 2.142a).

Figure 2.142 presents the typical characteristics of thin-film magnetoresistive sensor. Without the bias field ($H_b = 0$), the sample exhibits large hysteresis. But after applying sufficiently large additional bias field along the easy anisotropy axis, it is possible to reduce and even eliminate the hysteresis.

2.11.4 Preisach Model of Hysteresis

Preisach (Vajda and Della Torre 1995) proposed his model of hysteresis in 1935 (Preisach 1935). After this proposal, a huge number of papers and books extended this model (Everett and Whitton 1952, Everett 1954, 1955, Everett and Smith 1954, Woodward and Della Torre 1960, Del Vecchio 1980, Kadar 1987, Mayergoyz 1988, Della Torre 1994). The mathematical basis for the Preisach model was formulated by Krasnoselski et al. (1983). Also, several books on the subject were published (Ivanyi 1997, Bertotti 1998, Della Torre 1999, Mayergoyz 2003). Therefore, below only simplified description of the Preisach model is presented.

This model assumes that the ferromagnetic material consists of many elementary "domains" each of which is represented by a rectangular elementary hysteresis loop (Figure 2.143a). Each hysteresis loop is described by two parameters: critical field h_c and the field shift h_m representing interaction with neighbouring domains. The state of magnetization M is described by the operator γ (h_a, h_b) $H(t) = \pm 1$.

Moreover, all elementary domains are distributed in the plane called Preisach triangle (Figure 2.143). This triangle is divided by the line $L(t)$ into two parts—above this line there is the state T^{-1} (operator $\gamma = -1$), below the line the state T^{+1} (operator $\gamma = +1$). The line $L(t)$ represents the history of the material. If a magnetic field decreases, the horizontal part of the line L moves downward, while if a magnetic field increases, the vertical part of the line L moves rightward. After movement of the L line, we

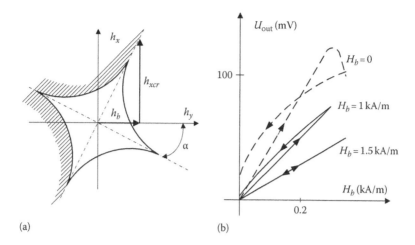

(a)   (b)

FIGURE 2.142
The mechanism of the demagnetization of a multidomain sample (a) and improvement of the magnetization process by applying the additional bias field (b).

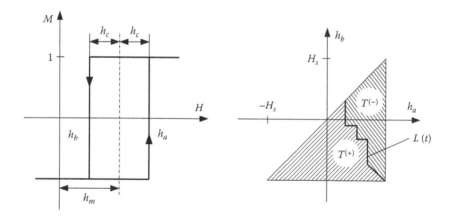

FIGURE 2.143
The elementary Preisach hysteresis loop and the Preisach triangle. (After Ivanyi, A., *Hysteresis Models in Electromagnetic Computation*, Akademiai Kiado, Budapest, Hungary, 1997.)

have nonreversible change of the state of magnetization—change of the γ operator. Previous position of the line L is erased (wiping out). The magnetization $M(t)$ of the sample can be determined as the sum of all elementary states*:

$$M(t) = \iint\limits_{h_a \geq h_b} P(h_a, h_b)\gamma(h_a, h_b)H(t)dh_a dh_b \quad (2.220)$$

where $P(h_a, h_b)$ is the Preisach function describing the material—usually determined experimentally.

Figure 2.144 presents the simplified method of determination of the loop using the Preisach model (Del Vecchio 1980, Ivanyi 1997). When we start, the material is demagnetized—both parts T^+ and T^- have the same

area and the line L is diagonal (Figure 2.144a). When magnetizing field increases, the line L is moved to the right (Figure 144b). In saturation, it reaches the right border and the whole triangle is filled by T^+ states. Then, if we decrease the field, the line L is moving down (Figure 2.144c). When it reaches zero (i.e., lies on the horizontal axis), the nonequilibrium of T^+ and T^- means remanence (Figure 2.144d). When both areas are the same, the line is at the position of coercive field H_c (Figure 2.144e). Further change of magnetic field results in the opposite saturation (Figure 2.144f).

The procedure presented in Figure 2.144 was proposed and described in more details by Del Vecchio (1980). After calculation of the complex hysteresis loops (Figure 2.145a), the experiment with non-oriented electrical steel has been performed (Figure 2.145b). The agreement of modeling and experimental results was quite satisfying.

---

* This operation can be performed numerically after mesh discretization of the triangle).

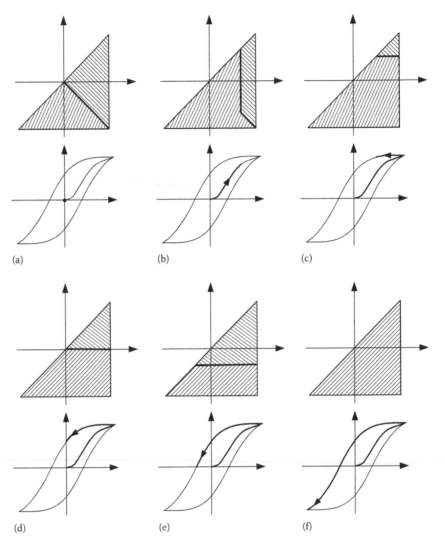

FIGURE 2.144
The illustration of the application of the Preisach model to determine the hysteresis loop. (After Ivanyi, A., *Hysteresis Models in Electromagnetic Computation*, Akademiai Kiado, Budapest, Hungary, 1997.)

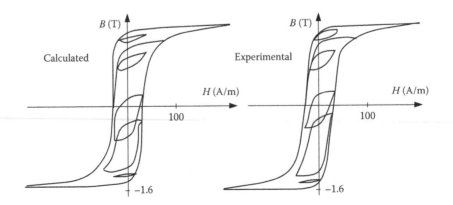

FIGURE 2.145
The hysteresis loops obtained using the model presented in Figure 2.144 and experimental results. (From Del Vecchio, R.M., 1980, *IEEE Trans. Magn.*, 16, 809, 1980.)

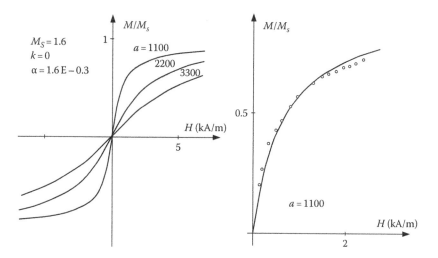

FIGURE 2.146
The example of computed anhysteric curves and comparison with experiment. (After Jiles, D.C. and Atherton, D.L., *J. Magn. Magn. Mater.*, 61, 48, 1986.)

2.11.5 Jiles–Atherton Model of Hysteresis

The Stoner–Wolhfarth model takes into account the rotation of magnetization. It can be used with success for the analysis of rather simple samples—for instance, thin-film elements.* On the other hand, Preisach's model does not take into account the rotation of magnetization. Although the Raleigh and Preisach models were based on physical phenomena, they were really just methods to mathematical curve fitting rather than truly physical model. The review of many other models of hysteresis is presented by Ivanyi (1997).

Beside various mathematical curve-fitting models, there were also attempts to construct models based more closely on real physics of the magnetization process. In 1971, Globus et al. presented method of determination of magnetization curve from size of crystallites and anisotropy field. Jiles and Atherton proposed model, which included mechanisms of nonreversible wall motion as well the reversible walls bending taking into account pinning of DWs (Jiles and Atherton 1983, 1986, Jiles 1998). The magnetization was determined from the energy analysis. This model was also later extended by many researchers.

The starting point for this model was the anhysteric magnetizing curve (see also Figure 2.125) because in this way, the nonreversible and reversible motion of DWs can be separated. By applying the Langevin expression, the following equation describing anhysteric magnetization M_{an} was proposed:

$$M_{an}(H_e) = M_s\left(\coth\left(\frac{H_e}{a}\right) - \left(\frac{a}{H_e}\right)\right) \quad (2.221)$$

where H_e is the effective field taking into account interdomain coupling (coefficient α) and $H_e = H + \alpha M$. The coefficient a is a parameter with a dimension of magnetic field characterizing the shape of the anhysteric magnetization. Figure 2.146 presents examples of calculation of anhysteric curve and comparison with experimental results.

Next, the energy of DW motion of rigid planar DWs and energy dissipated through pinning of the DW was analyzed. As a result, the following relationship describing the magnetization changes due to nonreversible motion of the walls was proposed:

$$\frac{dM_{nrev}}{dH} = \frac{1}{k - \alpha(M_{an} - M_{nrev})}(M_{an} - M_{nrev}) \quad (2.222)$$

where k is the pinning constant, and

$$k = M_{an}(H_c)\left\{\frac{\alpha}{1-c} + \left[\frac{1}{(1-c)\chi_{Hc} - cdM_{an}(H_c)/dH}\right]\right\} \quad (2.223)$$

where

$c = \chi_{in}/\chi_{an} = 3a\chi_{in}/M_s$ (χ_{in} is the initial susceptibility of the magnetization curve and χ_{an} is the differential susceptibility of the anhysteric curve at the origin)

χ_{Hc} is the slope of the hysteresis loop at H_c

---

* It can be also used for the analysis of certain cases of hard magnetic materials.

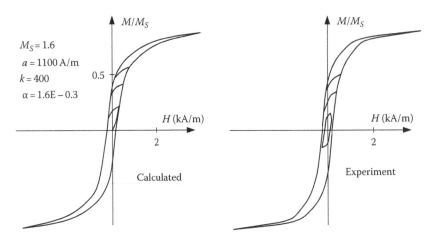

$M_S = 1.6$
$a = 1100 \text{ A/m}$
$k = 400$
$\alpha = 1.6E - 0.3$

FIGURE 2.147
The example of computed hysteresis loop and comparison with experiment (After Jiles, D.C. and Atherton, D.L., *J. Magn. Magn. Mater.*, 61, 48, 1986.)

For soft magnetic materials where $\chi_{an} \cong \chi_{H_c}$, Equation 2.223 can be simplified as

$$k = \frac{H_c}{1 - (\alpha M_s / 3a)} \quad (2.224)$$

From the expression (2.224) results that $k \cong H_c$, which means that coercivity depends mainly on the pinning of the DW motion.

The relationship (2.222) describes nonreversible component of DW motion. Taking into account also reversible changes $M_{rev} = c(M_{an} - M_{nrev})$, the final version of hysteresis model is

$$\frac{dM}{dH} = \frac{(M_{an} - M_{nrev})}{k - \alpha(M_{an} - M_{nrev})} + c\left(\frac{dM_{an}}{dH} - \frac{dM_{nrev}}{dH}\right) \quad (2.225)$$

Therefore, in order to construct the model of hysteresis, it is sufficient to determine four coefficients: a, α, k, and c from the anhysteric magnetizing curve and the hysteresis loop. Sometimes, it might be necessary to determine these coefficients by iteration. The first step is the preliminary determination of the coefficients from the experiment data. Further, computing for finding optimum values of these parameters might be required such that the calculated curve fits, in an optimum way, the experimental data. Nevertheless, after correct procedure, it is possible to obtain the mathematical model with a relatively good agreement with the experimental results (Figure 2.147). The advantage of this model is that a, α, k, and c coefficients have the physical meaning (for more details see Jiles 1998), which can be useful for further analysis.

2.11.6 Approximation of the Magnetizing Curve and Hysteresis

The presented models of hysteresis are rather complex. Often, when we want to design a magnetic device and compute the magnetic field distribution, we do not need to use a model of material, which expresses the physical phenomena. In such case, it is more important to use mathematically simple approximated model. For this purpose, there were proposed many various approximations of the *BH* magnetization curve (Trutt et al. 1968, Ivanyi 1997).

The simplest solution is to save into computer memory a lookup table with numerical data of the *BH* curve and to calculate a piecewise linear approximation—as it is illustrated in Figure 2.148. In a first approximation, the anisotropic material can be represented by two *BH* curves (one for the easy and one for the hard direction of magnetization). If necessary, it is also possible to introduce the parallelogram representation of the hysteresis loop—as it is shown in Figure 2.148b (Savini 1982, Lin et al. 1991).

Various other mathematical models can be used for approximation of the magnetization curve, from power

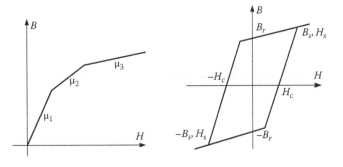

FIGURE 2.148
The example of the piecewise linear approximation of the magnetization curve or the hysteresis loop.

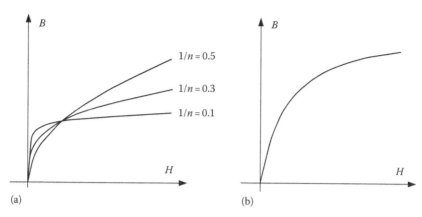

FIGURE 2.149
The example of the nth root approximation of the magnetization curve (a) and the anhysteric curve described by Fröhlih's equation (b).

series approximation and through to the Fourier harmonics representation (Trutt et al. 1968, El-Sherbiny 1973, Josephs et al. 1986):

$$B = a_0 + a_1 H + a_2 H^2 + a_3 H^3 + \cdots \qquad (2.226)$$

$$B = \mu_0 M_s \left(1 - \frac{a}{H} - \frac{b}{H^2} - \frac{c}{H^2} - \cdots \right) + \mu_0 H \qquad (2.227)$$

$$B = K H^{1/n} \qquad (2.228)$$

$$B = \frac{H}{\alpha + \beta |H|} \qquad (2.229)$$

$$B = B_0 \exp\left(\frac{H}{a + bH} \right) \qquad (2.230)$$

$$B = \frac{2}{\pi} B_s \tan^{-1}\left(\frac{H}{H_0} \right) \qquad (2.231)$$

Figure 2.149a presents the example of approximation of the magnetization curve by nth root approximation proposed by Nasar et al. (1994)—Equations 2.228. They found that the best fitting was for $n = 5$–14 and $K = 0.5$–1. The dependence (2.229) is known as *Fröhlih's relation* (Figure 2.149b) (Fröhlich 1881) and the relation (2.227) is derived from the *Fröhlih–Kennely equation* (Kennelly 1891) in a form $1/(\mu - \mu_0) = a + bH$.

2.12 Sources of a Magnetic Field

2.12.1 Helmholtz Coil

The *Helmholtz coils* is the most commonly used source of a uniform magnetic field (sometimes also as a standard

of magnetic field). A pair of Helmholtz coils (named in honor of the German physicist Hermann von Helmholtz) consists of two identical circular coils distanced by L equal to the radius r (Figure 2.150).

The magnetic field generated by the circular coils according to Equation 2.6 can be determined as

$$H_1 = \frac{nIr^2}{2}\left[r^2 + \left(\frac{L}{2} + x \right)^2 \right]^{-3/2} \quad \text{and}$$

$$H_2 = \frac{nIr^2}{2}\left[r^2 + \left(\frac{L}{2} - x \right)^2 \right]^{-3/2} \qquad (2.232)$$

The magnetic field between these two coils is therefore

$$H = H_1 + H_2 = \frac{nIr^2}{2}\left\{ \left[r^2 + \left(\frac{L}{2} + x \right)^2 \right]^{-3/2} \right.$$

$$\left. + \left[r^2 + \left(\frac{L}{2} - x \right)^2 \right]^{-3/2} \right\} \qquad (2.233)$$

This relation is presented in Figure 2.150 and we can see that in the central part between these two coils, the magnetic field is uniform. It can be proved that if $r = L$ this uniformity is the best and for $x = 0$, $r = L$ the magnetic field at the geometrical center is

$$H_0 = nIr^2(r^2 + 0.25r^2)^{-3/2} = 0.7155\frac{nI}{r} \qquad (2.234)$$

At a horizontal distance $1/4\,L$ from the center, the magnetic field value drops by less than 0.42% as compared with the center. Perpendicular to the coils axis at the same distance of $1/4\,L$, this difference is not larger than 0.75%.

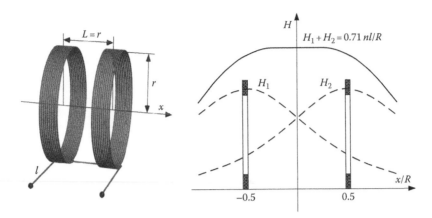

FIGURE 2.150
The Helmholtz coils and the magnetic field distribution inside the Helmholtz coils system.

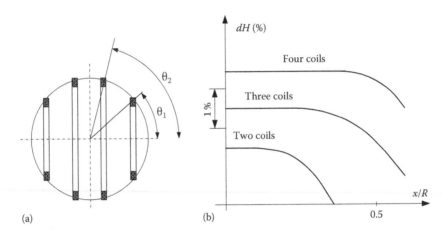

FIGURE 2.151
The double-pair Helmholtz coil system (a) and corresponding characteristics of the multicoil systems (b) (three coils—see also Figure 2.155c).

The field uniformity of the Helmholtz coil system can be significantly improved by using two pairs of coils connected in series (Garret 1951, Franzen 1962, Stamberger 1972, Kaminishi and Nawata 1981, Fiorillo 2004). It was estimated that the optimal configuration is $\theta_1 = 40.1°$, $\theta_2 = 73.4°$ (see Figure 2.151*), which corresponds with $L_1/L_2 = 2.685$, $r_1/r_2 = 0.672$, and $n_1/n_2 = 0.682$ (Fioriollo 2004).

Equation 2.232 was derived by neglecting the dimensions of the multilayer coil (Figure 2.152a). This problem was considered in many technical publications (Franzen 1962, Kaminishi and Nawata 1981, Fiorillo 2004). It was found that if the mean radius and mean distance fulfill the condition $\hat{r} = \hat{L}$, we obtain similar uniformity as in the one-wire coil but it is recommended that the ratio of the height h to the width w is $h/w = 1.078$.

The circular coils are not very convenient in practical applications due to the limited volume inside the coils. Therefore, a possibility of the square coil system was considered (Firester 1966). An example of a

three-axis square coil system, commonly used for magnetic field cancellation, is presented in Figure 2.152b. It was determined that the square coil system generates the magnetic field with uniformity not as excellent as the circular coils. For the square $a \times a$ coils distanced by

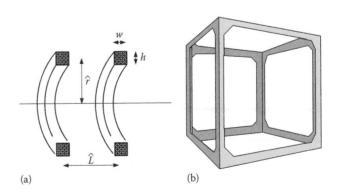

FIGURE 2.152
The multilayer Helmholtz coils (a) and rectangular three-axis Helmholtz coils system (b).

---

* For single-pair system, this optimal angle is $\theta = 63.4°$.

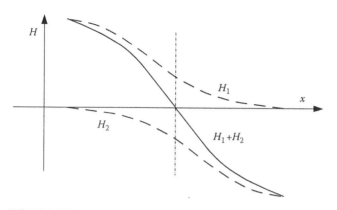

FIGURE 2.153
The inverse Helmholtz coils system as the source of a uniform magnetic field gradient.

L, the best coil arrangement is $L = 0.5445a$ and the magnetic field at the center of the cube is

$$H_0 = 0.648 \frac{nI}{a} \qquad (2.235)$$

For the two-coil system, if we connect the coils in series opposition, we obtain the source of magnetic field with a uniform gradient (Figure 2.153) (Murgatroyd and Bernard 1983). For such system, the optimum distance between coils is $L = 1.73r$ and the magnetic field gradient at the center is

$$\frac{dH_0}{dx} = 0.64 \frac{nI}{r^2} \qquad (2.236)$$

2.12.2 Long Solenoid

The solenoid is a long multi-turn coil (see also Figure 2.73) (Garret 1951, Underhill 2009). From Figures 2.73b and 2.154b, it can be concluded that inside the solenoid there is quite a long area where the magnetic field is uniform. That is why such coil can be used as a source of magnetic field. Moreover, such type of the coil is commonly used to magnetize the ferromagnetic cores when

they are placed inside—for example, in test devices (Epstein frame, single strip tester, etc.) as well as in electromechanical devices (transformers, relays).

The magnetic field inside the coil can be described by Equation 2.103. This equation was derived for one-layer coil. We can easily extend this relationship for the multilayer case with inner diameter $2r_1$ and outer diameter $2r_2$:

$$H = \frac{nI}{4L(r_2 - r_1)} \left[(L+x) \ln \frac{r_2 + \sqrt{r_2^2 + (L+x)^2}}{r_1 + \sqrt{r_1^2 + (L+x)^2}} \right.$$
$$\left. + (L-x) \ln \frac{r_2 + \sqrt{r_2^2 + (L-x)^2}}{r_1 + \sqrt{r_1^2 + (L-x)^2}} \right] \qquad (2.237)$$

At the center of the coil ($x=0$), the magnetic field is

$$H_0 = \frac{nI}{2\sqrt{r_0^2 + L^2}} \qquad (2.238)$$

For a very long coil, when $L \gg r$, Equation 2.238 can be presented in form

$$H_0 = \frac{nI}{2L} \qquad (2.239)$$

The relationship describing the magnetic field generated by solenoid can be conveniently presented in the following form:

$$H = kH_0 \qquad (2.240)$$

where k is the correction calculated as

$$k = \frac{1}{2} \left[\frac{L+x}{\sqrt{r_0^2 + (L+x)^2}} + \frac{L-x}{\sqrt{r_0^2 + (L-x)^2}} \right] \qquad (2.241)$$

Table 2.18 presents the values of the correction factor determined for typical dimensions of a solenoid. At the end of the coil, the magnetic field decreases to about 50% of the field at the center. This effect can be reduced to some extent (Figure 2.154b) by applying additional

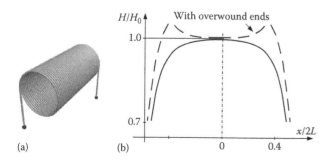

(a)   (b)

FIGURE 2.154
The solenoid and its characteristic.

TABLE 2.18

Correction Factor k for Typical Solenoids

$x/2L$	$L/r = 5$	$L/r = 12.5$	$L/r = 50$
0	0.981	0.997	1.000
0.1	0.978	0.996	1.000
0.2	0.969	0.995	1.000
0.3	0.943	0.990	0.999
0.4	0.850	0.964	0.997
0.5	0.498	0.500	0.500

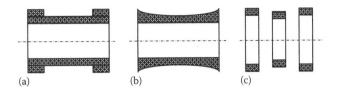

FIGURE 2.155
The various shapes of the modified solenoid.

coils at the ends (Figure 2.155a) (Gardener et al. 1960). It is also possible to change the distribution of magnetic field inside the solenoid by appropriate modification of the shape of the solenoid (Figure 2.155b) (Hak 1936, Barker 1950).

2.12.3 Sources of High Magnetic Field

In many experiments, it is necessary to apply very high magnetic fields, with the flux density up to 100 T (Montgomery 1963, Herlach et al. 1996). The modern, rare-earth-based hard magnets exhibit the coercivity as large as 5×10^6 A/m and similar magnetic fields are required to test them. Two main additional technical problems appear when we would like to generate such values of magnetic field. Special cooling system is required because the dissipated energy causes great growth of temperature of the magnetizing coils. Fortunately, in most experiments, only a pulsed magnetic field is necessary. But there is a second important problem—large mechanical forces accompanying the even short-timed but extremely large currents. These effects demand special design of the sources of such fields.

Bitter (1939) proposed a special design of the magnetizing coil composed of a number of discs (known as a *Bitter coil*, Figure 2.156). With such coil, it is possible to obtain, in the central bore of diameter 30 mm, a DC magnetic field capable of magnetizing paramagnetic air up to a flux density of 45 T. To obtain such value of a magnetic field, the current up to 67,000 A was necessary (which requires a power of 20 MW). The Bitter coil

(also called the Bitter magnet or electromagnet) consists of cooper disks with holes (for usually liquid cooling medium) and slits. These disks are overlapping each other (every disk is rotated with respect to its neighbor by 20°) what causes the radial current flow, which results with a smaller power dissipation. The Bitter coil concept is still used today as a high magnetic field source.

It is much easier to obtain the large magnetic field if the test area is small because the magnetostatic energy stored in a coil is proportional to its volume. The magnetic field generated by the disk coil can be expressed as (Montgomery 1963)

$$B = \frac{\mu_0 I}{r_1} \frac{F(\alpha,\beta)}{2(\alpha-1)\beta} \quad \text{with} \quad F(\alpha,\beta) = \beta \ln \frac{\alpha^2 + \sqrt{\alpha^2+\beta^2}}{1+\sqrt{1+\beta^2}}$$

(2.242)

where
  $F(\alpha,\beta)$ is F coefficient
  $\alpha = r_0/r_i$, $\beta = t/2r_i$, r_i and r_0 are the inside and outside radii
  t is the disk thickness

Mackay et al. (2000) proposed to use, as the magnetic field source, a flat copper coil with dimensions $r_i = 50$, $r_0 = 150$, and $t = 10\,\mu m$. After substitution to Equation 2.242, we obtain that in order to generate the flux density 50 T, a current of "only" 4000 A is required. Figure 2.157 presents the design of this microcoil.

The temperature of the coil can be determined from (Mackay et al. 2000)

$$\int_0^{\Delta t} B^2 dt = \mu_0^2 r_i F^2(\alpha,\beta) D \int_{T0}^{T1} \frac{C_p}{\rho} dT$$

(2.243)

where
  D is density of the coil
  ρ is resistivity
  C_p is specific heat capacity

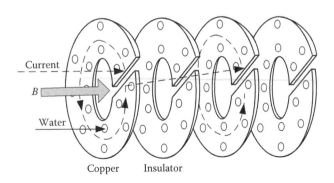

FIGURE 2.156
The Bitter coil.

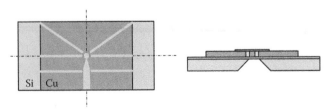

FIGURE 2.157
The micro-coil as a source of 50 T magnetic field. (After Mackay et al., 2000.)

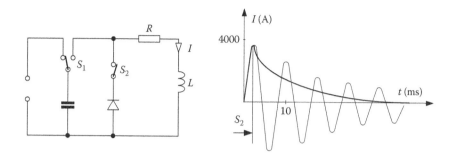

FIGURE 2.158
The capacity bank as a supply of pulse magnet device. (After *Measurement and Characterization of Magnetic Materials*, Fiorillo, F., Copyright (2004), from Elsevier.)

From Equation 2.243, it was estimated that for $B=50\,T$ after $\Delta t=30\,ns$, the temperature reaches the melting point of Cu. Thus, the pulses of magnetic field should be much shorter than 30 ns.

Theoretically, coils made from the superconducting materials could be a good candidate for such coils. But as it was described in Section 2.9, every superconducting material has limited current density and, even more important, limited critical magnetic field strength. Nevertheless, there have been developed some superconducting material based on Nb-Ti or Nb-Zr alloys capable of operation up to 8 T and current density 100,000 A/cm² (Kunzler et al. 1961). A large size superconducting NbTi coil for generation the magnetic field up to 6 T in 220 mm bore has been constructed (Watazawa et al. 1996). Hence, in the aspect of the maximum attainable magnetic field strength, the Bitter coil still looks favorable as compared to state-of-art superconducting technology.

As a supplying source for a large pulse magnetic field coils, the capacity banks (capacitor battery) (Figure 2.158) are commonly used. A bank of capacitors (with capacity exceeding sometimes 1000 µF) is charged by high voltage. When charged, the capacitors are connected directly to the coil of inductance L. In the resultant RLC circuit, oscillations are generated and they are stopped after the voltage attains the maximum value by a switching thyristor S2 (or ignitron) as shown in Figure 2.158. In this way, it is possible to supply the coil by a pulse exceeding 10,000 A and time up to several milliseconds.

The magnetic field can be generated by electromagnets or magnets or permanent magnets (Figure 2.159). The electromagnets are more versatile but on the other hand, the magnets do not require energy supply. Both types of devices have natural limitation of the magnetic circuit, usually not exceeding 1.3–1.4 T in the case of permanent magnet materials and about 2 T in the case of electromagnets. Nevertheless, by appropriate design of these devices, it was possible to obtain the flux density over 4 T in the case of magnets (Bloch et al. 1998). To increase the possible flux density range in electromagnets, often the ferromagnetic materials of the highest possible saturation polarization—Fe49:Co49:V2 with $J_s=2.35\,T$ are used for the poles. By correct design, it is possible to obtain up to 3 T in the air gap of electromagnet (Fiorillo 2004).

By appropriate design, it is possible to obtain a system of the permanent magnet with requested parameters (Abele 1993, Coey 1996, Skomski et al. 2000, Leupold and Potenziani 2007). Figure 2.160a presents a permanent magnet design numerically optimized to obtain a flux density value as high as possible in the geometrical center (Bloch et al. 1998). This magnet is composed of 192 individual magnets cut from NdFeB blocks as well as FeCo parts for the poles. All elementary magnets were magnetized in a hybrid coil in fields of up to 10 T. In the small gap at the center, a flux density of 4.3 T was obtained.

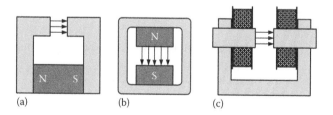

FIGURE 2.159
The permanent magnet or electromagnet as a source of large magnetic field.

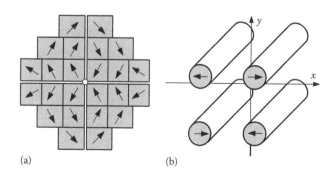

FIGURE 2.160
Two examples of complex permanent magnets systems.

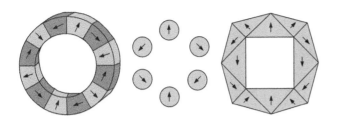

FIGURE 2.161
The Halbach cylinder and its modifications.

Another arrangement of magnets was proposed by Cugat et al. (1994) (Figure 2.160b). A set of four rod magnets when rotated can create the variable magnetic field depending on the angular position of the magnets (magic mangle), enabling to change the value and direction of the uniform magnetic field from −0.2 to +0.2 T.

Figure 2.161 presents the examples of a magnet array generating uniform magnetic field over a large area (used in magnetic resonance devices in medicine) and at the same time practically cancellation of external stray magnetic field. This configuration realizes the so-called *Halbach cylinder* principle (Halbach 1980, Shute et al. 2000).

The magnetic field generated by permanent magnet source in the gap l_g (Figure 2.159a) can be estimated as

$$H = \frac{J_m}{\mu_0}\left(1 - \frac{l_g/D}{\sqrt{1+(l_g/D)^2}}\right) \quad (2.244)$$

where
J_m is the polarization in the working point of the magnet (see Figure 2.24)
D is the diameter of the magnet pole
l_g is the air gap length

From (2.244), it is clear that to obtain uniform and large magnetic field in the gap the magnet pole diameter D should be at least a few times larger than the gap length (for $l_g/D \to 0$ magnetic field $H_{max} \to J_m/\mu_0$). The effective method to increase the value of magnetic field in the gap (at the cost of its area) is tapering of the poles. Dreyfus (1935) estimated that the optimal taper angle is about 60°—more exactly 54.74° (Fiorillo 2004).

In the case of tapered poles, it is possible to obtain the magnetic field slightly higher than the maximum value for the flat poles. The magnetic field in the gap can be estimated as (Fiorillo 2004)

$$H = \frac{2J_m}{3\sqrt{3}\mu_0}\ln\frac{r_0}{r_g} \quad (2.245)$$

Similar effect can be obtained in the case of electromagnet. As it is presented in Figure 2.162, tapering of the poles does not influence the magnetic field in the gap when the electromagnet is magnetized with moderate current. But if the yoke is in saturation, the tapering enables to obtain the magnetic field even higher than the saturation induction of the core. The magnetic field in the gap can be determined from the following simplified equation (Fiorillo 2004):

$$\mu_0 H \cong \mu_0 \frac{nI}{l_g} \quad (2.246)$$

Although the magnetic field in the gap of electromagnet depends mainly on the ampere-turns, the yoke plays important role as the flux concentrator. Figure 2.163 presents two examples of the electromagnets with yokes focusing the magnetic field in the test area. Such devices are used as magnetic lenses in synchrotron accelerators.

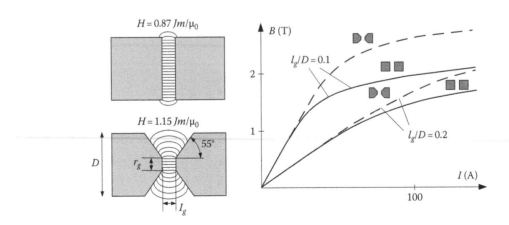

FIGURE 2.162
Flat and tapered poles of magnet or electromagnet (left: calculated for $l_g/D = 0.1$, right: characteristics determined for the yokes with $D = 250$ mm, $r_g = 50$ mm). (After *Measurement and Characterization of Magnetic Materials*, Fiorillo F., Copyright (2004), from Elsevier.).

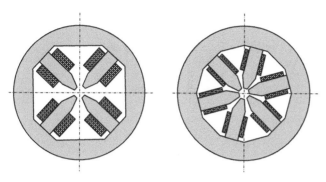

FIGURE 2.163
The quadrupole and sixtupole electromagnets.

2.12.4 Sources of a Magnetic Field with Arbitrary Waveform

Properties of magnetic materials and, most of all, losses depend strongly on the frequency of the magnetizing magnetic field and therefore on the presence of harmonics in the flux density waveform. Bertotti determined losses for the triangular and sinusoidal flux density as (Bertotti 1998)

$$P_{TRI} = \frac{4}{3}\sigma t^2 B_m^2 f^2 \quad \text{and} \quad P_{SIN} = \frac{\pi^2}{6}\sigma t^2 B_m^2 f^2 \qquad (2.247)$$

Thus, the difference between these losses is about 20% and direct comparison for values measured under such two different conditions cannot be made. For this reason, the International Standards require appropriate control of the magnetizing waveform. It was established to test all materials under sinusoidal waveform of the flux density. For example, the European standard EN 60404 recommends to control the form factor of the secondary (induced) voltage as $V_{rms}/V_{avg} = 1.111\% \pm 1\%$, which means pure sinusoidal waveform of the flux density.

On the other hand, it is reasonable to test the magnetic materials exactly in the same conditions as the future forecasted working conditions (Moses 1987, Fiorillo and Novikov 1990, Moses and Tutkun 1997). Therefore, in a special case, it should also be possible to control arbitrary waveform of magnetic field strength or magnetic flux density.

For relatively low values (up to 1 T), we can ensure the sinusoidal waveform of the flux density by supplying the winding from the source with small internal resistance. But because both magnetic field strength and magnetic induction normally are distorted (see Figure 2.135), it is usually necessary to use some form of negative feedback control. Sometimes it is sufficient to use a simple analog feedback circuit presented in Figure 2.164 (Mazzetti and Soardo 1966, Baldwin 1970, Blundell et al. 1980, Sankaran and Jagadeesh Kumar 1983).

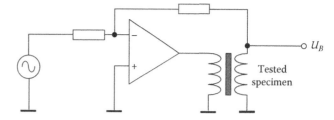

FIGURE 2.164
The sinusoidal waveform of flux density obtained by the analog feedback.

For more complex conditions (magnetizing close to saturation, material with square hysteresis loop, etc.), the digital control of waveform is more versatile and effective. By applying the digital method of control, we solve typical inverse problem: how to change the input (usually the magnetizing current) to obtain required waveform of the flux density. Because after the change of the input signal, response of the specimen changes (it depends on the waveform of excitation); therefore, a typical algorithm requires several iterative steps (successive approximation method). The realization of a given algorithm is stopped when appropriate waveform factor V_{rms}/V_{avg} (or THD factor) is obtained with sufficient accuracy. Recently computers are sufficiently fast to obtain requested conditions practically in real time, but for very low frequency (1 Hz and below) the iterative process can take even several minutes. The control process of iteration can be shortened by using predictive algorithms based on characteristics of the material (Matsubara et al. 1995).

One of the conventional methods of the waveform control is to use arbitrary waveform generator (as additional instrument, computer card, or software) (Birkenbach et al. 1986, Bertotti et al. 1993, Spornic et al. 1998). Using one cycle of measured signals, we can determine $h(t)$ and $b(t)$ and hence the hysteresis loop as the $b(t) = f(h(t))$ (Figure 2.165). From the hysteresis, we can compute appropriate waveform of the magnetizing current (similarly as in Figure 2.135).

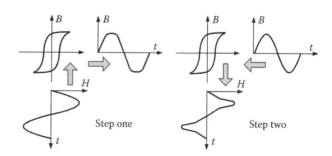

FIGURE 2.165
The sinusoidal waveform of flux density obtained by the digital analysis of the hysteresis curve.

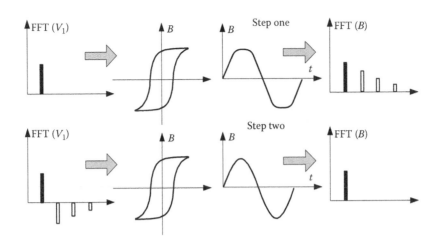

FIGURE 2.166
The sinusoidal waveform of the flux density obtained with the help of the FFT analysis of the signal.

It is also possible to use direct Fourier analysis to the signal of waveform control. Chatziilias et al. (2003) subtracted from the input signal the higher harmonics (taking into account also their phase shifts) responsible for the distortions. The inverse Fourier transform supplied the magnetic circuit with corrected input signal (Figure 2.166).

The fast digital methods apply the controller or adaptive filter between the generator and the magnetized specimen (De Wulf et al. 1998, Grote et al. 1998, Zurek et al. 2004, 2005, Zurek and Moses 2005). As a signal of error, the difference between required and measured signals can be used (Figures 2.167 and 2.168). These algorithms are versatile—they can cooperate with various test systems because it is not necessary to know properties of the magnetized specimen or the magnetizing yoke—the whole control circuit is treated as a "black box." After analog-to-digital conversion of the waveforms, the supplying voltage is corrected appropriately according to the calculated error signal. The main problem in such systems is to establish full synchronization between reference and measured signal to perform correction "sample by sample."

Figure 2.169 presents the principle of the waveform control by using the perceptron neural network (NN) (Baranowski 2008, Baranowski et al. 2009). The initial sinusoidal supplying signal is obtained from the virtual generator x in form of the n-number of samples. Initially, the weights of neuron are random. The y signal from NN is converted to analog signal by digital-to-analog converter and through a power amplifier (PA) supplies the magnetized specimen.

The voltage from the secondary winding proportional to dB/dt is integrated and converted again to a digital form b. The error signal (the difference between b signal and reference b_{ref} signal) is used for changing appropriately the weights of the neurons. Figure 2.170 presents the comparison of the initial signals y and b and the same signals after finishing the iteration. Number of iteration and the transition times depend on the specimen and the flux density value. For example, for 1.7 T and grain-oriented steel, it was necessary to repeat about 20 iteration cycles, which at 50 Hz required about 5 s.

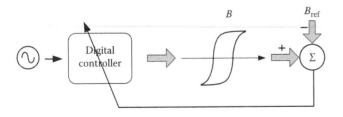

FIGURE 2.167
The sinusoidal waveform of flux density obtained with a digital controller.

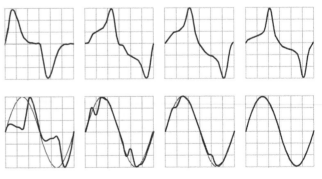

FIGURE 2.168
The last four steps of iterative process in the proportional controller (top row: magnetizing current, bottom: flux density and reference signals). (From Baranowski, 2008.)

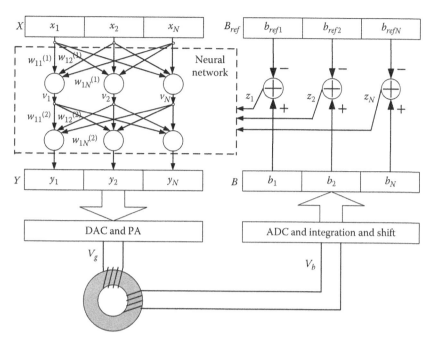

FIGURE 2.169
The sinusoidal waveform of induction obtained by applying of the perceptron neural network.

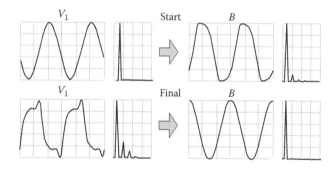

FIGURE 2.170
The sinusoidal waveform of flux density controlled by the perceptron neural network.

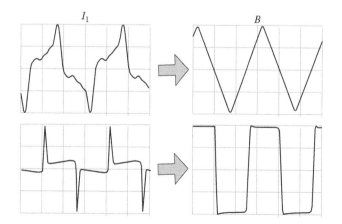

FIGURE 2.171
The triangular and trapezoidal waveform of the flux density (left side corresponding waveforms of the magnetizing currents).

The main advantage of the digital controllers (including artificial neural network applications) is that it is easy to obtain practically arbitrary waveform of the flux density or magnetic field strength. Figure 2.171 presents examples of triangular and trapezoidal waveform signals.

2.13 Samples and Circuits of the Material under Test

2.13.1 Measurements of *B* and *H* Values

Due to the demagnetizing field results of practically all tests of magnetic materials depend on the shape of the sample under test. Therefore, usually the properties are determined only of the specimen under test and not of the material. Moreover, during preparation of the the specimen, we also change somewhat its properties (e.g., by the process of cutting), and even de-stressing by annealing does not return the sample to exactly the same state as before the processing. The best would be the situation in which we test exactly the same specimen that is used in real application—such situation sometimes occurs for ring (toroidal) sample.

However, there is a need to compare properties of various materials with assumed reproducibility—for example, if we test the same material in different laboratories.

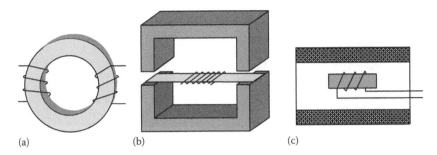

FIGURE 2.172
The closed magnetic sample (a), the sample closed by the yoke (b), and the open sample (c).

Therefore, international standards determine precisely the condition of the tests.*

As an example, we can point the IEC Standard 60404-2, which in detail describes the Epstein frame method. Although it is still under discussion whether the Epstein frame does determine the magnetic properties properly (in the physical meaning) (Sievert and Ahlers 1990), the reproducibility in inter-laboratory tests is on the level of 1% (which in magnetic measurements is quite satisfying). Manufacturers sometimes simply state in the data sheets: "Epstein data."

---

* Recently, IEC recommended the following standards for testing of magnetic materials:
  IEC 60404-1 "Magnetic materials. Part 1: Classification"
  IEC 60404-2 "Magnetic materials. Part 2: Methods of measurements of the magnetic properties of electrical steel sheet by means of an Epstein frame"
  IEC 60404-3. "Magnetic materials. Part 3: Methods of measurement of the magnetic properties of magnetic sheet and strip by means of a single sheet tester"
  IEC 60404-4 "Magnetic materials. Part 4: Methods of measurement of the dc magnetic properties of magnetically soft materials"
  IEC 60404-5 "Magnetic materials. Part 5: Permanent magnet (magnetically hard) materials. Methods of measurement of magnetic properties"
  IEC 60404-6 "Magnetic materials. Part 6: Methods of measurement of the magnetic properties of magnetically soft metallic and powder materials at frequencies in the range 20 Hz to 200 kHz by the use of ring specimen"
  IEC 60404-7 "Magnetic materials. Part 7: Method of measurement of the coercivity of magnetic materials in an open magnetic circuit"
  IEC 60404-8 "Magnetic materials. Part 8-1: Specifications for individual materials - Magnetically hard materials"
  IEC 60404-9 "Magnetic materials. Part 9: Methods of determination of the geometrical characteristics of magnetic steel sheet and strip"
  IEC 60404-10 "Magnetic materials. Part 10: Methods of measurement of magnetic properties of magnetic sheet and strip at medium frequencies"
  IEC 60404-11. "Magnetic materials. Part 11: Method of test for the determination of surface insulation resistance of magnetic sheet and strip"
  IEC 60404-12. "Magnetic materials. Part 12: Guide to methods of assessment of temperature capability of interlaminar insulation coatings"
  IEC 60404-13. "Magnetic materials. Part 13: Methods of measurement of density, resistivity, and stacking factor of electrical steel sheet and strip"
  IEC 60404-14. "Magnetic materials. Part 14: Methods of measurement of the magnetic dipole moment of a ferromagnetic material specimen by the withdrawal or rotation method"

It is optimal if our sample is in a form of closed magnetic circuit, for example, a ring core (Figure 2.172a). First of all it is the easiest to magnetize the closed sample uniformly to high value of the flux density. Therefore, even "open" samples such as sheets or strips are usually closed by an external yoke—the best symmetrical (Figure 2.172b). A completely open sample (Figure 2.172c) is used when it is not possible to close the magnetic circuit—for example, when we test a small mineral specimen, thin-film element, or a small magnet.

In the closed sample, it is possible to determine approximately the length of mean magnetic path l and then calculate the magnetic field strength H from Ampere's law knowing the magnetizing current I and number of turns n_1:

$$H = \frac{n_1 I}{l} \qquad (2.248)$$

The flux density B can be determined from Faraday's law knowing the cross-sectional area A of the sample, induced voltage $v(t)$, and number of turns n_2:

$$B(t) = -\frac{1}{An_2}\int v(t)\,dt \qquad (2.249)$$

Both main values B and H can be determined with error. Only in the case of the ring sample of recommended dimensions (described in the next section), we can assume that the $l = \pi d_{av}$. In other cases, we know the length of mean magnetic path l only approximately. Therefore, in some measurements, it is more reliable to measure the magnetic field strength directly as the tangential magnetic field component on the surface of the sample (according to Maxwell's law—see Equation 2.146) (Pfützner 1991, Nakata et al. 2000). For this goal, the Hall sensor and magnetoresistive sensors can be used (Pfützner 1980, Tumanski 1988). But, most commonly, for the measurement of tangential field component, a coil sensor is used—in a form of flat H-coil sensor (Tumanski 2002b) or as a Rogowski–Chattock coil (Rogowski–Chattock potentiometer [RCP]) (Iranmanesh et al. 1992, Shirkoohi and Kontopoulos 1994) (Figure 2.173).

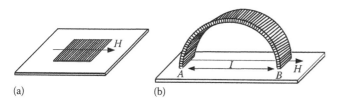

(a)   (b)

FIGURE 2.173
A flat tangential H-coil sensor (a) and Rogowski–Chattock coil (b) as a tool for measurement of magnetic field strength in the sample.

If the H-coil sensor is not exactly placed on the surface of the sample (e.g., due to the thickness of the sensor itself), it is possible to use more than one coil and then to extrapolate the result to the sample surface (Tumanski 2002). Nakata proposed to use a system of two H-coils for improvement of measuring accuracy of magnetic field strength (Nakata et al. 1987).

When we measure the flux density in the sample, we are usually able to determine the cross-sectional area quite accurately. Because the thickness of the sample can be varying in the length l, we can determine the average cross section from the mass m (by weighting the sample as it recommended in the Standard IEC 60404-2 and knowing the density ρ of the material):

$$A = \frac{m}{l\rho} \tag{2.250}$$

Other errors of measurement of the flux density value in the sample occur if the turns of the sensing coils are not wound close to the surface of the sample. In such case, the whole flux is divided into two parts—in the sample with area A_{Fe} and in the coil of the area A_C:

$$\Phi = nA_{Fe}B + n(A_c - A_{Fe})\mu_0H \tag{2.251}$$

If we assume that the flux density in the sample is Φ/nA_{Fe}, the error is

$$\Delta B = \left(\frac{A_C}{A_{Fe}} - 1\right)\mu_0H \tag{2.252}$$

Therefore, if possible, the B-coil should be wound as first directly on the sample as close to the surface as possible. It is better if the B-coil is wound on the whole length of the sample but sometimes we put the B-coil just in the center of the sample. This is not in the case for the primary magnetizing winding (wound usually as the second outer coil)—it should be wound on the whole sample length in order to minimize the additional errors caused by the stray or nonuniform fields.

If we would like to determine the localized values of B and H, we cannot use coils wound on the whole sample. For such measurements, the H-coil can be used. It is more difficult to determine the localized value of the flux density. Usually, for such measurements, it was necessary to drill small holes for the sensing coil (Figure 2.174a). Recently, destructive drilling of the small holes is sometimes substituted by the needle-probe method (Figure 2.174b) (Yamaguchi et al. 1998, Senda et al. 1999, Loisos and Moses 2001, Pfützner and Krismanic 2004).

In the case of 2D measurements, usually it is not possible to wind the coil around the sample (Figure 2.175a). Hence, the tangential H-coil and needle-probe method (or drilled holes method) are used (Figure 2.175b).

2.13.2 Ring Core

The *ring core* sample (a *toroidal sample*) is prepared by winding a ribbon or tape (Figure 2.176a), punching (2.176b), or bonding from a magnetic powder (Figure 2.176c). The first method is most frequently used because it enables to profit the best performances of anisotropic material in selected direction (usually the rolling direction). This kind of sample seems to be close to an ideal case—it is a closed magnetic circuit, with well-defined

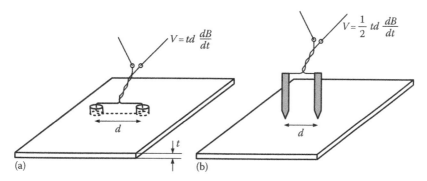

FIGURE 2.174
The measurements of the localized value of the flux density.

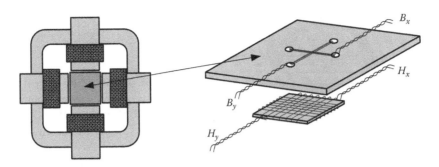

FIGURE 2.175
Two-dimensional measurements of the B and H values: the testing device (a) and the sensors (b).

FIGURE 2.176
Three examples of the toroidal cores, obtained by winding of the ribbon (a), punching the ring shaped laminations (b), or bonding from the magnetic powder (c).

dimensions—including the length of the magnetic path intuitively equal to $2\pi r_0$ (r_0 is the mean radius and $r_0 = (r_e + r_i)2$ where r_e is the external radius and r_i is internal radius of the toroid).

The main disadvantage of this type of sample is its tedious preparation—the magnetizing coils should be wound uniformly around the whole ring. Although recently there are commercial winding machines capable of winding such coils, nevertheless other samples like the strip sample and even the Epstein sample are easier and faster to prepare. Moreover, after winding of the ribbon, it is necessary to perform de-stressing annealing. For this reason, the ring core sample is mainly used in the case when also the final product is a toroid (e.g., in inductors or transformers).

Furthermore, the ring shape is in reality far from the "ideal" case. Even in the case of high permeability material, the magnetic flux is not completely concentrated in the sample—there is the stray magnetic field around the core. The magnetic field in the sample itself is not uniform—it is the largest near the inner perimeter and decreases not linearly with the distance to the external edge. That is why we can determine the length of the magnetic path as

$$l = 2\pi \left(\frac{r_e + r_i}{2} \right) \quad (2.253)$$

only if it has fulfilled the condition $r_e/r_i < 1.1$ (according to the IEC Standard 60404-4*). If this ratio is larger, we should determine the value of magnetic field as

$$\bar{H} = \frac{1}{r_e - r_i} \int_{r_i}^{r_e} H dr = \frac{nI}{2\pi(r_e - r_i)} \int_{r_i}^{r_e} \frac{dr}{r} = \frac{nI}{2\pi(r_e - r_i)} \ln \frac{r_e}{r_i} \quad (2.254)$$

Thus, if this ratio is between $1.1 < D_e/D_i < 1.6$, the length of magnetic path can be determined as

$$l = \pi \frac{r_e - r_i}{\ln(r_e/r_i)} \quad (2.255)$$

The ratio r_e/r_i higher than 1.6 is not recommended due to significant magnetic field nonuniformity in the sample. Nakata et al. (1992a,b) analyzed this error for various values of r_e/r_i ratio. If this ratio exceeds 1.5, the error of determination of B or H can be significant, especially at low magnetization. Nakata proposed a special numerical interpolation method that enables to reduce these errors when the sample has unconventional ratio r_e/r_i (for $r_e/r_i < 3$).

The magnetic field strength H_x at the distance x from the mean radius r_0 (where we assume the H_0 field) can be

---

* The Standard 60404-6 limits this ratio to 1.25.

determined from the following expression (Grimmond et al. 1989):

$$H_x = H_0 \frac{r_0}{r_0 + x} \qquad (2.256)$$

from which the magnetic field at the outer and inner surfaces are

$$H_i = \frac{r_i + r_e}{2r_i} H_0 \quad \text{and} \quad H_e = \frac{r_i + r_e}{2r_e} H_0 \qquad (2.257)$$

which means that $H_i / H_e = r_e / r_i$. As results from experiments, the distribution of H and B in the radial distance is more nonlinear.

The standards recommend the above-mentioned r_e / r_i value although other geometrical factors as the internal diameter $2r_i$ and height of the sample h also affect measured parameters. This problem was analyzed by Grimmond, Ling, and Moses (Moses 1987, Grimmond et al. 1989, Ling et al. 1990). They measured losses, permeability, and magnetizing current for the samples of various dimensions and estimated that the results can be different that obtained by using other methods. Figure 2.177 presents the results of experimental determination of the distribution of flux density (by using a special search coils) for various internal diameters and various heights of the sample.

From results presented in Figure 2.177, it is clear that the ring sample should have large inner diameter and large strip width (ring height). The geometrical factors influence following additional effects:

- The wound core is a spiral created from the thin strip. It means that the magnetic flux is forced to jump small interturn air gaps. This causes the additional normal flux component that produces additional eddy current losses.

- Small width of the strip is more influenced by the edge effects caused by the punching.

- Stress caused by the wounding strongly deteriorates material performance. Fortunately, correctly performed annealing greatly reduces this effect. For example, loss measured before annealing can be two times higher than after annealing of the ring sample (Moses 1987).

The IEC Standard 60404-4 also mentions another possible effect existing in the toroidal samples. The uniformly wound magnetizing coil is equivalent to a hypothetical single turn located at the mean circumference. This turn can be coupled with hypothetical single turn from uniformly wound secondary winding. This effect can be noticeable when the permeability of the sample is low—for example, near the saturation. Therefore, in exact measurements, an additional compensating arrangement should be included. The standard recommends reducing this effect by winding the magnetizing coil with even number of layers—alternately wound clockwise and anticlockwise. Also, at higher frequencies (kHz range and above), capacitive coupling between the windings plays increasing role and they have to be taken into account.

As we see from the above discussion, the real ring core sample is far from the "ideal" case and to obtain correct results of measurements, certain geometrical conditions should be fulfilled.

2.13.3 Epstein Frame

The Epstein frame is commonly used in electrical steel investigations due to the following advantages:

- Both sample and windings are described in detail by the IEC Standard 60404-2, which guarantee excellent repeatability and compliance in inter-laboratory comparison.

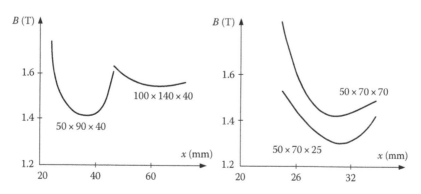

FIGURE 2.177
Experimentally determined flux density distribution in the toroid samples of various internal diameters and heights (samples with dimensions: $D_i \times D_e \times h$). (After Ling, P.C.Y. et al., *Ann. Fis.*, 86, 99, 1990.)

- Experienced laboratory worker can prepare the testing apparatus (assemble the multi-strip sample) within a time shorter than a minute.

- The sample is composed of the strips dimensioned 30 mm (±0.2 mm) × 280–320 mm (±0.5 mm), which is easy to prepare (although also, in this case, de-stress annealing is recommended).

The Epstein frame (Figures 2.178 and 2.179) was originally proposed as 50 cm square (Epstein 1900). Recently, smaller 25 cm version proposed by Burgwin (1941) is standardized and used.

The Epstein method is still controversial because the measured loss is determined only approximately (from purely physical point of view) (Sievert 1987, Sievert and Ahlers 1990). The main uncertainties are in the corner parts of the frame—it is not known what exactly happens in these parts of the magnetic circuit. Problematic is determination of the length of magnetic path especially when anisotropic material is tested (the flux can be directed along the critical 55°—see also Figure 2.16). The standard simply ignores all these issues and requires to assume that $l = 0.94$ m.

Another main drawback of the Epstein test is that it is a destructive, off-line method and therefore very difficult to automate. The relatively large weight and dimensions of the sample recognized often as an advantageous feature (due to averaging and taking into account the nonuniformity of the material) can be sometimes a drawback, for example, in the case of new amorphous or nanocrystalline materials.

In the Epstein frame, the windings are fixed and are composed of the two sets of four solenoids connected in series—every coil is 19 cm long. The first set of coils—measuring winding with 175 turns each (total 700 turns) is wound with the copper wire of the cross section 0.8 mm². Next, the magnetizing coils are wound with 1.8 mm² wire in three layers—also with total number of turns equal to 700.* Between the coils, a copper electrostatic shield can be inserted. Recommended internal cross-sectional area of the coils is a rectangle 10 × 32 mm. A mutual inductor (air-core transformer) can be inserted in the central part (see Figure 2.178) for compensation of the stray magnetic flux (resulting from the differences between the cross-sectional area of the coils and the sample). The primary winding of this inductor is connected in series with the magnetizing coil, the secondary with measuring coil in such a way that for device without ferromagnetic sample the secondary coil voltage is equal to zero (less than 0.1% of typical secondary voltage).

The sample under test is composed of multiple of four strips creating the square with mean length of side 25 cm. In the corners, these strips are double-overlapped as it is presented in Figure 2.179. It is allowed to use pressure up to 1N. The total mass of the specimen should be over 240 g. To compose the square, it is sufficient to use the strips 28 cm long. If this length l is larger, we determine effective mass of the sample m_e taking into account the total mass of the specimen m:

$$m_e = \frac{0.94}{4l} m \qquad (2.258)$$

From the total mass, we can also determine the sample cross-sectional area A knowing the density of material ρ (which is usually given by the manufacturer of the material):

$$A = \frac{m}{4l\rho} \qquad (2.259)$$

FIGURE 2.178
An example of the Epstein frame.

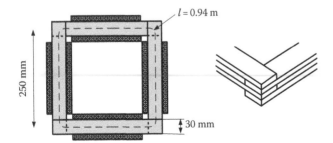

FIGURE 2.179
The Epstein frame: windings and sample arrangement and the corner strips arrangement.

The problem of reliability and accuracy of the Epstein method is still being debated. The main sources of uncertainty are the corner effects and, related to these effects, estimation of the magnetic path length. Dieterley

---

* The 60404-10 Standard allows 200 turns for medium frequency.

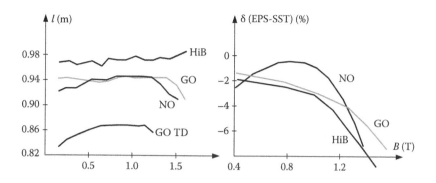

FIGURE 2.180
The results of tests of the Epstein frame: estimation of the mean length of the magnetic path and comparison with the strip tester by loss measurements (GO, grain-oriented steel, NO, non-oriented steel, HiB, HiB steel, GO TD, grain-oriented cut transversal to rolling direction). (After Marketos, P. et al., *IEEE Trans. Magn.*, 43, 2755, 2007; Ahlers, H. and Sievert, J., *J. Magn. Magn. Mater.*, 26, 176, 1982.)

tested different size of the Epstein frame and compared the results with the ring-core method for a large number of samples (Dieterly 1949). He obtained good correlation of the results for the fixed path length of 0.94 m. Similar investigations, under sinusoidal magnetization, repeated Marketos, Zurek and Moses for two Epstein frames of different sizes (which allowed estimation of the corner effect) (Marketos et al. 2007). The accurate effective mean path length was determined—the results are presented in Figure 2.180a. We can see that for conventional non-oriented and grain-oriented steel, the path length was indeed mostly equal to 0.94 m. But for high permeability steel and the grain-oriented steel cut perpendicular to the rolling direction, the differences were significant. Also, the magnetic path length was increasingly affected at measurement above 1.5 T.

Also, comparisons of the Epstein method results and the single strip tester (SST) results were carried out (Ahlers and Sievert 1982, Moses and Hamadeh 1983). Moses estimated that using the Epstein frame, the results were larger than in the case of SST and depending on the batch of the material, these differences can be as large as 20%. Figure 2.180b presents results obtained by Ahlers. Depending on the kind of material, the differences were reasonably small but they increase significantly with polarization (if we can really assume that the SST is a reliable standard).

Apart from the corner effects, the error of the Epstein method can also be influenced by the nonuniformity of magnetization. Figure 2.181 presents the results of measurements of the distribution of polarization performed with the help of a local search coil (Fiorillo 2004). We can see that although the magnetizing solenoid is uniformly wound on the length of 19 cm, only the central part of the strip is magnetized uniformly, which leads to an additional error. Nevertheless, even if the results of measurements are with errors, the main advantages of the Epstein method, repeatability and reproducibility

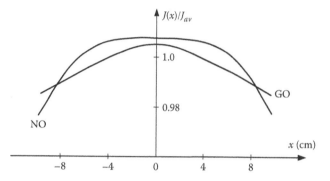

FIGURE 2.181
The distribution of the polarization in the length of Epstein strip. (After *Measurement and Characterization of Magnetic Materials*, Fiorillo, F., Copyright (2004), from Elsevier.)

are still valid due to high level of standardization and wide usage.

2.13.4 Single Sheet and Single Strip Testers

Figure 2.182 presents a comparison of the results of testing of two different samples of electrical steel performed using the SST and the Epstein frame (Tumanski and Baranowski 2007). We can see that the mean value calculated from more than 24 strips corresponds relatively well with the Epstein result. But reversely, the Epstein results lost the important information about the material nonuniformity.

Figure 2.182 presents the advantage of the single sheet/ strip tester in comparison with the Epstein method. The application of *SST* instead of classical ring core or Epstein frame offers several important advantages:

- Much less material needed for the test—much less wasted material.
- Possibility to test the nonuniformity of the material including localized measurements

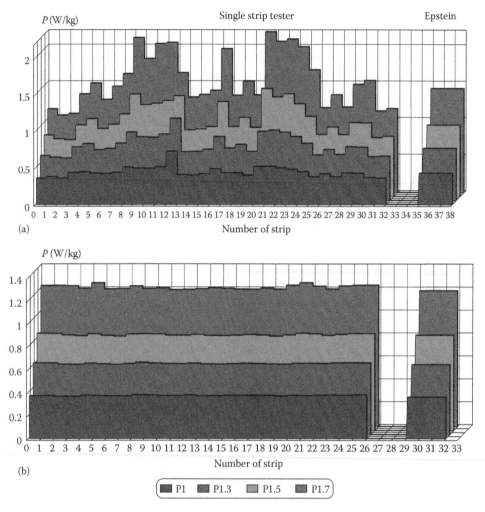

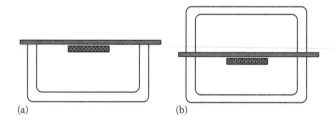

FIGURE 2.182
The comparison of results of the test of the same electrical steel samples obtained using single strip tester and the Epstein frame (a—sample of excellent uniformity, b—sample of non-homogeneity).

and, at the same time, possibility to test a large area by averaging the results obtained from many samples.

- Although the assembling of the Epstein frame is rather fast, it is still off-line test—by using as a sample, the sheet of material (SST), we can design the measuring system for online testing (Iranmanesh et al. 1992, Khanlou et al. 1995, Zemanek 2009).

- Correct designed strip or sheet tester can determine the material parameters with high accuracy.

As discussed earlier, when we have an open sample as sheet or strip, we close the magnetic circuit with a yoke (Figure 2.183). The asymmetrical yoke offers simplicity and easy access to the sample. But unfortunately as it was proved by Mikulec (Mikulec et al. 1985, Havlíček and Mikulec 1989), the asymmetrical yoke system

cannot be accepted for accurate AC measurements due to error caused by additional eddy currents above the yoke poles (Figure 2.184). The one-sided configuration is sometimes implemented in less accurate measurement systems where the access to the sample is necessary.

The error caused by the eddy currents can be significantly large when the sample is longer than the yoke

FIGURE 2.183
Asymmetrical (a) and symmetrical (b) yoke system for single sheet/strip testers.

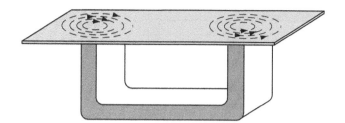

FIGURE 2.184
The eddy currents in an asymmetrical single strip tester.

(overhang effect), which was confirmed in experimental and in numerical simulations (Matsubara et al. 1989, Sievert et al. 1989, Enokizono and Sievert 1990, Nakata et al. 1990). If the sample is magnetized symmetrically, the eddy currents from both halves of the yoke (top and bottom) compensate each other.

Typically, both primary magnetizing coil and secondary B-coil are wound around the tested sample (Figure 2.185). For the H measurements, both techniques are used: direct by using H-coil and indirect where the magnetic field is calculated from the magnetizing current (Nakata et al. 2000, Pfützner 1991).

The main advantages of the H-coil method are as follows:

- Direct measurement of magnetic field strength (we avoid the problem of magnetic path length determination).
- If we put the sensor in the central part of the sample, we reduce the problem of nonuniformity of magnetization of the sample.

The main drawbacks of the H-coil method (in comparison with the magnetizing current method) are as follows:

- Much smaller signal additionally interferenced by the stray magnetic fields
- Necessity of using of the integrating amplifier
- Dependence of the result on the distance of the coil from the surface of the sample

This last problem can be overcome by using two or more H-coils and by extrapolation of the result toward the sheet surface (Nakata et al. 1987, Tumanski 2002). Instead of application of two H-coils, Pfützner (1991) proposed to use one coil in two positions. As was proved by Matsubara et al. (1989) in the system with two H-coils, we can also decrease the error caused by eddy currents in asymmetrical yoke. Figure 2.185a presents the typical design of SST with two H-coils.

The problem of uncertainty of determination of the length of magnetic path can be solved by using the compensation method (Figure 2.185b). This old principle

proposed by Iliovici (1913) was renewed by Mikulec (1981)* and later followed by many researchers (Nafalski et al. 1989, Iranmanesh et al. 1992, Khanlou et al. 1995). The RCP is used to measure the magnetic field between points A and B (Figure 2.174). The output signal of such sensor after amplification is fed back to the compensation winding to obtain the state of compensation—when its output signal is equal to zero, it means that $H_{A-B}=0$. In this state, the magnetic field in the air gaps H_a and magnetic field in the yoke H_{Fe} are compensated:

$$H_a l_a + H_{Fe} l_{Fe} = H_{AB} l_{AB} + nI = 0 \qquad (2.260)$$

and

$$H_{AB} l_{AB} = nI \qquad (2.261)$$

Thus, we can determine the magnetic field H_{AB} in the sample from the magnetizing current I because we know the length of magnetic path—it is, in this case, exactly equal to the distance between the points $A-B$.

The SST is described by the IEC standard 60404-3, although not as exactly as the Epstein frame. The dimensions of the yokes (see Figure 2.186) and of the sample are defined—length should not be shorter than 50 cm. Such large sample and the yoke made out mostly of electrical steel cause that the whole device is very heavy. Such large dimensions of the sample were decided upon in order to obtain mass comparable to the sample in the Epstein frame. The length of the former for the windings (with internal rectangular cross section 5 × 510 mm for the sample) is proposed as 445 mm (windings not shorter than 440 mm). The length of magnetic path is fixed as $l=450$ mm (exactly as the distance between poles of the yoke).

In the case of DC testing, the specimen seldom is in the form of a strip or sheet, more often as the bars, rods, or cylinders. Also, in this case, the sample is closed by the yoke and the whole device is known as a *permeameter*. Figure 2.187 presents two types of the permeameters recommended by IEC Standard 60404-4. The first one called permeameter type-A (Figure 2.187a) is based on the old Hopkinson permeameter (Hopkinson 1885). The magnetizing coil is placed around the specimen and for measurement of B and H, the flux-sensing coil and tangential coil can be used (this last can be substituted by MR or Hall sensor). The second device, shown in Figure 2.187b, called permeameter type-B is based on the old Neumann permeameter (Neumann

---

* In Iliovici permeameter, the magnetic field between points A and B was measured by special small yoke—in new permeameter, this yoke was substituted by RCP. In Iliovici permeameter, the state of compensation was obtained manually—in new device automatically by using feedback.

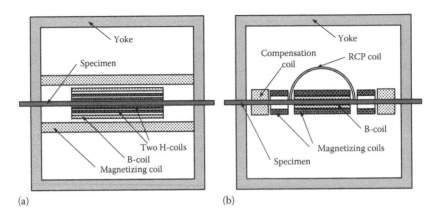

FIGURE 2.185
The single strip tester design: with H-coils method (a) and with compensation method (b).

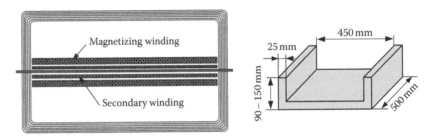

FIGURE 2.186
The single sheet tester recommended by IEC standard 60404-2.

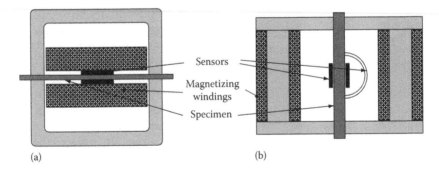

FIGURE 2.187
The permeameters recommended by IEC standard 60404-4: type A (a) and type B (design proposed by Magnet-Physik, www.magnet-physik.de) (b).

1934). The magnetizing coils are wound on the yoke and for measurement of *B* and *H* the flux-sensing coil and Rogowski–Chattock coil can be used. The coil sensors should be connected to the fluxmeter.

2.13.5 Samples for Two-Dimensional Measurements

In 2D measurements (e.g., rotational loss testing), we have two systems of magnetizing coils and two systems of measuring sensors—for *x* and *y* components. Two-dimensional measuring systems are still not standardized and therefore many different concepts of the testers are used in various laboratories. One of the most

popular is the square sample (typically 10 × 10 cm) presented in Figure 2.188 (Brix et al. 1984, Enokizono et al. 1990a,b, Sievert et al. 1990, Zhu and Ramsden 1993, Salz 1994).

Because the specimen is relatively small and rectangular, the distribution of magnetization is nonuniform (Enokizono and Sievert 1989, Nencib et al. 1995, Makaveev et al. 2000, Jesenik et al. 2003, 2004, Tumanski et al. 2004). Figure 2.189 presents the results of experimental results of the investigation of the magnetic field in the sample.

Although the tested area is usually not larger than 2 × 2 cm at the center of the specimen, the nonuniformity

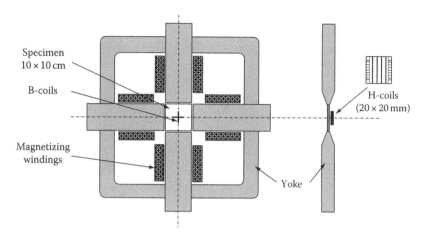

FIGURE 2.188
The two-dimensional tester of electrical steel with a square specimen.

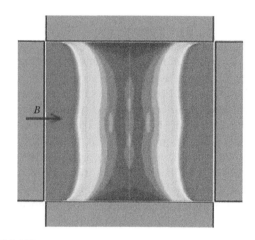

FIGURE 2.189
Experimentally determined distribution of magnetic field above a square sample in a RSST device (at the center, there are visible four micro-holes for flux density measurements). (From Tumanski, S., *Przegl. Elektrotech.*, 5, 32, 2005.)

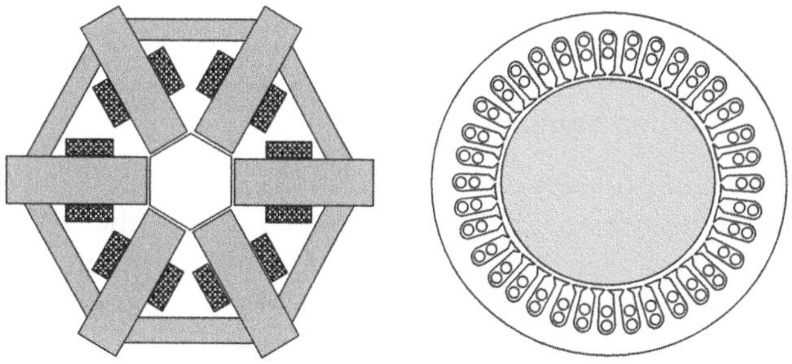

FIGURE 2.190
The rotational single sheet tester RSST with hexagonal and circular specimen.

of magnetization distribution is still significant. That is why other shapes of the samples were proposed: a hexagonal shape (Hasenzagl et al. 1996, 1998, Krell et al. 2000, Pluta et al. 2003) or even circular (Goričan et al. 2000, 2003) (Figure 2.190).

Following the idea of 2D testing, also samples and devices for 3D test were proposed. Figure 2.191 presents the example of 3D tester device with a cubic sample.

The measured results for all presented samples are shape dependent and the tests determine only properties of the selected samples of the magnetic material. No wonder that there are significant differences even in the case of only one type of tester (Sievert et al. 1996).

It would be interesting to find a sample and device for objective tests of magnetic materials. Probably, a good candidate could be the sheet sample much larger than the vertical type yoke (to eliminate the influence of

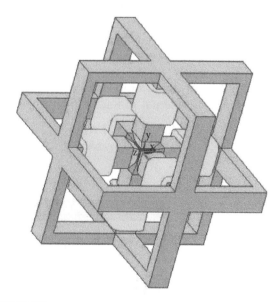

FIGURE 2.191
The example of the device for 3D tests of magnetic materials. (From Zhu, G.J. et al., *IEEE Trans. Magn.*, 39, 3429, 2003; Zhu, G.J. et al., *Przegl. Elektrotech.*, 1, 11, 2009.)

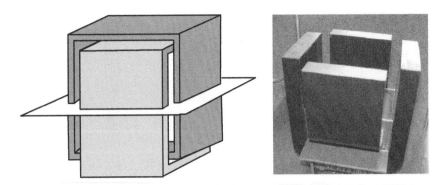

FIGURE 2.192
The example of a device with the sheet sample much larger than the vertical yoke system (photo on the right—only the bottom part of the yoke is shown).

the demagnetizing field) (Sievert et al. 1992, Tumanski 2002a, 2005). Figure 2.192 presents the example of such a 2D measuring system.

2.13.6 Open Samples

The main problem of the applications of the open samples is the influence of the demagnetizing field (the shape anisotropy—see Section 2.5.3). As a result, the magnetizing field H is weakened by contradicting demagnetizing field H_d and magnetic field in the sample H_m is

$$H_m = H - H_d = H - N_d M \qquad (2.262)$$

Theoretically, we can take into account the demagnetizing field knowing the demagnetizing factor N_d. But this factor is known only for ellipsoidal samples—for other shapes (as cylinders or rectangular prisms), it is known only approximately (Chen et al. 1991, Aharoni 1998). Moreover, only the ellipsoidal specimen can be magnetized uniformly.

Other problem of application of the open samples is the influence of external magnetic field, including

the Earth's magnetic field. Partially, this problem can be eliminated by placing the sample in West–East direction.

Nevertheless, open samples are used if it is difficult to use the closed magnetic circuit. The IEC Standard 60404-7 describes a method of measurement of the coercivity in an open magnetic circuit. The sample is placed inside a solenoid coil and first magnetized to saturation and then demagnetized monotonically until the resultant flux density decreases to zero. The state of demagnetization (coercive force) is detected using one or two differential sensors (Hall or flux-gate sensors) (Figure 2.193). The state of demagnetization does not depend on the shape of the specimen.

If we test an open ferromagnetic specimen in a solenoid coil, it is necessary to compensate the stray air flux. One method is to include the mutual inductance between the magnetizing and the measuring circuit (similarly as it is used in the case of the Epstein frame) (Figure 2.194a).

Another method is to connect in series the measuring coil n_1 with a second coil n_2 to obtain zero net signal in the absence of a specimen (Figure 2.194b). The flux of the second coil is $\Phi_2 = n_2 A_2 \mu_0 H$, while the first coil

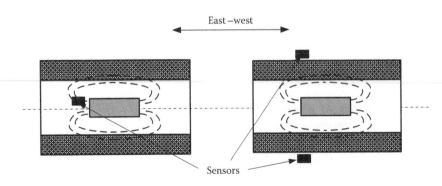

FIGURE 2.193
Two methods of determination of the coercivity in open sample system—according to IEC Standard 60404-7.

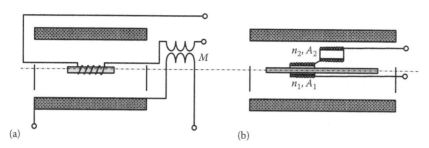

FIGURE 2.194
Two methods of air flux compensation.

is linked with the fluxes: $\Phi_1 = n_1 A_1 \mu_0 H + n_1 A_m J$ where A_m is the cross section of the specimen. If $n_1 A_1 = n_2 A_2$, the air flux is compensated and the resulting flux is (Fiorillo 2004)

$$\Phi = n_1 A_m J \left(1 - \frac{A_1}{A_m} N_d \right) \qquad (2.263)$$

In the case of magnetic thin film, it is difficult to realize magnetically closed sample and commonly the sample is used in form of the circle. The circular sample is also often used in magnetometric methods of investigations of magnetic materials (Humprey and Johnston 1963, Tejedor et al. 1985). For example, it is convenient to use the circular sample for anisotropy measurements using a torque magnetometer (Figure 2.195) (Kouvel and Graham 1957, Cornut et al. 2003).

In the torque magnetometer, the sample is placed in relatively high static magnetic field (typically of the magnet). Under the influence of this field, the sample is deflected and the deflection angle (a torque) depends on the magnetic properties of the sample. By rotating a magnet system, it is possible to measure the torque for various directions of magnetization, thus the analysis of anisotropy is possible. Usually, the torque is balanced

due to negative feedback, for instance, realized by using moving coil device (Figure 2.195).

2.14 Magnetic Shielding

2.14.1 Magnetic Field Pollution

The stray magnetic fields are practically always present. Beside the Earth's magnetic field (which in Europe is around $50\,\mu T$ with the horizontal component of $20\,\mu T$), there is also EM field generated mainly by human activity. This field can be especially higher in the urban or industrial environment, where it is very difficult to eliminate. It can be dangerous for health but it can also disturb performance of measuring devices. Therefore, a branch of electrical science called an *electromagnetic compatibility* (EMC) was developed which informs how to restrict the *electromagnetic interferences* (EMI) caused by emission of EM pollution. There are standards describing limitations of stray magnetic field (Figure 2.196). For

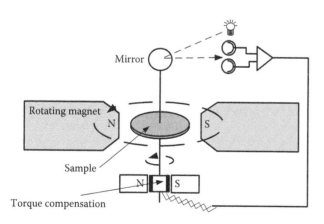

FIGURE 2.195
The torque magnetometer for anisotropy measurements.

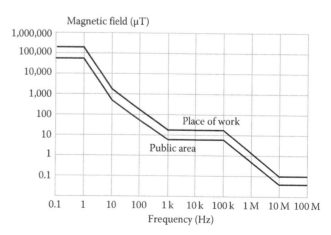

FIGURE 2.196
Safe levels of the magnetic field recommended by ICNIRP. (From ICNIRP, *Guidelines on Limiting to Non-Ionizing Radiation*, ICNIRP (International Commission on Non-Ionizing Radiation Protection) and WHO (World Health Organization), 1999.)

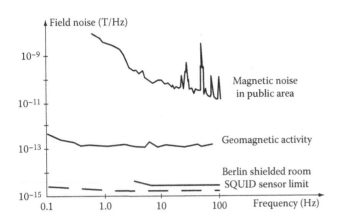

FIGURE 2.197
The typical magnetic noise. (After Romani, G.L. et al., *Rev. Sci. Instrum.*, 53, 1815, 1982.)

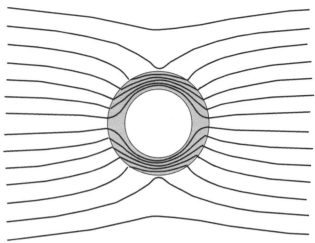

FIGURE 2.198
The principle of shielding by a ferromagnetic cylinder.

example, the industrial frequency magnetic field (50 or 60 Hz) is limited to 100 μT.

The EM compatibility also informs how to prepare the devices to obtain relative immunity to EMI. One of the commonly used methods is shielding.

Especially in biomagnetic investigations, it is necessary to lower the interference magnetic field because magnetic fields accompanying, for example, human brain activity are of the level of $10^{-12}–10^{-15}$ T. For such measurements, chamber free of magnetic field are built (Kelhä et al. 1982).

Figure 2.197 presents the characteristics of magnetic noise in human environment. The magnetic fields are not only of very different values but also the frequency bandwidth is very wide, from DC to GHz. The effectiveness of shielding depends on the permeability of the shielding material—which in turn depends on the value and frequency of magnetic fields. Therefore, correct design of magnetic shield is a complex knowledge (Mager 1970, Sumner et al. 1987, Schultz et al. 1988). Beside the *passive shielding* also *active shielding* methods (magnetic field cancellation) are developed.

2.14.2 Magnetic Shielding

A long time ago, it was discovered (della Porta 1589) that the magnetic field does not get through the ferromagnetic layer known as shield. First calculations and theory of a spherical shield were presented by Rücker (1894). The principle of operation of the magnetic shield is illustrated in Figure 2.198. If the shield is prepared from a material of high permeability, the magnetic field is concentrated in the shield and prefers the way through the shield bypassing the interior part of the shield.

The effectiveness of shielding is described by the shielding factor S as the relationship between external H_e and internal H_i magnetic field:

$$S = \frac{H_e}{H_i} \quad (2.264)$$

For simple shapes of the shield (sphere, cylinder, or cube), there were calculated, approximate formulas enabling determination of the shielding factor for DC magnetic field (Wills 1899, Mager 1970, Mager 1975, Gubser et al. 1979):

for cube with length a and wall thickness t

$$S = 1 + \frac{4}{5}\frac{\mu t}{a} \quad (2.265)$$

while for a long cylinder with diameter D

$$S = 1 + \frac{\mu t}{D} \quad (2.266)$$

We can therefore see that the most important factor is the relative permeability μ of the material. That is why usually high-permeability materials as NiFe (*mumetal*) with $\mu = 50{,}000–100{,}000$ or amorphous materials reaching even permeability as high as 800,000 are used. The second factor influencing the shielding effect is the wall thickness. However, it was proved that, instead of increasing the wall thickness, more effective is to use multiple shells (Wadey 1956, Freake and Thorp 1971, Paperno et al. 2000). For example, for a cylinder consisting of two layers with diameter D_1 and D_2 and shielding factors S_1 and S_2, the resultant shielding factor is

$$S = 1 + S_1 + S_2 + S_1 S_2 \left[1 - \left(\frac{D_2}{D_1} \right)^2 \right] \quad (2.267)$$

In the case of AC magnetic field, the eddy currents and skin-effect paradoxically improve the shielding effectiveness. For a long cylinder, the shielding factor is

$$S_{AC} = p(S_{DC} + 1) \qquad (2.268)$$

where $p \approx \delta/2t(\cosh 2t/\delta - \cos 2t/\delta)^{1/2}$ and $\delta = \sqrt{\rho / \pi\mu_0\mu f}$ is the penetration depth of the material with specific electrical resistivity ρ.

The equations presented above concern the situations when the external magnetic field is directed perpendicularly to plane of the shield (*transverse shielding*). More complex is the case when the magnetic field is directed along a cylinder shield (axial shielding). In such case, the demagnetizing factor N_d of the shield should be taken into account:

$$S \approx \frac{1 + 4N_d\mu t/D}{1 + 0.5D/L} \qquad (2.269)$$

Even more complex is the case when the shield is opened at the ends. The shielding factor in both cases is smaller than the transversal factor and depends on the L/D ratio (Mager 1970).

Obviously, the material of the shield strongly influences its performance. A mumetal alloy commonly used as the shield material is very sensitive to stress and mechanical shocks, so it should be carefully annealed. Both mumetal and amorphous materials are rather expensive and in many cases, the shield made from much cheaper SiFe electrical steel can be an alternative (Okazaki et al. 2009). At higher frequencies, it can also be effective to use a shield made from aluminum in which eddy currents aid the shielding. Also, for high frequencies (e.g., in microwave range), the *Faraday cage* (a mesh prepared from conducting material) is commonly used as a shield. Ideally, the shield should be designed in such a way as to make possible its periodic demagnetization, because magnetized shields exhibit lower performance.

In 1925, Spooner proved that by applying additional alternating magnetic field, it is possible to improve properties of ferromagnetic material, mainly by increasing the permeability (Spooner 1925). The material magnetized by a small AC field changes its domain mobility. This effect known as *shaking effect* is often used to improve the shielding factor (Cohen 1967, Kelhä et al. 1980) (Figure 2.199). When this effect was used with a shield made from amorphous Methglas 2705M, it was possible to increase the shielding factor from 4 to about 150 mainly by improving the incremental permeability from 35,000 to 500,000 (Sasada et al. 1988, 1990).

2.14.3 Active Magnetic Shielding—Magnetic Field Cancelling

It is also possible to cancel the magnetic field by applying feedback magnetic field—the method known as *active*

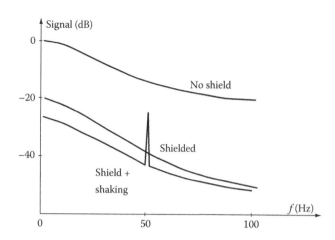

FIGURE 2.199
The typical dependence of shielding on frequency of magnetic field (the enhancement of the shielding by shaking with a 50 Hz magnetic field is also demonstrated). (After Kelhä, V.I. et al., *IEEE Trans. Magn.*, 16, 575, 1980.)

shielding or *magnetic field compensation*. Because even Earth's magnetic field is not stable, the compensating magnetic field should be controlled. The volume under test is usually surrounded by three-axis Helmholtz coils and magnetic field generated by these coils is proportional to the field detected by three-axis magnetic field sensors.

The active shielding technique can substitute the conventional ferromagnetic shields but more often it supports the conventional shield technique (Kelhä et al. 1982, ter Brake et al. 1993, Sergeant et al. 2004).

Figure 2.200 presents a system with the field cancellation offered by Spicerconsulting. This system enables active field compensation with a cancelling factor 50 (at 50 Hz) in the frequency bandwidth 0.5–5 kHz. Such systems are commonly used for protection of electron microscopes.

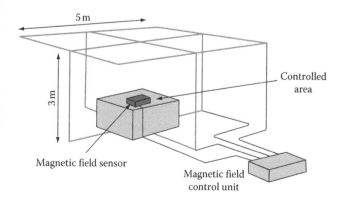

FIGURE 2.200
The principle of magnetic field cancellation—system of Spicerconsulting (Spicerconsulting, www.spicerconsulting.com).

2.14.4 Magnetically Shielded Room

Figure 2.197 presents typical magnetic field noise. We can see that in urban environment, the 50 (or 60 Hz) magnetic noises are on the level of 10^{-8} T, while the geomagnetic activity is on the level of 10^{-12} T. And meanwhile, the upper limit of SQUID sensor sensitivity is at the level of 10^{-15} T. Hence, in order to profit such sensor sensitivity, for example, in biomagnetic or paleomagnetic measurements, it is necessary to shield the area or sample under test with a shielding factor higher than 10^7. Therefore, in many scientific centers, there have been constructed special magnetically shielded rooms (Cohen 1967, Cohen 1970, Kelhä et al. 1982, Harakawa et al. 1996, Kajiwara et al. 1996, Yamazaki et al. 2006, Hirosato et al. 2009). One of the most known, PTB shielded room in Berlin enables to obtain the shielding factor of about 10^6, which is very near the SQUID sensor limit (Bork et al. 2001). Another application of magnetically shielded room is a protection during the tests performed by magnetic resonance imaging (MRI) device that uses magnetic field as high as 7 T. Although for such purposes, it is possible to design open-type room—without ferromagnetic wall shields (Yamazaki et al. 2006, Hirosato et al. 2009).

To obtain such large shielding factor, it is not sufficient to use a passive shield. Practically, all known shielding techniques (multilayer passive shield, active field cancellation, shaking, etc.) are commonly used to construct high-performance shielding rooms. Typically, several layers of mumetal separated by aluminum shield are used (Yamazaki et al. 2006) (see Figures 2.201 and 2.202).

Figure 2.202 presents the design of Japanese shielded room COSMOS (Harakawa et al. 1996, Kajiwara et al. 1996). The measured shielding factor for 1 Hz was determined as 420,000 (only passive shield). The new shielding room of PTB in Berlin MSR L1, built in 2000, is composed of seven mumetal layers ($D_1 = 3.2$ m, $t_1 = 4$ mm; $D_2 = 3.54$ m, $t_2 = 7$ mm; $D_3 = 3.88$ m, $t_3 = 6$ mm;

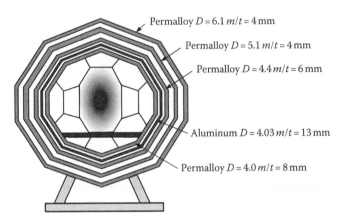

FIGURE 2.202
The design of COSMOS high-performance magnetically shielded room. (After Kajiwara, G. et al., *IEEE Trans. Magn.*, 32, 2582, 1996.)

$D_4 = 4.22$ m, $t_4 = 3$ mm; $D_5 = 4.93$ m, $t_5 = 3$ mm; $D_6 = 5.27$ m, $t_6 = 2$ mm; $D_7 = 5.61$ m, $t_7 = 2$ mm) and one aluminum layer ($D = 4.58$ m, $t = 10$ mm). The passive shielding factor for 0.01 Hz was 750,000 and for 5 Hz was 2×10^8. In 2000, it was a world record for passive shielding.

Performances of the shielded room can be further improved for low frequencies by applying of the active shielding. Figure 2.203 presents the comparison of passive and passive with active shielding of the magnetically shielded room constructed in Helsinki (Kelhä et al. 1982).

Figure 2.204 shows the comparison of the shielding factor versus frequency of main shielded rooms in Berlin BSR, MSR L1, and in Japan—COSMOS.

Modern shielding facilities and measuring system enable recently measurements of magnetic field on the level of fT. The following calculation demonstrates how small this value is. According to Biot–Savart law (Equation 2.96), the magnetic field generated by a current conducting wire is $H = I/2\pi x$. So a wire conducting a current of 1 mA generates the magnetic field of $1 fT$ at the distance of 200 km.

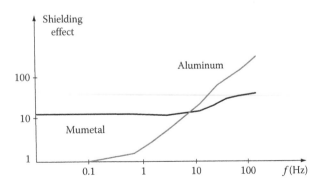

FIGURE 2.201
The comparison of shielding effectiveness of mumetal and aluminum layer. (After Yamazaki, K. et al., *IEEE Trans. Magn.*, 42, 3524, 2006.)

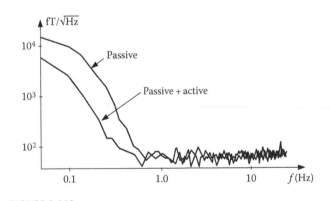

FIGURE 2.203
The comparison of the noise characteristics of passive shielding and passive/active shielding. (After Kelhä et al., 2001.)

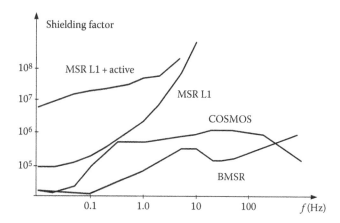

FIGURE 2.204
The comparison of the shielding effectiveness versus frequency of some shielding rooms. (After Bork, J. et al., The eight-layered magnetically shielded room of the PTB: Design and construction, in *Proceedings of 12th International Conferenvce on Bimagnetism*, Helsinki, pp. 970–973, 2001.)

References

Abele M.G., 1993, *Structures of Permanent Magnets*, Wiley, New York.

Aharoni A., 1998, Demagnetizing factors for rectangular ferromagnetic prisms, *J. Appl. Phys.*, **83**, 3432–3434.

Ahlers H., Sievert J., 1982, Comparison of a single strip tester and Epstein frame measurements, *J. Magn. Magn. Mater.*, **26**, 176–178.

Akulov N.S., 1929, Atomic theory of ferromagnetism, *Z. Phys.*, **54**, 582–587.

Anderson N.L., Lance T.M., 1922, Relation between magnetic hysteresis loss and coercivity, *Engineering*, **114**, 351–352.

Anuszczyk J.W., Pluta W., 2009, *Ferromagnetic Materials in Rotating Magnetic Fields* (in Polish), WNT, Warsaw, Poland.

Baibich M.N., 1988, Giant magnetoresitance of (100)Fe/(001) Cr magnetic superlattices, *Phys. Rev. Lett.*, **61**, 2472–2475.

Baldwin J.A., 1970, A controlled-flux hysteresis loop tracer, *Rev. Sci. Instrum.*, **41**, 468.

Bandyopadhyay S., Cahay M., 2008, *Introduction to Spintronics*, CRC Press, Boca Raton, FL.

Banerjee P.K., 1994, *The Boundary Element Methods in Engineering*, McGraw-Hill College, New York.

Baranowski S., 2008, Methods of investigation of magnetic materials in form of strips, PhD thesis, Warsaw University of Technology, Warsaw, Poland.

Baranowski S., Tumanski S., 2005, Comparison of various methods of the control of lux density waveform, Paper AB4 in *Proceedings of Soft Magnetic Materials Conference*, Bratislava, Slovakia

Baranowski S., Tumanski S., Zurek S., 2009, Comparison of digital methods of the control of flux density shape, *Przegl. Elektrotech.*, 85, 93–95

Barker J.R., 1950, An improved three-coil system for producing a uniform magnetic field, *J. Sci. Instrum.*, **27**, 197.

Barker W.A., 1962, Dynamic nuclear polarization, *Rev. Mod. Phys.*, **34**, 173–185.

Barkhausen H., 1919, Two phenomena uncovered with help of the new amplifier, *Z. Phys.*, **20**, 401–403.

Becker R., 1930, Theory of the magnetization curve, *Z. Phys.*, **62**, 253–269.

Becker R., Döring W., 1939, *Ferromagnetismus*, Springer, Berlin, Germany.

Becker R., Kersten M., 1930, Magnetization of Ni wire under large stress, *Z. Phys.*, **64**, 660–681.

Beljers H.G., van der Kint L., van Wieringen J.S., 1954, Overhauser effect in free radicals, *Phys. Rev.*, **95**, 1683

Bertotti G., 1998, *Hysteresis in Magnetism*, Academic Press, Sandiego, CA.

Bertotti G., Ferrara E., Fiorillo F., Pasquale M., 1993, Loss measurement on amorphous alloys under sinusoidal and distorted induction waveform using a digital feedback technique, *J.Appl. Phys.*, **73**, 5375–5377.

Birkenbach G., Hempel K.A., Schulte F.J., 1986, Very low frequency magnetic hysteresis measurements with well-defined time dependence of the flux density, *IEEE Trans. Magn.*, **22**, 505–507.

Birrs R.R., 1964, *Symmetry and Magnetism*, North Holland, Amsterdam, the Netherlands.

Bishop J.E.L., 1980, Understanding magnetization losses in terms of eddy currents dominated domain wall dynamics, *J. Magn. Magn. Mater.*, **19**, 336–344.

Bishop J.E.L., 1985, Enhanced eddy current loss due to domain displacement, *J. Magn. Magn. Mater.*, **49**, 241–249.

Bitter F., 1939, The design of powerful electromagnets, *Rev. Sci. Instrum.*, **10**, 373–381.

Blandford P., Szpunar J.A., 1989, On-line texture measurements for the prediction of the anisotropy of magnetic properties, *Texture. Microstruct.*, **11**, 249–260.

Bloch F., 1930, Zur Theorie des Ferromagnetismus, *Z. Phys.*, **61**, 206–219.

Bloch F., Cugat O., Meunier G., 1998, Innovating approaches to the generation of intense magnetic fields: Design and optimization of a 4 tesla permanent magnet, *IEEE Trans. Magn.*, **34**, 2465–2468.

Blundell M.G., Overshott K.J., Graham C.D., 1980, A new method of measuring power loss of magnetic materials under sinusoidal flux conditions, *J. Magn. Magn. Mater.*, **19**, 243–244.

Boll R., 1989, Introduction to magnetic sensors, Sensors—A comprehensive survey, in *Magnetic Sensors, Vol. 5*, Boll R. and Overshott K.J. (Eds.), VCH, Weinheim, Germany, Chapter 1.

Bork J., Halohm H.D., Klein R., Schnabe A., 2001, The 8-layered magnetically shielded room of the PTB: Design and construction, in *Proceedings of 12th International Conferenvce on Bimagnetism*, Helsinki, pp. 970–973.

Bozorth R.M., 1951, *Ferromagnetism*, Van Nostrand, New York.

Brailsford F., 1948, *Magnetic Materials*, Wiley & Sons, New York.

Bray D.E., McBride D. (Eds.), 1992, *Nondestructive Testing Techniques, Mössbauer Analysis METHOD*, John Wiley & Sons, New York, Chapter 9.

Brix W., Hempel K.A., Schulte F.J., 1984, Improved method for the investigation of the rotational magnetization process in electrical steel sheets, *IEEE Trans. Magn.*, **20**, 1708–1710.

Brown W.F., 1978, *Micromagnetics*, Wiley & Sons, New York.

Bunge H.J., 1982, *Texture Analysis in Material Science*, Butterworths, London, U.K.

Burgwin S.L., 1941, Measurements of core loss and ac permeability with the 25 cm Epstein frame, *Proc. ASTM*, **41**, 779.

Campbell W.H., 2001, *Earth Magnetism*, Harcourt/Academic Press, San Diego, CA

Carr W.J., 1969, Principles of ferromagnetic behavior, in *Magnetism and Metallurgy*, Berkowitz. A.E. and Kneller E. (Eds.), Academic Press, San Diego, CA, Chapter II.

Chatziilias N., Meydan T., Porter C., 2003, Real-time digital waveform control for magnetic testers, *J. Magn. Magn. Mater.*, **254–255**, 104–107.

Chen L.H., 1996, Colossal magnetoresistance in La-Y-Ca-Mn-O films, *IEEE Trans. Magn.*, **32**, 4692–4694.

Chen D.X., Brug J.A., Goldfrab R.B., 1991, Demagnetizing factors for cylinders, *IEEE Trans. Magn.*, **27**, 3601–3619.

Chikazumi S., 2009, *Physics of Ferromagnetism*, Oxford Scientific Publication, Oxford, U.K.

Clark J., 1989, Principles and applications of SQUIDs, *Proc IEEE.*, **77**, 1208–1223.

Coey J.M.D., 1996, *Rare Earth Iron Permanent Magnets*, Clarendon Press, Oxford, U.K.

Cohen D., 1967a, A shielded facility for low level magnetic measurements, *J. Appl. Phys.*, **38**, 1295–1296.

Cohen D., 1967b, Enhancement of ferromagnetic shielding against low-frequency magnetic fields, *Appl. Phys. Lett.*, **10**, 67–69.

Cohen D., 1970, Large volume conventional magnetic shields, *Rev. Phys. Appl.*, **5**, 53–58.

Comstock R.L., 1999, *Introduction to Magnetism and Magnetic Recording*, John Wiley & Sons, New York.

Cornut B., Cattelani S., Perrier J.C., Kedous-Lebouc A., Waeckerle T., Fraisse H., 2003, New compact and precise magnetometer, *J. Magn. Magn. Mater.*, **254–255**, 97–99.

Cracknell A.P., 1975, *Magnetism in Crystalline Materials*, Pergamon Press, Oxford, U.K.

Cugat O., Hansson P., Coey J.M.D., 1994, Permanent magnet variable flux source, *IEEE Trans. Magn.*, **30**, 4602–4604.

Cullity B.D., Graham C.D., 2009, *Introduction to Magnetic Materials*, IEEE Press/Wiley, Hoboken, NJ.

Curie P., 1895, Magnetic properties of bodies at various temperatures, *Ann. Chim. Phys.*, **5**, 289–405.

Daughton J., 2003, Spin-dependent sensors, *Proc. IEEE*, **91**(5), 681–686.

Debye P., 1926, *Ann. Physik.*, **81**, 1154

Defoug S., Kaczmarek R., Rave W., 1996, Measurements of local magnetization by Kerr effect on SiFe nonoriented sheets, *J. Appl. Phys.*, **79**, 6036–6038.

Del Vecchio R.M., 1980, An efficient procedure for modeling complex hysteresis processes in ferromagnetic materials, *IEEE Trans. Magn.*, **16**, 809–811.

Della Torre E., 1994, A Preisach model for accommodation, *IEEE Trans. Magn.*, **30**, 2701–2707.

Della Tore E., 1999, *Magnetic Hysteresis*, IEEE Press, New York.

De Wulf M., Dupre L.R., Melkebeek J., 1998, Real-time controlled arbitrary excitation for identification of electromagnetic properties of non-oriented steel, *J. Phys. IV*, **8**, 705–706.

Dieny B., 1991, Spin-valve effect in soft ferromagnetic sandwiches, *J. Magn. Magn. Mater.*, **93**, 101–104.

Dieny B., 1994, Giant magnetoresistance in spin-valve multilayers, *J. Magn. Magn. Mater.*, **136**, 335–359.

Dieterly D.C., 1949, DC permeability testing of Epstein samples with double-lap joints, *ASTM Spec. Tech. Publ.*, **85**, 39–62.

Dreyfus L., 1935, Grosse Elektromagnete für physikalischchemische Laboratorien, *E. u. M.*, **53**, 205–211.

Du Tremolet de Lacheisserie E., 1993, *Magnetostriction: Theory and Applications of Magnetoelasticity*, CRC Press, Boca Raton, FL.

Duret D., Bonzom J., Brochier M., Frances M., Leger J.M., Odru R., Salvi C., Thomas T., 1995, Overhauser magnetometer for the Danish Oersted Satellite, *IEEE Trans. Magn.*, **31**, 3197–3199.

El-Sherbiny L.K., 1973, Representation of magnetization characteristic by sum of exponentials, *IEEE Trans. Magn.*, **9**, 60–61.

Enokizono M., Sievert J., 1989, Magnetic field and loss analysis in an apparatus for the determination of rotational loss, *Phys. Scr.*, **39**, 356–359.

Enokizono M., Sievert J., 1990, Analytical studies on the yoke construction of a single sheet tester, *Ann. Fis.*, **86**, 123–125.

Enokizono M., Sievert J., Ahlers H., 1990a, Optimum yoker construction for rotational loss measurements apparatus, *Ann. Fis.*, **86**, 320–322.

Enokizono M., Suzuki T., Sievert J., Xu J., 1990b, Rotational power loss of silicon steel sheet, *IEEE Trans. Magn.*, **26**, 2562–2564.

Epstein J., 1900, Die magnetische Prüfung von Eisenblech, *ETZ*, **21**, 303–307.

Everett D.H., Whitton W.I., 1952, A general approach to hysteresis. Part I. *Trans. Faraday Soc.*, **48**, 749–757.

Everett D.H., Smith F.W., 1954, A general approach to hysteresis. Part II—Development of the domain theory, *Trans. Faraday Soc.*, **50**, 187–197.

Everett D.H., 1954, A general approach to hysteresis. Part III—A formal treatment of the independent domain model of hysteresis, *Trans. Faraday Soc.*, **50**, 1077–1096.

Everett D.H., 1955, A general approach to hysteresis. Part IV—An alternative formulation of the domain model, *Trans. Faraday Soc.*, **51**, 1551–1557.

Fert A., 2008, Origin, development and future of spintronics—Nobel lecture, *Angew. Chem. Int. Ed.*, **47**, 5956–5967.

Fiorillo F., 2004, *Measurement and Characterization of Magnetic Materials*, Elsevier, San Diego, CA.

Fiorillo F., Dupre L.R., Appino C., Rietto A.M., 2002, Comprehensive model of magnetization curve, hysteresis loops and losses in any direction in grain oriented Fe-Si, *IEEE Trans. Magn.*, **38**, 1467–1476.

Fiorillo F., Novikov A., 1990, Power losses under sinusoidal, trapezoidal and distorteed induction waveform, *IEEE Trans. Magn.*, **26**, 2559–2561.

Firester A.A., 1966, Design of square Helmholtz coil system, *Rev. Sci. Instrum.*, **37**, 1264–1265.

Fluitman J.H.J., 1973, The influence of the sample geometry on the magnetoresistance of NiFe films, *Thin Solid Films*, **16**, 269–276.

Franzen W., 1962, Generation of uniform magnetic fields by means of air-core coils, *Rev. Sci. Instrum.*, **33**, 933–938.

Freake S.M., Thorp T.L., 1971, Shielding of low magnetic fields with multiple cylinder shells, *Rev. Sci. Instrum.*, **42**, 1411–1413.

Freiser M.J., 1968, A survey of magnetooptic effects, *IEEE Trans. Magn.*, **4**, 152–161.

Fröhlich O., 1881, Investigations of dymoelectric machines and electric power transmission and theoretical conclusions therefrom, *Elektrotech. Z.*, **2**, 134–141.

Gao B.J., Schneider-Muntau H.J., Eyssa Y.M., Bird M.D., 1996, A new concept in Bitter disk design, *IEEE Trans. Magn.*, **32**, 2503–2506.

Gardener M.E., Jungerman J.A., Lichtenstein P.G., Patten C.G., 1960, Production of a uniform magnetic field by means of an end-corrected solenoid, *Rev. Sci. Instrum.*, **31**, 929–934.

Garret M.W., 1951, Axially symmetric systems for generating and measuring magnetic fields, *J. Appl. Phys.*, **22**, 1091–1107.

Giaugue W.F., 1927, Low temperature magnetic susceptibility of gadolinium sulfate, *J. Am. Chem. Soc.*, **49**, 1870–1877.

Giaugue W.F., MacDougal D.P., 1933, Attainment of temperature below 1° absolute by demagnetization of $Gd_2(SO_4)_3$ $8H_2O$, *Phys. Rev.*, **43**, 768–768.

Globus A., Duplex P., Guyot M., 1971, Determination of initial magnetization curve from crystallites size and effective anisotropy field, *IEEE Trans. Magn.*, **7**, 617–622.

Goričan V., Hamler A., Hribernik B., Jesenik M., Trlep M., 2000a, 2D measurements of magnetic properties using a round RSST, in *Proceedings of 2D Magnetic Measurements Workshop*, Vienna, Austria, pp. 66–75.

Goričan V., Jesenik M., Hamler A., Štumberger B., Trlep M., 2000b, Performance of round rotational single sheet tester at higher flux densities in the case of GO materials, in *Proceedings of 2D Magnetic Measurements Workshop*, PTB-Bericht PTB-E-81, pp. 143–149.

Goss N., 1934, Electrical sheet and method and apparatus for its manufacture and test, United States Patent 1965559.

Goss N., 1935, New development in electrical strip steels characterized by fine grain structure approaching the properties of a single crystal, *Trans. Am. Soc. Metals*, **23**, 511–531.

Graham C.D., 1969, Textured magnetic materials, in *Magnetism and Metallurgy*, Berkowitz A.E and Kneller E. (Eds), Academic Press, San Diego, CA, Chapter XV.

Grimmond W., Moses A.J., Ling P.C.Y., 1989, Geometrical factors affecting magnetic properties of wound toroidal cores, *IEEE Trans. Magn.*, **25**, 2686–2693.

Grote N., Denke P., Rademacher M., Kipscholl R., Bongards M., Siebert S., 1998, *Measurement of the Magnetic Properties of Soft Magnetic Materials Using Digital Real-Time Current Method*, Studies on Applications on Electromagnetics, IOS Press, pp. 566–569.

Gubser D.U., Wolf S.A., Cox J.E., 1979, Shielding of longitudinal magnetic fields with thin, closely spaced, concentric cylinders of high permeability material, *Rev. Sci. Instrum.*, **50**, 751–756.

Guy C., Ffyche D., 2005, *The Principles of Medical Imaging*, Imperial College Press, London, U.K.

Hak J., 1936, Eisenlose Zylinderspulen mit ungleichmässiger Windungsdichte zur Erzeugung von homogenen Feldern, *Arch. Elektrotech.*, **30**, 736.

Halbach K., 1980, Design of permanent multipole magnets with oriented rare earth cobalt material, *Nucl. Instrum. Methods*, **169**, 1–10.

Harakawa K., Kajiwara G., Kazami K., Ogata H., Kado H.,1996, Evaluation of a high performance magnetically shielded room for biomagnetic measurement, *IEEE Trans. Magn.*, **32**, 5256–5260.

Hartmann F., 1972, Resonance magnetometers, *IEEE Trans. Magn.*, **8**, 66–75.

Hasenzagl A., Pfützner H., Saito A., Okazaki Y., 1998, Field distribution in rotational single sheet testers, *J. Phys. IV*, **8**, 681–684.

Hasenzagl A., Weiser B., Pfützner H., 1996, Novel 3-phase excited single sheet tester for rotational magnetization, *J. Magn. Magn. Mater.*, **160**, 180–182.

Hasty T., 1963, Ferromagnetic resonance in thin film with perpendicular fields at radio frequencies, *J. Appl. Phys.*, **34**, 1079–1080.

Havlíček V., Mikulec M., 1989, On-line testing device using the compensation method, *Phys. Scr.*, **39**, 513–515.

Heisenberg W., 1928, On the theory of ferromagnetism, *Z. Phys.*, **49**, 619–636.

Herlach F., Li L., Harrison N., Von Bockstal N., Louis J.M., Frings P., Franse J. et al., 1996, Approaching 100 T with wire wound coil, *IEEE Trans. Magn.*, **32**, 2507–2510.

Herzer G., 1997, Nanocrystalline soft magnetic alloys, in *Handbook of Magnetic Materials*, Vol. 10, Elsevier, San Diego, CA, Chapter 3, pp. 415–462.

Hilzinger H.R., Kronmüller H., 1977, Pinning of curved domain walls by randomly distributed lattice defects, *Physica*, **86–88**, 1365–1366.

Hirosato S., Yamazaki K., Haraguchi Y., Muramatsu K., Haga A., Kamata K., Kobayashi K., Matsuura A., Sasaki H., 2009, Design and construction method of an open type magnetically shielded room for MRI composed of magnetic square cylinders, *IEEE Trans. Magn.*, **45**, 4636–4639.

Hopkinson J., 1885, Magnetization of iron, *Trans. R. Soc. A.*, **176**, 455–469.

Hoselitz K., 1952, *Ferromagnetic Properties of Metals and Alloys*, Oxford University Press, Amsterdam, the Netherlands.

Hubert A., Schäfer R., 1998, *Magnetic Domains*, Springer, Berlin, Germany.

Humprey F.B., Johnston A.R., 1963, Sensitive automatic torque balance for thin magnetic films, *Rev. Sci. Instrum.*, **34**, 348–358.

ICNIRP, 1999, *Guidelines on Limiting to Non-Ionizing Radiation*, ICNIRP (International Commission on Non-Ionizing Radiation Protection) and WHO (World Health Organization), July 1999, ISBN 3-9804789-6-3

Iliovici A., 1913, Sur un nouveau permeametre universal, *Bur. Soc. Int. des Electr.*, **3**, 581.

Iranmanesh H., Tahouri B., Moses A.J., Beckley., 1992, A computerised Rogowski-Chattock Potentiometer (RCP) compensated on-line power-loss measuring system for use on grain-oriented electrical steel production lines, *J. Magn. Magn. Mat.*, **112**, 99–102.

Ivanyi A., 1997, *Hysteresis Models in Electromagnetic Computation*, Akademiai Kiado, Budapest, Hungary.

Jesenik M., Goričan V., Trlep M., Hamler A., Štumberger B., 2003, Calculation of the rotational magnetic fields in the sample of the rotational single sheet testers, *Przegl. Elektrotech.*, **12**, 920–922.

Jesenik M., Goričan V., Trlep M., Hamler A., Štumberger B., 2004, Comparison of the rotational magnetic field homogeneity in 2D RRSST and 2D SRSST, in *Proceedings of Soft Magnetic Materials Conference*, pp. 829–833.

Jiles D., 1998, *Magnetism and Magnetic Materials*, Chapman Hall, London, U.K.

Jiles D.C., Atherton D.L., 1983, Ferromagnetic hysteresis, *IEEE Trans. Magn.*, **19**, 2183–2185.

Jiles D.C., Atherton D.L., 1986, Theory of ferromagnetic hysteresis, *J. Magn. Magn. Mater.*, **61**, 48–60.

Jin S., 1994, Colossal magnetoresistance in La-Ca-Mn-O ferromagnetic thin films, *J. Appl. Phys.*, **76**, 6929–6933.

Jin J., 2002, *The Finite Element Method in Electromagnetics*, Wiley & Sons, New York.

Josephs R.M., Crompton D.S., Krafft C.S., 1986, Characterization of magnetic oxide recording media using Fourier analysis of static hysteresis loop, *IEEE Trans. Magn.*, **22**, 653–655

Kadar G., 1987, On the Preisach function of ferromagnetic hysteresis, *J. Appl. Phys.*, **61**, 4013–4015.

Kajiwara G., Harakawa K., Ogata H., 1996, High-performance magnetically shielded room, *IEEE Trans. Magn.*, **32**, 2582–2585.

Kaleta J., Tumanski S., Zebracki J., 1996, Magnetoresistors as a tool for investigating the mechanical properties of ferromagnetic materials, *J. Magn. Magn. Mater.*, **160**, 199–200.

Kaminishi K., Nawata S., 1981, Practical method of improving the uniformity of magnetic fields generated by single and double Helmholtz coils, *Rev. Sci. Instrum.*, **52**, 447–453.

Keithley J.F., 1999, *The Story of Electrical and Magnetic Measurements*, IEEE Press, New York.

Kelhä V.I., Peltonen R.S., Rantala B., 1980, The effect of shaking on magnetic shields, *IEEE Trans. Magn.*, **16**, 575–578.

Kelhä V.I., Pukki J.M., Peltonen R.S., Penttinen A.J., Ilmoniemi R.J., Heino J.J., 1982, Design, construction and performance of a large volume magnetic shield, *IEEE Trans. Magn.*, **18**, 260–270.

Kennelly A.E., 1891, Magnetic reluctance, *Trans. Am. Inst. Electr. Eng.*, **8**, 485–517.

Kernevez N., Duret D., Moussavi M., Leger J.M., 1992, Weak field NMR and RESR spectrometers and magnetometers, *IEEE Trans. Magn.*, **28**, 3054–3058.

Kernevez N., Glenat H., 1991, Description of a high sensitivity CW scalar DNP-NMR magnetometer, *IEEE Trans. Magn.*, **27**, 5402–5404.

Kerr J., 1877, On rotation of the plane of polarization by reflection from the pole of magnet, *Philos. Mag.*, **3**, 321–343.

Kersten M., 1943, Theory of ferromagnetic hysteresis and initial permeability, *Z. Phys.*, **44**, 63–77.

Khanlou A., Moses A.J., Meydan T., Beckley P., 1995, A computerised on-line power loss testing system for the steel industry, based on the RCP compensation technique, *IEEE Trans. Magn.*, **31**, 3385–3387.

Kittel C., 1949, Physical theory of ferromagnetic domains, *Rev. Mod. Phys.*, **21**, 541–583.

Kittel C., 2004, *Introduction to Solid State Physics*, John Wiley & Sons, New York

Kiwi M., 2001, Exchange bias theory, *J. Magn. Magn. Mater.*, **234**(3), 584–595.

Koch H. 1989, SQUID sensors, in *Magnetic Sensors*, W. Göpel (Ed.), VCH Verlageselschaft, Weinheim, Germany, Chapter 10.

Kouvel J.S., Graham C.D., 1957, On the determination of magnetocrystalline anisotropy constants from torque measurements, *J. Appl. Phys.*, **28**, 340–343.

Krasnoselski M.A., Pokrovski A.V., 1983, *Systems with Hysteresis*, Nauka, Moscow (in Russian).

Krell C., Mehnen L., Leiss E., Pfützner H., 2000, Rotational single sheet testing on samples with arbitrary size and shape, in *Proceedings of 2D Magnetic Measurements Workshop*, Vienna, Austria, pp. 96–103.

Kunzler J.E., Buehler E., Hsu F.S.L., Wernick J.H., 1961, Superconductivity in NbSn at high current density ina magnetic field of 88 kgauss, *Phys. Rev. Lett.*, **6**, 89–91.

Kwiatkowski W., Stabrowski M., Tumanski S., 1983, Numerical analysis of the shape anisotropy and isotropy dispersion in thin film permalloy magnetoresistors, *IEEE Trans. Magn.*, **19**, 2502–2505.

Landau L.D., Lifshitz E.M., 1935. Theory of the dispersion of magnetic permeability in ferromagnetic bodies, *Phys. Z. Sowietunion*, **8**, 153.

Langevin P., 1905, Magnetism and electron theory, *Ann. Chim. Phys.*, **5**, 70–127.

Lee E.W., 1955, *Rep. Prog. Phys.*, **18**, 184.

Leupold H.A., Potenziani E., 2007, *Design of Rare-Earth Permanent Magnets*, Wexford College, Palm Springs, CA.

Lin C.E., Cheng C.L., Huang C.L., 1991, Hysteresis haracteristic analysis of transformer under different excitations using real time measurements, *IEEE Trans. Power Delivery*, **6**, 873–879.

Ling P.C.Y., Moses A.J., Grimmond W., 1990, Investigations of magnetic flux distribution in wound toroidal cores taking into account of geometrical factors, *Ann. Fis*, **86**, 99–101.

Loisos G., Moses A.J., 2001, Critical evaluation and limitations of the localized flux density measurements in electrical steels, *IEEE Trans. Magn.*, **37**, 2755–2757.

Mackay K., Bonfim M., Givord D., Fontaine A., 2000, 50 T pulsed magnetic fields in microcoils, *J. Appl. Phys.*, **87**, 1996–2002.

Mager A., 1952, Über der Einfluss der Korngrosse auf die Koerzitivkraft, *Ann. Phys.*, **11**, 15–16.

Mager A.J., 1970, Magnetic shields, *IEEE Trans. Magn.*, **6**, 67–75.

Mager A., 1975, Magnetostatische Abschirmfaktoren von Zylindern mit recheckigen Querschnittformen, *Physica*, **80B**, 451–463.

Makaveev D., von Rauch M., de Wulf M., Melkebeek J., 2000, Accurate field strength measurement in rotational single sheet testers, *J. Magn. Magn. Mater.*, **215–216**, 673–676.

Marketos P., Zurek S., Moses A. J., 2007, A method for defining the mean path length of the Epstein frame, *IEEE Trans. Magn.*, **43**, 2755–2757.

Matsubara K., Nakata T., Takahashi N., Fujiwara K., Nakano M., 1989, Effects of the overhang of a specimen on the accuracy of a single sheet tester, *Phys. Scr.*, **40**, 529–531.

Matsubara K., Takahashi N., Fujiwara K., Nakata T., Nakano M., Aoki H., 1995, Accelaration technique of waveform control for single sheet tester, *IEEE Trans. Magn.*, **31**, 3400–3402.

Mayergoyz I.D., 1988, Dynamic Preisach models of hysteresis, *IEEE Trans. Magn.*, **24**, 2925–2927.

Mayergoyz I.D., 1991, *Mathematical Models of Hysteresis*, Springer, Berlin, Germany.

Mayergoyz I.D., 2003, *Mathematical Models of Hysteresis and Their Applications*, Academic Press, San Diego, CA.

Mazzetti P., Soardo P., 1966, Electronic hysteresisgraph holds dB/dt constant, *Rev. Sci. Instrum.*, **37**, 548–552.

McCurrie R.A., 1994, *Ferromagnetic Materials—Structure and Properties*, Academic Press, San Diego, CA.

Meiklejohn W.H., Bean C.P., 1957, New magnetic anisotropy, *Phys. Rev.*, **105**(3), 904–913.

Mikulec N., 1981, Kompenzačni ferometr (compensation ferrometer), *Elektrotech. Obzor*, **70**, 435–441.

Mikulec M., Havlíček V., Wiglasz V., Čech D., 1985, Present problems of the AC magnetic measurements on sheets with respect to standardization, in *Proceedings of Soft Magnetic Materials Conference*, pp. 16–18.

Montgomery D.B., 1963, The generation of high magnetic fields, *Rep. Prog. Phys.*, **26**, 69–104.

Morrish A.H., 2001, *The Physical Principles of Magnetism*, IEEE Press, New York

Moses A.J., 1990, Electrical steel—Past, present and future developments, *IEE Proc.*, **137**, 233–245.

Moses A.J., 1992, Importance of rotational losses in rotating machines and transformers, *J. Mater. Eng. Perform.*, **1**, 235–244.

Moses A.J., Hamadeh S., 1983, Comparison of the Epstein square and single strip tester for measuring the power loss of nonoriented electrical steel, *IEEE Trans. Magn.*, **19**, 2705–2710.

Moses A.J., Leicht J., 2004, Measurement and prediction of iron loss in electrical steel under controlled magnetization condition, *Przegl. Elektrotech.*, **12**, 1181–1187.

Moses A.J., Ling P.C.Y., 1987, Dimensional factors affecting magnetic properties of wound cores, in *Proceedings of Soft Magnetic Materials Conference*, Badgastein, Austria, pp. 257–258.

Moses A.J., Shirkochi G.H., 1987, Iron loss in non-oriented electrical steels under distorted flux conditions, *IEEE Trans. Magn.*, **23**, 3217–3220.

Moses A.J., Tutkun N., 1997, Investigations of the power loss in wound toroidal cores under PWM excitation, *IEEE Trans. Magn.*, **33**, 3763–3765.

Mössbauer R.L., 1958, Gammastrahlung in Ir[197], *Z. Phys. A*, **151**, 124–143.

Murgatroyd P.N., Bernard B.E., 1983, Inverse Helmholtz pairs, *Rev. Sci. Instrum.*, **54**, 1736–1738.

Nafalski A., Moses A.J., Meydan T., Abousetta M.M., 1989, Loss measurement on amorphous materials using a field compensated single strip tester, *IEEE Trans. Magn.*, **25**, 4287–4291.

Nakata T., Fujiwara K., Nakano M., Kayada T., 1990, Effects of the construction of yokes on the accuracy of a single sheet tester, *Ann. Fis.*, **86**, 190–192.

Nakata T., Ishihara Y., Nakaji M., Todaka T., 2000, Comparison between the H-coil method and the magnetizing current method for the single sheet tester, *J. Magn. Magn. Mater.*, **215**, 607–610.

Nakata T., Kawase Y., Nakano M., 1987, Improvement of measuring accuracy of magnetic field strength in single sheet testers by using two H-coils, *IEEE Trans. Magn.*, **23**, 2596–2598.

Nakata T., Takahashi N., Fujiwara K., Nakano M., Ogura Y., 1992a, An improved method for determining the DC magnetization curve using a ring specimen, *IEEE Trans. Magn.*, **28**, 2456–2458.

Nakata T., Takahashi N., Fujiwara K., Nakano M., Ogura Y., 1992b, Accurate measurement method of magnetization curve using ring specimen, *J. Magn. Magn. Mater.*, **112**, 71–73.

Nasar S.A., Xiong G.Y., Fu Z.X., 1994, Eddy-current losses in a tubular linear induction motor, *IEEE Trans. Magn.*, **30**, 1437–1445.

Nawrocki W., 2007, *Introduction to Quantum Metrology* (in Polish), Poznan University of Technology, Poznan, Poland.

Néel L., 1932, Influence of fluctuations of molecular field on magnetic properties of bodies, *Ann. Phys.*, **17**, 5–105.

Néel L., 1944a, Effect of cavities and inclusions on the coercive force, *Cahiers Phys.*, **25**, 21–44.

Néel L., 1944b, Some properties of boundaries between ferromagnetic domains, *Cahiers Phys.*, **25**, 1–20.

Néel L., 1944c, Laws of magnetization and subdivision of elementary domains in iron, *J. Phys. Radium*, **5**, 241–276.

Néel L., 1946, Principles of a new general theory of the coercive field, *Ann. Univ. Grenoble*, **22**, 290–343.

Nencib N., Kedous-Lebouc A., Cornut B., 1995, 2D analysis of rotational loss tester, *IEEE Trans. Magn.*, **31**, 3388–3390.

Neumann H., 1934, Das Spannungsmesserjoch, ein neues Prüfgerät für Dauermagnetstähle und fertige Magnete, *Z. Tech. Phys.*, **15**, 473–477.

Nogués, J., Schuller I.K., 1999, Exchange bias, *J. Magn. Magn. Mater.*, **192**(2), 203–232.

Nozawa T., Mizogami M., Mogi H., Matsuo Y., 1996, Magnetic properties and dynamic domain behavior in grain-oriented 3% SiFe, *IEEE Trans. Magn.*, **32**, 572–589.

O'Handley R.C., 1987, Physics of ferromagnetic amorphous alloys, *J. Appl. Phys.*, **62**, R15–R49.

O'Handley R.C., 2000, *Modern Magnetic Materials*, John Wiley & Sons, New York.

Okazakio Y., Yanase S., Nakamura Y., Maeno R., 2009, Magnetic shielding by grain-oriented electrical steel sheet under alternating fields up to 100 kHz, *Przegl. Elektrotech.*, **85**(1), 55–59.

Osborn J.A., 1945, Demagnetizing factors of the general ellipsoid, *Phys. Rev.*, **67**, 351–357.

Overhauser A., 1953, Polarization of nuclei in metals, *Phys. Rev.*, **92**, 411–415.

Panina L.V., 1995, Giant magnetoimpedance in Co-rich amorphous wires and films, *IEEE Trans. Magn.*, **31**, 1249–1260.

Paperno E., Koide H., Sasada I., 2000, A new estimation of the axial shielding factors for multi-shell cylinder shields, *J. Appl. Phys.*, **87**, 5959–5961.

Parkin S., Jiang X., Kaiser C., Panchula A., Roche K., Samant M., 2003, Magnetically engineered spintronic sensors and memory, *Proc IEEE*, **91**(5), 661–680.

Parkin S., Kaiser C., Panchula A., Rice P., Hughes B., Samant M., See-Hun Y., 2004, Giant tunneling magnetoresistance at room temperature with MgO(100) tunnel barriers, *Nat. Mater.*, **31**, 862–867.

Pauli W., 1926, Paramagnetism of a degenerate gas, *Z. Phys.*, **41**, 81–102.

Pershan P.S., 1967, Magneto-optical effects, *J. Appl. Phys.*, **38**, 1482–1490.

Pfützner H., 1980, Anwendungen von Hallgeneratoren im Vergleich zu anderen Methoden der Feldstärkeerfassung bei den Prüfung von Elektroblechen, *Z. Elektrotech. Inform. Energietech.*, **10**, 534–546.

Pfützner H., 1991, On the problem of the field detection for single sheet tester, *IEEE Trans. Magn.*, **27**, 778–785.

Pfützner H., 1994, Rotational magnetization and rotational losses of grain oriented silicon steel sheets—Fundamental aspects and theory, *IEEE Trans. Magn.*, **30**, 2802–2807.

Pfützner H., Krismanic G., 2004, The needle method for induction tests—Sources of error, *IEEE Trans. Magn.*, **40**, 1610–1616.

Phan M.-H., Yu S.-C., 2007, Review of the magnetocaloric effect in manganite materials, *J. Magn. Magn. Mater.*, **308**, 325–340

Pluta W., Kitz E., Krell C., Rygal R., Soinski M., Pfützner H., 2003, Practical relevance of rotational loss measurement of laminated machine cores, *Przegl. Elektrotech.*, **3**, 151–154.

Preisach F., 1935, Magnetic investigations of precipitations hardening FeNi alloys, *Z. Phys.*, **93**, 245–269.

Pry R.H., Bean C.P., 1958, Calculation of the energy losses in magnetic sheet material using a domain model, *J. Appl. Phys.*, **29**, 532–533.

Polik Z., Kuczmann M., 2009, Potential functions for solving TEAM Problem 27, *Przegl. Eleketrotech.*, **85**(12), 137–140.

Rave W., Reichel P., Brendel H., Leicht M., McCord J., Hubert A., 1993, Progress in quantitative magnetic domain observation, *IEEE Trans. Magn.*, **29**, 2551–2552.

Rayleigh L., 1887, The behavior of iron and steel under the operation of feeble magnetic forces, *Philos. Mag.*, **23**, 225–245.

Richter R., 1910, Proposal for description of hysteresis loss, *Elektrotech. Z.*, **31**, 1241–1246.

Romani G.L., Willimason S.J., Kaufman L., 1982, Biomagnetic instrumentation, *Rev. Sci. Instrum.*, **53**, 1815–1845.

Rossignol M., Schlenker M., Samson Y., 2005, Ferromagnetism of an ideal system, in *Magnetism—Fundamentals*, E. du Trémolet de Lacheisserie (Ed.), Springer, Berlin, Germany, Chapter 5.

Rouessac F., Rouessac A., 2000, *Chemical Analysis*, John Wiley & Sons, New York.

Rücker A.W., 1894, On the magnetic shielding on concentric spherical shells, *Philos. Mag.*, **37**, 95–130.

Salz W., 1994, A two-dimensional measuring equipment for electrical steel, *IEEE Trans. Magn.*, **30**, 1253–1257.

Sankaran P., Jagadeesh Kumar V., 1983, Use a voltage follower to ensure sinusoidal flux in a core, *IEEE Trans. Magn.*, **19**, 1572–1573.

Sasada I., Kubo S., O'Handley R.C., Harada K., 1990, Low-frequency characteristic of the enhanced incremental permeability by magnetic shielding, *J. Appl. Phys.*, **67**, 5583–5585.

Sasada I., Kubo S., Harada K., 1988, Effective shielding for low-level magnetic fields, *J. Appl. Phys.*, **64**, 5696–5698.

Savini A., 1982, Modelling hysteresis loops for finite element magnetic field calculation, *IEEE Trans. Magn.*, **18**, 552–557.

Schultz R.B., Plantz V.C., Brush D.R., 1988, Shielding theory and practice, *IEEE Trans. Electromagn. Compat.*, **30**, 187–201.

Seeger A., Kronmuller H., Rieger H., Trauble H., 1964, Effect of lattice defects on the magnetization curve of ferromagnets, *J. Appl. Phys.*, **35**, 740–748.

Seiden P.E., 1969, Magnetic resonance, in *Magnetism and Metallurgy*, Berkowitz A.E., Kneller E. (Eds.), Academic Press, San Diego, CA, Chapter III.

Senda K., Ishida M., Sato K., Komatsubara M., Yamaguchi T., 1999, Localized magnetic properties in grain-oriented silicon steel measured by stylus probe method, *Trans. IEEE Jpn.*, **117-A**, 941–950.

Sergeant P., Adriano U., Dupre L., Bottauscio O., de Wulf M., Zucca M., Melkebeek J.A.A., 2004, Passive and active electromagnetic shielding of inductive heaters, *IEEE Trans. Magn.*, **40**, 675–678.

Shiling J.W., Houze G.L., 1974, Magnetic properties and domain structure in grain oriented 3% SiFe, *IEEE Trans. Magn.*, **10**, 195–223.

Shirkoohi G.H., Kontopoulos A.S., 1994, Computation of magnetic field ion Rogowski-Chattock potentiometer compensated magnetic testers, *J. Magn. Magn. Mater.*, **133**, 587–590.

Shute H.A., Mallinson J.C., Wilton D.T., Mapps D.J., 2000, One-sided fluxes in planar, cylindrical and spherical magnetized structures, *IEEE Trans. Magn.*, **36**, 440–451.

Shull C.G., Smart J.S., 1949, Detection of antiferromagnetism by neutron diffraction, *Phys. Rev.*, **76**, 1256–1257.

Sievert J., 1987, Determination of AC magnetic power loss of electrical steel sheet: Present status and trends, *IEEE Trans. Magn.*, **20**, 1702–1707.

Sievert J., 2005, On the metrology of the magnetic properties of electrical sheet steel, *Przegl. Elektrotech.*, **5**, 1–5.

Sievert J., Ahlers H., 1990, Is the Epstein frame replaceable? *Ann. Fis.*, **86**, 58–63.

Sievert J., Ahlers H., Birkfeld M., Cornut B., Fiorillor F., Hempel K.A., Kochmann T. et al., 1996, European intercomparison of measurements of rotational power loss in electrical sheet steel, *J. Magn. Magn. Mater.*, **160**, 115–118.

Sievert J., Ahlers H., Enokizono M., Kauke S., Rahf L., Xu J., 1992, The measurement of rotational power loss in electrical sheet steel using a vertical yoke system, *J. Magn. Magn. Mater.*, **112**, 91–94.

Sievert J., Enokizono M., Woo B.CH., 1989, Experimental studies on single sheet testers, in *Proceedings of Soft Magnetic Materials Conference*, Madrid, Spain, Paper 2.17.

Sievert J., Xu J., Rahf L., Enokizono M., Ahlers H., 1990a, Studies on the rotational power loss measurement problem, *Ann. Fis.*, **86**, 35–37.

Slater J.C., 1936, The ferromagnetism of nickel, *Phys. Rev.*, **49**, 537–545.

Slonczewski J.C., 1989, Conductance and exchange coupling of two ferromagnets separated by a tunneling barrier, *Phys. Rev.*, **B39**, 6995–7002.

Skomski R., Coey J.M.D., 2000, *Permanent Magnetism*, IOP Publishing, Bristol, U.K.

Smit J., Wijn H.P.J., 1959, *Ferrites*, John Wiley & Sons, New York.

Spooner T., 1925, Effect of a superposed alternating field on apparent magnetic permeability and hysteresis loss, *Phys. Rev.*, **25**, 527–540.

Spornic S.A., Kedous-Lebouc A., Cornut B., 1998, Numerical waveform control for rotational single sheet testers, *J. Phys. IV*, **8**, 741–744.

Stamberger G.A., 1972, *Devices for Generating Weak Magnetic Field* (in Russian), Nauka.

Steinmetz C.P., 1891, Note on the law of hysteresis, *Electrician*, **26**, 261–262.

Steinmetz C.P., 1892, On the law of the hysteresis, *Trans. Am. Inst. Electr. Eng.*, **9**, 3–51.

Stodolny J., 1995, Alternative materials and development of conventional electrical steel, *Metall. Foundry Eng.*, **21**(4), 307–318.

Stoner E.C., 1933, Atomic moments in ferromagnetic alloys with non-ferromagnetic elements, *Philos. Mag.*, **15**, 1018–1034.

Stoner E.C., Wohlfarth E.P., 1948, A mechanism of magnetic hysteresis in heterogeneous alloys, *Phil. Trans. R. Soc. London*, **A240**, 599–642.

Stoner E.C., Wohlfarth E.P., 1991, A mechanism of magnetic hysteresis in heterogeneous alloys, a reprint from 1948, *IEEE Trans. Magn.*, **27**, 3475–3518.

Sumner T.J., Pendlebury J.M., Smith K.F., 1987, Conventional magnetic shielding, *J. Phys. D.*, **20**, 1095–1101.

Szymczak H., Szymczak R., 2008, Magnetocaloric effects, *Phys. Appl. Mater. Sci. (Poland)*, **26**, 807–814.

Tejedor M., Fernandez A., Hernando B., Carrizo J., 1985, Very simple torque magnetometer for measuring magnetic thin films, *Rev. Sci. Instrum.*, **56**, 2160–2161.

ter Brake H.J.M., Huonker R., Rogalla H., 1993, New results in active noise compensation for magnetically shielded rooms, *Meas. Sci. Technol.*, **4**, 1370–1375.

Thomson W., 1857, On the electrodynamic qualities of metals: effect of magnetization on the lectric conductivity of nickel and iron, *Proc. R. Soc.*, **8**, 546–550.

Tishin A.M., Spichkin Y.L., 2003, *The Magnetocaloric Effect and Its Applications*, Institute of Physics Publishing, Philadelphia, PA.

Todaka T., Maeda Y., Enokizono M., 2009, Counterclockwise/clockwise rotational losses under high magnetic field, *Przegl. Elektrotech.*, **85**(1), 20–24.

Trutt F.C., Erdelyi E.A., Hopkins R.E., 1968, Representation of magnetization characteristic of DC machines for computer use, *IEEE Trans. Power Appl.*, **87**, 665–669.

Tumanski S., 1988, The application of permalloy magnetoresistive sensors for nondestructive testing of electrical steel sheets, *J. Magn. Magn. Mater.*, **75**, 266–272.

Tumanski S., 2001, *Thin Film Magnetoresistive Sensors*, IOP Publishing, Bristol, U.K.

Tumanski S., 2002a, Which magnetizing circuit is suitable for two-dimensional measurements, in *Proceedings of 2D Magnetic Measurements Workshop*, PTB-Bericht PTB-E-81, pp. 151–157.

Tumanski S., 2002b, A multi-coil sensor for tangential magnetic field investigations, *J. Magn. Magn. Mater.*, **242–245**, 1153–1156.

Tumanski S., 2005, New design of the magnetizing circuit for 2D testing of electrical steel, *Przegl. Elektrotech.*, **5**, 32–34.

Tumanski S., Bakon T., 2001, Measuring system for two-dimensional testing of electrical steel, *J. Magn. Magn. Mater.*, **223**, 315–325.

Tumanski S., Baranowski S., 2007, Single strip tester of magnetic materials with array of magnetic sensors, *Przegl. Elektrotech.*, **83**, 46–49.

Tumanski S., Pluta W., Soinski M., 2004, Analysis of magnetic field distribution in the sample of RSST device, in *Proceedings of Soft Magnetic Materials Conference*, pp. 85–864.

Underhill C.R., 2009, *Solenoids, Electromagnets and Electromagnets Windings*, University of Michigan Library, Ann Arbor, MI.

Vajda P., Della Torre E., 1995, Ferenc Preisach, in memoriam, *IEEE Trans. Magn.*, 31, i–ii.

Velazquez J., 1994, Giant magnetoimpedance in non-magnetostrictive amorphous wires, *Phys. Rev.*, **B50**, 16737–16740.

Wadey W.G., 1956, Magnetic shielding with multiple cylindrical shields, *Rev. Sci. Instrum.*, **27**, 910–916.

Waeckerle T., Alves F., 2006, Aliages magnetiques amorphes, in *Materiaux magnetiques en genie electrique 2*, Kedous-Lebous A. (Ed.), Lavoisier, Chapter 1.

Wakoh S., Yamashita J., 1966, Band structure of ferromagnetic iron self-consistent procedure, *J. Phys. Soc. Jpn.*, **21**, 1712–1726.

Warburg E., 1881, Magnetic investigations, *Ann. Phys.*, **13**, 141–164.

Watazawa K., Sakuraba J., Hata F., Hasebe T., Chong C.K., Yamada Y., Watana-be K., Awaji S., Fukase T., 1996, A cryocooler cooled 6T NbTi superconducting magnet with room temperature bore of 220 mm, *IEEE Trans. Magn.*, **32**, 2594–2597.

Webster W.L., 1925, Magnetostriction in iron crystals, *Proc. R. Soc.*, **109A**, 570–584.

Weiser B, Pfützner H., Anger J., 2000, Relevance of magnetostriction and forces for the generation of audible noise of transformer cores, *IEEE Trans. Magn.*, **36**(5), 3759–3777.

Weiss P., 1907, Hypothesis of the molecular field and ferromagnetic properties, *J. Phys.*, **6**, 661–690.

Wenk H.R., van Houtte P., 2004, Texture and anisotropy, *Rep. Prog. Phys.*, **67**, 1367–1428.

White R.L., 1992, Giant magnetoresistance: A primer, *IEEE Trans. Magn.*, **28**, 2482–2486.

Wilczynski W., Schoppa A., Schneider J., 2004, Influence of the different fabrication steps of magnetic cores on their magnetic properties, *Przegl. Elektrotech.*, 2, 118–122.

Williams H.J., Shockley W., 1949, A simple domain structure in an iron crystal showing a direct correlation with the magnetization, *Phys. Rev.*, **75**, 183–187.

Williams H.J., Shockley W., Kittel C., 1950, Studies of the propagation velocity of a ferromagnetic domain boundary, *Phys. Rev.*, **80**, 1090–1094.

Wills A.P., 1899, On the magnetic shielding effect of trilamellar cylindrical shields, *Phys. Rev.*, **9**, 193–213.

Wolf S.A., Chtchelkanova A.Y., Treger D.M., 2006, Spintronics—A retrospective and perspective, *IBM J. Res. Dev.*, **50**, 101–110.

Woodward J.G., Della Torre E., 1960, Particle interaction in magnetic recording tape, *J. Appl. Phys.*, **31**, 56–62.

Wrobel L.C., Aliabadi M.H., 2002, *The Boundary Element Method*, John Wiley & Sons, New York.

Wrona J., 2002, Magnetometry of the thin film structures, PhD thesis, AGH University of Science and Technology, Cracow, Poland.

Yabumoto M., 2009, Review of techniques for measurement of magnetostriction in electrical steels and progress towards standardization, *Przegl. Elektrotech.*, **1**, 1–6.

Yamaguchi T., Senda K., Ishida M., Sato K, Honda A., Yamamoto T., 1998, Theoretical analysis of localized magnetic flux measurement, *J. Phys. IV*, **8**, 712–720.

Yamazaki K., Muramatsu K., Hirayama M., Haga A., Torita F., 2006, Optimal structure of magnetic and conductive layers of a magnetically shielded room, *IEEE Trans. Magn.*, **42**, 3524–3526.

Yuasa S., Nagahama T., Fukushima A, Suzuki Y., Ando K., 2004, Giant room-temperature magnetoresitance in single-crystal Fe/MgO/Fe magnetic tunel junctions, *Nat. Mater.*, **31**, 868–871.

Zeeman P., 1897, The effect of magnetization on the nature of light emitted by a substance, *Nature*, **55**, 347.

Zemanek I., 2009, Single sheet and on-line testing based on MMF compensation method, *Przegl. Elektrotech.*, **85**(1), 79–83.

Zheng Y., Zhu J.G., 1996, Micromagnetics of spin valve memory cell, *IEEE Trans. Magn.*, **32**, 4237–4239.

Zhu G.J., Lin Z.W., Guo Y.G., Huang Y., 2009, 3D measurement and modelling of magnetic properties of soft magnetic composite, *Przegl. Elektrotech.*, **1**, 11–15.

Zhu J.G., Ramsden V.S., 1993, Two dimensional measurement of magnetic field and core loss using a square specimen tester, *IEEE Trans. Magn.*, **29**, 2995–2997.

Zhu G.J., Zhong J.J., Lin Z.W., Sievert J., 2003, Measurement of magnetic properties under 3D magnetic excitations, *IEEE Trans. Magn.*, **39**, 3429–3431.

Zienkiewicz O.C., Taylor R.L., Zhu J.Z., 2005, *The Finite Element Method. Its Basis and Fundamentals*, Butterworth Heinemann, St. Louis, MO

Zurek S., 2009, Static and dynamic rotational losses in non-oriented electrical steel, *Przegl. Elektrotech.*, **85**, 89–92.

Zurek S., Marketos P., Meydan T., 2004, Control of arbitrary waveform in magnetic measurements by measns of adaptive iterative digital feedback algorithm, *Przegl. Elektrotech.*, **80**, 122–125

Zurek S., Marketos P., Meydan T., Moses A.J., 2005, Use of novel adaptive digital feedback for magnetic measurements under controlled magnetizing conditions, *IEEE Trans. Magn.*, **41**, 4242–4249.

Zurek S., Meydan T., 2006, Rotational power losses of magnetic steel sheets in circular rotational magnetic field in ccw/cw direction, *IEE Proc. Meas. Sci. Technol.*, **153**, 147–157.

Zurek S. Moses A.J., 2005, Adaptive iterative feedback algorithm for measurements of magnetic properties under controlled magnetizing conditions over wide frequency range, *Przegl. Elektrotech.*, **81**, 5–7.

3

Magnetic Materials

3.1 Soft Magnetic Materials: General Information

3.1.1 Properties and Classification

Commonly, ferromagnetic or ferrimagnetic materials are considered as magnetic materials although other materials (diamagnetic and paramagnetic) also exhibit some magnetic properties, as discussed earlier. The magnetic materials can be further classified into two clearly separate categories: soft magnetic materials and hard magnetic materials. Coercivity is assumed as the main criterion, and IEC Standard 404-1 recommends the coercivity of 1000 A/m as a value to distinguish both groups. This border is rather symbolic because both classes are completely different. From soft magnetic materials, we require the coercivity to be as small as possible (usually much less than 100 A/m) while hard magnetic materials should have coercivity as high as possible (commonly above 100,000 A/m). There is also a subclass of hard magnetic materials called semi-hard magnetic materials (with coercivity between 1,000 and 100,000 A/m). Figure 3.1 presents magnetic materials taking into account their coercivity available Vacuumschemlze who is one of the main manufacturers.

Soft magnetic materials cover huge market of various products: about 7×10^6 tons annually and about 10^{10} Euro (Moses 2003). We can divide these products taking onto account their magnetic performance, applications, cost, and other properties. For example, grain-oriented silicon steel is mechanically much harder than the nonoriented, so the same punching die will wear off after producing smaller quantity of elements. Even in the case of SiFe electrical steel, the best grade can be 10 times more expensive than ordinary grades of steel. And between cheap ferrites and high-quality soft magnetic materials, these differences in cost can be much larger.

Therefore, selection of appropriate kind and quality of material for a given application is an important knowledge. For example, the best quality steel after preparation of the product can be much more deteriorated than cheaper material that after the same technology can exhibit better performance (Schneider et al. 1998, Schoppa et al. 2000, Wilczynski et al. 2004). Figure 3.2 presents a comparison of the main parameters of typical soft magnetic materials including their cost.

It would be nice to be able to find the soft magnetic material with all excellent properties (high saturation polarization, small losses, small coercivity, small magnetostriction, good mechanical properties, etc.) even at much higher price. But such material simply does not exist. We have to accept always some compromises—high permeability at the cost of saturation polarization (Figure 3.3), small power loss at the cost of saturation polarization, better magnetic parameters at the cost of mechanical properties, etc. Fortunately, there is a plethora of various magnetic materials and appropriate technology often helps to find desirable material (Fish 1990, Moses 1990, 1992, 2003, Pfützner 1992, Arai and Ishiyama 1994, McCurrie 1994, Kronmüller 1995, Stodolny 1995, Fiorillo 1996, Schneider et al. 1998, Goldman 1999, O'Handley 2000, Beckley 2000, 2002, Geoffroy and Porteseil 2005, Peuzin 2005, Degauque et al. 2006, De Wulf 2006, Lebourgeois and Guyot 2006, Waecklerle 2006, Waeckerle and Alves 2006a,b, Kazimierczuk 2009).

Taking into account the main applications of soft magnetic materials, it should be noted that this situation continues to change and develop. For example, it was traditionally assumed that the main area of application of silicon steel is electric power industry. But recently, more and more power electric and power electronics devices use higher frequency signals, up to MHz. In high frequency range, electrical steel exhibits prohibitively high power loss and should be substituted by nanocrystalline and even ferrite materials (Figure 3.4). Consequently, in such applications, other accompanying devices, for example, measuring transformers, should be also made from high-frequency materials. In turn, the progress in nanocrystalline/amorphous materials resulted in development of new classical electrical steel (e.g., thinner gauge of even 0.15 mm).

Taking into account the importance of various groups of soft magnetic materials, it should be noted that almost 80% of the market is occupied by SiFe electrical steel (Figure 3.5). With ferrites and permalloys (NIFe), it is more than 95% and we can see that other materials, including amorphous and nanocrystalline are marginal in value.

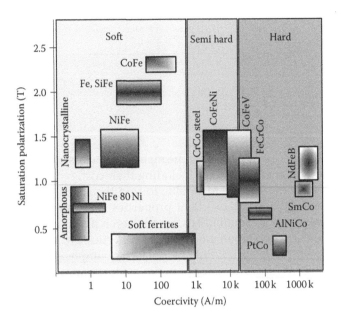

FIGURE 3.1
Ranges of commercially available magnetic materials (as an example of products offered by Vacuumschmelze).

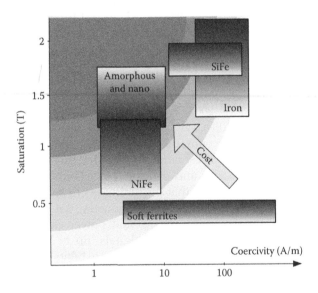

FIGURE 3.2
Comparison of the coercivity, saturation, and cost of typical soft magnetic materials.

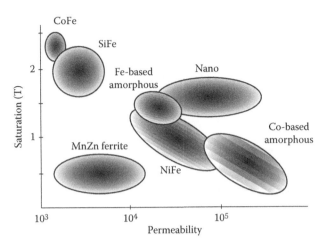

FIGURE 3.3
Comparison of the permeability and coercivity of the typical soft magnetic materials. (After Moses, A.J., *Przegl. Elektr.*, 79, 457, 2003.)

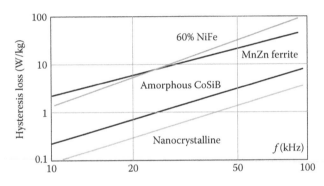

FIGURE 3.4
Hysteresis power loss versus frequency of high-frequency materials. (From Kolano, R. and Kolano-Burian, A., *Przegl. Elektr.*, 78, 241, 2002.)

Depending on application, various properties are required. In the case of electric power devices (power and distribution transformers, electric machines), the most important factors are low power loss and high saturation polarization. If we would like to choose only between silicon steel and amorphous materials (neglecting other factors), we arrive at a contradiction—amorphous materials exhibit smaller power loss but also significantly smaller saturation polarization and vice versa. Table 3.1 presents the comparison of parameters for the main soft magnetic materials.

If a material is used for magnetic shielding, then its losses are not as important as the permeability, and hence amorphous materials or permalloy is advisable. In the case of high-frequency applications, apart from the losses, deterioration of magnetic properties (e.g., permeability) with frequency is important, so from Table 3.1 we can see that, in this case, the materials would be ordered as follows: SiFe, NiFe, amorphous/nanocrystalline, MnZn ferrite, NiZn ferrite (and in microwave range, garnets).

Especially important are the CoFe alloys because they exhibit high saturation polarization with the highest known value of 2.46 T. Table 3.2 presents the typical applications of soft magnetic materials.

Figure 3.6 presents a diversity of soft magnetic materials currently available commercially. The properties of such materials will discussed in more detail in the following sections.

3.1.2 Pure Iron

Pure iron has excellent magnetic properties: large saturation polarization $J_S = 2.15\,T$, low coercivity

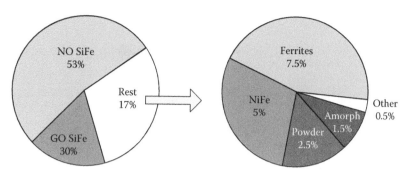

FIGURE 3.5
Annual value of world production of soft magnetic materials. (After Schneider, J. et al., *J. Phys.*, 8, Pr2-755, 1998.)

TABLE 3.1

Comparison of Parameters for the Main Soft Magnetic Materials

Parameter	3% SiFe GO	FeSiB Metglas	Ni80Fe20 Permalloy	Co50Fe50 Permendur	MnZn Ferrite
B_s (T)	2.03	1.56	0.82	2.46	0.2–0.5
H_c (A/m)	4–15	0.5–2	0.4–2	160	20–80
$P1.5\,T/50\,Hz$ (W/kg)	0.83	0.27		1	
$P1\,T/1\,kHz$ (W/kg)	20	5	10	20	
$\mu_{max} \times 1000$	20–80	100–500	100–1,000	2–6	3–6
Frequency range (kHz)	3	250	20	up to 1 kHz	2,000 NiZn—100,000

TABLE 3.2

Typical Applications of the Main Soft Magnetic Materials

Application	Electrical Steel	Fe-Based Amorphous	Powder	CoFe	Ferrite
Power transformers					
Distribution transformers					
Lamp ballasts					
Induction motors					
Generators					
Reactors					
Other motors					
Special transformers					
Chokes					
Power electronics					
Instrumentation					
Pulsed power					
Shielding					

Source: After Moses, A.J., *Przegl. Elektr.*, 80, 1181, 2004. With permission.

$H_c = 3$–$12\,A/m$, and high permeability $\mu_{max} = 280,000$ (single crystal magnetically annealed even up to 1,400,000). But the main problem is that such performance is displayed only by pure iron: even small quantities of impurities cause significant deterioration of magnetic properties (Figure 3.7). In practice, such extremely pure material is expensive and possible to use only in laboratory.

Commercially available pure iron has much smaller permeability $\mu_{max} = 10,000$–$20,000$ and larger coercivity $H_c = 20$–$100\,A/m$ because impurities such as C, Mn, P, S, N, and O impede the domain wall motion. By annealing such material in hydrogen at 1200°C–1500°C, it is possible to remove most of these impurities but such process is also quite expensive.

Pure iron has low resistivity $\rho = 10\,\mu\Omega$ cm (in comparison with $45\,\mu\Omega$ cm of GO SiFe and $140\,\mu\Omega$ cm of amorphous material). Such good conductivity causes large eddy current loss and practically precludes pure iron from AC application.

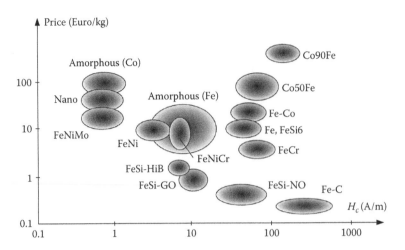

FIGURE 3.6
Diversity of soft magnetic materials. (After Waecklerle, T., Materiaux magnetiques doux speciaux et applications, in *Materiaux magnetiques en genie electrique I*, Kedous-Lebouc, A. (Ed.), Lavoisier, Chapter 3, pp. 153–223, 2006.)

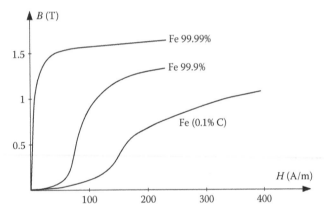

FIGURE 3.7
Magnetization curves of iron. (After McCurrie, R.A., *Ferromagnetic Materials*, Academic Press, London, U.K., 1994.)

Both problems—expensive manufacturing and limited frequency application—can be solved if we use the same material in form of a powder iron. This material is manufactured by grinding iron (or iron alloys*) into powder with dimension of particles 5–200 μm and next by pressing this powder with insulating material. Resistivity of such material is of the order of $\rho = 10^4\ \mu\Omega$ cm and therefore it can be used in high-frequency range to 100 kHz (specially prepared NiFe powder to 100 MHz) (Kazimierczuk 2009).

Instead of cheap pressing, most often sintering technology is used, which results in better magnetic performances of the powder materials (Table 3.3).

As a material for powder iron cores, most commonly, carbonyl iron is used. Technology of obtaining extra pure iron powder from iron pentacarbonyl, Fe(CO)$_5$, was developed in 1925 by BASF Company. By thermal decomposition of iron pentacarbonyl, it is possible to produce 99.8% pure iron powder with spherical particles ranging from 1 to 8 μm.

Although the presence of carbon significantly deteriorate magnetic properties of iron (Figures 3.6 and 3.7), low-carbon steel is widely used as magnetic material mainly due to its low price (Figure 3.6) and good mechanical properties. As "low-carbon" steel it is assumed the material with following: C, 0.04%–0.06%; P, 0.05%–0.15%; Mn, 0.35%–0.8%; S, 0.006%–0.025%; and Si, 0.05%–0.25%. Although the magnetic properties of low-carbon electrical steel are rather poor, they are acceptable for many cheap devices, like small motors, relays, or electromechanical mechanisms. Figure 3.9 presents the part of phase diagram of Fe-C alloys.

Iron exists in two allotropic forms: α-Fe (ferrite Fe-C) ferromagnetic body-centered cubic and γ-Fe (austenite Fe-C) paramagnetic face-centered cubic. Above 0.008% of C in ferrite appears as impurity cementite (iron carbide, Fe$_3$C) that above 210°C is nonmagnetic. Transition between α-Fe and γ-Fe is at 910°C,[†] but also ferrite is paramagnetic above Curie temperature 768°C.

TABLE 3.3

DC Magnetic Properties of Powder Material

	Pressed	Sintered
B at 8000 A/m	1.65 T	1.75 T
B_r from 8000 A/m	0.34 T	0.93 T
H_c from 8000 A/m	192 A/m	80 A/m
μ_{max}	800	7000

Source: Bularzik, J.H. et al., *J. Phys.*, 8, Pr2-747, 1998.

---

* In powder materials also other substances, like NiFe or Sendust are used.

† Above 1538°C iron again is ferromagnetic in body-centered cubic structure known as δ-Fe.

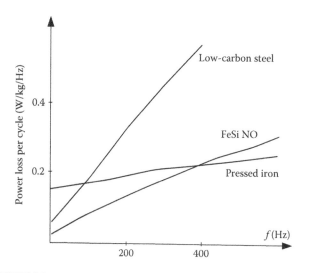

FIGURE 3.8
Losses of various iron-based materials. (After Bularzik, J.H. et al., *J. Phys.*, 8, Pr2-747, 1998.)

Especially important are Fe-Co-based alloys Fe50Co50 (known as Permendur) that exhibit the largest possible saturation polarization $J_S = 2.46\,T$ and very high Curie temperature ($T_c = 930°C$). To improve mechanical properties of FeCo alloy (and increase resistivity $\rho = 40\,\mu\Omega\,cm$), a small part of vanadium is added: Fe49Co49V2. The alloy Fe6Co94 has very high Curie temperature ($T_c = 950°C$) (pure cobalt has $T_c = 1130°C$). Table 3.4 collects magnetic properties of iron and some of its alloys.

3.2 Silicon Iron Electrical Steel

3.2.1 Conventional Grain-Oriented SiFe Steel

As presented in Figure 2.25, two inventions were ground breaking in history of improvement of electrical steels:

addition of Si and Goss texture. The first invention was proposed by Robert Hadfield in 1902 (Barret et al. 1902) and patented in 1903 (Hadfield 1903). The second invention was patented by Norman Goss in 1934 (Goss 1934) and described in 1935 (Goss 1935).

As discussed earlier, one of the most important drawbacks of pure iron is its relatively low resistivity and hence large eddy current loss. Figure 3.10 presents the resistivity of different iron alloys. We can see that good candidates for resistivity improvement are silicon and aluminum. Addition of silicon influences also saturation polarization and Curie temperature (both of which decrease with silicon content; see Figure 3.11).

From Figure 3.11, the best would be 6.5% content of Si because resistivity increases almost sevenfold and the material is non-magnetostrictive. Unfortunately, this material is very hard and brittle, what is disadvantageous in rolling process as well as in punching the final product. In practice, punching can be carried out only for the steel with up to 3%–4% silicon content. Large content of Si causes also decrease of saturation polarization as well as the permeability. Therefore, the GO silicon steel is manufactured mostly often with 2.7%–3.3% of silicon although also 6.5% SiFe is offered in the market (in small volume and for higher price).

Figure 3.12 presents part of a phase diagram of iron–silicon. The transition between α-Fe and γ-Fe is at 911°C. For silicon content higher than 1.86%, this transition no longer takes place and it is possible to anneal the material to high temperatures for removal of parasitic impurities. The most unwanted components in SiFe steel are carbon, oxide, sulfur, and nitrogen because even small amounts of this element cause increase of hysteresis loss. Therefore, the starting material should be as pure as possible and after manufacturing, the content of these elements can be smaller than 10 ppm.

Figure 3.13 presents typical route of production of GO SiFe. The Goss invention is a method of developing of a grain texture. By suitable combination of annealing and

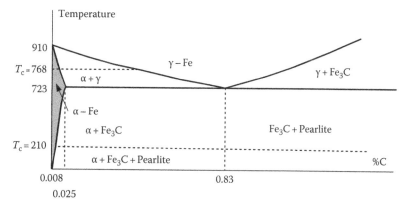

FIGURE 3.9
Iron–carbon phase diagram.

TABLE 3.4

Performances of Iron and Iron-Alloy Materials

	H_c (A/m)	μ_{max} ×1000	J_1 (T)
99.95 Iron	4	230	2.15
Iron (commercially)	20–100	4–20	2.15
Carbonyl (powder)	6	20	2.15
CoFe2%V (Permendur)	200	3	2.4
Low-carbon steel C, 0.04%–0.06%	Power loss 5.5–10 W/kg at 1.5 T, 50 Hz		

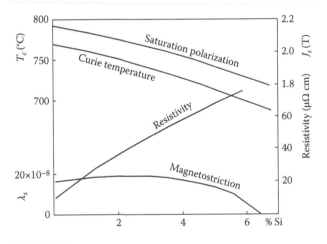

FIGURE 3.10
Resistivity of different iron alloys.

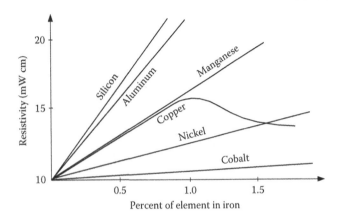

FIGURE 3.11
Magnetic and electrical parameters as a function of silicon content.

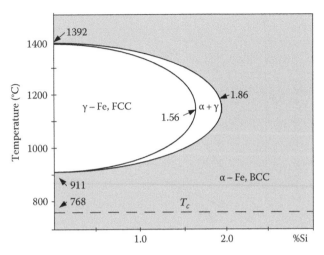

FIGURE 3.12
Iron–silicon phase diagram.

cold rolling, the grains having [001] direction in the rolling direction and (110) plane close to the sheet plane are privileged to grow. In the meantime, the inhibitor manganese sulfide (MnS) suppresses the growth of other grains. Figure 3.14 presents the example of grain structure of GO electrical steel.

In a grain-oriented steel, we profit from its anisotropic properties of the fact that the iron crystal have the best magnetic properties in "easy" [100]

direction (Figure 2.14). Therefore, the main effort is made to obtain the best Goss texture with relatively large grains ordered in one direction. Figure 3.15 presents dependence of flux density and loss on the tilt angle. Surprisingly, the minimum of the loss occurs when the grain are slightly misoriented from the perfect direction. The best results are obtained for the grain orientation of about 2°.

The data from Figure 3.15 can be partially explained by results of domain investigations presented in Figures 3.16 and 3.17. Perfectly oriented grains (0°) have wider spacing of the 180° domain walls than for a tilt angle 2°. This domain width strongly influences excess loss (see Figure 2.118). The domain wall spacing can be significantly decreased by applying a stress, which is one of the methods known as a domain refinement.

After the first annealing in production process (Figure 3.13), the grain dimensions are only around 0.02 mm. After second annealing (secondary recrystallization), the Goss-oriented grains grow through the thickness of the sheet to diameter 3–7 mm with average misorientation of around 6°.

Theoretically, as larger grains as better, but measurements of loss versus grain dimensions did not confirm such a simple relationship. Indeed domain observations confirm that large grains have a wide domain spacing (Beckley 2000). This is why usually grain diameter in conventional GO steel does not exceed about 8 mm.

Excellent properties along rolling direction are advantageous when we can guarantee that magnetization is applied only in this direction. But this advantage can be a problem when a part is magnetized not exactly in the rolling direction (e.g., corners of a square core). In such a case, we have to expect significant deterioration of the material performance. Figure 3.18 presents the

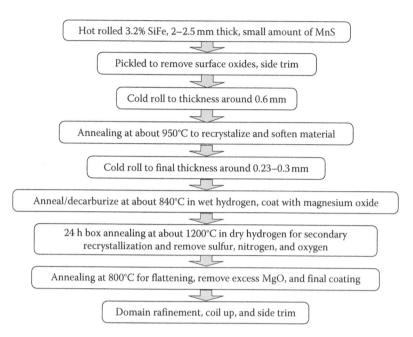

FIGURE 3.13
Production route of grain-oriented silicon iron. (From Moses, A.J., *IEE Proc.*, 137, 233, 1990.)

FIGURE 3.14
The example of typical grain structure of GO steel: in the left part, an incompletely recrystallized line is visible.

magnetization curve and losses determined for various directions of magnetization (Tumanski 2002).

Electrical GO steel is classified according to international standards based on the power loss. European standard EN 10107 uses the following nomenclature for the steel grades:

a. First letter M for electrical steel.

b. Three digits after the first letter denote value of specific loss measured at 1.5 or 1.7 T.

c. Two further digits represent the thickness.

d. Last letter describes type of material: N, normal (loss measured at 1.5 T), S, reduced loss (loss at 1.7 T); P, high permeability (loss at 1.7 T).

For example, M097-30N means electrical steel (M) of normal grade (N) with material thickness 0.3 mm (30) and power loss at 1.5 T notexceeding 0.97 W/kg.* Table 3.5 presents examples of GO electrical steel according to EN Standard 10107.

Figure 3.19 presents parameters of a typical GO SiFe steel grade M089–27N and Table 3.6 presents an example of the measured results for a similar steel sample (as measured by Epstein method).

3.2.2 HiB Grain-Oriented Electrical Steel

The conventional GO SiFe steel has Goss texture [001] (110) with grain orientation dispersion (tilt angle) of about 6°. In 1965, Nippon Steel Corporation developed new technology for production of improved GO SiFe steel (Taguchi and Sakakura 1964, 1969, Yamamoto et al. 1972). After addition of around 0.025% aluminum to the starting melt, the recrystallization process was

---

* Unfortunately different countries often use different standards: ASTM A876M (American Society for Testing and Materials), JIS C2553 (Japanese Industrial Standard), AISI classification (American Iron and Steel Institute), IEC 60404-8-7 (International Electrotechnical Commission). But all standards of classification use the same parameters: power loss, thickness, and type of material. For example, *Japanese Standard* classify GO steel as: Z, normal; ZH, HiB steel; ZDKH, with laser domain refinement. For example, 30ZH100 means HiB steel of thickness 0.3 mm and power loss at 1.7 T/50 Hz not exceeding 1 W/kg. According to American Standard 30P154M means steel of thickness 0.3 mm and power loss at 1.7 T /60 Hz not exceeding 1.54 W/kg (or according to Standard ASTM A876 it is equivalent to 30P070 what means thickness 0.3 mm and power loss 1.7 T/60 Hz not exceeding 0.7 W/ lb). Moreover, there are many different Company names, as ORSI (Thyssen), ORIENTCORE (Nippon Steel), etc. which can use their own classification.

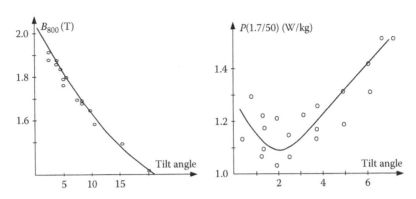

FIGURE 3.15
Flux density and core loss versus the tilt angle. (After Littmann, M.F., *J. Appl. Phys.*, 38, 1104, 1967; Littmann, M.F., *IEEE Trans. Magn.*, 7, 48, 1971; Littman, M.F., *J. Magn. Magn. Mater.*, 26, 1, 1982; Nozawa, T. et al., *IEEE Trans. Magn.*, 14, 252, 1978.)

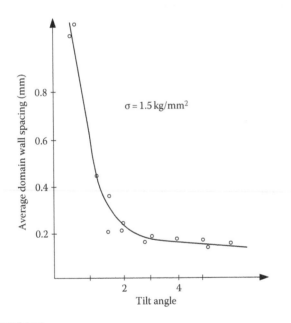

FIGURE 3.16
Dependence of average domain wall spacing on the tilt angle. (After Nozawa, T. et al., *IEEE Trans. Magn.*, 14, 252, 1978.)

enhanced due to aluminum nitride AlN acting as inhibitor. The production route was simplified; hot rolled material was initially annealed at 1100°C in N_2 and in one cycle cold rolled to final thickness. After that conventional procedure was performed, decarburization, batch annealing for recrystallization at 1200°C, and final annealing at 800°C are carried out.

As a result, a new class of material with more perfect texture was obtained: a tilt angle was on average 2°–3° and grain dimensions exceeding 10 mm. This material

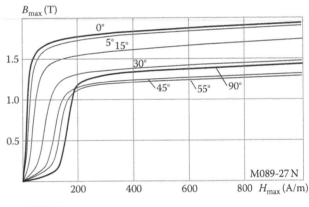

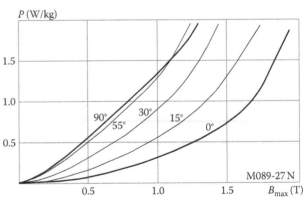

FIGURE 3.18
Properties of the GO SiFe steel in different directions of magnetization (with respect to the rolling direction). (From Tumanski, S., 2002.)

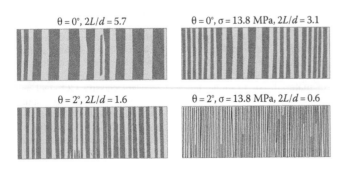

FIGURE 3.17
Domain structure of 3% FeSi crystals (θ, tilt angle; $2L$, domain wall spacing; d, thickness). (After Shilling, J.W. et al., *IEEE Trans. Magn.*, 14, 104, 1978; 1978a.)

TABLE 3.5

Classification of Electrical Steel according to Standard EN 10107

Name	Thickness (mm)	Loss at 1.5 T, 50 Hz (W/kg)	Loss at 1.7 T, 50 Hz (W/kg)	Polarization for 800 A/m (T)
Normal material				
M080-23N	0.23	0.80	1.27	1.75
M089-27N	0.27	0.89	1.40	1.75
M097-30N	0.30	0.97	1.50	1.75
M111-35N	0.35	1.11	1.65	1.75
Material with reduced loss				
M120-23S	0.23	0.77	1.20	1.78
M130-27S	0.27	0.85	1.30	1.78
M140-30S	0.30	0.92	1.40	1.78
M150-35S	0.35	1.05	1.50	1.78
Material with high permeability (HiB)				
M100-23P	0.23		1.00	1.85
M103-27P	0.27		1.03	1.88
M105-30P	0.30		1.05	1.88
M111-30P	0.30		1.11	1.88
M117-30P	0.30		1.17	1.85

TABLE 3.6

Example of the Measured Results for a Typical GO SiFe Steel—Grade: M089-27N

J_{max} (T)	B_r (T)	H_{max} (A/m)	H_c (A/m)	P (W/kg)	μ
0.1	0.048	4.2	2		18,800
0.2	0.136	6.7	4	0.016	23,600
0.3	0.189	8.9	6	0.034	26,700
0.4	0.262	10.7	8	0.060	29,500
0.5	0.343	12.4	9	0.091	31,900
0.6	0.433	14.1	11	0.129	33,700
0.7	0.530	15.6	12	0.174	35,700
0.8	0.629	17.0	13	0.224	37,400
0.9	0.735	18.4	15	0.281	38,900
1.0	0.846	19.6	16	0.345	40,500
1.1	0.959	21.3	17	0.417	41,000
1.2	1.076	23.0	18	0.495	41,400
1.3	1.180	25.8	19	0.582	40,000
1.4	1.302	31.1	20	0.684	35,800
1.5	**1.407**	**40.4**	**21**	**0.802**	**29,500**
1.6	1.511	62.0	22	0.952	20,500
1.7	**1.620**	**132.8**	**24**	**1.189**	**10,200**
1.8	1.685	455.8	27	1.590	3,100
1.9	1.726	1767.3	33	2.026	900

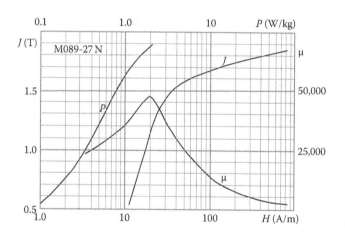

FIGURE 3.19
Parameters of the typical GO SiFe steel-grade: M089-27N.

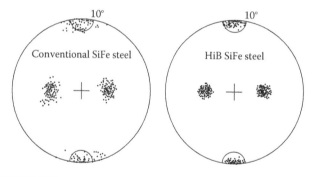

FIGURE 3.20
Pole figures for the (100) plane. (From Yamamoto, T. et al., *IEEE Trans. Magn.*, 8, 677, 1972.)

exhibits significantly lower losses at higher polarization (above 1.7 T), which in turn could be achieved at significantly lower magnetic field strength. This steel is known as high-permeability material (HiB). Figure 3.20 presents comparison of pole figures determined for HiB and conventional steel.

Similar new technology was introduced in 1973 by Kawasaki Steel with inhibitor MnSe + Sb and in 1975 by Allogheny Ludlum Steel Corporation used boron as inhibitor. The Kawasaki technology has two cycles of cold rolling (Goto et al. 1975, Fiedler 1977).

It should be noted that the better performance of the HiB steel is especially evident for high polarization (as the name suggests). For small polarization, the HiB steel does not really offer any advantages; sometimes this material can be comparable or even worse than a conventional steel. Table 3.7 presents an example of the measurement results for HiB steel sample (by Epstein method). Figure 3.21 presents the typical parameters of HiB steel.

As discussed above in the previous section (see Figures 3.15 through 3.17), the effect of perfect grain orientation can be weakened by wide domain wall spacing. Therefore, the HiB steel is often produced with the support of the domain refinement tools: laser scratching, plasma jest irradiation, spark ablation, groove making, chemical treatment, or coating stress (Nozawa et al. 1978, 1979, 1996, Fukuda et al. 1981, Iuchi et al. 1982,

TABLE 3.7

Example of the Results of Measurements for a Typical HiB
GO SiFe Steel—Grade: M100-23P

J_{max} (T)	B_r (T)	H_{max} (A/m)	H_c (A/m)	P (W/kg)	μ
0.1	0.054	2.9	1.9		27,300
0.2	0.142	5.3	3.7	0.014	29,800
0.3	0.210	7.2	5.1	0.031	33 200
0.4	0.287	8.9	6.6	0.054	35,600
0.5	0.378	10.6	8.0	0.083	37,500
0.6	0.479	11.7	9.4	0.117	40,700
0.7	0.572	13.2	10.9	0.157	42,200
0.8	0.676	14.5	12.3	0.203	43,900
0.9	0.774	15.5	13.6	0.256	46,000
1.0	0.885	17.0	14.9	0.314	46,800
1.1	0.999	18.4	15.9	0.379	47,700
1.2	1.069	18.5	18.0	0.467	51,700
1.3	1.205	20.8	18.2	0.534	49,800
1.4	1.302	23.4	18.80	0.622	47,700
1.5	**1.418**	**28.1**	**19.8**	**0.720**	**42,500**
1.6	1.527	38.4	20.6	0.839	33,200
1.7	**1.622**	**64.6**	**21.8**	**0.999**	**20,900**
1.8	1.732	185.5	24.2	1.301	7,700
1.9	1.783	1155.0	30.3	1.827	1,300

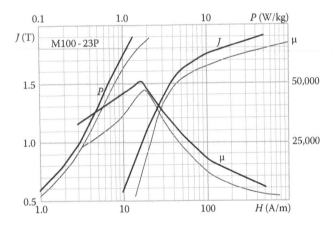

FIGURE 3.21
Parameters of the HiB GO SiFe steel-grade: M100-23P (thin curves; the same parameters of conventional steel for comparison).

Ichijima et al. 1984, Krause et al. 1984, Beckley et al. 1985, Takahashi et al. 1986, Yabumoto et al. 1987, Sato et al. 1998). Recently, the most commonly used technique is laser scratching (sometimes supported by stress coating or chemical etching).

Figure 3.22 presents the effect of domain refinement on losses. It is clearly visible that the most successful domain refinement occurs when the sample has grain orientation close to perfect. This corresponds with results of domain investigation presented in Figures 3.16 and 3.17. Therefore, domain refinement usually is used in the case of HiB steel.

Usually, the high-power laser beam focused into a 150 μm spot scribes a line transversely to the rolling direction with scratch distance of about 5 mm. Figure 3.23 presents the example of result of laser scratching technique. After application of domain refinement, it is possible to obtain the core loss lower by 5%–10% than the untreated high-permeability steel. It can be also deduced from Figure 3.23 that the domain refinement practically does not change the static hysteresis loss. Because after laser scribing the dependence $P/f = f(f)$ is more linear (thin straight lines shown for comparison), this means that the domain refinement mainly decreases excess loss (related to domain wall).

3.2.3 SiFe Non-Oriented Electrical Steel

As presented in Figure 3.18, grain-oriented SiFe steel is strongly anisotropic so magnetic properties in direction different than the rolling are poor. Thus if we cannot guarantee that direction of magnetic flux is the same as the rolling direction (e.g., in rotating machines), we can expect worst performance of the designed device as compared with the raw material. In such a case, the non-oriented material is more advisable.

Figure 3.24 presents the magnetization curves of typical NO steel determined for various angles of magnetization. It should be noted that this material is not purely isotropic but in comparison with grain-oriented steel, the change of properties with the change of direction of magnetization are acceptably small. That is why in rotating machines the NO material is much more often used than the grain-oriented steel (see Figure 3.5).

Figure 3.25 presents a collection of the main properties of typical non-oriented SiFe steel. For comparison on the same scale, there are presented the same properties of conventional grain-oriented (CGO) steel. We can see that generally the NO steel exhibits lower quality in comparison with the CGO. But in many devices, such poor properties are acceptable mainly taking into account economic point of view: NO steel is much cheaper. Moreover, because this SiFe steel contains less silicon (0%–3%), it is more ductile, which means that more parts can be punched with the same stamping tool.

NO steel is delivered in one of the two possible forms: fully finished or semifinished. In the case of semifinished steel, the customer has to anneal the material after stamping. Semifinished steel is only a partly decarburized to obtain better punchability (final decarburization is obtained in final annealing). Because it is not necessary to perform 24 h annealing for recrystallization and grain growth, the production process is much simpler (faster and therefore cheaper) (Figure 3.26).

During annealing, care should be taken to not allow the extra grain growth. For this reason, the duration

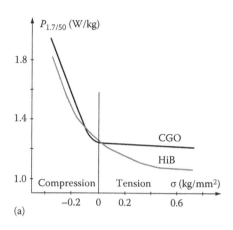
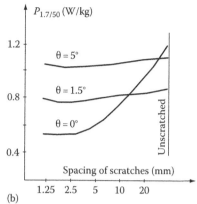

FIGURE 3.22

Influence of stress or laser scratching on losses of conventional (CGO) and HiB steel (a) or single crystal sample ($t = 0.2$ mm) of different orientation (b). (After Yamamoto, T. et al., *IEEE Trans. Magn.*, 8, 677, 1972; Nozawa T. et al., *IEEE Trans. Magn.*, 15, 972, 1979.)

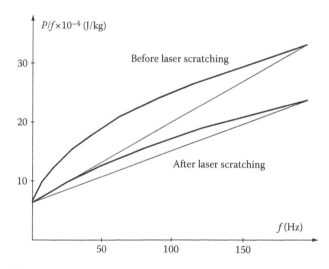

FIGURE 3.23

Core loss per cycle before and after laser scribing (0.23 mm GO 3% SiFe). (After Nozawa, T. et al., *IEEE Trans. Magn.*, 32, 572, 1996.)

of annealing and of cooling should be precisely controlled. The optimal grain size is around 100–200 μm (Figure 3.27).

In NO SiFe, the presence of impurities is directly related to the hysteresis loss, which constitutes almost 75% of the total loss. Hence it is a crucial for magnetic performances is presence of impurities (Figure 3.28). Therefore, the starting material and technology of removing carbon (decarburization), oxide, sulfur, and nitride are very important. In modern NO steel, these impurities do not exceed 10 ppm. NO SiFe steel is manufactured with different contents of silicon (Table 3.8).

The influence of Si on steel parameters is described by following empirical rules:

$$K_1 = 5.2 - 0.5\, \text{Si}\%\, [10^4\, \text{J/m}^3]$$

$$\rho = 12 + 11\, \text{Si}\%\, [\mu \text{Wcm}] \qquad (3.1)$$

$$J_s = 2.16 - 0.05\, \text{Si}\%\, [\text{T}]$$

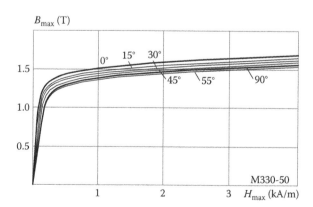

FIGURE 3.24

Magnetization curves of typical non-oriented steel determined for various angles of magnetization. (From Tumanski S., 2002.)

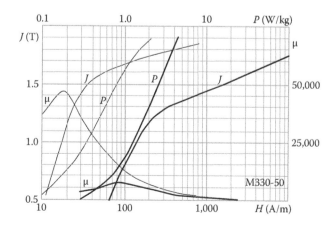

FIGURE 3.25

Properties of typical NO steel (thin curves; the same properties of typical grain-oriented steel for comparison).

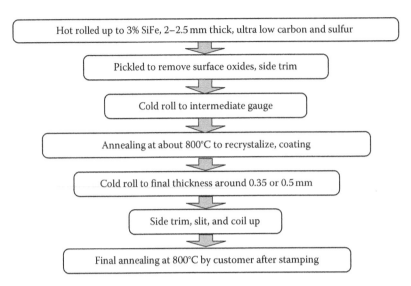

FIGURE 3.26
Production route of NO silicon iron (in fully processed steel only one cold roll to final gage is used). (From Beckley, P., *Electrical Steels, European Electrical Steel*, Orb Works, New Port, U.K., 2000.)

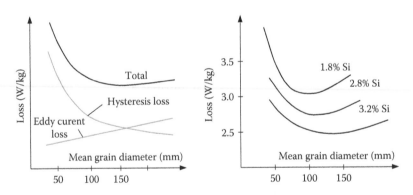

FIGURE 3.27
Influence of mean grain diameter on power loss. (After Matsumara, K. and Fukuda, B., *IEEE Trans. Magn.*, 20, 1533, 1984; Shimanaka, H. et al., *J. Magn. Magn. Mater.*, 26, 57, 1982.)

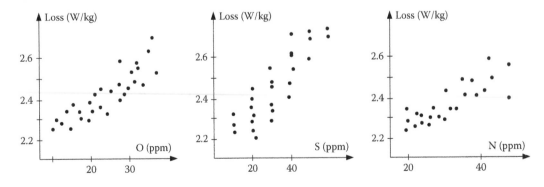

FIGURE 3.28
Effect of impurities on total loss. (After Brissonneau, P., *J. Magn. Magn. Mater.*, 41, 38, 1984; Shimanaka, H. et al., *Trans. Met. Soc. AIME*, 193, 1980.)

TABLE 3.8

Classification of NO SiFe Steel

	% Si	Thickness (mm)	Loss$_{1.5}$ (W/kg)
Top grades	2.5–3.2	0.35–0.5	2.2–2.5
Medium grades	1.5–2.5	0.35–0.5	4–6
Low grades	0.5–1.5	0.5–0.65	7–9
Non-Si		0.65–1	8–12

Source: Brissonneau, P., *J. Magn. Magn. Mater.*, 19, 52, 1980.

NO SiFe steel is classified by international standards in a similar way as the GO SiFe steel, mainly according to the power loss.* Table 3.9 presents the classification according to European Standard EN 10106.

Standards determine maximum anisotropy of loss described as

$$T = \frac{P_1 - P_2}{P_1 + P_2} \times 100 \qquad (3.2)$$

where

P_1 is the loss in a sample cut perpendicular to the rolling direction

P_2 is the loss in a sample cut parallel to the rolling direction

Thus, the NO SiFe steel should be tested using two sets of samples. Table 3.10 presents an example of the results of measurement for a typical NO SiFe steel.

NO SiFe steel is usually delivered in coated form. Various coating materials, organic or inorganic, are used. Coating plays important role because it is not only the insulating layer but also protects against oxidation, aids in punchability, and in certain cases it can introduce tensile stress to improve the quality. Comprehensive review on this subject is presented in a review article of Coombs et al. (2001).

3.2.4 Unconventional Iron-Based Alloys

It can be concluded from data presented in Figure 3.11 that by increasing the contents of Si in the SiFe steel, we can obtain attractive material in comparison with 3% SiFe. Although saturation polarization decreases with silicon content, the 6.5% SiFe exhibits almost two times higher resistivity, which makes it a good candidate for medium-frequency applications (due to significantly smaller eddy current losses). Moreover, the 6.5% SiFe exhibits very small magnetostriction.

The main problem is that steel with this amount of silicon is very hard and brittle and it is not possible to manufacture it by the conventional cold rolling. Therefore, several other manufacturing technologies of

---

* See comments on p. 123.

TABLE 3.9

Classification of Fully Processed (Letter A) Non-Oriented Electrical Steel according to Standard EN 10106

Name	Thickness (mm)	Loss at 1.5 T, 50 Hz (W/kg)	Polarization for 5000 A/m (T)	Anizotropy of Loss (%)
M235-35A	0.35	2.35		
M250-35A		2.50		
M270-35A		2.70	1.60	17
M300-35A		3.00		
M330-35A		3.30		
M250-50A	0.50	2.50	1.60	17
M270-50A		2.70	1.60	17
M290-50A		2.90	1.60	17
M310-50A		3.10	1.60	14
M330-50A		3.30	1.60	14
M350-50A		3.50	1.60	12
M400-50A		4.00	1.63	12
M470-50A		4.70	1.64	10
M530-50A		5.30	1.65	10
M600-50A		6.00	1.66	10
M700-50A		7.00	1.69	10
M800-50A		8.00	1.70	10
M940-50A		9.40	1.72	8
M310-65A	0.65	3.10	1.60	15
M330-65A		3.30	1.60	15
M350-65A		3.50	1.60	14
M400-65A		4.00	1.62	14
M470-65A		4.70	1.63	12
M530-65A		5.30	1.64	12
M600-65A		6.00	1.66	10
M700-65A		7.00	1.67	10
M800-65A		8.00	1.70	10
M1000-65A		10.00	1.71	10
M600-100A		6.00	1.63	10
M700-100A		7.00	1.64	8
M800-100A	1.00	8.00	1.66	6
M1000-100A		10.00	1.68	6
M1300-100A		13.00	1.70	6

this material are proposed: by powder rolling processing (Yuan et al. 2008), by rapid solidification (Fiorillo 2004, Bolfarini et al. 2008), or chemical vapor deposition (CVD) (Crottier-Combe et al. 1996, Haiji et al. 1996).

The rapid solidification technology is used to avoid formation of FeSi (B$_2$) and Fe$_3$Si (DO$_3$) phases because they make the material brittle. The molten metal stream is ejected onto a rotating metallic drum. Thin (30–100 μm) ribbon is obtained as result with acceptable ductility, coercivity lower than 10 A/m, and maximum permeability higher than 10,000. The average grain diameter is 5–10 μm and therefore sometimes this material is called as a *microcrystalline material*.

The magnetic properties can be improved by annealing and recrystallization to grain with

TABLE 3.10

Example of the Results of Measurements for Typical NO
SiFe Steel—Grade: M400-50AP

J_{max} (T)	B_r (T)	H_{max} (A/m)	H_c (A/m)	P (W/kg)	μ
0.1	0.039	35.24	14.2	0.026	2280
0.2	0.106	48.00	27.2	0.102	3310
0.3	0.188	56.55	36.3	0.210	4220
0.4	0.275	64.00	43.8	0.345	4980
0.5	0.360	70.34	49.1	0.495	5650
0.6	0.443	77.27	53.8	0.664	6180
0.7	0.529	84.35	57.7	0.850	6600
0.8	0.617	92.78	61.3	1.055	6860
0.9	0.709	103.23	65.7	1.279	6940
1.0	0.790	116.80	69.4	1.530	6820
1.1	0.878	134.64	72.0	1.810	6500
1.2	0.962	165.32	75.9	2.122	5780
1.3	1.043	224.16	80.3	2.479	4610
1.4	1.123	363.38	83.6	2.910	3070
1.5	**1.157**	**782.52**	**86.7**	**3.427**	**1530**
1.6	1.210	1990.36	91.5	4.002	640
1.7	**1.230**	**4413.73**	**98.2**	**4.485**	**310**

average diameter of around 300 μm (Roy et al. 2009).
But unfortunately annealing leads to poor mechanical
properties.

The main advantage of microcrystalline 6.5% SiFe is
much wider frequency bandwidth in comparison with
conventional SiFe (Figure 3.29). From Figure 3.29, it is
clear that 6.5% SiFe is superior in comparison with 3.2%
SiFe above around 100 Hz.

More convenient for large volume manufacturing is
CVD technology. In this technology, conventional 3%

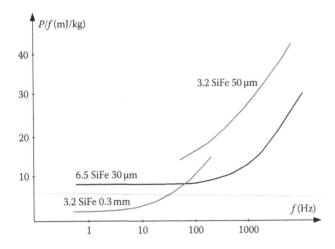

FIGURE 3.29
Core loss per cycle of microcrystalline 6.5% SiFe (and comparison
with conventional 3.2% SiFe). (After Degauque, J. and Fiorillo, F.,
Alliages magnetiques doux enrichis en silicium, in *Materiaux magnetiques en genie electrique I*, Kedous-Lebouc, A. (Ed.), Lavoisier, Chapter 4,
pp. 227–286, 2006.)

SiFe is chemically polished and next is held at 1000°C for
1 h under a flowing gaseous mixture of $SiCl_4$ and argon
(Figure 3.30). Fe_3Si forms near the surface of the sheet
and the sheet loses Fe as a gaseous $FeCl_2$. During 13 h
diffusion process at 1000°C, a homogeneous FeSi solution is obtained. In this way, it is possible to manufacture various thickness grain-oriented or non-oriented
6.5% SiFe sheet.

As result of the CVD siliconization, a significant
improvement of most of all frequency-dependent properties is observed. Figure 3.31 presents the typical reduction of energy loss after the CVD process.

In 1993, NKK Corporation started a commercial
scale production of 6.5% SiFe by CVD siliconizing process. Several grades of 6.5% SiFe called NK E-core 0.05,
0.1, 0.2, and 0.3 mm thick were available. Table 3.11
presents properties of NKK 6.5% SiFe. For medium
frequency, this material is much better than conventional steel and can be an alternative for amorphous
material.

From time to time, the idea of cubic texture is investigated. Indeed, looking at the Goss texture (Figure 3.32),
the positioning of crystals seems to be rather extrava-
gant. More natural order, like cobbles, seems to be the
cube order (Figure 3.32). Moreover, in the Goss texture,
there is just one easy direction and perpendicular to it is
the hard direction: in cube texture, both directions are
easy (see Figure 2.18). Such feature could be profitable in
rotating machines.

First information about possibility of obtaining SiFe
electrical steel with cube texture was published by
Sixtus in 1935 (Sixtus 1935). In 1957, Assmus described
condition of processing and annealing necessary for a
cubic orientation (Assmus et al. 1957). By applying this
technology, Kohler confirmed it practically by applying
annealing of 3% SiFe in slightly oxidizing atmosphere
(Kohler 1960).

Since then, other technological possibilities were
demonstrated, for example, by rolling the sheet two
times in perpendicular directions. In 1988, a patent was
published describing technology where starting point
to cubic texture was (114) [401] texture (Sakakura et al.
1988). In the 1970s, two manufacturers, Armco Steel and
Vacuumschelze, offered electrical steel with cubic texture in laboratory scale.

But this material was not accepted by the market.
Cubic SiFe steel was expensive (due to very complex
technology of producing), with large magnetostriction
and therefore also recently is prepared only on the laboratory scale.

Other alternative to SiFe steel is iron–aluminum alloy.
As can be seen from Figure 3.10, that aluminum changes
the resistivity similarly to the addition of silicon. Indeed
16% Al-Fe exhibits large resistivity ($\rho = 140 \, \mu\Omega m$), almost
four times larger than 3% SiFe. Additionally, it exhibits

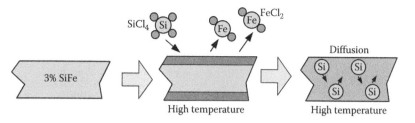

FIGURE 3.30
CVD technology of 6.5% SiFe production. (After Degauque, J. and Fiorillo, F., Alliages magnetiques doux enrichis en silicium, in *Materiaux magnetiques en genie electrique I*, Kedous-Lebouc, A. (Ed.), Lavoisier, Chapter 4, pp. 227–286, 2006.)

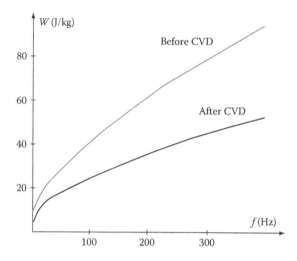

FIGURE 3.31
Energy loss versus frequency of 6.5 SiFe. (After Crotier-Combe, S. et al., *J. Magn. Magn. Mater.*, 160, 151, 1996.)

reasonably small coercivity and high permeability (after annealing it can be up to 50,000). The drawback of AlFe steel is that this material is more susceptible to oxidation (Figure 3.33).

The other commercially available steel is 16% AlFe known as *alperm* (also as alfenol). Due to its high resistivity, it can be used for medium frequency applications (Adams 1962). Another alloy, 13% AlFe, known as *alfer* exhibits large magnetostriction and is used in magnetoelastic sensors.

In 1936, Masumoto developed new Fe-Si Al alloy known as *sendust*. This alloy (84.9 Fe–9.5 Si–5.6 Al) has very high permeability (even 140,000), low coercivity, and large resistivity (Masumoto 1936). Additionally, it exhibits a unique feature; simultaneous zero magnetostriction and zero magnetocrystalline anisotropy constant K_1. It can be used in medium- or high-frequency applications as competitive to NiFe alloys. Due to its hardness, it was used as material for magnetic reading heads. Because it is very brittle, often it is manufactured as powder cores. Table 3.12 present properties of main iron–aluminum alloys.

3.3 Nickel- and Cobalt-Based Alloys

3.3.1 NiFe Alloys (Permalloy)

Permalloy* (NiFe alloy) was for many years the symbol of the highest quality soft magnetic material with extremely high permeability ($\mu_{max} = 1,000,000$ in the case of Supermalloy) and small coercivity ($H_c = 0.2\,A/m$ in the case of Supermalloy) (Chin 1971, Pfeifer and Radeloff 1980, Couderchon et al. 1982). Superiority of permalloy illustrates impressive Figure 3.34 from Fiorillo book (Fiorillo 2004). Recently, the importance of NiFe alloys is shadowed by new stars: amorphous and nanocrystalline materials having similar or better properties.

By an appropriate selection of the alloy composition (supported by additional elements like Mn, Cu, Cr, V) and annealing/cooling treatment, it is possible to obtain an alloy exhibiting various, often extraordinary properties, for example,

- Alloy with close to zero coefficient of thermal expansion (FeNi36 known as invar or Fe59Ni36Cr5 known as elinvar)
- Alloy with the coefficient of thermal expansion the same as glass (Fe53.5Ni29Co17Mn0.3Si0.2 known as *kovar*)
- Alloy with very high permeability (Fe16Ni79Mn4 known as supermalloy or Fe16Ni77Cu14 known as mumetal)
- Alloy with constant permeability (Fe55Ni36Cu9 known as isoperm)
- Alloy with permeability linearly dependent on temperature (Fe70Ni30 known as thermoperm or Ni67Fe2Cu30 known as thermalloy)
- Alloy with square hysteresis loop (Fe50Ni50 known as hipernik or permenorm)

---

* The name Permalloy is formally related to only one type of NiFe alloys: the nonmagnetostrictive NiFe alloy with Ni content around 80%, but it is commonly used to all types of NiFe alloys.

TABLE 3.11

Properties of 6.5% SiFe (NKK E-Core)

	t (mm)	J_s (T)	Loss (W/kg)					μ_{max} 1000	λ_s 10^{-6}
			1 T	1 T	0.5 T	0.2 T	0.1 T		
			50 Hz	0.4 kHz	1 kHz	5 kHz	10 kHz		
NK E-core	0.05	1.25	0.85	7.0	5.2	8.0	5.1	16	0.1
6.5%	0.1	1.25	0.72	7.3	6.2	11.8	9.7	18	
	0.2	1.27	0.55	8.1	8.4	19.0	16.8	19	
	0.3	1.30	0.53	10.0	11.0	25.5	24.5	25	
3% SiFe	0.05	1.79	0.68	7.2	7.6	19.5	18.0		−0.8
	0.1	1.85	0.72	7.2	7.6	19.5	18.0	24	
	0.23	1.92	0.29	7.8	10.4	33.0	30.0	92	
	0.35	1.93	0.40	12.3	15.2	49.0	46.0	94	
Amorph Fe-Si-B	0.03	1.38	0.13	1.5	2.2	4.0	4.0	300	27

Source: After Degauque, J. and Fiorillo, F., Alliages magnetiques doux enrichis en silicium, in *Materiaux magnetiques en genie electrique I*, Kedous-Lebouc, A. (Ed.), Lavoisier, Chapter 4, pp. 227–286, 2006.

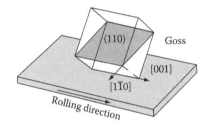

FIGURE 3.32
Goss (110)[001] and cubic (001)[100] texture.

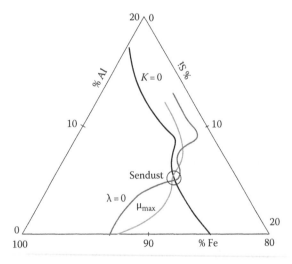

FIGURE 3.33
Parameters of FeAlSi alloy.

By appropriate heat treatment, it is possible to obtain various types of texture, including isotropy. Figure 3.35 demonstrates the possibility of change of the hysteresis loop by appropriate annealing: from square loop to linear.

Flexibility and versatility of NiFe as a function of temperature treatment and composition are advantageous but under some circumstances they could be a drawback. Their properties can be easily altered during operation of a given device and especially after cutting and stamping or mechanical shocks, it is necessary to anneal the material.

Figure 3.36 presents dependence of NiFe alloy properties on the nickel contents. There are three main areas of application:

- Close to 80% of Ni where the permeability is very high and coercivity is small
- Close to 50% where the saturation polarization is the largest
- Close to 36% where resistivity is large

TABLE 3.12

Properties of Al–Fe Alloys

		J_s (T)	H_c (A/m)	μ_{max}
Alfer	13% Al	1.28	53	4,000
Alperm	16% Al	0.8	3	55,000
Sendust	10% Si, 5% Al	1	4	140,000

Source: After Adams, E., *J. Appl. Phys.*, 33, 1214, 1962.

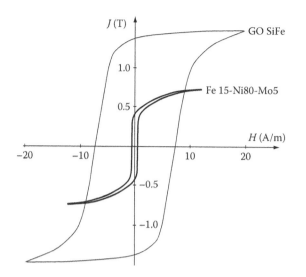

FIGURE 3.34

Comparison of the hysteresis loops of NiFe and FeSi alloys. (After *Measurement and Characterization of Magnetic Materials*, Fiorillo, F., Copyright (2004).)

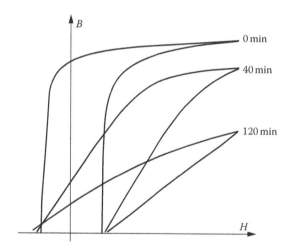

FIGURE 3.35

Changes of the shape of hysteresis loop with time of annealing in magnetic field. (After Pfeifer, F. and Radeloff, C., *J. Magn. Magn. Mater.*, 19, 190, 1980.)

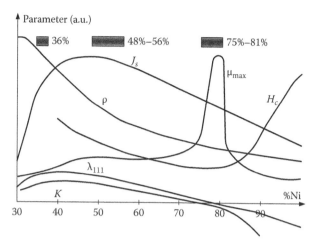

FIGURE 3.36

NiFe alloy parameters versus the nickel content.

The anisotropy constant and magnetoelasticity constant are close to zero around 75%–91% Ni. Unfortunately, it is not possible to obtain zero value of both parameters simultaneously (as it is the case with Sendust) but with small addition of other components (molybdenum, copper), it is possible to obtain nonmagnetostrictive material with zero anisotropy constant.

Table 3.13 presents the parameters of the main NiFe alloys. In application of NiFe alloys, two types of alloys are important. The first is around 80% of Ni with high permeability, small coercivity, and close to zero magnetostriction. These alloys are most of all used as magnetic shields (mumetal) and for sensors (supermalloy). To obtain high permeability, the special heat treatment called permalloy treatment (or double treatment) is done. This annealing consists of heating for 1 h at 900°C–950°C and cooling not faster than 100°C/h to room temperature. Then, heating is continued to 600°C and the alloy is cooled in the open air to room temperature by placing on the copper plate.

High-permeability alloys have low resistivity and saturation polarization. Therefore, for other purposes (pulse transformers, audiofrequency transformers, null

TABLE 3.13

Properties of NiFe and CoFe Alloys

	ρ ($\mu\Omega$ cm)	J_s (T)	H_c (A/m)	$\mu_{max} \times 1000$	
Fe64Ni36	75	1.3	40	20	Invar
Fe50Ni50	45	1.6	7	15	Isoperm
Fe52Ni48	45	1.6	8	180	
Fe44Ni56	35	1.5	1	300	
Fe20Ni80	16	1.1	0.4	100	Permalloy
Fe16Ni79Mo5	60	0.8	0.4	550	Supermalloy
Fe16Ni77Cu5Cr2	56	0.75	0.8	500	Mumetal
Fe50Co50	7	2.45	160	5	Permendur
Fe49Co49V2	40	2.4	400	17	Hiperco, vacoflux

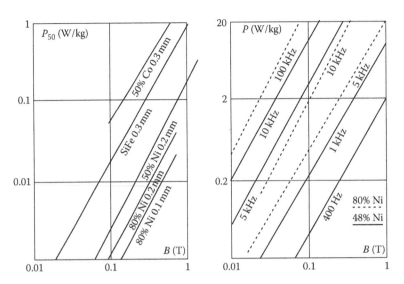

FIGURE 3.37
Power loss of typical NiFe and CoFe alloys. (Based on the data from Vacuumschmelze and Magnetics.)

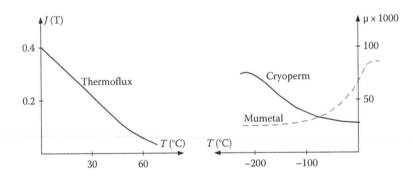

FIGURE 3.38
Special purposes NiFe alloys (Vaccumschmelze).

balance transformers, switches, chokes, etc.), alloys with around 50% Ni are used. Figure 3.37 presents the dependence of core loss on a peak magnetic flux density.

As discussed earlier, NiFe alloys are very versatile in their performances. Figure 3.38 presents the special purposes alloys of Vacuumschmelze. Thermoflux is an alloy with practically linear dependence of flux density on the temperature. It can be used as a magnetic shunt to compensate temperature effects in permanent magnet systems. Cryoperm is an alloy developed for low-temperature applications.

NiFe alloys are also available in form of the powder cores known as MPP (molybdenum permalloy powder, Ni81Fe17Mo2).

3.3.2 CoFe and CoFeNi Alloys (Permendur and Perminvar)

Alloys with around 50% Co is known as permendur—the alloy of the highest available saturation

polarization—up to 2.46 T. Data and characteristics of such alloy are presented in Table 3.12 and Figure 3.39.

To improve the material properties, a small amount of vanadium (2%) is added. In this way, resistivity and 10 A/m ductibility are improved partly at the cost of small decrease of saturation polarization. There is also a special kind of permendur alloy with carefully controlled purity and magnetically annealed known as supermendur. It exhibits much lower coercivity around 10 A m and large permeability around 80,000.

CoFe alloys are used in applications where high saturation flux density is necessary, for example, as poles of electromagnets. Mechanical force depends on the square of flux density, hence alloys with higher saturation flux density allow miniaturization of the cores. For this reason, the CoFe alloys are used when the weight is at premium, in the aerospace industry.

There is also the group of NiFeCo alloys, in which the most well known is perminvar Ni45Fe30Co25, the alloy with permeability constant over a wide range of applied field values.

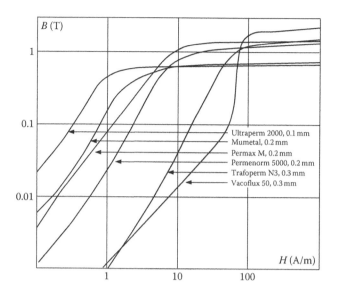

FIGURE 3.39

Magnetization curves of various NiFe and CoFe alloys (Vaccumschmelze data: *ultraperm* and *mumetal* alloys from 72% to 83% NiFe group, *permax* from 54% to 68% NiFe group, *permenorm* from 45%–50% NiFe group, *trafoperm* from nearly isotropic SiFe, *vacoflux* from 50% Co alloy).

3.4 Amorphous and Nanocrystalline Alloys

3.4.1 Amorphous Soft Magnetic Materials (Metallic Glass)

Figure 3.40 illustrates technology of the amorphous ribbon production. The material in liquid state is rapidly cooled on a rotating copper drum. The speed of cooling should be fast enough not to allow the formation of crystal structure. To help in formation of the amorphous state, small addition of metalloid (mostly boron) is made in order to improve viscosity of the molten metal. Since the ribbon should be cooled very quickly, it is thin with the thickness not exceeding 50 μm. Also the width of the ribbon is limited, usually not exceeding 20 cm.

Table 3.14 presents data of main amorphous ribbons. Exceptional performances are exhibited by cobalt-based

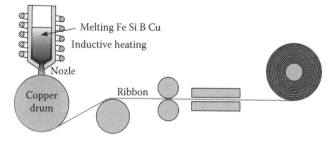

FIGURE 3.40

Technology of the amorphous ribbon production.

alloys: very high permeability, small coercivity, and possibility to obtain material with negligible magnetostriction. But these alloys suffer from low polarization saturation and low Curie temperature. On the other hand, iron-based alloys exhibit higher saturation but at the cost of permeability and coercivity. The highest polarization is possible to obtain by adding of about 20% of expensive cobalt to iron-based alloy. Intermediate properties exhibit FeNi-Mo alloys that make them suitable for shielding purposes.

The magnetic parameters of amorphous ribbon can be improved by annealing (Figure 3.41), especially by annealing in longitudinal magnetic field. Because metallic glass is brittle, usually heat treatment is performed on the final form of the core (e.g., wound toroid). This way mechanical stress is removed, but this is also caused by winding of the toroid. The temperature should not exceed crystallization temperature of the amorphous state and is around 400°C for NiFe alloys, 480 for Fe alloys, and 550 for Co alloys.

The amorphous alloys have been introduced in 1970s by Allied Signals Inc. Metglas Products. Recently, market available products are offered by Metglas Inc. Hitachi Metals and by Vaccumschmelze. Table 3.15 presents data of commercially available soft magnetic amorphous materials. Typical thickness of the ribbon is around 25 μm, and its width does not exceed 20 cm.

The most versatile is Metglas SA1 with relatively large saturation flux density and permeability and low power loss. Figure 3.42 presents the core power loss of this alloy.

It can be seen from Figures 3.42 and 3.43 that amorphous alloys can work in high-frequency bandwidth. It is the main application of this material—pulse transformers, high-frequency transformers, current transformers, and ground fault interrupters. Indeed, as illustrated in Figure 3.44, amorphous materials are superior up to 1 MHz in comparison with other materials.

Another important application of amorphous materials is profiting on exceptionally high permeability: in Co-based alloy even 1,000,000. Such large permeability is recommended especially in shielding and sensors application. Figure 3.45 presents typical magnetization curves of Co-base amorphous alloy.

The possibility of application of amorphous cores in power transformers is still under discussion (Moses 1994, Hasegawa and Azuma 2008). Indeed, a comparison of power loss at 1.3 T–0.64 W/kg in the case of silicon steel and 0.11 W/kg in the case of amorphous ribbon is rather impressive.

After introduction of amorphous alloys, it was expected that they should substitute conventional distribution transformer due to smaller loss and better efficiency. Up to 1990, around 20,000 amorphous metal distribution transformers (AMDTs) have been

TABLE 3.14

Properties of Main Amorphous Alloys

	J_s (T)	H_c (A/m)	$\mu_{max} \times 1000$	ρ ($\mu\Omega$ cm)	$\lambda_s \times 10^{-6}$	T_c (°C)
FeSiBC	1.6	2.2	300	135	30	370
FeSiBCo	1.8	4	400	123	35	415
FeNiMoB	0.9	1.2	800	138	12	350
CoNiFeBSi	0.6	0.3	1 000	142	~0	225

Source: After Waecklerle, T., Materiaux magnetiques doux speciaux et applications, in *Materiaux magnetiques en genie electrique I*, Kedous-Lebouc, A. (Ed.), Lavoisier, Chapter 3, pp. 153–223, 2006.

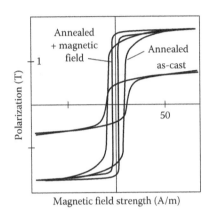
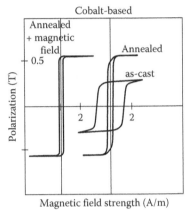

FIGURE 3.41
Effect of annealing of amorphous ribbon. (After Metglas Inc. data.)

installed in the United States (in Japan about 32,000 units) (Moses 1994). Although amorphous material has much lower power loss in comparison with GO SiFe steel (due to smaller coercivity and larger resistivity), there are several practical obstacles in expansion of the AMDT units.

Amorphous ribbon is very thin (about 25 μm) and relatively narrow (10–20 cm), which makes difficult the design and manufacture large transformers. Practically, only wound corers can be made. In the case of one-phase distribution (dominant in the United States and Japan), the design of wound amorphous core transformer is

relatively easy. In the case of three-phase distribution (dominant in Europe), the design of AMDT unit is still a challenge.

The very significant obstacle is lower saturation flux density 1.56 T for most popular Metglas 2605 SA1 ribbon in comparison with 2.03 for GO SiFe. That is why Metglas Inc. introduced a new-grade Metglas 2605 HB1 with

TABLE 3.15

Commercially Available Amorphous Materials

		J_s (T)	$\mu_{max} \times 1000$	$\lambda \times 10^{-6}$
Metglas 2605 SA1	$Fe_{78}B_{13}Si_9$	1.56	600	27
Metglas 2605 SC	$Fe_{81}B_{13.5}Si_{3.5}C_2$	1.61	300	30
Metglas 2605CO	$Fe_{66}Co_{18}B_{15}Si_1$	1.8	400	35
Metglas 2705 M	$Co_{69}Fe_4Ni_1Mo_2Si_{12}B_{12}$	0.77	800	<0.5
Metglas 2714 A	$Co_{66}Fe_4B_{14}Si_{15}Ni_1$	0.57	1000	<0.5
Matglas 2826 MB	$Fe_{40}Ni_{38}B_{18}Mo_4$	0.88	800	12
Vitrovac 6025	$Co_{66}Fe_4B_{12}Si_{16}Mo_2$	0.55	600	0.3
Vitrovac 6030	$Co_{70}(FeMo)_2Mn_5(SiB)$	0.8	300	0.3

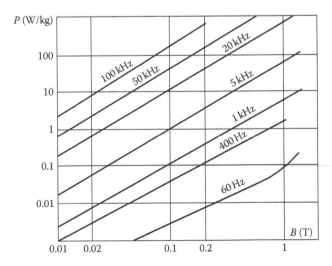

FIGURE 3.42
Core power loss of Metglas 6025 SA1 alloy. (After Metglas Inc. data.)

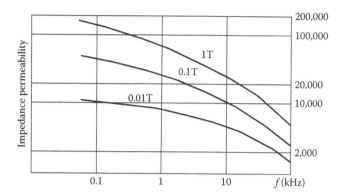

FIGURE 3.43
Core power loss of Metglas 6025 SA1 alloy (After Metglas Inc. data.)

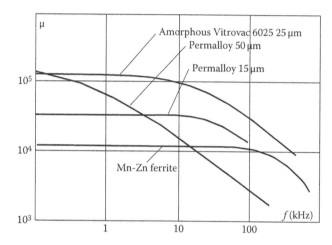

FIGURE 3.44
Permeability versus frequency of different magnetic materials (as measure on toroidal cores). (After Hilzinger H.R., *J. Magn. Magn. Mater.*, 83, 370, 1990.)

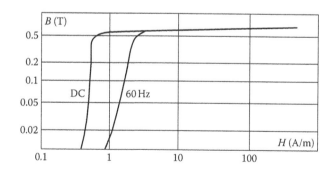

FIGURE 3.45
Magnetization curve of cobalt Metglas 2705M. (After Metglas Inc. data.)

improved properties: $B_s = 1.64\,T$ and $H_c = 2.4\,A/m$ (in the case of Metglas 2605 SA1 $H_c = 3.4\,A/m$) (Hasegawa 2006). Table 3.16 presents performance of a 50 Hz three-phase 500 kVA transformer and Figure 3.46 the efficiency of distribution transformer.

TABLE 3.16

Performances of Three-Phase 500 kVA Transformer

	SiFe	Metglas SA1	Metglas HB1
Weight (ratio)	1	1.23	1.17
No load loss (W)	665	215	215
40% load loss (W)	1353	1207	1207
Audible noise (dB)	53	58	55

Source: Hasegawa, R. and Azuma, D., *J. Magn. Magn. Mater.*, 320, 2451, 2008.

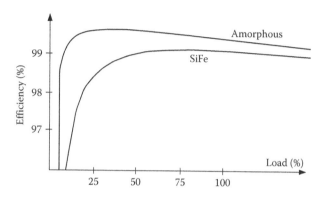

FIGURE 3.46
Efficiency of 2000 kVA dry-type transformer. (From Hasegawa, R., *J. Magn. Magn. Mater.*, 215, 240, 2000.)

It was estimated that in the United States the annual energy lost in transformer is around 140 TWh. Introduction of AMDT transformers resulted in saving of around 80 TWh, which corresponds with annual reduction of 60 tons of CO_2[*] (Hasegawa and Azuma 2008). In 2009, ABB Corp. introduced production of AMDT transformers in New Jersey.

Amorphous materials are described in detail in many review papers or books (Chen 1980, Luborsky et al. 1980, Hasegawa 1983, Egami 1984, Moorjani and Coey 1984, McHenry et al. 1999, Waeckerle et al. 2006). Lately, we observe that new papers on this subject are low in number. Even introduction of the new grade Metglas HB1 is a rather cosmetic improvement. Recently, more research papers are devoted to nanocrystalline materials and they can be considered as further development of amorphous materials (because as starting material often amorphous material is used for devitrification process).

3.4.2 Nanocrystalline Soft Magnetic Materials

In 1988, Yoshizawa et al. from Hitachi Metals Company proved that after appropriate annealing of Fe-based amorphous ribbon, it is possible to create very small grains of α-FeSi (average diameter around 10 nm) embedded in an amorphous matrix (see Figure 2.115)

---

[*] For Europe, the same data are estimated as follows: annual transformer power loss 55 TWh, saving 22 TWh, CO_2 reduction 15 tons.

(Yoshizawa et al. 1988). This new material called FINEMET exhibited excellent magnetic properties, close to those possible to obtain in Co-based, much more expensive amorphous ribbon. As the initial amorphous material $Fe_{73.5}Cu_1Nb_3Si_{13.5}B_9$ ribbon with thickness 20 μm was used (note that also expensive boron was used in smaller amount than in the typical amorphous material), copper was added to enhance the nucleation of α-Fe grains and niobium was used to lower the growth of the grains. The temperature of annealing for about 1 h was above the crystallization temperature, around 550°C. Moreover, the starting material had saturation magnetostriction $\lambda_s = 20 \; 10^{-6}$ but after annealing the nanocrystalline material exhibited close to zero magnetostriction. Thus, the paper of Yoshizawa was the starting point for the new class of soft magnetic materials known as a nanocrystalline material. Figure 3.47 shows the properties of material presented in paper of Yoshizawa.

In nanocrystalline materials, the direction of anisotropy of the grains is randomized and therefore reduced in the overall performance. The properties of a nanocrystalline material are similar to the best grades of permalloy. However, the NiFe alloys can be used in frequency only up to 100 kHz but the nanocrystalline materials can work correctly in the frequency bandwidth similar to best grades of ferrites.

After the initial information about FINEMET, many other publications about similar or other nanocrystalline materials appeared. The most widely known are NANOPERM and HITPERM (see Figure 3.50). Team from Tohoku University in 1981 proposed to substitute Nb by other transition metal like zirconium (Zr) or hafnium (Hf) (Suzuki et al. 1991, Kawamura et al. 1994). Typical materials $Fe_{87}Zr_7B_5Cu_1$ or $Fe_{86}Zr_7B_6Cu_1$ are known as NANOPERM. Such materials were obtained by rapid solidification of melting mixtures or by compaction of powder by hot-pressing machine (Kawamura et al. 1994). Figure 3.48 presents properties of NANOPERM.

Figure 3.49 presents changes of amorphous/nanocrystalline material during annealing.

From the data presented in Figure 3.48, it can be seen that the nanocrystalline NANOPERM has lower power loss and higher permeability in comparison with the Fe-based amorphous alloy. In comparison with the FINEMET, this material exhibits higher saturation polarization (1.2 T for FINEMET compared to 1.5–1.8 T

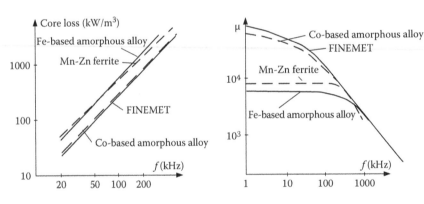

FIGURE 3.47
Properties of the FINEMET material. (After Yoshizawa, Y. et al., *J. Appl. Phys.*, 64, 6044, 1988.)

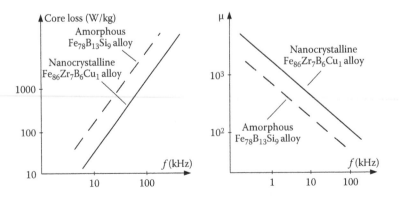

FIGURE 3.48
Properties of the $Fe_{86}Zr_7B_6Cu_1$ soft magnetic compacts material. (After Kawamura, Y. et al., *J. Appl. Phys.*, 76, 5545, 1994.)

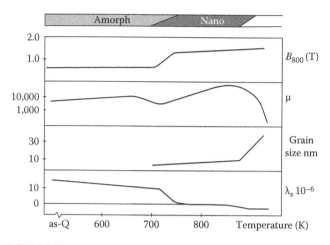

FIGURE 3.49

Change of material parameters during annealing of amorphous $Fe_{86}Zr_7B_6Cu_1$ material. (After Suzuki, K. et al., *J. Appl. Phys.*, 70, 6232, 1991.)

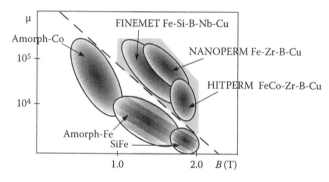

FIGURE 3.50

Permeability and polarization range of main nanocrystalline materials.

for NANOPERM). The main drawback of amorphous and nanocrystalline materials is their relatively small (compared to SiFe) polarization. Similarly, as in the case of NiFe alloys, in order to obtain higher polarization, special alloys with an addition of cobalt known as HITPERM (Figure 3.50) were developed.

The FeCo nanocrystalline alloys with composition (FeCo)-M-B-Cu (where M—Zr, Hf, Nb, etc.) enable obtaining the saturation polarization close to the polarization of SiFe (Villard et al. 1999a,b). These alloys are known as HITPERM. They can be used in elevated temperatures up to 650°C. Table 3.17 presents the main features of nanocrystalline alloys.

The most commonly used nanocrystalline material is close to $Fe_{73.5}Cu_1Nb_3Si_{13.5}B_9$ alloy and is available under various trade names: Finemet, Vitroperm, Nanophy.

Usually nanocrystalline material is delivered as ready-to-use toroidal core, for example, choke or transformer because nanocrystalline material is very sensitive to annealing conditions. Figure 3.51 presents the influence of the annealing temperature on coercivity and permeability. Incorrect temperature of annealing leads to significant increase of coercivity due to overgrowth of grain diameter (coercivity strongly depends on grain diameter proportional roughly to D^6). Similarly small changes of alloy composition cause significant changes of parameters. Figure 3.52 presents the annealing temperature recommended for Finemet alloy.

By applying the magnetic field during annealing, it is also possible to change the magnetic parameters, for example, the shape of the hysteresis loop. Figure 3.53 presents various types of hysteresis for the same Finemet type alloy annealed under various conditions.

Materials with high permeability are used in EMI filters, shielding sheets, current sensors, and magnetic sensors. Materials with square hysteresis are used in pulsed power cores and surge absorbers. Low magnetostriction is required for high-frequency applications such as transformers, filters, and chokes. Table 3.18 presents the properties of the main FINEMET materials.

Nanocrystalline materials are used mainly in high-frequency applications as competition to ferrites (see Figure 3.54 and Table 3.18).

TABLE 3.17

Properties of the Main Nanocrystalline Alloys

	J_s (T)	H_c (A/m)	μ (1 kHz)	$P_{0.2}$,100 kHz (W/kg)	D (nm)
$Fe_{73.5}Cu_1Nb_3Si_{13.5}B_9$	1.24	0.5	100,000	38	13
$Fe_{73.5}Cu_1Nb_3Si_{15.5}B_7$	1.23	0.4	110,000	35	14
$Fe_{84}Nb_7B_9$	1.49	8	22,000	76	9
$Fe_{86}Cu_1Zr_7B_6$	1.52	3.2	48,000	116	10
$Fe_{91}Zr_7B_3$	1.63	5.6	22,000	80	17
$Fe_{67}Co_{18}Si_1B_{14}$	1.8	5	1,500		

Source: From *Handbook of Magnetic Materials*, Vol. 10, Herzer, G., Nanocrystalline soft magnetic alloys, pp. 415–462, Copyright (1997), with permission from Elsevier.

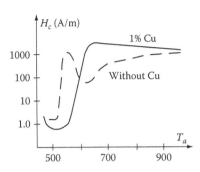
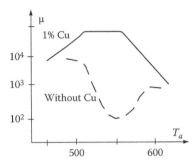

FIGURE 3.51

Changes of magnetic parameters for different temperatures of 1 h annealing of Finemet. (From Herzer G., *IEEE Trans. Magn.*, 26, 1397, 1990.)

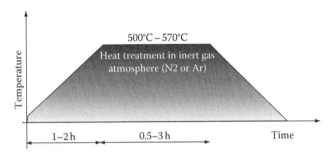

FIGURE 3.52

Typical process of annealing of Finemet-type nanocrystalline alloy. (From Hitachi Metals data.)

3.5 Soft Ferrites

3.5.1 MnZn and NiZn Ferrites

Ferrites are the ceramic ferrimagnetic materials with a spinel* structure. Investigating various possible compositions of ferrites, Snoek predicted that the two main classes of materials, MnZn and NiZn ferrites, should exhibit the best performances (Snoek 1936, Sugimoto 1999). And indeed although various soft ferrites compositions are being investigated (Stoppels 1996), only these two main mutually complementing families of materials are commercially available (Figure 3.55). Table 3.19 presents the main properties of both classes of soft ferrites. MnZn ferrites exhibit higher permeability and saturation polarization but NiZn ferrites have much higher resistivity (close to electric insulators) and therefore can be used at higher frequencies (1–500 MHz while MnZn up to a few MHz).

The two classes of ferrites can be described by formulas $Mn_xZn_{(1-x)}Fe_2O_4$ or $Ni_xZn_{(1-x)}Fe_2O_4$. By changing the composition, it is possible to influence the properties

of ferrite (Figure 3.56). There are three main groups of ferrites: with optimized permeability (for EMI filters, chokes and shield), with optimized saturation polarization (for signal processing), and with optimized loss (quality factor Q) for power application. Generally, magnetic properties of soft ferrites are poor in comparison with those of SiFe and additionally these parameters strongly depend on temperature. But, there are a plethora of commercially available ferrite products due to their dominating performance in the high and very high frequency range. Ordinary, inexpensive ferrites are also used in many applications where requirements are not high.

The manufacturing process of ferrites is not very complex. The raw materials (the oxides or carbonates of the constituent metals) in form of a powder are mixed. Then this mixture is calcined (presintered at approximately 1000°C), where apart from other processes, calcination of carbonates $MCO_3 \rightarrow MO + CO_2$ occurs (in ordinary materials, sometimes this stage is omitted). After calcination, the material is milled to the powder with grains smaller than 2 μm. Finally, the product is formed by pressing. The main process of sintering is performed at temperature between 1150°C and 1300°C. Depending on the composition (including small addition of other elements), technology market offers a diversity of various ferrite materials (Figure 3.57).

In assessment of the quality of ferrites, slightly different parameters are used than it is in the case of low-frequency materials. By considering permeability, both components (real and imaginary) are taken into account (Figure 3.58). The complex permeability

$$\mu = \mu' + j\mu'' \tag{3.3}$$

describes in a better way the energy dissipation represented by a phase shift δ between polarization $J(t)$ and magnetic field $H(t)$:

$$\tan\delta = \frac{\mu''}{\mu'} \tag{3.4}$$

---

* Spinel is a mineral described by a formula $MgAl_2O_4$ ($A^{2+}B_2^{3+}O_4^{2-}$). In the crystal, the oxide anions are arranged in a cubic close-packed lattice while the cations A and B are occupying octahedral or tetrahedral sites in the lattice (see Figure 2.110).

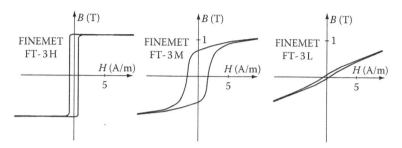

FIGURE 3.53
Various types of hysteresis loop of Finemet material. (From Hitachi Metals data.)

TABLE 3.18

Properties of the Main FINEMET Cores

	J_s (T)	B_r/B_s (%)	H_c (A/m)	μ (1 kHz)	μ (100 kHz)	P (kW/m³)	λ_s (10⁻⁶)
FT-3H	1.23	89	0.6	30,000	5,000	600	0
FT-3M		50	2.5	70,000	15,000	300	0
FT-3L		5	0.6	50,000	16,000	250	0
Mn-Zn ferrite	0.44	23	8.0	5,300	5,300	1,200	−0.6

Source: Hitachi Metals data.

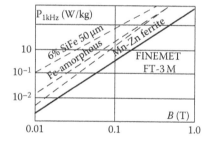
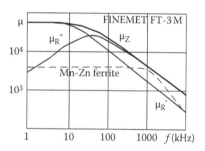

FIGURE 3.54
Power loss and impedance permeability of Finemet material. (From Hitachi Metals data.)

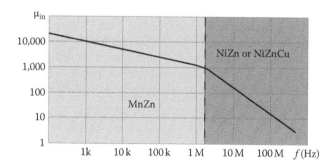

FIGURE 3.55
Initial permeability versus frequency of MnZn and NiZn ferrites.

The excess power loss is expressed as (Fiorillo 2004)

$$W = \pi JH \frac{\tan\delta}{\sqrt{1+\tan^2\delta}} \quad (3.5)$$

For inductors used in filter applications, a quality factor $Q = 1/\tan\delta$ often is used as the figure of merit. Another quality and performance indicator in form of the factor $f \cdot B_{max}$ for a fixed loss sometimes is used. The example of such dependence is presented in Figure 3.59.

Figure 3.60 presents the hysteresis loop of typical ferrite core and Table 3.20 presents properties of typical grades of commercially available ferrites.

3.5.2 Ferrites for Microwave Applications

New generations of electronic devices work at ever increasing frequencies. Therefore, there is a need for magnetic materials in the GHz range, up to 150 GHz. In many microwave devices, magnetic part is very important, taking into account antennas, circulators, isolators, phase shifters, and filters. Fortunately, there are ferrites that work in the microwave range.

In the GHz frequency range, ferrites exhibit ferromagnetic resonance phenomenon (more exactly ferrimagnetic resonance; see Section 2.7.4). Therefore, an important factor of a microwave material is ΔH, the

TABLE 3.19

Properties of MnZn and NiZn Ferrites

	μ_{in}	J_s (T)	H_c (A/m)	ρ (Ωm)	T_c (°C)	f (MHz)
MnZn	500–20,000	0.3–0.5	4–100	0.02–20	100–250	DC-1
NiZn	10–2,000	0.1–0.36	16–1,600	10–10^7	100–500	1–500

Source: McCurrie, R.A., *Ferromagnetic Materials*, Academic Press, London, U.K., 1994.

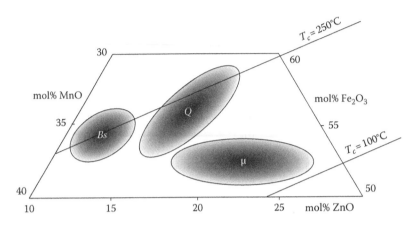

FIGURE 3.56
Phase diagram of MnO-ZnO-Fe$_2$. (From *Ferrite Users Guide* edited by Magnetic Materials Producers Association MMPA.)

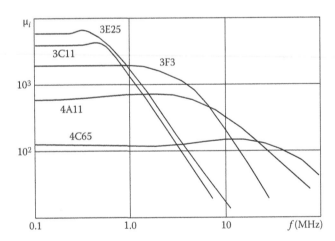

FIGURE 3.57
Diversity of market available ferrites ("3" means MnZn and "4" are NiZn ferrites). (From data of Ferroxcube.)

ferromagnetic resonance linewidth. The permeability of material is described by the Polder tensor

$$B = \begin{bmatrix} \mu & -j\kappa & 0 \\ j\kappa & \mu & 0 \\ 0 & 0 & \mu_0 \end{bmatrix} H \qquad (3.6)$$

where $\mu = \mu_0(1 + \omega_0\omega_m/(\omega_0^2 - \omega^2))$, $\kappa = \mu_0\omega\omega_m/(\omega_0^2 - \omega^2)$, $\omega_0 = \gamma\mu_0 H_0$, $\omega_m = \gamma\mu_0 M$

and γ is ferromagnetic resonance (gyromagnetic) ratio, $\gamma = g\cdot 17.6$ [MHz/(kA/m)] and g is Landé factor (between 1.9 and 2.4 for various ferrite materials).

Thus, permeability depends on the frequency (ω), resonance frequency (ω_0), and internal bias magnetic field (H_0).

In the microwave range, three most important materials can be used: garnets ferrite (in the range 1–10 GHz), spinel ferrites (in the range 3–30 GHz), and hexagonal ferrites (in the range 1–100 GHz). Table 3.21 presents properties of the main microwave materials.

Microwave technique and microwave materials are described in many books and review papers (Dionne 1975, Nicolas 1980, Pardavi-Horvath 2000, Adam et al. 2002, Özgür et al. 2009).

3.6 Hard Magnetic Materials

3.6.1 General Remarks

Since the introduction of rare-earth metals such as neodymium and samarium in the production of permanent magnets (and after wider exploitation of deposits of the main minerals containing these elements—monazite, bastnäzite in China and Brasil), significant progress in the quality of permanent magnets is observed. It is illustrated in Figure 3.61.

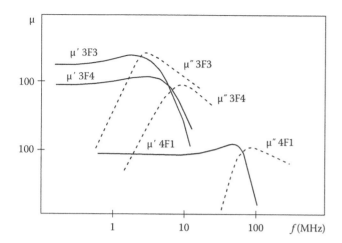

FIGURE 3.58
Real and imaginary parts of permeability for selected ferrites. (After Stoppels, D., *J. Magn. Magn. Mater.*, 160, 323, 1996.)

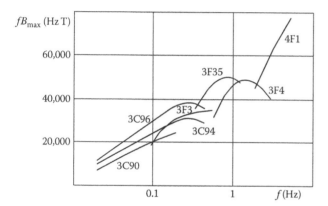

FIGURE 3.59
Performance factor of various grades of ferrites. (From data of Ferroxcube.)

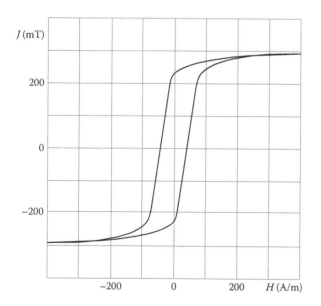

FIGURE 3.60
Typical hysteresis loop of ferrite: *3R1*, $f = 100$ kHz, $T = 25°C$. (From data of Ferroxcube.)

In the case of soft magnetic materials, we require as small as possible coercivity because hysteresis power loss strongly depends on this value. Conversely, in the case of hard magnetic materials, we need to have as large as possible coercivity and remanence because stored magnetic energy approximately depends on the $H_c \cdot B_r$ value (Figure 3.62).

Historically, as the first hard magnetic materials, simply different kinds of steel were used. In 1917, in Japan, cobalt steel ($Fe_{55}Co_{35}W_7Cr_2C_{0.6}$) known as Honda alloy was developed. In 1931, also in Japan, Mishima invented the Alnico alloy ($Fe_{58}Ni_{30}Al_{12}$), which gave better performance in comparison with steel. Recently, cobalt steel has practically vanished from the market as a material for permanent magnets. Table 3.22 presents parameters of the main hard magnetic materials and Figure 3.63 the market segmentation of permanent magnet materials.[*]

From the data presented in Figure 3.63, it can be seen that hard magnetic ferrites are still dominating the market (almost 60% of the market). The rare-earth-metal-based magnets overshadowed the earlier very popular Alnico and their applications are growing (Fastenau and van Loenen 1996).

For many years, the evaluation of magnetic materials was commonly based on the relationship $B = f(H)$. Recently, this relation is commonly substituted by $J = f(H)$ as more reliable and describing physics of the magnetic phenomena.[†] In the case of soft magnetic materials, both relations are very similar and differences are detectable only for extremely large H values.[‡] In the case of hard magnetic materials, hysteresis $B = f(H)$ is significantly different than $J = f(H)$ (Figure 3.64) and therefore we need to distinguish between coercivity for the $B(H)$ loop known as ${}_BH_c$ and coercivity for the $J(H)$ loop known as ${}_JH_c$. The coercivity ${}_JH_c$ is usually larger than ${}_BH_c$ and the difference is a measure of internal possibility of material to store the magnetic energy.

Another difference is that the soft magnetic materials are usually used in a form of a closed magnetic circuit. But in the case of permanent magnets, we never use the hard magnetic materials in the form of a closed circuit (we are interested in generation of the external magnetic field between poles of the magnet). An example of a permanent magnet with a gap of the length l_g (and cross-section A_g) is presented in Figure 3.65.

After opening of the closed circle, the remanence flux density B_r is decreased to the value B_m as result of the demagnetizing field. In the analysis of permanent magnets, we usually take into account only the second

---

[*] Data from 1995. Recently, there are probably more rare-earth-metal-based magnets.
[†] In relationship $B = f(H)$, it is not possible to identify saturation.
[‡] Between flux density B and polarization J (or magnetization M), there is the following difference: $B = J + \mu_0 H$ or $B = \mu_0 M + \mu_0 H$.

TABLE 3.20

Properties of Typical Power Ferrites

	3C90	3F3	3F4	4F1
Type	MnZn	MnZn	MnZn	NiZn
Frequency range	25–200 kHz	0.1–0.7 MHz	0.5–3 MHz	2–20 MHz
ρ (Ωm) 25° DC	5	7	10	2×10^5
ρ (Ωm) 100° DC	1.3	2	3	3×10^4
ρ 25° 100 kHz	4	6	7	1×10^5
ρ 25° 1 MHz	1	1.5	3	5×10^4
ρ 25° 10 MHz			0.6	2×10^4
Average grain size (μm)	7	5	2	3
μ (10 kHz, 0.1 mT)	2000	1800	800	80
B_s (T)	0.5	0.5	0.44	0.32
Loss P (mW/cm³)				
25 kHz, 200 mT	60	100		
100 kHz, 100 mT	70	70		
500 kHz, 50 mT		200	180	
1 MHz, 30 mT		390	150	
3 MHz, 10 mT			220	150
5 MHz, 10 mT				300
10 MHz, 5 mT				150

Source: After Stoppels, D., *J. Magn. Magn. Mater.*, 160, 323, 1996.

TABLE 3.21

Properties of the Main Microwave Materials

	J_s (T)	ΔH (kA/m)	T_c (°C)
Spinels			
$Mn_{0.1}Mg_{0.9}Fe_2O_4$	0.25	56	290
$Li_{0.2}Zn_{0.6}Fe_{2.2}O_4$	0.5	14	450
$MgFe_2O_4$	0.25	70	320
$NiFe_2O_4$	0.3	24	575
Garnets			
$Y_3Fe_5O_{12}$ (YIG)	0.18	4	280
$(Y,Al)_3Fe_5O_{12}$	0.12	6	250
$(Y,Gd)_3Fe_5O_{12}$	0.06	12	250
Hexaferrites			
$BaFe_{12}O_9$	0.45	1.5	430
$Ba_3Co_2Fe_{24}O_{41}$	0.34	12	410
$Ba_2Zn_2Fe_{12}O_{22}$	0.28	25	130

quadrant of the hysteresis loop, known as *demagnetization curve* (Figure 3.65). The value of B_m is determined as the crossing point P of the hysteresis loop and the demagnetization line (or the load line) as

$$H = -N_d \frac{B}{\mu_0} \quad \text{or} \quad \frac{B}{\mu_0 H} = -\frac{1}{N_d} \tag{3.7}$$

where N_d is the demagnetizing coefficient depending on the geometry of the sample.

For the magnetic circuit presented in Figure 3.65, we can write the following relationships (neglecting the leakage magnetic field around the gap of the area A_g):

$$H_g l_g - H_m l_m = 0 \tag{3.8}$$

$$\Phi = B_g A_g = \mu_0 H_g A_g = B_m A_m \tag{3.9}$$

From these equations, we obtain

$$H_g^2 = B_m H_m \frac{l_m A_m}{l_g A_g} \tag{3.10}$$

or

$$H_g^2 V_g = B_m H_m \cdot V_m \tag{3.11}$$

From Equation 3.11, we can conclude that

- The possibility of obtaining the magnetic field of the magnet depends directly on the (BH) value, called often as an *energy product*
- For the assumed volume of an air gap V_g and assumed maximum energy BH_{max} of hard magnetic materials, we can determine the necessary volume of the magnet V_m (see Figure 3.61)
- We should design the dimensions of the magnet to ensure that the working point P (Figure 3.65) corresponds with the maximum of $(BH)_{max}$

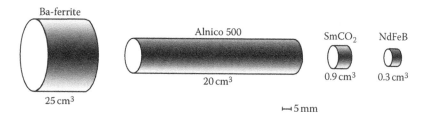

FIGURE 3.61

Permanent magnet producing the magnetic field of the flux density 0.1 T at the distance of 5 mm from the face. (After Boll, 1989.)

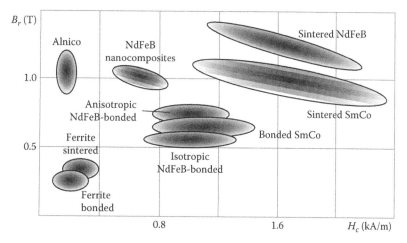

FIGURE 3.62

Comparison of the main hard magnetic materials. (From Leonowicz, 2005.)

TABLE 3.22

Properties of the Main Hard Magnetic Materials

	$(BH)_{max}$ (kJ/m³)	$_JH_c$ (kA/m)	$_BH_c$ (kA/m)	B_r (T)	T_c (°C)	Price (Relative)
FeCoCr	7	20	20	0.9		—
Ferrite	28	275	265	0.4	450	1
Alnico	40	124	124	1.2	850	10
NdFeB	320	1200	750	1.0	310	20
SmCo	160	1500	750	0.95	720	100

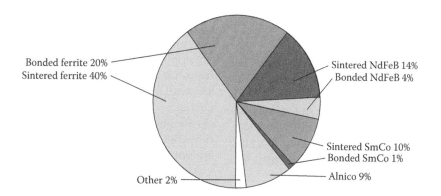

FIGURE 3.63

Permanent magnet market (1995). (After Jiles, D., *Magnetism and Magnetic Materials*, Chapman & Hall, New York, 1998.)

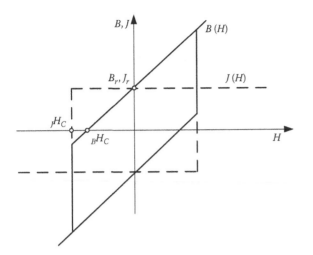

FIGURE 3.64
Ideal $B(H)$ and $J(H)$ hysteresis loops.

Figure 3.66 illustrates that we can obtain various $BH(H)$ curves for the same values of B_r, H_c.

For a theoretical ideal magnet with square-shape hysteresis loop, the optimum working point P is in the middle of the line $B_r - {}_BH_c$ and maximum energy product is

$$(BH)_{max} = \frac{1}{\mu_0}\left(\frac{B_r}{2}\frac{\mu_0\,{}_BH_c}{2}\right) = \frac{B_r^2}{4\mu_0} = \frac{J_s^2}{4\mu_0} \qquad (3.12)$$

For NdFeB magnets, this theoretical limit is estimated as 485 kJ/m^3. Usually, real magnets exhibit less than 60% of this theoretical value.*

Usually on the graph presenting the demagnetizing curve of a material, the lines B/μ_0H representing $1/N$ value (Equation 3.7) as well as those representing the BH values (Figure 3.67) are also indicated. This way the designer of the permanent magnet can at once determine possible performances of the material and recommended geometry of the magnet.

The energy stored by the magnet also depends on the energy product BH and volume of the magnet V_m (Fiorillo 2004):

$$E_{max} = \frac{1}{2}(B_mH_m)V_m \qquad (3.13)$$

In many applications (e.g., in electric machines), the gap is not fixed; sometimes a magnet armature can move up to the state of closed magnetic circuit (full contact with the poles). The flux density does not return to the remanence B_r but to other point B_{rc} (Figure 3.68). Moreover, this movement creates a small hysteresis loop. Because this loop is often very narrow, it is substituted by a line with a slope called the *recoil permeability* μ_r.

---

* 1 T = kg/A/s^2; 1 T·1 A/m = kg/s^2/m; and 1 J = kg m^2/s, thus (BH) is represented by J/m^3.

3.6.2 Alnico Alloys

Alnico was discovered in 1930 by Mishima and the first classical alloy contains 58% Fe, 30% Ni, and 12% Al. It exhibited coercivity of over 30 kA/m, which was almost double of the best steel magnets then available. Until introduction of the rare earth magnets in 1970s, Alnico was the main hard magnetic material. Recently, its importance is reduced (see Figure 3.63); its main advantage is still high working temperature—up to 500°C. Alnico was evaluated with modified composition (addition of cobalt and titanium) known as Alnico 2, Alnico 3, ..., Alnico 9 (Figure 3.69).

Various types of Alnico differ not only by the composition but also by heat treatment (including annealing in magnetic field); they are isotropic or anisotropic, non-oriented, or grain oriented. The properties of Alnico alloys are presented in Table 3.23.

Alnico are recognized as Alnico 2, ..., Alnico 9 or by their various trade names: Alni, Alcomax, Hycomax, Ticonal. In manufacturing of Alnico alloy heat treatment, often in presence of magnetic field is very important. After the heat treatment, Alnico transforms to a composite material with rich iron and cobalt precipitates in a NiAl matrix. Starting from Alnico 5, anisotropy is improved due to grain orientation—after heating in a presence of magnetic field and appropriate cooling the Fe-Co particles are long and ordered in one direction. Starting from Alnico 6, titanium is added which increases the coercivity. In Alnico 9, after appropriate heat treatment, a columnar crystallization is obtained. Figure 3.70 presents a typical annealing process of Alnico material.

The Alnico material is hard and brittle hence it is usually cast or sintered into a final shape. Alnico 5 and higher versions contain cobalt and titanium and therefore this material is more expensive. Figure 3.71 presents the demagnetization curves of the Alnico alloys.

3.6.3 Hard Magnetic Ferrites

Hard ferrites exhibit rather modest performance, especially relatively small saturation. But because of their low price, they are commonly used in less demanding applications. Hard ferrites are prepared as sintered from the powder but also composite, bonded materials are available. They are bonded with rubber, PVC, polyethylene, or polyester resins. Although bonded materials are inferior in comparison with sintered (as result of reduction of magnetic part in the whole volume), they are popular due to flexibility and low price.

Almost exclusively, two types are manufactured as hard magnetic ferrites: barium ferrite (BaO·6Fe$_2$O$_3$) and strontium ferrite (SrO·6Fe$_2$O$_3$). Both groups of hard ferrites have similar properties although strontium ferrites

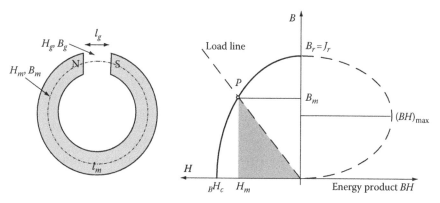

FIGURE 3.65
Open magnetic circuit of permanent magnet and its demagnetization curve.

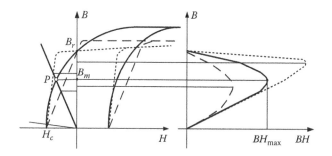

FIGURE 3.66
Energy product of the magnet for different shapes of the hysteresis loop (and the same values of B_r and H_c).

have slightly better coercivity and saturation. The properties of these ferrites are presented in Figure 3.72 and Table 3.24.

Barium and strontium ferrites have exceptional properties due to two main reasons: they exhibit very large anisotropy and the powder grain is small (diameter around 1 m), so that every grain is close to a one-domain structure. The one-domain sample is magnetized by coherent rotation of magnetization and according to Stoner-Wohlfarth model, this leads to a rectangular hysteresis loop (see Figure 2.138).

The manufacturing process of hard ferrites includes calcination in a furnace at about 1200°C, formation of the powder, and then pressing and sintering at around 1200°C. The anisotropic properties are obtained by pressing the powder in the presence of magnetic field.

3.6.4 Rare Earth Hard Magnetic Materials

High anisotropy is a crucial for properties of hard magnetic materials. Hence, many researchers looked for materials with exceptionally large magnetocrystalline anisotropy. In 1966, Hoffer and Strnat announced that they obtained material based on yttrium with very large anisotropy $K_1 = 5.7 \, MJ/m^3$ (Hoffer and Strnat 1966).

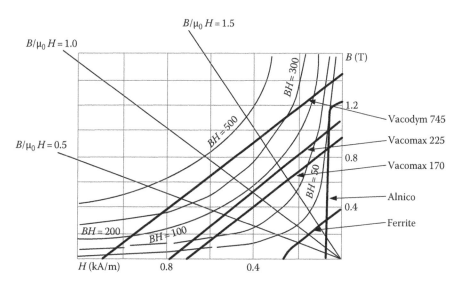

FIGURE 3.67
Demagnetization curves of various materials. (From Vacuumschmelze catalogue.)

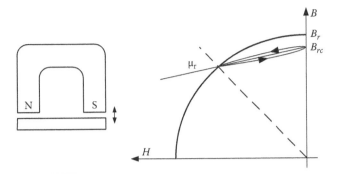

FIGURE 3.68
Dynamic work of the permanent magnet.

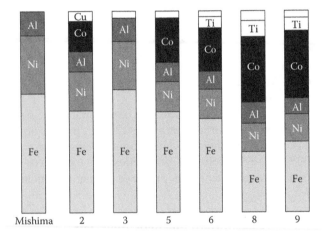

FIGURE 3.69
Various compositions of alnico alloys.

TABLE 3.23

Properties of Alnico Alloys

	B_r (T)	$_BH_c$ (kA/m)	$(BH)_{max}$ (kJ/m³)	
Alnico 2	0.72	45	12.7	Isotropic, 12%–20% Co
Alnico 3	0.7	38	11.1	Isotropic, Co-free
Alnico 4	0.55	57	10.7	Isotropic
Alnico 5	1.32	240	51.7	Anisotropic, oriented
Alnico 6	1.08	240	31	Anisotropic, titanium
Alnico 8	0.93	480	47.7	Anisotropic, high H_c
Alnico 9	1.12	480	83.6	Columnar, oriented

Source: Arnold Magnetic Technologies.

Later, other compositions—rare earth metal/ferromagnetic metals as Co or Fe were investigated. The most widely studied materials have the structure $RECo_5$, $RECo_{17}$, $REFe_{14}B$. Light lanthanides were found to be most promising. Indeed, these elements crystallized in complex hexagonal or rhombohedral structures exhibited not only high anisotropy but also large saturation polarization (Table 3.23) (Strnat et al. 1967, Strnat 1988, 1990, Buschow 1991, Kirchmayr 1996).

From data presented in Table 3.25, particularly interesting are $SmCo_5$ with exceptionally large anisotropy, $Nd_2Fe_{14}B$ with high saturation, and Sm_2Co_{17} with high Curie temperature. Indeed, in 1967, the first rare-earth permanent magnet material $SmCo_5$ was proposed as exhibiting the best properties and new era in magnet technology started (Strnat et al. 1967).

Rare-earth-based materials are prepared similarly to the ferrite materials: fine powder (diameter of grains to a few µm to obtain conditions close to one domain) is pressed (sometimes in a presence of magnetic field) and then sintered at about 1150°C for 3h. Sometimes, this material is further heat-treated at a temperature of 850°C–900°C. There are also bonded materials with polymer or epoxy resin.

Both components of the $SmCo_5$ material are expensive. Therefore, other similar materials were tested. One of them is MMCo5 or $Sm_{0.2}MM_{0.8}Co_5$ where samarium is substituted by mischmetal MM composed from various other lanthanides, for example, 55% Ce, 24% La, 16% Nd, 5% Pr, and 2% of additional lanthanide metal. In another material known as Sm_2Co_{17}, there is less cobalt and samarium and other components (e.g., Fe, Mn, Cu, Cr) are added for the improvement of magnetic properties. Figure 3.73 presents typical demagnetization characteristics of samarium–cobalt materials.

In 1984, two teams reported new hard magnetic material based on neodymium (Croat et al. 1984, Sagawa et al. 1984). The new material has much larger saturation polarization, larger coercivity, and therefore higher energy product $(BH)_{max}$ (Figure 3.74). Moreover, the new material used less expensive components. For this reason, recently the neodymium magnets slightly overshadowed the samarium-based materials.

First reported neodymium materials were obtained using classical technique (sintering) (Sagawa et al. 1984) but later also rapid solidification was employed (similar to the technology used in amorphous alloys) (Croat et al. 1984). Rapid solidification enables to obtain small grains of about 30nm while in sintered materials they are around 3µm. Theoretically (Kneller and Hawig 1991), small grains should result in higher coercivity but to date nanomaterials and nanocomposites exhibit worst parameters than sintered materials.

In other technology known as magnetquench, the alloy is rapidly quenched. After grinding, the flakes are bonded into an epoxy resin.

Figure 3.75 and Table 3.26 presents properties of typical rare-earth hard magnetic materials. The best performance is demonstrated by NdFeB material, but its Curie temperature is relatively low at 312°C. Therefore, samarium-based materials with Curie temperature 820°C are also commercially available.

Table 3.26 presents parameters of commercially available rare-earth hard magnetic materials, VACODYM

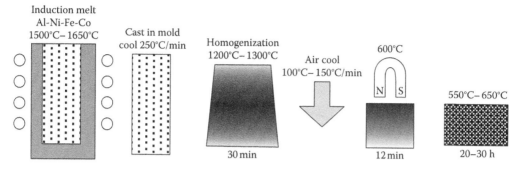

FIGURE 3.70
Typical process of heat treatment of alnico material.

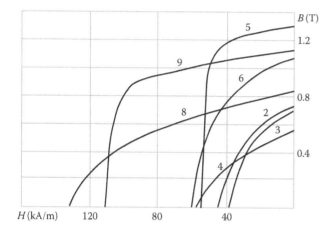

FIGURE 3.71
Demagnetization curves of different grades of alnico alloy. (From Arnold Magnetic Technologies catalogue.)

(neodymium based) and VACOMAX (samarium based), the best grades of Vacuumschmelze. Vacodym 722 with $(BH)_{max}$ equal to $415\,kJ/m^3$ is close to the theoretical limit ($485\,kJ/m^3$) (see Equation 3.12). Figure 3.76 presents characteristics of this material. As can be seen, the temperature changes of the parameters are significant. This influence can be decreased by changing geometry of the magnet (compare movement of the working point with temperature for $B/\mu_0 H = 1.0$ and $B/\mu_0 H = 2.0$).

3.7 Special Magnetic Materials

3.7.1 Thin Magnetic Films

Thin film is a nanometer range layer deposited onto a substrate. Such layer can be only a few atoms thick: fraction of a nanometer* or up to several hundred nanometer (typical thickness is several dozen of nanometers). Almost every material can be deposited, including soft

* The diameter of iron atom is about 0.15 nm.

and hard magnetic materials. Thin films are usually polycrystalline but they can be also amorphous or monocrystalline. In electronics, the most important are, of course, semiconductor thin films. In this section of the book, mainly ferromagnetic thin films are described.

Up to 1970, ferromagnetic thin films were mainly the object of interest of physicists due to their unique features different than those of bulk materials. In 1970, Hunt patented application of thin film magnetoresistor as a magnetic reading head (Hunt 1970). From that time, thin films gained rapid attention in computer industry as reading heads of discs. This development was crowned by a Nobel prize in 2007 (Fert and Grünberg) for GMR thin film for magnetoresistive effect (Fert 2008). Recently, thin ferromagnetic films are the object of interest in nanotechnology and a new scientific branch spintronics (Wolf et al. 2006, Bandyopadhyay and Cahay 2008).

One of unique features of thin ferromagnetic films (in comparison with a bulk material) is that they magnetize similarly to one domain. Properties of one domain are described by Stoner-Wohlfarth model of hysteresis (see Section 2.11.3). Thin film exhibits uniaxial anisotropy induced in process of deposition. In the initial state, a thin film is magnetized along the anisotropy axis called also *easy axis of magnetization*. Usually, we apply the magnetic field directed perpendicularly to anisotropy axis and the thin film is magnetized by coherent rotation of magnetization (hysteresis-free) up to the saturation (when the magnetic field reaches value equal to the anisotropy field H_k). If we apply the magnetic field along the easy axis, but apposite to initial magnetization, the direction of magnetization changes suddenly when the magnetic field reaches value equal to anisotropy field H_k. Thus in the hard direction of magnetization, thin film is described by linear dependence $M = f(H)$ while in the easy direction it is described by rectangular hysteresis (see Figure 2.138).

Theoretically one-domain sample cannot be demagnetized. The real thin films exhibit dispersion of

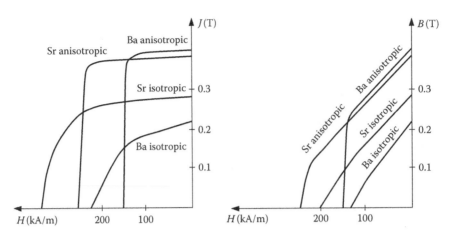

FIGURE 3.72
Demagnetization curves of different hard ferrites. (From McCurrie, R.A., *Ferromagnetic Materials*, Academic Press, London, U.K., 1994.)

TABLE 3.24

Properties of Typical Hard Ferrites

	B_r (T)	$_BH_c$ (kA/m)	$_JH_c$ (kA/m)	$(BH)_{max}$ (kJ/m³)
BaO 6Fe₂O₃				
Sintered, isotropic	0.21	145	240	7.4
Sintered, anisotropic	0.4	160	165	29.5
Bonded isotropic	0.12	88	190	2.8
SrO 6Fe₂O₃				
Sintered, anisotropic	0.38	280	320	26.8
Bonded	0.27	196	260	14.0

Source: McCurrie, R.A., *Ferromagnetic Materials*, Academic Press, London, U.K., 1994.

TABLE 3.25

Anisotropy Constant, Anisotropy Field, Saturation, and Curie Temperature of Rare Earth-Based Materials

	B_r (T)	K_1 (MJ/m³)	H_A (MA/m)	T_c (°C)
YCo₅	1.06	5.5	10.4	630
LaCo₅	0.91	6.3	14.0	567
PrCo₅	1.20	8.1	13.5	620
NdCo₅	1.22	0.24	0.4	637
SmCo₅	1.14	11–20	20–35	727
Nd₂Fe₁₄B	1.69		5.4	312
Sm₂Fe₁₄B	1.52		11.9	343
Sm₂Co₁₇	1.25	3.2	5.2	920

Source: From *Ferromagnetic Materials*, Vol. 4, Strnat, K.J., Copyright (1988); Herbst, J.F. and Croat, J.J., *J. Magn., Magn. Mater.*, 100, 57, 1991.

anisotropy and therefore they are not in ideal one-domain state. Dispersion of anisotropy means that the direction (angle dispersion of anisotropy) and magnitude (magnitude dispersion of anisotropy) can vary on the plane from one grain to another (not perfect texture).

Figure 3.77 presents comparison of theoretical and real hysteresis loops of a thin film element. We can see that also in hard direction of magnetization, there is a narrow hysteresis. But if magnetizing field is sufficiently small ($H_x \ll H_k$), the magnetization is linear and hysteresis free (Figure 3.77c). This area is of particular interested for sensor applications (Tumanski 2001) because between direction of magnetization φ and value of external field H_x is the dependence close to linear (see Figure 2.39):

$$\sin \varphi = \frac{H_x}{H_y + H_k} \tag{3.14}$$

If external magnetic field is higher than the anisotropy field, the thin film can be demagnetized (see Figure 2.141). We can avoid such a situation by artificial increase of anisotropy—by additional biasing the film with the field H_B directed along the easy anisotropy axis:

$$\sin \varphi = \frac{H_x}{H_y + H_k + H_B} \tag{3.15}$$

Thin film can be deposited with various technologies (Figure 3.78) such as by electroplating and chemical methods, but most commonly, sputtering technique or vacuum evaporation is used. In the evaporation method, the source of the material is heated by electron beam and the vapor of metal is transferred to the substrate in vacuum. Glasses, oxidized silicon, or ceramic material can be used as the substrate. To improve adhesion to the substrate, often predeposition of the buffer layer, for example, titanium or tantalum is used. The anisotropy of thin film is induced by deposition in presence of the magnetic field.

The evaporation technique is used mostly in laboratory. In industry, often the sputtering technique is more efficient. The chamber of deposition is filled by a

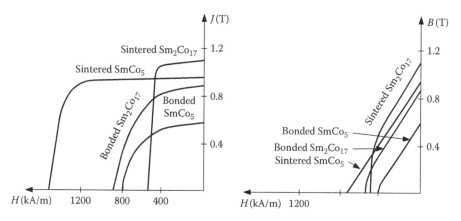

FIGURE 3.73
Typical demagnetization characteristics of samarium–cobalt materials. (From McCaig, M. and Clegg, A.G., *Rare Earth Permanent Magnets in Theory and Practice*, Pentech Press, London, U.K., 1987; Jiles, D., *Magnetism and Magnetic Materials*, Chapman & Hall, New York, 1998.)

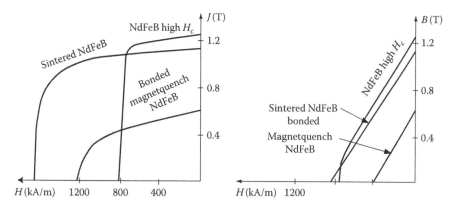

FIGURE 3.74
Typical demagnetization characteristics of neodymium–iron–boron materials. (From McCaig, M. and Clegg, A.G., *Rare Earth Permanent Magnets in Theory and Practice*, Pentech Press, London, U.K., 1987; Jiles, D., *Magnetism and Magnetic Materials*, Chapman & Hall, New York, 1998.)

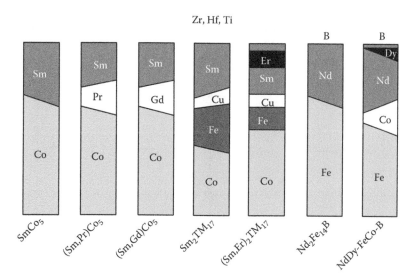

FIGURE 3.75
Various compositions of rare earth hard magnetic materials. (After Strnat, K.J., *Proc. IEEE*, 78, 923, 1990.)

TABLE 3.26

Properties of Typical Rare-Earth Hard Magnetic Materials

	B_r (T)	$_BH_c$ (kA/m)	$_JH_c$ (kA/m)	$(BH)_{max}$ (kJ/m³)
SmCo$_5$	1.00	790	1500	196
MMCo$_5$	0.80	620	700	124
Sm$_{0.4}$Pr$_{0.6}$Co$_5$	1.03	800	1300	208
Sm$_2$(Co$_{0.8}$Fe$_{0.14}$Mn$_{0.04}$Cr$_{0.02}$)$_{17}$	1.13	880	1000	248
Sm$_2$TM$_{17}$ (bonded)	0.89	540	1200	136
NdFeB sintered	1.22	840	1280	280
NdFeB magnetquench	1.17	840	1040	256
Sm$_2$Fe$_{17}$N$_3$	1.22	750	2400	160
NdFeB nano (30 nm)	1.12		440	157
Vacodym (Vacumschelze)	1.47	875	915	415
Vacomax (Vacuumschelze)	1.12	640	730	240

Source: After McCurie (1994).

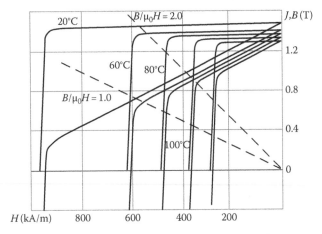

FIGURE 3.76
Demagnetization characteristics of VACODYM 722 HR hard magnetic material. (From Vacuumschmelze catalogue.)

noble gas, for example, argon. Under the electrical field between the anode (attributed to the substrate) and the cathode (attributed to the source), the glow discharge appears. The ionized gas ejects the atoms of the material from the source and moves them into the substrate.

By changing sources, it is possible to deposit various materials and in this way form a multilayer structure.

In the thin film technology, molecular beam epitaxy (MBE) technique is very important (Figure 3.79). In this technique, it is possible to obtain almost perfect crystal structure because deposited film mimics the structure of substrate acting as a seed crystal. Modern MBE apparatus is equipped with the testing method, for example, reflection high energy electron diffraction (RHEED). This enables to control the deposition process, crystal structure, and composition. MBE method utilizes very high vacuum.

As a ferromagnetic thin film material, most commonly permalloy 81–19 is used. Figure 3.80 presents the main properties of NiFe alloys. The main advantage of 81/19 alloy (permalloy) is that its magnetostriction coefficient (λ) is close to zero (non-magnetostrictrive material). Moreover, this material also exhibits relatively small anisotropy ($H_k = 250$ A/m), large magnetoresistance ($\Delta\rho/\rho = 2.2\%$), and small dispersion of anisotropy (α).

In GMR technology, thin film is created in a form of multilayer: typically two ferromagnetic layers separated by a conducting layer (spacer). This conducting layer

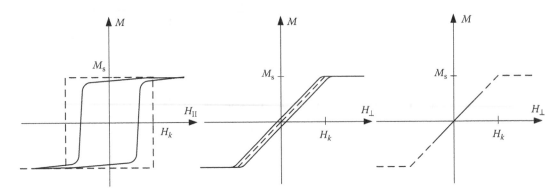

FIGURE 3.77
Hysteresis lops of thin film magnetized along the anisotropy axis $H_{||}$ (a) and perpendicular to the anisotropy axis $H_\perp$ (b and c) (dashed line, theoretical model).

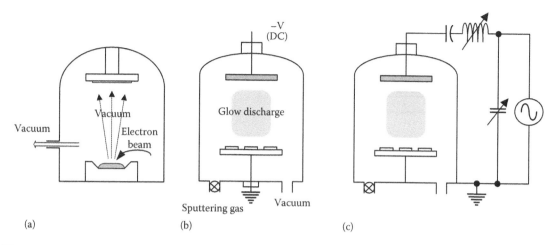

FIGURE 3.78
Various methods of thin film deposition: evaporation (a), DC sputtering (b), RF sputtering.

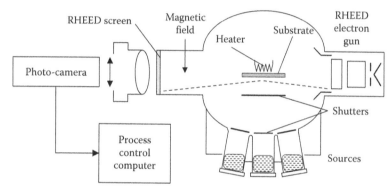

FIGURE 3.79
Principle of operation of the MBE deposition system.

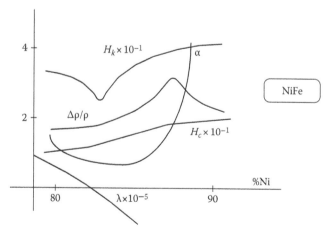

FIGURE 3.80
Performance of NiFe thin film.

should be very thin (less than a few nanometers) and therefore deposition of such thin layer without defects (especially pin-holes) is crucial for this technology. Even more difficult is deposition of the spacer layer for a magnetic tunnel junction. In this case, the intermediate layer should be an insulator; commonly, it is aluminum layer that is later oxidized.

After deposition, final shape of the thin film element is formed. This includes etching of the final form (most often a strip), leads or terminals, additional bias layers, additional planar coils, cover layer, etc. This process can change the properties of the thin film due the shape anisotropy (see Equation 2.214). It is not recommended to form a shape inclined at some angle with respect to anisotropy axis of material because the resultant anisotropy axis is then different than the axis of the material. The most critical case is when the strip is perpendicular to easy anisotropy axis because it causes significant increase of the dispersion of anisotropy.

The main application of thin magnetic films is the reading head for disk drives. Such a head is a complex device in which a magnetoresistive reading head is usually merged with thin film writing head and both are inserted between thick film shields (Figure 3.81). Also the magnetoresistive element of the head is a complex multilayer design. Figure 3.82 presents a surface view of a spin valve head. The GMR sensor is composed of

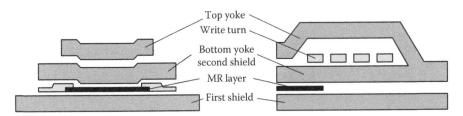

FIGURE 3.81
Typical design of a shielded reading head.

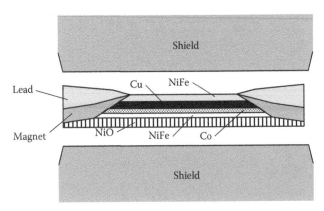

FIGURE 3.82
Surface view of typical spin valve sensing element of the reading head.

four layers: two permalloy layers separated by a Cu layer and additional Co layer for improvement of the performances. This element is additionally magnetized by antiferromagnetic layer of NiO. Sometimes, a magnet part for biasing the permalloy layer is added. More information about spin-valve sensors is given later in the part devoted to magnetic sensors.

Overcoat/lubricant	$\sim$ 4 nm	
Recording layer	$\sim$ 15 nm	
Intermediate layer	$\sim$ 20 nm	
Soft magnetic underlayer	$\sim$ 80 nm	
Adhesion layer	$\sim$ 10 nm	
Substrate	$\sim$ few mm	

FIGURE 3.83
Typical functional layers in perpendicular recording media (in the case of longitudinal recording soft magnetic underlayer is not necessary). (After Piramanayagam, S.N., *J. Appl. Phys.*, 102, 011301, 2007.)

Progress in the density of information recording (due to miniaturization or writing/reading heads) forces further improvements in recording. Materials should follow these improvements. Previously as most commonly used recording material for tape recorders was γFe_2O_3 with coercivity of about 30 kA/m. Recently, hard disks composed of a few platters are used for data recording and storage. Onto such platters quite complex multilayer structure is deposited by sputtering (Figure 3.83).

Typical recording media consist of six functional layers (Figure 3.83)—each of which can be composed from further several sub-layers, for example, two antiferromagnetic coupled layers separated by third layer as recording layer can be used as a recording layer. Of course for quality of recording the recording layer is crucial—it is a polycrystalline layer. Such material should exhibits high coercivity, high anisotropy, large saturation magnetization, and small grains (not coupled magnetically to each other). Recently, CoCrPt-oxide-based material is used commonly as magnetic recording medium. This material exhibits coercivity of about $H_c \sim 300$ kA/m and anisotropy $H_k \sim 1000$ kA/m. The average diameter of the grains is around 8 nm.

Typical hard disk platter is manufactured as follows. A disk from aluminum with several additions (magnesium silicon, copper, and iron to improve mechanical properties) is used as a substrate and its thickness is 0.635, 0.8, 1.0, 1.27, 1.5, or 1.8 mm. This aluminum platter is carefully polished to decrease the roughness to below 0.1 nm. Then an adhesion layer (typically NiP) is deposited and followed by a magnetically soft underlayer (e.g., amorphous CoTaZr- or NiFe-based alloys). CoCrPt:SiO$_2$ layer is typically used as a recording layer.

Recently longitudinal recording practically reached limit of its possibilities. According to "superparamagnetic limit" $KV > 40$ kBT (K is magnetic anisotropy and V is volume of a grain), which means that the grain cannot be smaller than about 8 nm. Below this value, there is a risk of unstable reading with the temperature fluctuations. This limit for longitudinal recording is about 130 Gbit/in.² and this value is recently demonstrated. Further improvement of density recording, up to about 500 Gbit/in.², is possible in slightly more complicated perpendicular recording. In this recording, the reading head is a "monopole" instead of "ring"

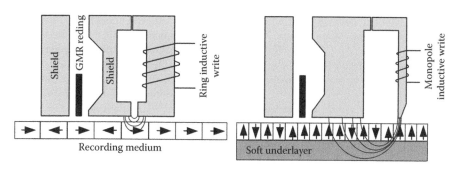

FIGURE 3.84
Reading/writing head and recorder medium in longitudinal and perpendicular recording.

(Figure 3.84). Also the recording medium should be modified. The anisotropy of the grains should be directed perpendicularly to the surface. Moreover, additional soft magnetic underlayer should be applied to close the magnetic flux beneath the recording layer. It is estimated that in order to achieve the density 500 Gbit/in.^{2}, the recording medium should exhibit coercivity of about 500 kA/m and the grain diameters should be about 6 nm (Piramanayagam 2007).

3.7.2 Ferrofluids and Magnetorheological Liquids

Magnetic liquids consist of small ferromagnetic or ferrimagnetic particles (most often γFe_2O_3, Fe_3O_4, or Co) suspended in a carrier liquid (usually mixture of water and oil). There are two classes of these materials: magnetorheological liquid with particle dimensions at the order of μm and ferrofluid with much smaller particles, typically of dimension around 10 nm.

In both cases, viscosity depends on external magnetic field, what can be used in lubricants, dampers, shock absorbers, etc. From magnetic point of view, the two materials differ; the magnetorheological liquid behaves like a fluid ferromagnetic material but the ferrofluid with very small particles suspended by Brownian thermal motion is closer to a paramagnetic state. Although we can assume the nanometer range particles as one-domain particles (thus elementary magnets) due to thermal movements, they are distributed randomly. This effect is known as superparamagnetism. In the absence of magnetic field, the average resultant magnetization of ferrofluid is not observed but in the presence of magnetic field, it can be polarized magnetically according to Langevin law with susceptibility much higher than in classical paramagnetic materials.

Ferrofluids are commonly known from their fantastic geometrical effects in the presence of magnetic field (Figure 3.85) and are used in art (there are readily available clips on YouTube). But due to their unique features, they are used in many applications: their birefringence effect is employed in light polarizers, telescopes, antiradar systems, in medicine for improvement contrast in

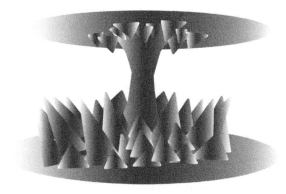

FIGURE 3.85
Surface structure of ferrofluid in presence of magnetic field.

NMR method of tumor detection, as domain and magnetic field viewers. The magnetorheological liquids are commonly used in mechatronics and robotic devices.

At the end of this chapter devoted to special magnetic materials it is worth to say several words about extraordinary magnetic material known as *Heusler alloy*. The curiosity of this alloy discovered in 1903 by Heusler (Heusler 1903, 1912, Webster 1969) is that it is a ferromagnetic material although it is composed of non-magnetic elements, for example Cu_2MnSn, Cu_2MnAl, Pd_2MnAl. The ferromagnetic features are obtained due to special crystal structure of these alloys with double-exchange mechanism between neighboring ions. Many years these materials were studied mainly by physicians but recently application in spintronics elements is strongly reported (Ishikawa 2006).

References

Adam J.D., Davis L.E., Dionne G.F., Schloeman E.F., Stitzer S.N., 2002, Ferrite device and materials, *IEEE Trans. Microwave Theory Tech.*, **50**, 721–737.

Adams E., 1962, Recent developments in soft magnetic alloys, *J. Appl. Phys.*, **33**, 1214–1220.

Arai K.I., Ishiyama K., 1994, Recent development of new soft magnetic materials, *J. Magn. Magn. Mater.*, **133**, 233–237.

Assmus P., Detent K., Ibe G., 1957, *Z. Metallk.*, **48**, 341.

Bandyopadhyay S., Cahay M., 2008, *Introduction to Spintronics*, CRC Press, Boca Raton, FL.

Barret W.F., Brown W., Hadfield R.A., 1902, Researches on the electrical conductivity and magnetic properties of upwards of one hundred alloys of iron, *J. IEE*, **31**, 674–729.

Beckley P., 2000, *Electrical Steels, European Electrical Steel*, Orb Works, New Port, U.K.

Beckley P., Snell D., Lockhart C., 1985, Domain control by spark ablation, *J. Appl. Phys.*, **57**, 4212–4213

Beckley P., 2002, *Electrical Steels for Rotating Machines*, IEE.

Bolfarini C., Silva M.C.A., Jorge A.M., Kiminami C.S., Botta W.J., 2008, Magnetic properties of spray-formed Fe-6.5% Si and Fe-6.5%Si-1%Al after rolling and heat treatment, *J. Magn. Magn. Mater.*, **320**, e653–e656.

Boll R., 1989, Sensors, a comprehensive survey, in *Magnetic Sensors*, Vol. 5, Boll R., Overshott K.J. (Eds.), VCH Publishers Inc., Deerfield Beach, FL, Chapter 1.

Brissonneau P., 1980, Non-oriented SiFe sheets, *J. Magn. Magn. Mater.*, **19**, 52–59.

Brissonneau P., 1984, Non-oriented electrical sheets, *J. Magn. Magn. Mater.*, **41**, 38–46.

Bularzik J.H., Krause R.F., Kokal H.R., 1998, Properties of new soft magnetic material for AC and DC motor applications, *J. Phys.*, **8**, Pr2-747–Pr-2853.

Buschow K.H.J., 1991, New developments in hard magnetic materials, *Rep. Prog. Phys.*, **54**, 1123–1213.

Chen H.S., 1980, Glassy metals, *Rep. Prog. Phys.*, **43**, 353–432.

Chin G.Y., 1971, Review of magnetic properties of FeNi alloys, *IEEE Trans. Magn.*, **7**, 102–113.

Coombs A., Lindemno M., Snell D., Power D., 2001, Review of the types, properties, advantages and latest developments in insulating coating on non-oriented electrical steels, *IEEE Trans. Magn.*, **37**, 544–557.

Couderchon G., Tiers J.F., 1982, Some aspects of magnetic properties of NiFe and CoFe alloys, *J. Magn. Magn. Mater.*, **26**, 196–214.

Croat J.J., Herbst J.F., Lee R.W., Pinkertyon F.E., 1984, Pr-Fe and Nd-Fe-based materials: a new class of high-performance permanent magnets, *J. Appl. Phys.*, **55**, 2078–2082.

Crottier-Combe S., Audisio S., Degaugue J., Porteseil J.L., Ferrara E., Pasquale M., Fiorillo F., 1996, The magnetic properties of FeSi 6.5 wt% alloys obtained by a SiCl4based CVD process, *J. Magn. Magn. Mater.*, **160**, 151–153.

Degauque J., Fiorillo F., 2006, Alliages magnetiques doux enrichis en silicium, in *Materiaux magnetiques en genie electrique I*, Kedous-Lebouc A. (Ed.), Lavoisier, Chapter 4, pp. 227–286.

De Wulf M., 2006, Aciers electriques non orientes pour machines electriques et autres applications, in *Materiaux magnetiques en genie electrique I*, Kedous-Lebouc A. (Ed.), Lavoisier, Chapter 2, pp. 87–148.

Dionne G., 1975, A review of ferrites for microwave applications, *Proc. IEEE*, **63**, 777–790.

Egami T., 1984, Magnetic amorphous alloys, *Rep. Prog. Phys.*, **47**, 1601–1725.

Fastenau R.H.J., van Loenen E.J., 1996, Applications of rare earth permanent magnets, *J. Magn. Magn. Mater.*, **157–158**, 1–6.

Fert A., 2008, Origin, development and future of spintronics (Nobel lecture), *Angew. Chem. Int.*, **47**, 5956–5967.

Fiedler H.C., 1977, A new high induction grain oriented 3% silicon iron, *IEEE Trans. Magn.*, **13**, 1433–1436.

Fiorillo F., 1996, Advances in Fe-Si properties and their interpretation, *J. Magn. Magn. Mater.*, **157–158**, 428–431.

Fiorillo F., 2004, *Measurement and Characterization of Magnetic Materials*, Elsevier, Amsterdam, the Netherlands.

Fish G.E., 1990, Soft magnetic materials, *Proc. IEEE*, **78**, 947–972.

Fukuda B., Satoh K., Ichida T., Itoh Y., Shimanaka H., 1981, Effects of surface coating on domain structure in grain oriented 3% SiFe, *IEEE Trans. Magn.*, **17**, 2878–2880.

Geoffroy O., Porteseil J.L., 2005, Soft materials for electrical engineering and low frequency electronics, in *Magnetism—Materials and Applications*, du Tremolet de Lacheisserie E., Gignoux D., Schlenker M. (Eds.), Springer, New York, Chapter 16, pp. 89–153.

Goldman A., 1999, *Handbook of Modern Ferromagnetic Materials*, Kluwer Academic Publishers, Boston, MA.

Goss N., 1934, Electrical sheet and method and apparatus for its manufacture and test, U.S. Patent 1965559.

Goss N., 1935, New development in electrical strip steels characterized by fine grain structure approaching the properties of a single crystal, *Trans. Am. Soc. Metals*, **23**, 511–531.

Goto I., Matoba I., Imanaka T., Gotoh T., Kan T., 1975, Development of a new grain oriented silicon steels RG-H, in *Proceedings of 2nd Soft Magnetic Conference*, pp. 262–268.

Hadfield R., 1903, Magnetic composition and method of making some, U.S. Patent 745829.

Haiji H., Okada K., Hiratani T., Abe A., Ninomiya M., 1996, Magnetic properties and workability of 6.5% Si steel sheet, *J. Magn. Magn. Mater.*, **160**, 109–114.

Hasegawa R. (Ed.), 1983, *Glassy Metals*, CRC Press, New York.

Hasegawa R., 2000, Present status of amorphous soft magnetic alloys, *J. Magn. Magn. Mater.*, **215–216**, 240–245.

Hasegawa R., 2006, Advances in amorphous and nanocrystalline magnetic materials, *J. Magn. Magn. Mater.*, **304**, 187–191.

Hasegawa R., Azuma D., 2008, Impacts of amorphous metal-based transformers on energy efficiency and environment, *J. Magn. Magn. Mater.*, **320**, 2451–2456.

Herbst J.F., Croat J.J., 1991, Neodymium-iron-boron permanent magnets, *J. Magn., Magn. Mater.*, **100**, 57–78.

Herzer G., 1990, Grain size dependence of coercitity and permeability in nanocrystalline ferromagnets, *IEEE Trans. Magn.*, **26**, 1397–1310.

Herzer G., 1997, Nanocrystalline soft magnetic alloys, in *Handbook of Magnetic Materials*, Vol. 10, Elsevier, Amsterdam, the Netherlands, Chapter 3, pp. 415–462.

Heusler F., 1903, Über magnetische manganlegierungen, *Verhand. Deutsch. Phys. Ges.*, **5**, 219.

Heusler F., Take E., 1912, Nature of the Heusler alloy, *Trans. Faraday Soc.*, **8**, 169–184.

Hilzinger H.R., 1990, Recent advances in rapidly solidified soft magnetic materials, *J. Magn. Magn. Mater.*, **83**, 370–374.

Hoffer G., Strnat K., 1966, Magnetocrystalline anisotropy of YCo5 and Y2Co17, *IEEE Trans. Magn.*, **2**, 487–489.

Hunt R.P., 1970, Magnetoresistive head, U.S. Patent 3493694.

Ichijima I., Nakamura M., Nozawa T., Nakata T., 1984, Improvement of magnetic properties in thinner HiB with domain-refinement, *IEEE Trans. Magn.*, **20**, 1557–1559.

Ishikawa T., Marukame T., Kijima H., Matsuda K.I., Uemura T., Arita M., Yamamoto M., 2006, Spin-dependent tunneling characteristics of fully epitaxial magnetic tunneling junctions with a full-Heusler alloy Co_2MnSi thin film and a MgO tunnel barrier, *Appl. Phys. Lett.*, **89**, 192505.

Iuchi T., Yamaguchi S., Ichiyama T., 1982, Laser processing for reduction core loss of grain oriented silicon steel, *J. Appl. Phys.*, **53**, 2410–2412.

Jiles D., 1998, *Magnetism and Magnetic Materials*, Chapman & Hall, New York.

Kawamura Y., Inoue A., Kojima A., Masumoto T., 1994, Fabrication of nanocrystalline Fe86Zr7B6Cu1 soft magnetic compacts with high saturation magnetization, *J. Appl. Phys.*, **76**, 5545–5551.

Kazimierczuk M., 2009, *High-Frequency Magnetic Components*, John Wiley & Sons, New York.

Kirchmayr H.R., 1996, Permanent magnets and hard magnetic materials, *J. Phys. D.*, **29**, 2763–2778.

Kneller E.F., Hawig R., 1991, The exchange-spring magnet: A new material principle for permanent magnets, *IEEE Trans. Magn.*, **27**, 3588–3600.

Kohler D., 1960, Promotion of cubic grain growth in 3% silicon iron by control of annealing atmosphere conditions, *J. Appl. Phys.*, **31**, 408s–409s.

Kolano R., Kolano-Burian A., 2002, Soft magnetic amorphous and nanocrystalline materials—New generation of materials for electrical engineering, *Przegl. Elektr.*, **78**, 241–248.

Krause R.F., Rauch G.C., Kasner W.H., Miller R.A., 1984, Effect of laser scribing on the magnetic properties and domain structure of high-permeability 3% SiFe, *J. Appl. Phys.*, **55**, 2121–2123.

Kronmüller H., 1995, Recent development in high-tech magnetic materials, *J. Magn. Magn. Mater.*, **140–144**, 25–28

Lebourgeois R., Guyot M., 2006, Les ferrites doux, in *Materiaux magnetiques en genie electrique II*, Kedous-Lebouc A. (Ed.), Lavoisier, Chapter 3, pp. 149–198.

Leonowicz M., Wyslocki J., 2005, *Modern Permanent Magnets* (in Polish), WNT, Warsaw, Poland.

Littmann M.F., 1967, Structures and magnetic properties of grain-oriented 3.2% silicon iron, *J. Appl. Phys.*, **38**, 1104–1108.

Littmann M.F., 1971, Iron and silicon-iron alloys, *IEEE Trans. Magn.*, **7**, 48–61.

Littman M.F., 1982, Grain-oriented silicon steel sheets, *J. Magn. Magn. Mater.*, **26**, 1–10.

Luborsky F.E., Frischmann P.G., Johnson L.A., 1980, Amorphous materials—A new class of soft magnetic alloys, *J. Magn. Magn. Mater.*, **19**, 130–137.

Masumoto H., 1936, On a new alloy Sendust and its magnetic and electric properties, *Sci. Rep. Tohoku Imp. Univ.*, **95**, 388–402.

Matsumara K., Fukuda B., 1984, Recent developments of non-oriented electrical steel sheets, *IEEE Trans. Magn.*, **20**, 1533–1538.

McCaig M., Clegg A.G., 1987, *Rare Earth Permanent Magnets in Theory and Practice*, Pentech Press, London, U.K.

McCurrie R.A., 1994, *Ferromagnetic Materials*, Academic Press, London, U.K.

McHenry M.E., Willard M.A., Laughlin D.E., 1999, Amorphous and nanocrystalline materials for applications as soft magnets, *Prog. Mater. Sci.*, **44**, 291–443.

Moorjani K., Coey J.M.D., 1984, *Magnetic Glasses*, Elsevier, Amsterdam, the Netherlands.

Morito N., Kamatsubara M., Shimizu Y., 1998, History and recent development of grain oriented electrical steel at Kawasaki Steel, *Kawasaki Steel Technical Report*, **39**, 3–12.

Moses A.J., 1990, Electrical steels: Past, present and future developments, *IEE Proc.*, **137**, 233–245.

Moses A.J., 1992, Development of alternative magnetic core materials and incentives for their use, *J. Magn. Magn. Mater.*, **112**, 150–155.

Moses A.J., 1994, Steel restructured, *IEEE Rev.*, **40**, 11–14.

Moses A.J., 2003, Soft magnetic materials for future power applications, *Przegl. Elektr.*, **79**, 457–460.

Moses A.J., 2004, Measurement and prediction of iron loss in electrical steel under controlled magnetization condition, *Przegl. Elektr.*, **80**, 1181–1187.

Nicolas J., 1980, Microwave ferrites, in *Ferromagnetic Materials*, Wohlfarth E.P. (Ed.), North Holland, Amsterdam, the Netherlands, Chapter 4.

Nozawa T., Mizogami M., Mogi H., Matsuo Y., 1996, Magnetic properties and dynamic domain behavior in grain-oriented 3% SiFe, *IEEE Trans. Magn.*, **32**, 572–589.

Nozawa T., Yamamoto T., Matsuo Y., Ohya Y., 1978, Relationship between total losses under tensile stress in 3 percent DiFe single crystal and their orientations near (110)[001], *IEEE Trans. Magn.*, **14**, 252–257.

Nozawa T., Yamamoto Y., Matsuo Y., Ohya Y., 1979, Effect of scratching on losses in 3-percent SiFe single crystals with orientation near (110)[001], *IEEE Trans. Magn.*, **15**, 972–981.

O'Handley R.C., 2000, *Modern Magnetic Materials*, John Wiley & Sons, New York.

Özgür U., Alivov Y., Morkoc H., 2009, Microwave ferrites—Fundamental properties, *J. Mater. Sci.*, **20**, 789–834.

Pardavi-Horvath M., 2000, Microwave applications of soft ferrites, *J. Magn. Magn. Mater.*, **215–216**, 171–183.

Peuzin J.C., 2005, Soft materials for high frequency electronics, in *Magnetism—Materials and Applications*, du Tremolet de Lacheisserie E., Gignoux D., Schlenker M. (Eds.), Springer, New York, Chapter 17, pp. 155–210.

Pfeifer F., Radeloff C., 1980, Soft magnetic NiFe and CoFe alloys—Some physical and metallurgical aspects, *J. Magn. Magn. Mater.*, **19**, 190–207.

Pfützner H., 1992, Performance of new materials in transformer cores, *J. Magn. Magn. Mater.*, **112**, 399–405.

Piramanayagam S.N., 2007, Perpendicular recording media for hard disk drives, *J. Appl. Phys.*, **102**, 011301.

Roy R.K., Pada A.K., Ghosh M., Mitra A., Ghosh R.N., 2009, Effect of annealing treatment on soft magnetic properties of Fe-6.5 Si wide ribbon, *J. Magn. Magn. Mater.*, **321**, 2865–2870.

Sagawa M, Fujimura S., Togawa N., Yamamoto H., Matsuura Y., 1984, New material for permanent magnets on a base of Nd and Fe, *J. Appl. Phys.*, **55**, 2083–2087.

Sakakura A., Hoshina K., Uematsu Y., Igawa T., Fujimoto H., 1988, Process for producing electrical steel sheet, U.S. Patent 4762575.

Sato K., Ishida M., Hina E., 1998, Heat-proof domain-refined grain-oriented electrical steel, *Kawasaki Steel Technical Report*, **39**, 21–28.

Schneider J., Schoppa A., Peters K., 1998, Magnetic application choice among different nonoriented electrical steels, *J. Phys.*, **8**, Pr2-755–Pr2-762.

Schoppa A., Schneider J., Wuppermann C.D., 2000, Influence of manufacturing process on the magnetic properties of non-oriented electrical steels, *J. Magn. Magn. Mater.*, **215–216**, 74–78.

Shiling J.W., Houze G.L., 1974, Magnetic properties and domain structure in grain oriented 3% SiFe, *IEEE Trans. Magn.*, **10**, 195–223.

Shilling J.W., Morris W.G., Osborn M.L., Prakash Rao, 1978, Orientation dependence of domain wall spacing and losses in 3-percent SiFe single crystals, *IEEE Trans. Magn.*, **14**, 104–111.

Shimanaka H., Ito Y., Matsumara K., Fukuda B., 1982, Recent development of non-oriented electrical steel sheets, *J. Magn. Magn. Mater.*, **26**, 57–64.

Shimanaka H., Ito Y., Matsumara K., Irie T., Nakamura H., Shono Y., 1980, *Trans. Met. Soc. AIME*, 193.

Sixtus K.J., 1935, Magnetic anisotropy in silicon steel, *Physica*, **6**, 105–111.

Snoek J.L., 1936, Magnetic and electrical properties of the binary systems MO Fe_2O_3, *Physica*, **3**, 463–483.

Stodolny J., 1995, Alternative materials and development of conventional electrical sheets, *Metall. Foundry Eng.*, **21**, 307–318.

Stoppels D., 1996, Development in soft magnetic power ferrites, *J. Magn. Magn. Mater.*, **160**, 323–328.

Strnat K., Hoffer G., Olson J., Ostertag W., Becker J.J., 1967, A family of new cobalt-base permanent magnet materials, *J. Appl. Phys.*, **38**, 1001–1002.

Strnat K.J., 1988, Rare-earth cobalt permanent magnets, in *A Handbook of the Properties of Magnetically Ordered Substances*, Vol. 4, Wohlfarth E.P., Buschow K.J.H. (Eds.), North-Holland, Amsterdam, the Netherlands, pp. 131–209.

Strnat K.J., 1990, Modern permanent magnets for applications in electro-technology, *Proc. IEEE*, **78**, 923–946.

Sugimoto M., 1999, The past, present and future of ferrites, *J. Am. Ceram. Soc.*, **82**, 269–280.

Suzuki K., Makino A., Inoue A., Masumoto T., 1991, Soft magnetic properties of nanocrystalline bcc FeZrB and FeMBCu (M—transitiom metal) alloys with high saturation magnetization, *J. Appl. Phys.*, **70**, 6232–6237.

Taguchi S., Sakakura A., 1964, Process of producing single-oriented silicon steel, U.S. Patent 3159511.

Taguchi S., Sakakura A., 1969, Characteristics of magnetic properties of grain oriented silicon iron with high permeability, *J. Appl. Phys.*, **40**, 1539–1541.

Takahashi N., Ushigami Y., Yabumoto M., Suga Y., Kobayashi H., Nakayama T., Nozawa T., 1986, Production of very low core loss grain-oriented silicon steel, *IEEE Trans. Magn.*, **22**, 490–495

Tumanski S., 2001, *Thin Film Magnetoresistive Sensors*, IOP Publishing, Briatol, U.K.

Tumanski S., 2002, Investigations of two-dimensional properties of selected electrical steel samples by means of the Epstein method, in *Proceedings of the 7th Workshop on 2D Measurements*, Lüdenscheid, Germany, September 16–17, 2002, pp. 151–157.

Villard M.A., Huang M.Q., Laughlin D.E., McHenry M.E., Cross J.O., Harris V.G., Franchetti C.,1999a, Magnetic properties of HITPERM (Fe,Co)88Zr7B4Cu1 magnets, *J. Appl. Phys.*, **85**, 4421–4423.

Villard M.A., Gingras M., Lee M.J., Harris V.G., Laughlin D.E., McHenry M.E., 1999b, Magnetic properties of HITPERM (Fe,Co)88Zr4B4Cua nanocrystalline magnets, *Mat. Res. Soc.*, **577**, 469–479.

Waecklerle T., 2006, Materiaux magnetiques doux speciaux et applications, in *Materiaux magnetiques en genie electrique I*, Kedous-Lebouc A. (Ed.), Lavoisier, Chapter 3, pp. 153–223.

Waeckerle T., Alves F., 2006a, Alliages magnetiques amorphes, in *Materiaux magnetiques en genie electrique II*, Kedous-Lebouc A. (Ed.), Lavoisier, Chapter 1, pp. 17–75.

Waeckerle T., Alves F., 2006b, Alliages magnetiques nanocritallins, in *Materiaux magnetiques en genie electrique II*, Kedous-Lebouc A. (Ed.), Lavoisier, Chapter 2, pp. 79–145.

Webster P.J., 1969, Heusler alloys, *Cont. Phys.*, **10**, 559–577.

Wilczynski W., Schoppa A., Schneider J., 2004, Influence of different fabrications steps of magnetic core on their magnetic properties, *Przegl. Elektr.*, **80**, 118–122.

Wolf S.A., Chtchelkanova A.Y., Treger D.M., 2006, Spintronics—A retrospective and perspective, *IBM J. Res. Dev.*, **50**, 101–110.

Yabumoto M., Kobayashi H., Nozawa T., Hirose K., Takahashi N., 1987, Heatproof domain refining method using chemically etched pits on the surface of grain oriented SiFe, *IEEE Trans. Magn.*, **23**, 3062–3064.

Yamamoto T., Taguchi S., Sakakura A., Nozawa T., 1972, Magnetic properties of grain-oriented silicon steel with high permeability orientcore HiB, *IEEE Trans. Magn.*, **8**, 677–681.

Yoshizawa Y., Oguma S., Yamauchi K., 1988, New Fe-based soft magnetic alloys composed of ultrafine grain structure, *J. Appl. Phys.*, **64**, 6044–6046.

4

Magnetic Sensors

4.1 General Remarks

4.1.1 Magnetic Field Sensors—Classification

Most of magnetic sensors are really sensors of flux density of magnetic field*—there are some sensors that measure other values by magnetic principle, for example, magnetostrictive sensors of pressure. Figure 4.1 presents sensitivity ranges of the most commonly used *magnetic field sensors* (MFS).

Author of this book presented in the past a review of MFS (Tumanski 2004). There are also other reviews presented by other authors (Lenz 1990, Heremans 1993, Lenz and Edelstein 2006, Ripka and Janosek 2010) as well as review books (Boll and Overshott 1989, Ripka 2001). Looking at these reviews, the general conclusion is that the overall state is rather stable. During the last years, the same sensors (superconducting quantum interference device [SQUID], resonance, fluxgate, magnetoresistive, Hall sensors, induction sensors) are dominating. Among new sensors, only tunnel magnetic junction (TMJ) and giant magneto-impedance (GMI) introduced new possibilities. Of course, all traditional sensors are improved through miniaturization, microelectronics, etc.

Figure 4.1 presents the typical ranges of application of the main sensors. Very small magnetic fields, usually biomagnetic fields, are measured using SQUID sensors. Above 0.1 nT, two classes of sensors can be used—resonance and fluxgate. The resonance methods (optical pumping and nuclear resonance) exhibit very good resolution but the sensors are rather large and detect mainly the scalar value of magnetic field. Therefore, they are usually used in geophysics and for outer space investigations.

More versatile in a similar range is the fluxgate sensor. It is much smaller and detects the vector value. But in order to obtain the best parameters, it should be very carefully manufactured. In the range around the Earth's magnetic field, much cheaper are the anisotropic magnetoresistance (AMR) and giant magnetoresistivity

(GMR) sensors. Large magnetic field is almost exclusively measured by Hall effect sensors.[†]

Special features exhibit inductive sensors (search coil). They can be used over a very wide range of values—from very small pT range to very large magnetic fields, from DC (when sensor is moved) to GHz bandwidth.

4.1.2 Specification of the Performances of Magnetic Sensors

When a new sensor principle is announced, it is usually a long way to practical applications. It is not sufficient just to find the relation $H \rightarrow X$ (X—sensor output signal)—it is more important to ensure appropriate measuring parameters, as follows:

- Noise level and resolution
- Offset and long-term stability
- Sensitivity, nonlinearity, and range of measured values
- Frequency bandwidth
- Temperature stability
- Immunity to orthogonal excitation
- Hysteresis
- And, of course, cost of material and fabrication

One of the most important parameters is the *power spectral density* (PSD) *of noise* because it is a crucial parameter if we consider the resolution of the sensor. The main sources of noise are internal, for example, resistance of the sensor is the source of *thermal Johnson noise* V_{nT} while semiconductor junction is the source of *shot noise* I_{ns}:

$$V_{nT} = \sqrt{4kTR\Delta f} \tag{4.1}$$

$$I_{ns} = \sqrt{2qI\Delta f} \tag{4.2}$$

where
k is the Boltzman constant
q is the electron charge

---

* When the sensor of magnetic field is used to measure the magnetic field in the air, it is not important if it measures a flux density B or a magnetic field strength H because both values are connected by linear relationship $B = \mu_0 H$. Hence measuring instruments are usually scaled in flux density units tesla, but it is possible to meet instruments scaled in magnetic field strength units A/m (or both).

[†] Although, for scaling of these sensors, the resonance methods are used.

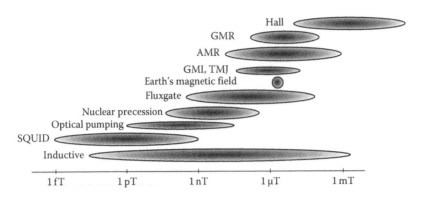

FIGURE 4.1
The range of application of the main magnetic field sensors.

Because the noise depends on the frequency range Δf, usually the spectral density $S(f)$ of noise is presented:

$$S(f) = \frac{V_n^2}{\Delta f} = \left(\frac{V_n}{\sqrt{\Delta f}} \right)^2 \qquad (4.3)$$

Hence, a "unit" of noise can be described, for example, $\mu V/\sqrt{Hz}$ or a noise equivalent magnetic field, for example, $nT/\sqrt{Hz}$. Figure 4.2 presents the typical comparison of noise spectrum of two magnetic sensors and Table 4.1 presents noise levels of the main MFS.

Electronic circuit is usually an additional source of noise. But the signal processing can also help in suppression of noise. One of the best tools for this task is a lock-in amplifier utilizing the principle of synchronous detector. A lock-in amplifier passes only signals whose frequency corresponds with reference frequency.[*] In some sensors, the reference signal is easy to obtain. For example, in a fluxgate sensor, the output signal has double frequency of the magnetizing source[†] (Figure 4.3). If we connect the sensor into a bridge circuit, the output signal has the same frequency as the supplying source.[‡] Sometimes, we have to generate such signals artificially—for example, in DC SQUID, we additionally magnetize SQUID using modulation oscillator (Figure 4.3).

The second important limitation of the sensor resolution is offset, in particular a temperature zero drift. Such problem is especially difficult in the case of magnetoresistive sensors where changes of resistance caused by magnetic field and changes of resistance caused by the temperature are not easy to separate. The commonly used method for rejection of the offset is a bridge circuit

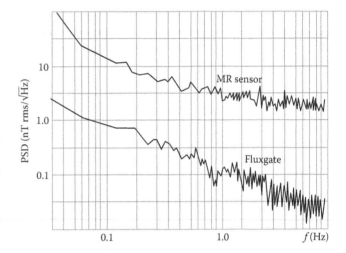

FIGURE 4.2
Comparison of noise spectrum of two magnetic field sensors. (After Ripka, P., *J. Magn. Magn. Mater.*, 157–158, 424, 1996.)

TABLE 4.1

Noise or Resolution of Various Magnetic Sensors

	$S(f)$	Resolution
SQUID	$5\,fT/\sqrt{Hz}$	
High T_c SQUID	$100\,fT/\sqrt{Hz}$	
Air coil	$0.3\,pT/\sqrt{Hz}$ at 20 Hz	
Ferromagnetic core coil	$2.5\,pT/\sqrt{Hz}$ at 1 Hz	
Fluxgate	$2\,pT/\sqrt{Hz}$–$5\,nT/\sqrt{Hz}$	
AMR	$15\,nT/\sqrt{Hz}$	
Hall	$0.5\,\mu T/\sqrt{Hz}$	
Optically pumped		1 pT
Overhauser		10 pT
Proton precession		100 pT

Source: **After Ripka, P.,** *J. Magn. Magn. Mater.*, 157–158, 424, 1996.

---

[*] Noise signal (white noise) is a signal composed of all possible frequency components randomly distributed.

[†] This signal is modulated by the signal proportional to measured magnetic field.

[‡] Also, in this case, the measured signal modulates the carrier signal of frequency the same as supplying voltage.

with one sensor active (with signal proportional to measured value x) and second sensor passive (Figure 4.4a). If both sensors (active and passive) are influenced by some other value, for instance, temperature T, this common value is rejected because

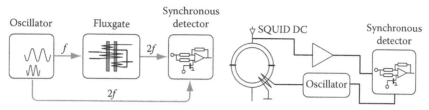

FIGURE 4.3
Application of the lock-in amplifier for magnetic sensors.

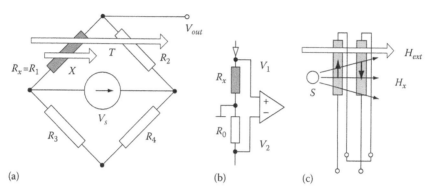

FIGURE 4.4
Examples of the common component rejection.

$$V_{out} \approx \frac{1}{4}\left(\frac{\Delta R_1}{R_1} - \frac{\Delta R_2}{R_2}\right)V_s$$

$$= \frac{1}{4}\left(\frac{\Delta R_1(x)}{R_1} + \frac{\Delta R_1(T)}{R_1} - \frac{\Delta R_2(T)}{R_2}\right)V_s = \frac{1}{4}\frac{\Delta R_1(x)}{R_1}V_s$$

$$(4.4)$$

The same effect can be obtained by using a differential amplifier because $V_{out}=K(V_1-V_2)$ (Figure 4.4b). Figure 4.4c illustrates rejection of common influence of external magnetic field in gradiometer sensors, commonly used in SQUID measurements. Both sensors are influenced by the same value H_{ext}, but the sensor that is nearer the source S of an unknown magnetic field H_x exhibits slightly greater output signal.

Figure 4.5a presents the method of compensation of the temperature influence of magnetoresistive GMR sensors. Both sensors (of the pair of sensors are influenced by the same temperature, but only one (active) is

influenced by magnetic field because second one (passive) is covered by shield. Much better is the possibility of rejection of temperature influence in the case of AMR sensors (Figure 4.5b) because by appropriate design, it is possible to obtain sensors of differential transfer characteristics.

Of course, it is easier is to separate low-frequency (or DC) zero drift if the output signal of the sensor is at much higher frequency. This is in the case of fluxgate sensor because the low-frequency measured magnetic field causes modulation of the high-frequency output signal. Figure 4.6 presents the same effect obtained artificially in an AMR sensor. If we bias this sensor by alternating "flipping" field perpendicular to measured field (along the anisotropy axis), we obtain alternating output signal that can be easily separated from the low-frequency temperature zero drift.

The best case is when the transfer characteristic of the sensors is linear:

$$X_{out} = S \cdot B \quad (4.5)$$

where
S is the sensitivity coefficient
X is the output signal

Nonlinearity of the transfer characteristic appears when sensitivity depends on the measured value:

$$X'_{out} = S(B) \cdot B \quad (4.6)$$

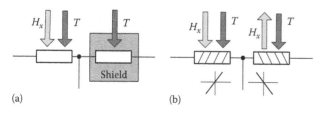

(a)　　　(b)

FIGURE 4.5
Compensation of the temperature influence in GMR sensors (a) and AMR sensors (b).

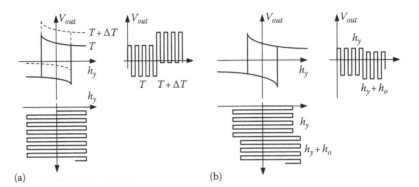

(a)   (b)

FIGURE 4.6
Compensation of the temperature influence (a) and orthogonal field influence (b) in AMR sensors by using the flipping bias magnetic field. (From Tumanski, S., *Thin Film Magnetoresistive Sensors*, IOP Publishing, Bristol, U.K., 2001.)

The error of nonlinearity is

$$\delta_{nl} = \frac{X_{out} - X'_{out}}{X_{out}} = 1 - \frac{S(B)}{S} \quad (4.7)$$

For example, sensitivity coefficient of a Hall sensor of thickness t is (Popovic 2004)

$$S_i = G_H \frac{R_H}{t} \quad (4.8)$$

Both the geometrical factor G_H and the Hall effect coefficient R_H depend not only on the carrier mobility μ_H but also on the measured magnetic field B:

$$R_H = R_{HO}\left(1 - \alpha\mu_H^2 B^2\right) \quad \text{and} \quad G_H = G_{H0}\left(1 - \beta\mu_H^2 B^2\right) \quad (4.9)$$

Fortunately, the coefficients α and β have opposite signs and therefore, it is possible to design a Hall sensor with a transfer characteristic close to linear.

In the case of AMR sensors (Figure 4.7), the transfer characteristic is very nonlinear because the output signal depends on the magnetoresistivity coefficient $\Delta\rho/\rho$ and on square of the magnetic field h_x^*:

$$\frac{\Delta R}{R} = -\frac{\Delta\rho}{\rho} h_x^2 \quad (4.10)$$

This relation is additionally influenced by nonhomogeneous demagnetizing field. We can significantly improve the linearity by an appropriate design of the sensor (Tumanski 2001) such that the final transfer characteristic becomes

---

* h_x is the value of magnetic field H_x related to anisotropy field H_k and perpendicular component H_y and $h_x = H_x/(H_k + H_y)$.

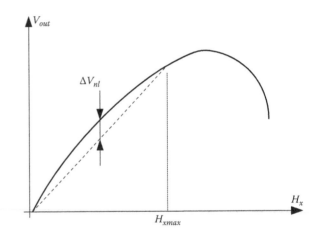

FIGURE 4.7
Nonlinearity error of typical AMR sensor.

$$\frac{\Delta R}{R} = \pm\frac{\Delta\rho}{\rho} h_x \sqrt{1 - h_x^2} \cong \pm\frac{\Delta\rho}{\rho} h_x \quad (4.11)$$

Thus, the sensor is very close to linear if the measured magnetic field h_x is sufficiently small. Generally, radical improvement of linearity can be obtained by employing the negative feedback, because the sensor works then only as a zero field detector (Figure 4.8).

One of the main disadvantages of the feedback is that the magnetic field of feedback destroys the magnetic field distribution at the measured location. This problem can be overcome by using a very small feedback coil as to generate magnetic field in a very small volume. Figure 4.9 presents a design of Philips MR sensor with two small planar coils—one for feedback and the second for flipping (see also Figure 4.6).

By using the feedback, it is possible to extend the *range of magnetic field* for all types of sensors (see Figure 4.8). But practically, the *lower limit* of measured magnetic field is determined from resolution (noise level and offset) and the *upper limit* often depends on competitive

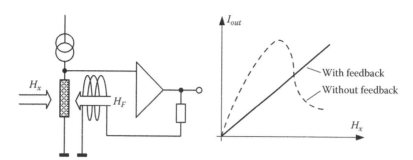

FIGURE 4.8
Improvement of linearity by using a feedback.

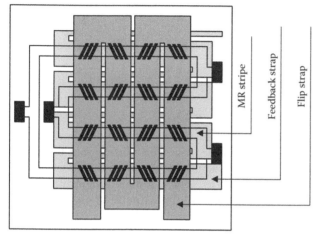

FIGURE 4.9
AMR sensor with two additional planar coils for feedback and flipping.

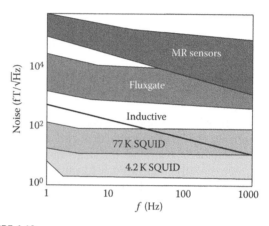

FIGURE 4.10
Typical noise level (lower limit of the range) for various sensors. (After Schilling, M., Rauscheigenschaften magnetoresistiver sensoren, in *Proceedings of the VII Symposium. "Magnetoresistive Sensoren"*, Sensitec, 2003.)

sensors. Let us look at Figure 4.10 with noise spectrum of various sensors. The range 1–10 fT can be measured only by using 4.2K SQUID. But in the range between 10 fT and 1 pT, it is possible to perform similar measurements using cheaper 77K SQUIDs. Moreover, toward the top of this range, it is also possible to use inductive sensors.

Between 1 pT and about 1 nT, we can use much cheaper fluxgate sensors. Above 1 nT, we meet even more cheaper (than fluxgate) AMR sensors. And later, above 10 mT, more convenient is to use Hall sensors.

Magnetoresistive and Hall sensors exhibit excellent frequency behavior—they can operate from DC to GHz and more importantly, the output signal depends directly on the magnetic field value. Inductive sensors can operate also up to GHz (Tumanski 2007), but signal depends on the derivative of flux density *dB/dt*—hence an integrating circuit is needed. Moreover, often the frequency bandwidth of coil sensors is limited by the resonance. Fluxgate sensors have limited upper frequency by the magnetizing frequency because the output 2*f* signal is modulated by measured value. The carrier frequency should be at least about five times higher than

frequency of the sensor signal. Hence, the fluxgate sensors usually operate to about 1 kHz (it means that the frequency of the magnetizing field should be not lower than 5 kHz).

Practically, all magnetic sensors are influenced by temperature. But by an appropriate design, it is possible to make temperature-compensated sensors. For example, temperature error of AMR sensor depends on the temperature changes of magnetoresistivity $\alpha_{\Delta\rho}$ and the temperature changes of anisotropy α_{Hk}:

$$\alpha_t = \alpha_{\Delta\rho} - \frac{H_k}{H_k + H_y + tM/w}\alpha_{Hk} \qquad (4.12)$$

For permalloy $\alpha_{\Delta\rho} \approx -0.018\,\mathrm{K^{-1}}$ and $\alpha_{Hk} \approx -0.022\,\mathrm{K^{-1}}$ and by appropriate design of thickness *t* and width *w* of the magnetic strip, we can obtain temperature self-compensating sensor.

Vector MFS should measure only one component of magnetic field. Fluxgate sensor, Hall and GMR sensors can exhibit errors caused by the orthogonal component of magnetic field—the *crossfield effect*. This effect is especially harmful in the case of AMR sensor (an important drawback of this type of sensor) because its sensitivity depends on H_y component:

$$\frac{\Delta R}{R} \cong \frac{\Delta \rho}{\rho} \frac{1}{H_k + H_y} H_x \qquad (4.13)$$

and if this perpendicular field component is large (larger than anisotropy field H_k), the performance of the sensor can be destroyed (large hysteresis). Fortunately, it is possible to restore its correct operation by biasing the sensor with large field in the easy axis of anisotropy.

The crossfield effect error in AMR sensor can be decreased by applying AC flipping field as it is demonstrated in Figure 4.6. The best method to avoid problems with orthogonal component in all sensors is applying the feedback because then the sensor works only as zero field detector.

In magnetic sensors, especially containing ferromagnetic materials, the *perming error* can occur. It is the effect when the sensor is overload by large magnetic field (much larger than the measured field range). Therefore, after magnetic shock, the demagnetization process can be required. In fluxgate sensor, the perming effect can be suppressed by applying sufficiently large excitation field. In the magnetoresistive sensors, the flipping field can be helpful to remove the perming effect.

4.2 Induction Sensors*

4.2.1 Introduction

Induction coil sensors (Dehmel 1989, Ripka 2001, Prance et al. 2006, Tumanski 2007) (also called search coil, pickup coil, B-coil, H-coil, magnetic antenna, etc.) are one of the oldest and most well-known types of magnetic sensors. Its transfer function $V = f(B)$ results from the fundamental Faraday's law of induction:

$$V = -n \cdot \frac{d\Phi}{dt} = -n \cdot A \cdot \frac{dB}{dt} = -\mu_0 \cdot n \cdot A \cdot \frac{dH}{dt} \qquad (4.14)$$

where Φ is the magnetic flux passing through a coil with an area A and a number of turns n.

The operating principles of coil sensors are generally known but the technical details and practical

FIGURE 4.11
The simplest coil-based sensor for flux density (or magnetic fields strength).

implementation of such devices are only known to specialists. For example, it is well known that the output signal, V, of a coil sensor depends on the rate of change of flux density, dB/dt, which requires integration of the output signal. However, there are other useful methods that enable the obtaining of results proportional directly to flux density B.

The properties of a coil sensor were described in detail many years ago (Zijlstra 1967) (Figure 4.11). However, sensors based on the same operating principle are still widely used in many applications, especially for detection of stray magnetic field (potentially dangerous for health). The induction sensor is practically the only one that can be manufactured directly by its users (in comparison to Hall, magnetoresistive, or fluxgate sensors). The method of coil manufacture is simple and the materials (winding wire) are readily available. Thus, almost everyone can perform such investigations using simple and very low-cost, yet quite accurate, induction coil sensors.

There are many historical examples of the renaissance of various coil sensors. For example, the Rogowski coil and Rogowski–Chattock coil were first described in 1887 (Chattock 1887, Rogowski and Steinhaus 1912). Today, these sensors have been rediscovered: Rogowski coil as excellent current transducer (Murgatroyd 1996) and Rogowski–Chattock as sensor used in measurement of magnetic properties of soft magnetic materials (Shirkoohi and Kontopoulos 1994). An old Austrian patent from 1957 (Werner 1957) describing the use of a needle sensor for the investigation of local flux density in electrical steel was revived some years ago for magnetic measurements (Senda et al. 1997, Pfützner and Krismanic 2004).

* Based on the paper "Induction coil sensors—A review" published in *Measurements, Science and Technology*, **18** (2007), R31–R46 (permission of Editor—Institute of Physics Publishing).

4.2.2 Air Coils versus Ferromagnetic Core Coils

The relatively low sensitivity of an air coil sensor and problems with its miniaturization can be partially overcome by incorporation of a ferromagnetic core, which acts as a flux concentrator inside the coil. For a coil with a ferromagnetic core, Equation 4.14 can be rewritten as

$$V = -\mu_0 \cdot \mu_r \cdot n \cdot A \cdot \frac{dH}{dt} \quad (4.15)$$

Modern soft magnetic materials exhibit a relative permeability μ_r, which can be higher than 10^5, so this can result in a significant increase of the sensor sensitivity. However, it should be taken into account that the resultant permeability of the core μ_c can be much lower than the actual material permeability. This is due to the demagnetizing field effect defined by the demagnetizing factor N_d, which is dependent on the geometry of the core:

$$\mu_c = \frac{\mu_r}{1 + N_d \cdot (\mu_r - 1)} \quad (4.16)$$

If the permeability μ_r of a material is relatively large (which is generally the case), the resultant permeability of the core μ_c depends mainly on the demagnetizing factor N (mic = 1/N). Thus, in the case of high permeability material, the sensitivity of the sensor depends mostly on the geometry of the core.

The demagnetizing factor N_d for an ellipsoidal core depends on the core length l_c and core diameter D_c according to an approximate equation, given as

$$N_d \cong \frac{D_c^2}{l_c^2} \cdot \left(\ln \frac{2l_c}{D_c} - 1 \right) \quad (4.17)$$

It can be derived from Equation 4.17 that in order to obtain small value of N_d (and a large resultant permeability μ_c), the core should be long and with small diameter. Let us consider the dimensions of a search coil sensor optimized for a large sensitivity as described by Prance (Prance et al. 2003).

The core was prepared from amorphous ribbon (Metglas 2714AF) with dimensions $l_c = 300\,\text{mm}$ and $D_c = 10\,\text{mm}$ (aspect ratio equal to 30). Substituting these values into Equation 4.17, we obtain $N \cong 3.5 \times 10^{-3}$, which means that the sensitivity is about 300 times larger in comparison with the air coil sensor.

The use of a core made of a soft magnetic material leads to a significant improvement of the sensor sensitivity. However, this enhancement is achieved with the sacrifice of one of the most important advantages of the air coil sensor—the linearity. The core, even if made from the best ferromagnetic material, introduces to the transfer function of the sensor some nonlinear factors, which depend on temperature, frequency, flux density, etc. Additional magnetic noise (e.g., Barkhausen noise) also decreases the resolution of the sensor. Moreover, the ferromagnetic core alters the distribution of the investigated magnetic field, which can have important consequences.

4.2.3 Design of the Air Coil Sensor

A typical design of an air coil sensor is presented in Figure 4.12. The resultant area of a multilayer coil sensor can be calculated using integration (Zijlstra 1967)

$$A = \frac{\pi}{4} \cdot \frac{1}{D - D_i} \cdot \int_{D_i}^{D} (y^2) dy = \frac{\pi}{12} \cdot \frac{D^3 - D_i^3}{D - D_i} \quad (4.18)$$

However, Equation 4.18 is of limited accuracy. Thus, in practice, it is better to determine the resultant area of the coil experimentally by means of calibration in a known field. The area of the coil can be calculated with a simplified formula based on the assumption that its diameter is equal to mean value $D_m = (D + D_i)/2$, thus

$$A = \frac{\pi}{8} \cdot (D + D_i)^2 \quad (4.19)$$

If we assume that flux density to be measured is a sine wave $b = B_m \cdot \sin(\omega \cdot t)$, and the sensing coil is a ring with diameter D, then Equation 4.14 can be rewritten in a form

$$V = 0.5 \cdot \pi^2 \cdot f \cdot n \cdot D^2 \cdot B \quad (4.20)$$

where
f is a frequency of the measured field
n and D are number of turns and diameter of the coil, respectively
B is the measured flux density

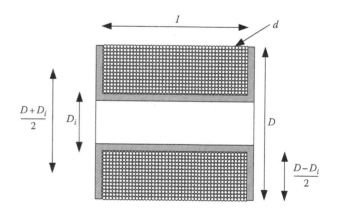

FIGURE 4.12
Typical design of an air coil sensor (*l*, length of the coil; *D*, outer diameter of the coil; D_i, inner diameter of the coil; *d*, diameter of the wire).

If we would like to determine the magnetic field strength H instead of flux density B, then Equation 4.20 can be easily transformed knowing that for a non-ferromagnetic medium $B = \mu_0 \cdot H$ ($\mu_0 = 4 \cdot \pi \cdot 10^{-7}$ H/m) and

$$V = 2 \times 10^{-7} \cdot \pi^3 \cdot f \cdot n \cdot D^2 \cdot H \qquad (4.21)$$

Taking into account (4.19), Equation 4.21 can be presented as

$$V = \frac{10^{-7}}{2} \cdot \pi^3 \cdot f \cdot n \cdot (D + D_i)^2 \cdot H \qquad (4.22)$$

The number of turns depends on the diameter, d, of the wire that is used, the packing factor k (typically $k \approx 0.85$ (Dehmel 1989)), and the dimensions of the coil

$$n = \frac{l \cdot (D - D_i)}{2 \cdot k \cdot d^2} \qquad (4.23)$$

Thus, the sensitivity $S = V/H$ of an air coil sensor can be calculated as

$$S = \frac{10^{-7}}{4} \cdot \frac{\pi^3 \cdot f \cdot l}{kd^2} \cdot (D - D_i) \cdot (D + D_i)^2 \qquad (4.24)$$

The resolution of the coil sensor is limited by thermal noise, V_T, which depends on the resistance R of the coil, the temperature T, the frequency bandwidth Δf with coefficient equal to the Boltzman factor $k_B = 1.38 \times 10^{-23}$ W·s/K

$$V_T = 2\sqrt{k_B \cdot T \cdot \Delta f \cdot R} \qquad (4.25)$$

The resistance of the coil can be calculated as (Richter 1979)

$$R = \frac{\rho \cdot l}{d^4} (D - D_i) \cdot (D + D_i) \qquad (4.26)$$

and the signal-to-noise ratio, SNR, of the air coil sensor is

$$SNR = \frac{\pi^3 \cdot 10^{-7}}{8} \cdot \frac{f}{\sqrt{\Delta f}} \cdot \frac{\sqrt{l} \cdot (D + D_i) \cdot \sqrt{D^2 - D_i^2} \cdot H}{\sqrt{k^2 \cdot k_B \cdot T \cdot \rho}} \qquad (4.27)$$

As can be seen from the above equation, the sensitivity increases roughly proportionally to D^3 and SNR increases with D^2, so the best way to obtain maximum sensitivity is to increase the coil diameter D. Increasing of the coil length is less effective, because the sensitivity increases with l, while the SNR increases only

with $\sqrt{l}$. The sensitivity can also be improved by an increase of the number of turns. For example, by using wire with smaller diameter, the sensitivity increases with d^2 and SNR ratio does not depend on the wire diameter.

Various publications discussed other geometrical factors of the air coil sensor. For instance, the optimum relation between the length and the diameter of the coil can be determined taking into account the error caused by inhomogeneous field. It was found (Herzog and Tischler 1953, Zijlstra 1967) that for $l/D = 0.866$, undesirable components are eliminated at the center of the coil. This analysis was performed for the coil with one layer. For a multilayer coil, the recommended relation is $l/D = 0.67 - 0.866$ (0.67 for $D_i/D = 0$ and 0.866 for $D_i/D = 1$). The same source (Zijlstra 1967) recommends that D_i/D to be less than 0.3.

It can be concluded from the analysis presented above that in order to obtain high sensitivity, the air coil sensor should be very large. For instance, the induction coil magnetometer used for measurements of micropulsations of the Earth's magnetic field in the bandwidth 0.001–10 Hz with resolution 1 pT–1 nT (Stuart 1972), had meter-range dimensions and even hundreds of kilograms in weight. An example of a design of such coil sensor is presented by Campbell (1969). A coil with diameter 2 m (16,000 turns of copper wire 0.125 mm in diameter) detected micropulsation of flux density in the bandwidth 0.004–10 Hz. For 1 pT field, the output signal was about 0.32 μV, while the thermal noise level was about 0.1 μV.

The air coil sensor with 10,000 turns and diameter 1 m was used to detect flux density of magnetic field in pT range for magnetocardiograms (Estola and Malmivuo 1982).

Coil sensors are sensitive only to the flux that is perpendicular to their main axis. Therefore, in order to determine all directional components of the magnetic field vector, three mutually perpendicular coils should be used (Figure 4.13). An example of such a low-noise, three-axis search coil magnetometer is described by Macintyre (Macintyre 1980). Such a "portable" magnetometer consists of three coils of between 19 and 33 cm in diameter and 4100–6500 turns with a weight 14 kg. is capable of measuring magnetic field between 20 Hz and 20 kHz with a noise level lower than 170 dB/100 μT.

On the other hand, there are examples of extremely small air-cored sensors. Three orthogonal coil system with dimensions less than 2 mm and a weight about 1 mg (40 turns) have been used for detection of position of small, fast-moving animals (Schilstra and van Hateren 1998).

Air coil sensors are widely used as eddy-current proximity sensors or for eddy-current sensors for nondestructive testing (e.g., for detection of cracks). In such array has been developed (Uesaka et al. 1998). The array consists of 16 microloop sensors with an area $14 \times 14\,mm^2$ and a thickness of $125\,\mu m$. Each coil has 40 turns within an area of $2 \times 2\,mm^2$.

Sometimes, the frequency response of the air-core sensor is a more important factor than the sensitivity or the spatial resolution. The dimensions of the coil can be optimized for better frequency performance. These factors are discussed later.

4.2.4 Design of Ferromagnetic Core Coil Sensor

High-permeability core coil sensors are often used in the case when high sensitivity or dimension limitations are important. A typical geometry of such a sensor is presented in Figure 4.15.

The optimal value of core diameter D_i has been determined as $D_i \cong 0.3D$ (Richter 1979). The length of the coil l is recommended to be about 0.7–0.9 of the length l_c of the core. For such coil dimensions, the output signal V and SNR ratio at room temperature can be described as (Tumanski 1986)

$$V \cong 0.9 \times 10^{-5} \cdot f \cdot \frac{l^3}{d^2} \cdot D_i \cdot \frac{1}{\ln\left(2 \cdot l/D_i\right)-1} \cdot H \quad (4.28)$$

$$SNR \cong 1.4 \times 10^8 \cdot \frac{f}{\sqrt{\Delta f}} \cdot l^2 \cdot \sqrt{l} \cdot \frac{1}{\ln\left(2 \cdot l/D_i\right)-1} H \quad (4.29)$$

It can be concluded from relationships (4.28) and (4.29) that in the case of coil sensor with a ferromagnetic core, the most efficient method of improving the sensor performance is to make the length of the core (or rather the ratio l/D_i) as large as possible, since the sensitivity is proportional to l^3. Figure 4.16 presents the dependence of the resultant permeability of the core μ_c on the aspect ratio l/D and the core material permeability μ_r (as described in Richter (1979)).

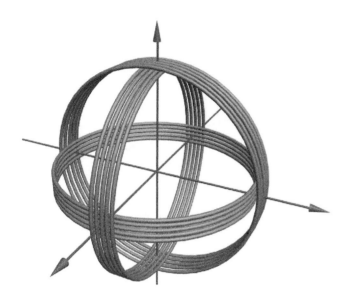

FIGURE 4.13
The concept of three mutually perpendicular coils for three-axis magnetic field measurements.

cases, sensitivity is not as important as the spatial resolution and compactness of the whole device. Such sensors are often manufactured as a flat planar coil (made in PCB or thin film technology (Hirota et al. 1993, Sadler and Ahn 2001, Gatzen et al. 2002, Mukhopadhyay et al. 2002, Zhao et al. 2004) connected to an on-chip complementary metal-oxide semiconductor (CMOS) electronic circuit (Sadler and Ahn 2001). An example of such a sensor with dimensions $400 \times 400\,\mu m$ (7 turns) is presented in Figure 4.14.

For testing of the spatial distribution of magnetic field by means of coil sensors, the flexible microloop sensor

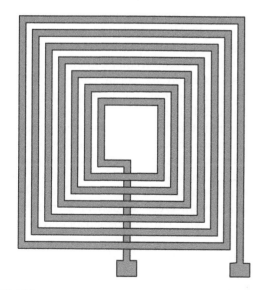

FIGURE 4.14
An example of a pickup planar thin-film coil designed for eddy-current sensors. (From Gatzen, H.H. et al., *IEEE Trans. Magn.*, 38, 3368, 2002.)

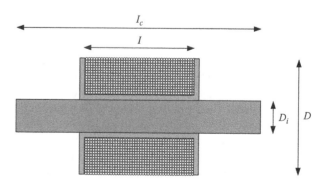

FIGURE 4.15
Design of a typical ferromagnetic core coil sensor (l, length of the coil; l_c, length of the core; D, diameter of the coil; D_i, diameter of the core).

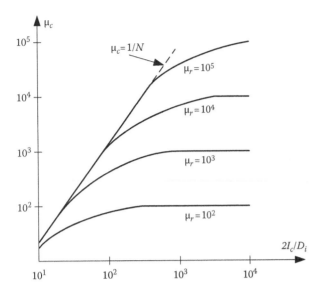

FIGURE 4.16
The dependence of the resultant permeability of the core on the dimensions of the core and the permeability of the material. (From Richter, W., *Exp. Technik Phys.*, 27, 235, 1979.)

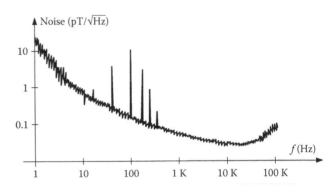

FIGURE 4.17
An example of the noise performance of the induction magnetometer. (After Prance, R.J. et al., *Sens. Actuat.*, 85, 361, 2000.)

The choice of the aspect ratio of the core is very important. The length should be sufficiently large to benefit from the permeability of the core material. On the other hand, if the aspect ratio is large, the resultant permeability depends on the material permeability. This may cause error resulting from the instability of material permeability due to the changes of temperature or applied field frequency. For large values of material permeability, the resultant permeability μ_c practically does not depend on material characteristic because relation (4.16) is then

$$\mu_c \approx \frac{1}{N_d} \tag{4.30}$$

Higher values of material permeability allow the use of longer cores without the risk of the resultant permeability depending on the magnetic characteristic of the material used.

As an example, let us consider a low-noise induction magnetometer that is described by Prance (Prance et al. 2000). Here, the core was prepared from amorphous ribbon (Metglas 2714AF) with temperature-independent properties and dimensions: length 150 mm, cross section of the order 5×5 mm (aspect ratio of around 27). A coil of 10,000 turns was wound with 0.15 mm diameter wire. The noise characteristic of this sensor is presented in Figure 4.17. The obtained minimum noise level around 0.05 pT/√Hz was found to be comparable with the values reported for SQUID sensors.

The same authors compared the influence of the core material. For an amorphous Metglas core, the noise was

found to be 0.05 pT/√Hz, while for the same sensor with a permalloy supermumetal core, it exhibited larger noise of 2 pT/√Hz. Also, a comparison of air coil and ferromagnetic core sensors has been reported (Hayashi et al. 1978, Ueda and Watanabe 1980). Experimental results show that well-designed ferromagnetic core induction sensors exhibited linearity comparable with air-core sensors.

Sensors with ferromagnetic core are often used for magnetic investigations in space research (Ness 1970, Korepanowv 2003). Devices with a core length of 51 cm and the weight of 75 g (including preamplifier) exhibited the resolution (noise level) of 2 fT/√Hz (Korepanow 2001). In an analysis of Earth's magnetic field (OGO Search coil experiments), the following three-axis sensors have been used: coil 100,000 turns of 0.036 mm in diameter, core made from nickel-iron alloy 27 cm long and square (0.6×0.6 cm) cross section. Each sensor weighted 150 g (with half of the weight being the core). The sensitivity in this case was 10 μV/nT Hz (Frandsen et al. 1969).

A detailed design and optimization of an extremely sensitive, three-axis search coil magnetometer for space research is described by Seran (Seran and Fergeau 2005). The coil magnetometer developed for the scientific satellite DEMETER had a noise level of 4 fT/√Hz at 6 kHz. To obtain the desired resonance frequency and a resistance noise above the preamplifier voltage noise, the diameter of copper wire of 71 μm and number of turns of 12,200 were selected. The core was built from 170 mm long 50 μm thick annealed FeNiMo 15-80-5 permalloy strips, with cross section of 4.2 mm × 4.2 mm. The mass of the whole three-axis sensor and the bracket was only 430 g.

There are commercially available search coil sensors. For example, MEDA Company offers sensors with a sensitivity of 25 mV/nT and noise at 10 kHz equal to 10 fT/√Hz (MGCH-2 sensor with core length about 32 cm) or at 0.2 Hz equal to 2.5 pT/√Hz (MGCH-3 sensor with core length of about 1 m).

4.2.5 Frequency Response of Search Coil Sensor

It is obvious from Equation 4.14 that in order to obtain any output voltage signal from the sensor, the flux density must be varying with time. Therefore, the coil sensors are capable of measuring only dynamic (AC) magnetic fields. In the case of the DC magnetic fields, the variation of the flux density can be "forced" in the sensors by moving the coil. However, the term "DC magnetic field" can be understood as a relative one. By using a sensitive amplifier and a large coil sensor, it is possible to determine low frequency (mHz) magnetic fields (Campbell 1969, Stuart 1972). Thus, it is also possible to investigate quasi-static magnetic fields with fixed-coil (stationary) sensors.

AC magnetic fields with a frequency up to several MHz can be investigated by means of coil sensors (Cavoit 2006). In special designs, this bandwidth can be extended to GHz range (Yabukami et al. 1997, Yamaguchi et al. 2000). An example of typical frequency characteristic of a coil sensor is presented in Figure 4.18.

According to Equation 4.20, the output signal depends linearly on frequency, but due to the internal resistance R, inductance L, and self-capacitance C of the sensor, the dependency $V = f(f)$ is more complex. The equivalent electric circuit of an induction sensor is presented in Figure 4.19.

The output signal increases, initially almost linearly with the frequency of measured field, up to the resonance frequency:

$$f_0 = \frac{1}{2 \cdot \pi \cdot \sqrt{L \cdot C}} \tag{4.31}$$

Above the resonance frequency, the influence of self-capacitance causes the output signal to decrease with frequency.

Analyzing the equivalent circuit of the sensor, the sensitivity $S = V/H$ can be expressed in the form (Ueda and Watanabe 1975)

$$S = \frac{S_0}{\sqrt{(1+\alpha)^2 + \left(\beta^2 + \frac{\alpha^2}{\beta^2} - 2\right) \cdot \gamma^2 + \gamma^4}} \tag{4.32}$$

where
$\alpha = R/R_0$
$\beta = R \cdot \sqrt{C/L}$
$\gamma = f/f_0 = 2 \cdot \pi \cdot f \cdot \sqrt{L \cdot C}$

The absolute sensitivity S_0 can be described as $S_0 = 2 \times 10^{-7} \cdot \pi^3 \cdot n \cdot D^2$. The graphical form of the relation (4.32) is presented in Figure 4.20.

The sensor loaded with a small resistance R_0 exhibits frequency characteristic with a plateau between the low corner frequency,

$$f_l = \frac{R + R_0}{2 \cdot \pi \cdot L} \tag{4.33}$$

and the high corner frequency,

$$f_h = \frac{1}{2 \cdot \pi \cdot R_0 \cdot C} \tag{4.34}$$

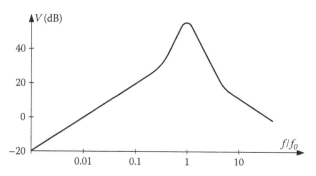

FIGURE 4.18
Typical frequency characteristic of an induction coil sensor.

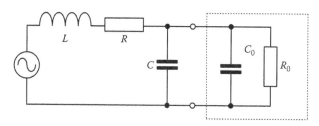

FIGURE 4.19
Equivalent circuit of induction sensor loaded with capacity C_0 and resistance R_0.

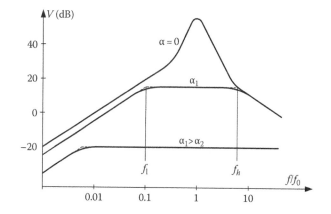

FIGURE 4.20
Frequency characteristics of the induction coil loaded by the resistance R_0 (the coefficient $\alpha = R/R_0$).

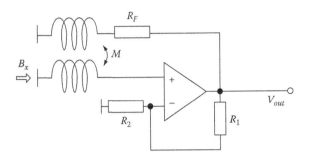

FIGURE 4.21
Induction coil sensor connected to the amplifier with negative transformer type feedback.

An integrating transducer connected to the sensor output is a commonly used method for the improvement of the frequency characteristic. Another method is a load of a sensor with very low resistance (current-to-voltage converter). For low value of load resistance R_0 (high value of α coefficient—see Figure 4.20), we can operate on the plateau of frequency characteristic (in so-called self-integration mode).

The inductance of the sensor depends on the number of turns, permeability, and the core dimensions, according to the following empirical formula (Ueda and Watanabe 1975):

$$L = n^2 \cdot \frac{\mu_0 \cdot \mu_c \cdot A_c}{l_c} \cdot \left(\frac{l}{l_c}\right)^{-3/5} \tag{4.35}$$

The self-capacitance of the sensor strongly depends on the construction of the coil (the application of the shield between the coil layers significantly changes the capacitance).

The frequency characteristic below f_l can be additionally improved by incorporation of a PI correction circuit (more details are given in the next section).

A feedback circuit (as presented in Figure 4.21) is another method of improvement of the frequency characteristic of the sensor (Clerc and Gilbert 1964, Micheel 1987).

The output signal of the circuit presented in Figure 4.21 can be described as (Clerc and Gilbert 1964)

$$V_{out} = 2 \cdot \pi \cdot f \cdot n \cdot \mu_0 \cdot \mu_c \cdot A_c$$

$$\cdot \frac{R_F}{2 \cdot \pi \cdot M} \cdot \frac{j \cdot \omega / \omega_l}{1 + j \cdot \omega / \omega_l} \cdot \frac{1}{1 + j \cdot \omega / \omega_h} \cdot B_x \tag{4.36}$$

The circuit described by Equation 4.36 represents a high-pass filter working under $\omega \ll \omega_0$, and a low-pass filter for $\omega \gg \omega_0$. As a result, the frequency characteristic is flat between low frequency,

$$\omega_l = \frac{R_F}{M} \cdot \frac{1}{1 + R_1 / R_2} \tag{4.37}$$

and high frequency,

$$\omega_h = \frac{1 + R_1 / R_2}{R_F} \cdot M \cdot \omega_0 \tag{4.38}$$

When improvement of the low-frequency characteristic due to the feedback is not sufficient, additional suppression of low-frequency noise is possible by introduction of RC filters, as proposed by Sklyar (Sklyar 2006).

The air-core sensors, due to their relatively low inductance, are used as large frequency bandwidth current transducer (in the Rogowski coil configuration described later), typically up to 1 MHz with an output integrator and up to 100 MHz with current-to-voltage output.

4.2.6 Electronic Circuits Connected to the Coil Sensor

Because the output signal of an induction coil is dependent on the derivative of the measured value (*dB/dt* or *dI/dt* in the case of Rogowski coil), one of the methods of recovering the original signal is application of an integrating transducer (Scholes 1970) (Figure 4.22).

Figure 4.22 presents a typical analog integrating circuit (IC). The presence of the offset voltage and associated zero drift are a significant problem in correct design of such transducers. For this reason, an additional potentiometer is sometimes used for offset correction and resistor R' is introduced for the limitation of the low-frequency bandwidth. The output signal of an integrating transducer is

$$V_{out} = -\frac{1}{R \cdot C} \cdot \int_{t_0}^{T+t_0} (V_{in}) dt + V_0 \tag{4.39}$$

where $R = R_1 + R_{coil}$. The resistance R should be sufficiently large (as not to load the coil) as well as the capacitance C—typical values are $R_1 = 10\,k\Omega$ and $C = 10\,\mu F$ (Scholes 1970).

The amplifier can introduce several limitations at higher frequencies. A passive IC (Figure 4.23) can exhibit

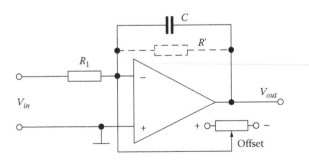

FIGURE 4.22
Typical integrating circuit for the coil signal.

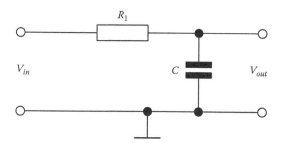

FIGURE 4.23
The passive integrating circuit.

somewhat better performance at those frequencies. Combinations of various methods of integration (active and passive) can be used for large bandwidth—as proposed by Pettinga (Pettinga and Siersema 1983).

Problems with correct design of a measuring system with an integrating transducer are often overcome by applying low resistance loading to the coils (self-integration mode presented in Figure 4.20). Usually, a current-to-voltage converter is used as an output transducer, and is additionally supported by a low-frequency correction circuit. An example of such a transducer is presented in Figure 4.24.

In the current-to-voltage transducer, presented in Figure 4.24, the correction circuit R_1C is introduced to adjust the characteristic of low-resistance loaded circuit at lower frequencies, as illustrated in Figure 4.25.

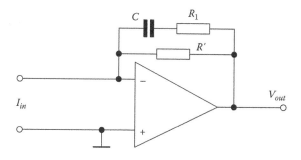

FIGURE 4.24
Current-to-voltage transducer with additional frequency correction circuit. An example described with elements: $R_1 = 100\,\mathrm{k\Omega}$, $C = 0.22\,\mathrm{\mu F}$, $R' = 47\,\mathrm{M\Omega}$.

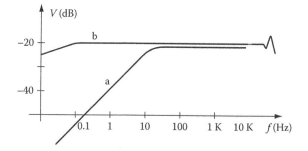

FIGURE 4.25
Frequency characteristic of a coil sensor with current-to-voltage transducer (a) and with additional correction circuit (b).

The above-described circuits were equipped with analog transducers at the output of the coil sensor. However, it is also possible to convert the output signal to a digital form and then to perform digital integration—several examples of such approaches (with satisfactory results) have been described by Smith and Annan (2000) and D'Antona et al. (2002, 2003).

Accurate digital integration of the coil sensor signal is not a trivial task. Also, in digital processing, there are integration zero drifts. The most frequently used method eliminating this problem is subtraction of the calculated average value of the signal. The integrating period, sampling frequency, and triggering time should be carefully chosen, especially when the exact frequency of the processed signal is unknown (which is often the case). Moreover, cost of a good quality data acquisition circuit (analog-to-digital converter) is relatively high and a use of a PC is necessary. Therefore, in some cases, the integrating process is performed digitally by relatively simple hardware (without a PC), consisting of analog-to-digital (ADC) and digital-to-analog (DAC) circuits and registers (AR and MR) as it is presented in Figure 4.26.

It is worth noting that almost all companies manufacturing the equipment for magnetic measurements (as Brockhaus, LakeShore, Magnet Physik, Walker Scientific) offer digital integrator instruments, called fluxmeters*—often equipped with the coil sensors. Figure 4.27 presents a coil sensor used in this instrument.

An amplifier connected to the coil sensor introduces additional noise, voltage noise, and current noise, as

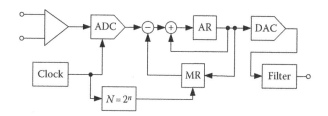

FIGURE 4.26
Digital transducer of the coil sensor signal. (After D'Antona, G. et al., *IEEE Instrum. Meas. Technol. Conf., Vail*, 1185, 2003.)

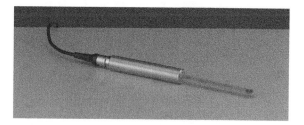

FIGURE 4.27
An example of the coil sensor used in a digital fluxmeter of Brockhaus.

---

* Digital and analog fluxmeters are described in Section 5.2.1.

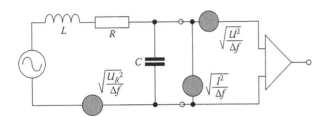

FIGURE 4.28
The sources of noise in an equivalent circuit of a coil connected to an amplifier.

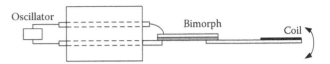

FIGURE 4.30
A vibrating cantilever magnetic field sensor. (After Hetrick, R.E., *Sens. Actuat.*, 16, 197, 1989.)

$$V = -B_x \cdot n \cdot A \cdot \sin(\omega \cdot t) \qquad (4.40)$$

illustrated in Figure 4.28. Each noise component is frequency dependent. Analysis of the coil sensor connected to an amplifier indicates that at low frequency, the thermal noise of the sensor dominates (Dehmel 1989). Above the resonance frequency, the amplifier noise dominates.

The reduction of the amplifier noise can be achieved by application of a HTS SQUID picovoltmeter, as it has been reported by Eriksson (Eriksson et al. 2002). Indeed, the noise level of the preamplifier was decreased to a level of 110 pV/√Hz, but dynamic performance of such circuit was deteriorated.

4.2.7 Moving Coil Sensor

One of the drawbacks of the induction coil sensor—sensitivity only to varying magnetic fields can be overcome by introducing movement to the coil. For example, if the coil is rotating (Figure 4.29) with quartz-stabilized speed rotations, it is possible to measure DC magnetic fields with very good accuracy. The main condition of the Faraday's law (the variation of the flux) is fulfilled, because the sensor area varies as $a(t) = A \cdot \cos(\omega \cdot t)$ and the induced voltage is

Instead of rotation, it is possible to move the sensor in other ways, the most popular being a vibrating coil. One of the first such ideas was applied by Groszkowski, who in 1937 demonstrated the moving coil magnetometer (Groszkowski 1937). The coil was forced to vibrate by connecting it to a rotating eccentric wheel.

One of the best excitation methods for vibrating a coil is by connecting it with an oscillating element, such as piezoelectric ceramic plate (Hagedorn and Mende 1976, Hetrick 1989). Due to relatively high frequency of vibration (in the kHz range), it is possible to make very small sensor with a relatively good geometrical resolution. Figure 4.30 presents an example of a pickup coil sensor mounted onto a piezoelectric bimorph cantilever. The 10-turn coil, 30 μm wide and 0.8 μm thick, excited to a vibration frequency of around 2 kHz (mechanical resonance frequency) exhibited sensitivity of around 18 μV/100 μT (Hetrick 1989).

It is also possible to perform measurements by quick removal of the coil sensor from a magnetic field (or quick insertion into a magnetic field). Such extraction coil methods (used a digital fluxmeter with large time constant) enable the measurement of DC magnetic field according to the following relationship (Jiles 1998):

$$\int V dt = -n \cdot A \cdot (B_x - B_o) \qquad (4.41)$$

The moving coil methods are currently rarely used, because generally there is tendency to avoid any moving parts in measuring instruments. For measurements of DC magnetic field, Hall sensors and fluxgate sensors are most frequently used.

4.2.8 Gradiometer Sensor

The gradiometer sensors (gradient sensors) are commonly used in SQUID magnetometers for elimination of the influence of ambient field (Karp and Duret 1980). These sensors can be also used in other applications where the ambient fields disturb the measurements, or determination of the magnetic field gradient itself (Senaj et al. 1998).

The operating principle of the gradiometer sensor is shown in Figure 4.31. The external magnetic field H_{ext}

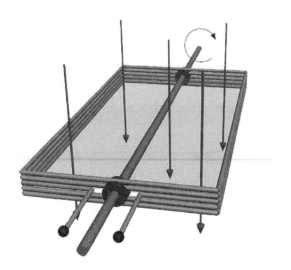

FIGURE 4.29
The sensor of a DC magnetic field with a rotating coil.

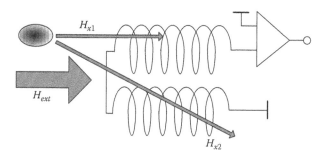

FIGURE 4.31
Operating principle of gradiometer sensor.

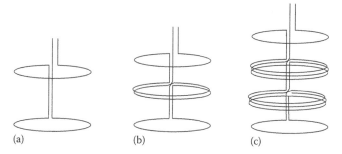

FIGURE 4.33
Gradiometer design: (a) first-order, (b) second-order, (c) third-order.

is generated by a large and distant source (e.g., Earth's magnetic field), so it can be assumed that this field is practically uniform. If two coil sensors (with small distance between them) are inserted in such a field, then both will sense the same magnetic field. As both coils are connected differentially, the influence of the external field is eliminated. If at the same time, there is a smaller source of magnetic field (e.g., due to the human heart investigated in magnetocardiograms) near both of the coils, then the magnetic field in the coil placed nearer the source is larger than in the other coil. This small difference, hence the gradient of magnetic field, is therefore detected by the gradiometer sensor. In this way, it is possible to measure relatively small magnetic field from a local source in the presence of magnetic field generated by more distanced source.

Figure 4.32 presents typical arrangements of gradiometer coils: vertical, planar, and asymmetric. A well-designed gradiometer coil should indicate zero output signal when inserted into a uniform field. In the asymmetric arrangement, the sensing coil is smaller, hence has more turns in order to compensate the magnetic field detected by the larger coil.

It is possible to improve the rejection of the common component by employing several gradiometers. Figure 4.33 presents a first-order, second-order (consisting of two gradiometers of a first order), and third-order gradiometers.

Figure 4.34 presents a typical response of the gradiometer coils to the source distant from the sensor (Fagaly

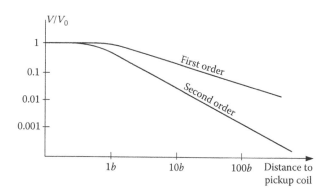

FIGURE 4.34
A typical response of a gradient coil sensor to the source with a distance from the pickup coil smaller than 0.3b (b the distance between coils of gradiometer). (After Fagaly, R.L., Superconducting quantum interference devices, in *Magnetic Sensors and Magnetometers*, Chapter 8, Artech House, Boston, MA, 2001, pp. 305–347.)

2001). The first-order gradiometer rejects more than 99% of the source at a distance of 300b (where b is the distance between the single coils of the gradiometer). For the second-order gradiometer, this distance is reduced to about 30b. It is possible to arrange even higher order of the gradiometer (Figure 4.33c presents the gradiometer of the third order), but gradiometers of higher order have reduced sensitivity and SNR ratio.

The quality of a gradiometer sensor is often described by the formula

$$V \propto G + \beta \cdot H \qquad (4.42)$$

where
   G is the gradient of the measured field
   H is the magnitude of the uniform field

In superconducting devices, it is possible to obtain the β factor as small as part-per-million. This means that it is possible to detect magnetic fields of tens of femtotesla in a presence of millitesla uniform ambient field.

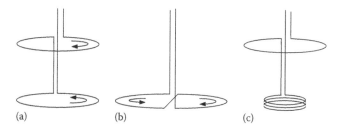

FIGURE 4.32
Gradiometer coils arranged: (a) vertically, (b) horizontally, (c) asymmetrically.

4.2.9 Rogowski Coil

The special kind of helical coil sensor uniformly wound on a relatively long nonmagnetic circular or rectangular strip, usually flexible (Figure 4.35), is commonly known as the Rogowski coil, after the description given by Rogowski and Steinhaus in 1912 (Rogowski and Steinhaus 1912). In certain applications, this coil, known as *Rogowski–Chattock potentiometer RCP*, is used for measurement of magnetic field strength. Indeed, the operating principle of such a coil sensor was as first described by Chattock in 1887 (Chattock 1887) (it is not clear if Rogowski knew the disclosure of Chattock, because in Rogowski's article the Chattock was not cited).

The induced voltage is used as the output signal of the Rogowski coil. But the principle of operation of this sensor is based on the Ampere's law rather than the Faraday's law. If the coil of the length l is inserted into a magnetic field, then the output voltage is the sum of voltages induced in each turn (all turns are connected in series):

$$V = \sum \left(-n \cdot \frac{d}{dl} \cdot \frac{d\varphi}{dt} \right) = \mu_0 \cdot \frac{n}{l} \cdot A \cdot \frac{d}{dt} \int_A^B (H) dl \cdot \cos(\alpha) \quad (4.43)$$

The output signal of the Rogowski coil depends on the number of turns per unit length n/l and the cross-sectional area, A, of the coil. A correctly designed and manufactured Rogowski coil inserted in the magnetic field at fixed points A–B should give the same output signal independent of the shape of the coil between the points A and B.

One of the important applications of the Rogowski–Chattock coil is the device for testing of magnetic materials known as a single sheet tester (SST) (Shirkoohi and

Kontopoulos 1994, Nafalski et al. 1989, Khanlou et al. 1995). In such a device, it is rather difficult to use the Ampere's law ($H \cdot l = I \cdot n$) for determination of the magnetic field strength H (from the magnetizing current I), because the mean length l of the magnetic path is not exactly known (in comparison to a closed-circuit system). But if the RCP coil is used, we can assume that the output signal of this coil is proportional to the magnetic field strength between the points A–B:

$$V = \mu_0 \cdot \frac{n}{l} \cdot A \cdot \frac{d}{dt} (H \cdot l_{AB}) \quad (4.44)$$

The RCP coil can be used to determine the $H \cdot l$ value, that is, the difference of magnetic potentials. The application of the coil to direct measurements of H (for fixed value of length l_{AB}) is not convenient, because the output signal is relatively small and integration of the output voltage is required. For this reason, the compensation method shown in Figure 4.36 is more often used. In such a method, the output signal of the RCP coil is utilized as the signal for the feedback circuit for the current exciting the correction coils. Due to the negative feedback circuit, the output signal of the coil is equal to zero, which means that all magnetic field components in the air gaps and the yoke are compensated and

$$H \cdot l_{AB} - n \cdot I = 0 \quad (4.45)$$

Thus, the magnetic field strength can be determined directly from the magnetizing current because the other parameters (l_{AB} and n) are known.

The most important application of the Rogowski coil is for current measurements (Stoll 1975, Nassisi and Luches 1979, Pellinen et al. 1980, Bellm et al. 1985, Ward and Exon 1993). When the coil wraps around a current conducting wire (Figure 4.37), the output signal of the sensor is

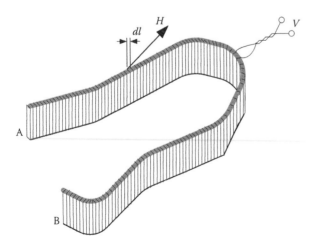

FIGURE 4.35
An example of the Rogowski coil (A, B—ends of the coil).

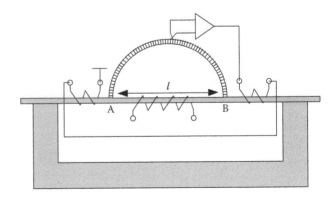

FIGURE 4.36
The application of RCP sensor for determination of $H \cdot l$ value.

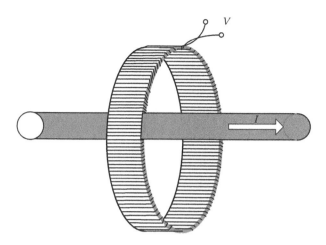

FIGURE 4.37
The Rogowski coil as a current sensor.

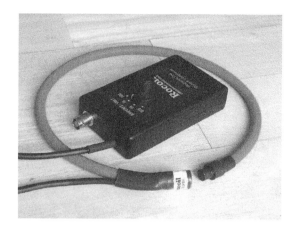

FIGURE 4.38
An example of the Rogowski coil current sensor—model 8000 of Rocoil.

$$V = \mu_0 \cdot \frac{n}{l} A \cdot \frac{dI}{dt} \qquad (4.46)$$

The Rogowski coil used as the current sensor enables measurements of very high values of current (including the plasma current measurements in space (Korepanow 2003)). Due to relatively small inductance of such sensors, they can be used to measure transient current of pulse times down to several nanoseconds (Bellm et al. 1985). Other advantages of the Rogowski coil as the current sensor in comparison with current transformers are as follows: excellent linearity (lack of the saturating core), no danger of opening of the second winding of current transformer, and low construction costs. Therefore, the Rogowski current sensor in many applications has effectively substituted current transformers. The application of the Rogowski coil for current measurements is so wide that recently analog devices equipped the energy transducer (model AD7763) with an integrating amplifier for connecting the coil current sensors.

Although the Rogowski coil seems to be relative simple in design, careful and accurate preparation is strongly recommended to obtain expected performance (Murgatroyd 1992, Murgatroyd and Woodland 1994, Murgatroyd et al. 1991). First of all, it is important to ensure uniformity of the winding (for perfectly uniform winding, the output signal does not depend on the path the coil follows around the current carrying conductor or on the position of the conductor). Special methods and machines for manufacture of Rogowski coils were proposed (Murgatroyd 1992, Murgatroyd and Woodland 1994, Ramboz 1996). The output signal, thus also the sensitivity, can be improved by increasing of the area of the turn, but for correct operation (according to the Ampere's law), it is required to ensure the homogeneity of the flux in each turn. For this reason, the coil is wound on a thin strip with small cross-sectional area.

Also, the positioning of the return loop is important—both terminals should be at the same end of the coil. When the coil is wound on a coaxial cable, the central conductor can be used as the return path (Murgatroyd et al. 1991).

There are several manufacturers offering various types of Rogowski coils or Rogowski coil current sensors. Figure 4.38 shows an example of a Rogowski coil transducer (coil and integrator transducer) of Rocoil Rogowski Coils Ltd.

4.2.10 Other Induction Sensors

4.2.10.1 Flux Ball Sensor

It is quite difficult to manufacture a coil sensor with sufficient sensitivity and small dimensions for measurements of local magnetic fields. If the investigated magnetic field is inhomogeneous, then the sensor averages the magnetic field over the area of the coil.

Brown and Sweer (Brown and Sweer 1945) demonstrated that the volume averaging of the field over the interior of any sphere centered at a point is equal to the value of magnetic field at this point. Thus, a spherical coil measures the field value at its center. An example of a design of such a spherical coil sensor is shown in Figure 4.39.

4.2.10.2 Tangential Field Sensor (H-Coil Sensor)

Measurements of magnetic field strength in magnetic materials (e.g., electrical steel sheet) utilize the fact that the tangential magnetic field inside of the magnetic materials is the same as the tangential field component directly above this material. Thus, a flat coil called an H-coil is often used for the measurements of the magnetic field strength (Pfützner and Schönhuber 1991, Nakata et al. 2000) (Figure 4.40).

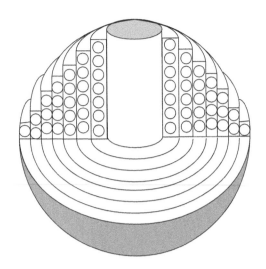

FIGURE 4.39
An example of the spherical coil sensor.

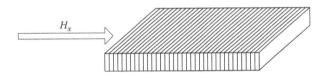

FIGURE 4.40
An example of a typical H-coil sensor.

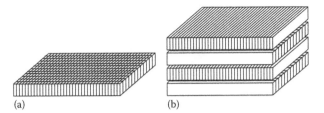

FIGURE 4.41
Two examples of a coil sensor for two components of magnetic field: (a) two perpendicular H-coils wound around the same former and (b) four separate coils allowing double-coil measurements.

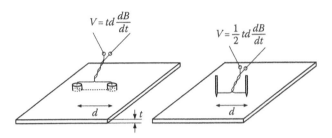

FIGURE 4.42
Two methods of local flux density measurements: micro-holes method and needle method.

The coil sensor for measurement of the tangential component should be as thin as possible. This is conflict with the requirement for the optimum sensitivity of the sensor, which depends on the cross-sectional area of the coil. A typical coil sensor with thickness 0.5 mm and area 25 mm × 25 mm wound with wire of 0.05 mm in diameter (about 400 turns) exhibits a sensitivity of around 3 μV/(A/m) (Tumanski 2002).

In order to obtain sufficient sensitivity, the H-coil is manufactured with a thickness of about 0.5 mm or more. Thus, the axis of the coil is somewhat distanced from the magnetic material surface under investigation. Nakata (Nakata et al. 1987) proposed a two-coil system where the value of the magnetic field directly at the surface can be extrapolated from two results. This idea has been tested numerically and experimentally (Tumanski 2002). It was shown that such method enables the determination of the magnetic field at the magnetic material surface even if the sensor is at a distance from this surface of more than 1 mm. Additionally, it was proposed to enhance the two-coil method by application of up to four parallel (or perpendicular) sensors. An example of such multi-coil sensor is shown in Figure 4.41b.

4.2.10.3 Needle Sensors (B-Coil Sensors)

When it was necessary to determine the local value of the flux density in electrical steel sheet, practically only one method was available (apart from the optical Kerr method): to drill two micro-holes (with diameter 0.2–0.5 mm) and to wind one- or more-turn coil (Figure 4.42a).

However, drilling of such small holes in relatively hard material is not easy. Moreover, this method is destructive. In order to avoid these problems, several years ago, Japanese researchers (Senda et al. 1997, 1999, Yamaguchi et al. 1998) returned to an old Austrian patent of Werner (Werner 1957). Werner proposed to form a one-turn coil by using two pairs of needles, but the application of his invention was difficult, due to relatively small (less than mV) output signal of the sensor. Experimental and theoretical analysis (Senda et al. 1997, 1999, Yamaguchi et al. 1998, Loisos and Moses 2001, Krismanic et al. 2000) proved that today this method can be used with satisfactory results. It is sufficient to use one pair of needles to form a half-turn coil sensor where the induced voltage is described by the following relationship:

$$V \approx \frac{1}{2} \cdot t \cdot d \cdot \frac{dB}{dt} \qquad (4.47)$$

where
t is the thickness of the steel sheet
d is the distance between the needles

Although the needle method is not as accurate as the micro-holes method, it is recently used for the flux density measurements, especially in two-dimensional (2D) testing of electrical steel sheets. To ensure correct contact with insulated surface of the sample, the needle tip should be specially prepared (Krismanic et al. 2000).

4.2.11 Coil Sensor Used as a Magnetic Antenna

Similarly as other MFS, the coil sensors are often used for measurements of nonmagnetic values. They are used in nondestructive testing (NDT), as proximity sensors, current sensors, reading heads, etc. However, the detection of magnetic fields is their primary application, the importance of which increased due the need for stray field investigations, especially for electromagnetic (EM) compatibility and protection against dangerous magnetic fields in human environment.

University laboratories are often visited by ordinary people, who live near transformer stations or power transmission lines. They might be wondering if a close proximity to such sources of magnetic fields can be potentially dangerous for their health, and so they have a genuine interest in measurement of magnetic field strength.

It is possible to do such measurements as a "do it yourself" project. It is sufficient, for example, to wind 100 turns of a copper wire on a nonmagnetic tube (5–10 cm diameter) (Figure 4.11) and to measure the induced voltage. Then, if for instance $f = 50\,\text{Hz}$, $n = 100$ and $D = 10\,\text{cm}$, we obtain (according to Equations 4.20 and 4.21, a simple expression for output voltage): V [V] $\approx 246.5\,\text{B}$ [T] or V [V] $\approx 3.096 \times 10^{-3}$ [A/m]. Of course, there are many commercially available professional measuring instruments using similar methods.

Moreover, starting from 2006, all products indicated with a CE sign should fulfill European Standard EN 50366:2003 "Household and similar appliances—electromagnetic fields—methods for evaluation and measurements." Figure 4.43 presents an example of professional measuring instrument of Narda designed for determination of EM compatibility conditions according to the European Standard EN 50366.

The European Standard EN 50366 requires that magnetic field value should be isotropic. As the coil sensor detects magnetic field only in one direction, a three-coil system (shown in Figure 4.13) should be used and then the value of magnetic field should be determined as follows:

$$b(t) = \sqrt{b_x^2(t) + b_y^2(t) + b_z^2(t)} \tag{4.48}$$

For investigations of magnetic pollution (magnetic smog), the author designed and constructed a magnetometer consisting of an air coil sensor (Figure 4.44) and

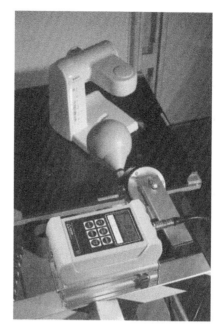

FIGURE 4.43
An example of a search coil magnetometer used for testing the electromagnetic compatibility. (Courtesy of Narda Safety Test Solution Company).

FIGURE 4.44
The air coil sensor designed for investigations of magnetic pollution.

amplifier/frequency correction system (Figure 4.45). The coil was wound onto a ring with diameter 60 mm and width 35 mm using wire 0.1 mm in diameter. There were 70,000 turns and the coil was divided in seven sections with 10,000 turns each.

The block diagram of the constructed magnetometer is presented in Figure 4.45. Two notch filters (50 and 150 Hz) were used for elimination of the power frequency components. An integrating amplifier (for voltage coil input) or current-to-voltage converter (for current coil input) can be selected by the user. The best attainable sensitivity of the magnetometer was found to be 0.1 pT.

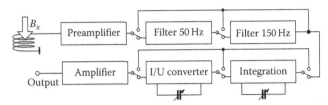

FIGURE 4.45
Block diagram of an induction magnetometer.

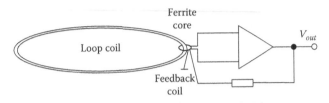

FIGURE 4.46
A loop sensor with feedback coil for MHz range application. (From Cavoit, C., *Rev. Sci. Instrum.*, 77, 064703, 2006.)

FIGURE 4.48
An induction search coil as a tool for detection of metal objects. (Courtesy of Minelab Electronics).

By applying of a special design of a sensor and an electronics, it is possible to extend the working range toward high frequencies, up to MHz range (Figure 4.46). Cavoit (Cavoit 2006) designed a large primary shorted coil (with small capacity and high resonance frequency) coupled to a toroidal ferrite core with a winding (4.46). Prototypes of such sensors, used for radar and space probes, worked at frequency range of 0.1–50 MHz.

The interest in measurements of high-frequency magnetic fields (up to several GHz) grew with the latest rapid development of high-frequency applications, for example, mobile phones and their antennas. Inductive sensors can also be used in this range of frequencies, although several additional problems can occur (Grudzinski and Rozwalka 2004). The dimensions of a sensor should be appropriately smaller than the wavelength of the measured field. Therefore, typically, measuring instruments are equipped with several sensors. An example of a high-frequency field sensor is presented in Figure 4.47. Additional problems arise from the fact that in a high-frequency EM field, it is rather difficult to separate the electric and magnetic components (especially near the source of the field).

Coil sensor antennas can be used as pickup coil sensors for detection of metal objects. An example of such a device is shown in Figure 4.48. Practically, almost all mine detectors use coil sensors (Clem 1997, Riggs et al. 2002, Black et al. 2005). Especially important are applications of magnetic sensors in water environment, for submarine communication and location of submarines (Ioannindis 1977, Murray and McAuly 2004).

Induction coil sensors are commonly used in geophysics, for observation of magnetic anomaly and low-frequency fields. Interesting application of such investigations is the possibility of prognosis of the earthquake events, probably due to the generation of the signals generated by piezoelectric rock formation before the earthquake (Fraser-Smith et al. 1990, Fenoglio 1995, Karakelian et al. 2002, Surkov and Hayakawa 2006).

Great commercial success can be noted in the application of magnetic antennas in the magnetic article surveillance systems (Figure 4.49). Two main types of such systems are used.

In the first one (Gregor et al. 1984), thin strips cut from amorphous materials are magnetized by the field emitted from the transmitter antenna. Due to the nonlinearity of the magnetic properties, the output signal contains harmonics of the predetermined frequency detectable by the receiver antenna.

Recently, a more efficient magneto-acoustic system was proposed by Herzer (Herzer 1998). The label is prepared as a strip made from magnetostrictive amorphous material. A transmitter produces pulses of frequency of 58 kHz, 2 ms on and 20 ms off. The vibrating amorphous strip generates signal of a frequency of 58 kHz,

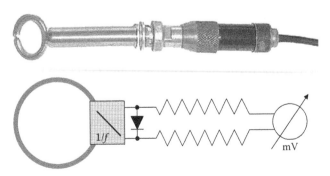

FIGURE 4.47
Magnetic field sensor for the frequency bandwidth 0.6–2 GHz.

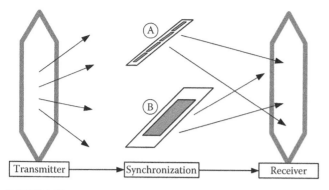

FIGURE 4.49
Magnetic article surveillance system: (A) magnetic harmonic system and (B) magneto-acoustic system.

which is detected by the receiving antenna. The system is synchronized, so the antenna is active only during the pause, and therefore, the influence of a background noise is decreased.

In both systems, the activation or deactivation of the marker is realized by an additional strip made from a semi-hard magnetic material. After demagnetization of this initially magnetized element, the frequency generated by the label changes and it is not detected by the receiver antenna, thus the deactivation is achieved.

The induction sensors used for the magnetic field measurements (and indirectly other quantities such as, e.g., current) have been known for many years. Today, they are still in common use for their important advantages: simplicity of operation and design, wide frequency bandwidth and large dynamics.

The performance of an air-cored induction coil sensor can be precisely calculated due to the simplicity of the transfer function $V = f(B)$. All factors (number of turns and cross-sectional area) can be accurately determined and because function $V = f(B)$ does not include material factors (potentially influenced by external conditions such as temperature), their dependence is excellently linear without upper limit (without saturation).

Because of the absence of any magnetic elements and excitation currents, the sensor practically does not disturb the measured magnetic field (as compared for instance with fluxgate sensors).

The case of a coil sensor with ferromagnetic core is more complicated, because the permeability depends on the magnetic field value and/or temperature. Despite this, if the core is well designed, then these influences can be significantly reduced.

However, the coil sensors also exhibit some shortcomings. First of all, they are sensitive only to AC magnetic fields, although quasi-static magnetic field (of frequencies down to mHz range) can be measured. One notable inconvenience is that the output signal does not depend on the magnetic field value but on the derivative of this field dB/dt or dH/dt. Therefore, the output signal is frequency dependent and it is necessary to connect an integrating amplifier to the sensor, which can introduce additional errors of signal processing.

It is rather difficult to miniaturize the induction coil sensors because their sensitivity depends on the sensor area (or the length of the core). Nevertheless, there are reported micro-coil sensors with dimensions less than 1 mm, which were prepared with the help of thin-film techniques.

Two methods are used for output electronic circuit: the integrator circuit and the current-to-voltage transducer (self-integration mode). Although digital integration techniques are developed and commonly used, the analog techniques, especially current-to-voltage transducers, are often applied due to their simplicity and good dynamics.

Certain old inventions, such as Rogowski coil, Rogowski–Chattock coil, or needle method are today successfully reutilized. The importance of the coil sensor has also increased recently, because it is easy to measure stray fields generated by electrical devices (to meet EM compatibility requirements).

The induction coil sensors with ferromagnetic core prepared from modern amorphous materials exhibit sensitivity comparable to the sensitivity of SQUID sensors. But low magnetic field applications (at less than pT level) can be better served by using SQUID methods. On the other hand, it is more convenient to use the Hall sensors for large magnetic fields (above 1 mT).

4.3 Fluxgate Sensor

4.3.1 Principle of Operation

The fluxgate sensor known from 1936 (Aschennbrener et al. 1936) is still commonly used due to many advantages: resolution starting from 1 pT, alternating signal modulated by low-frequency signal proportional to measured magnetic field—this way noise and offset can be relatively easy suppressed (Gordon and Brown 1972, Primdahl 1979, Bornhöfft and Trenkler 1989, Ripka 1992, 2003, Ripka et al. 2001). Operating of such sensor is based on the fact that the magnetizing curve (or hysteresis loop) is nonlinear (with saturation) and symmetrical. Figure 4.50 illustrates the principle of operation of a fluxgate sensor. If we magnetize the ferromagnetic material periodically with magnetic field H significantly larger than "knee" of the magnetization curve, we obtain the distorted flux density (signal A—Figure 4.50). But due to symmetry of the magnetization curve, this signal contains only odd harmonics. When additional external magnetic field H_x appears, then the magnetization curve

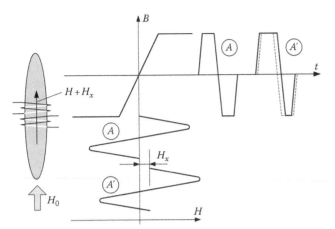

FIGURE 4.50
Principle of operation of the fluxgate sensor.

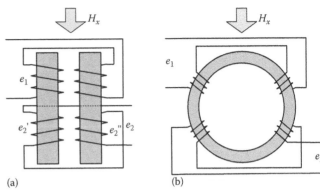

(a)   (b)

FIGURE 4.51
A fluxgate sensor with two rod cores (a) and ring core fluxgate sensor (b) (bright).

becomes asymmetrical-top half of the output signal (in secondary winding) is wider than the bottom half (signal A′—Figure 4.50). This causes presence of even harmonics in the output signal. Usually, the second harmonic is used as the signal proportional to external magnetic field H_x.

Although the principle of operation seems to be rather simple, in the practical realization, many details are crucial to obtaining a satisfying performance of the sensor. Most of all, it is important that the material of the core exhibits small hysteresis and small noise level. Also the construction should be prepared carefully to avoid mechanical stresses and to ensure perfect symmetry.

The small second harmonic signal is accompanied by a large common odd harmonics signal. For this reason, the sensor is commonly composed of two strip cores (Figure 4.51a). The driving magnetic fields are of opposing directions and the external magnetic field is added to the magnetizing field in one core and is subtracted in the second core. So in the pick-up coils connected in series, the common components are compensated out while the even harmonic components are added.* It is also possible to achieve the same effect with a single ring core (Figure 4.51b), where in the left half the external magnetic field is added to the magnetizing field while in the right half it is subtracted. The main advantage of a ring core sensor is the much smaller power necessary to magnetize the core to saturation.

Let us consider the sensor with the core represented by simplified hysteresis shape and the exciting signal by a simplified triangular shape (Figure 4.52). The output signal in secondary coil with n_2 turns wound on the core with area A is

---

* Sometimes a reversed configuration is used—magnetization is in the same direction (sometimes by a common coil) and the pick-up coils are connected in opposing directions.

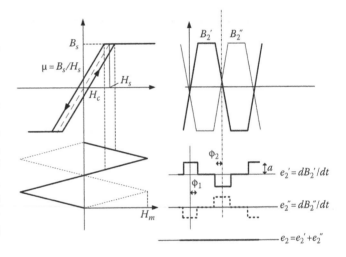

FIGURE 4.52
Signals in fluxgate sensor for $H_x=0$.

$$e_2 = -An_2 \frac{dB}{dt} \qquad (4.49)$$

According to Figure 4.52, the main parameters of a rectangular signal are

$$\varphi_1 = \frac{\pi}{2} \frac{H_s + H_c}{H_m}; \quad \varphi_1 = \frac{\pi}{2} \frac{H_s - H_c}{H_m}; \quad a = \frac{B_s}{\varphi_1 + \varphi_2} = 4\mu f H_m \qquad (4.50)$$

Expanding to a Fourier series, we obtain

$$e_2' = e_2'' = \frac{16}{\pi} n_2 f A \mu H_m \sin\frac{\pi}{2}\frac{H_s}{H_m} \cos\left(\omega t - \frac{\pi}{2}\frac{H_c}{H_m}\right)$$

$$+ \frac{16}{3\pi} n_2 f A \mu H_m \sin\frac{3\pi}{2}\frac{H_s}{H_m} \cos\left(3\omega t - \frac{3\pi}{2}\frac{H_c}{H_m}\right) \qquad (4.51)$$

Thus, the output signal is composed from odd harmonics and the resultant signal is $e_1' - e_2'' = 0$.

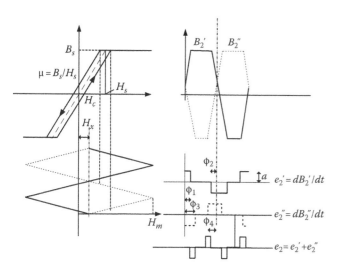

FIGURE 4.53
Signals in a fluxgate sensor for some external field H_x.

For presence of external field H_x, we have (Figure 4.53)

$$\varphi_1 = \frac{\pi}{2}\frac{H_s + H_c - H_x}{H_m} ; \quad \varphi_2 = \frac{\pi}{2}\frac{H_s - H_c - H_x}{H_m} ;$$

$$\varphi_3 = \frac{\pi}{2}\frac{H_s + H_c + H_x}{H_m} ; \quad \varphi_4 = \frac{\pi}{2}\frac{H_s - H_c + H_x}{H_m} \quad (4.52)$$

and the output signals are

$$e_2' = \frac{16}{\pi} n_2 f A \mu H_m \sin\frac{\pi}{2}\frac{H_s}{H_m}\cos\left(\omega t - \frac{\pi}{2}\frac{H_c}{H_m}\right)$$

$$-8n_2 f A \mu H_x \sin\pi\frac{H_s}{H_m}\sin\left(2\omega t - \pi\frac{H_c}{H_m}\right)+\cdots$$

$$e_2'' = -\frac{16}{\pi} n_2 f A \mu H_m \sin\frac{\pi}{2}\frac{H_s}{H_m}\cos\left(\omega t - \frac{\pi}{2}\frac{H_c}{H_m}\right)$$

$$-8n_2 f A \mu H_x \sin\pi\frac{H_s}{H_m}\sin\left(2\omega t - \pi\frac{H_c}{H_m}\right)+\cdots$$

and

$$e_2 = e_2' + e_2'' = 16 n_2 f A \mu H_x \sin\pi\frac{H_s}{H_m}\sin\left(2\omega t - \pi\frac{H_c}{H_m}\right)+\cdots$$

$$(4.53)$$

Thus, the odd harmonics are eliminated and the output comprises only even harmonics. The magnitude of the second harmonic output signal is

$$E_2 = 16 n_2 f A \mu H_x \sin\pi\frac{H_s}{H_m} \quad (4.54)$$

Other authors present similar results, for example, Afanasjev (Afanasjev 1986) determined similar relationship for sinusoidal excitation as

$$E_2 = 32 n_2 f A \mu H_x \frac{H_s}{H_m}\sqrt{1 - \left(\frac{H_s}{H_m}\right)^2} \quad (4.55)$$

Instead of using two cores, it is also possible to eliminate the zero-field component of the output signal in orthogonal-type fluxgate sensors (Figure 4.54). In these sensors, the two coils: exciting and pick-up are perpendicular to each other and in the absence of external magnetic field the output signal is equal to zero. Any external magnetic field is added to the exciting field and the resultant magnetic field,

$$H_w = \sqrt{H_x^2 + \left(H_m \sin\omega t\right)^2} \quad (4.56)$$

saturates the core periodically. As a result, the output second harmonic component is proportional to the external measured field.

Sensors of this type were proposed many years ago (Alldredge 1951, Primdhal 1970, 1979, Gise and Yarbrlough 1977), but recently they are studied again because they can be very small and fabricated with thin-film technology (Fan et al. 2006, Zhao et al. 2007).

The principle of operation described above shows a very simplified picture of real sensors. Especially, the ring core in practice is far from idealized picture because it is magnetized nonuniformly by external measured field and the halves of the ring do not work separately. Between the magnetizing and pick-up coils, signals can be carried through the mutual inductance. Also a parasitic capacitance and inductance of the coil influences the shape of the magnetizing current wave as well as output parameters.

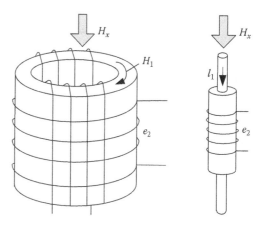

FIGURE 4.54
Orthogonal type fluxgate sensors.

4.3.2 Performances of Fluxgate Sensors

The relation (4.54) can be presented as

$$E_2 = 16n_2 f A \mu_0 \frac{\mu_r}{1 + N_d \mu_r} H_x \sin \pi \frac{H_s}{H_m} \quad (4.57)$$

Thus, if permeability μ_r of the core material is very large, the performance of the sensor does not depend on the μ_r value and hence does not depend on the instability of material properties (for instance, variation with temperature). Therefore, the permeability of the core material should be high and then the output signal depends practically only on the demagnetizing factor N_d:

$$E_2 = 16n_2 f A \mu_0 \frac{1}{N_d} H_x \sin \pi \frac{H_s}{H_m} \quad (4.58)$$

After calculation of the sensitivity factor $S = dE_2/dH_s$, we obtain the following relation (Afanasjev 1986):

$$S = 16n_2 \mu A f \left(\frac{H_s}{H_m} \right) \sqrt{1 - \left(\frac{H_s}{H_m} \right)^2} \quad (4.59)$$

Figure 4.55 presents dependence of sensitivity factor S on the magnetizing field H_m. The optimum value of the magnetizing field is $H_m = \sqrt{2}H_s$. Usually, the magnetizing field is higher (at the cost of sensitivity) because it reduces noise (Figure 4.55) (Scouten 1972, Afanasjev 1986). Assuming that H_m/H_s is around 2, we can rewrite Equation 4.58 as

$$E_2 \cong 10n_2 f A \mu_0 \frac{1}{N_d} H_x \quad (4.60)$$

The core material is magnetized nonuniformly and therefore, the demagnetizing factor can be only determined experimentally or numerically (Primdahl et al. 1989, How et al. 1997, Clarke 1999). But for sensitivity assessment, the simplified relationship can be used. Assuming that the demagnetizing factor for a strip core is approximately $N_d \approx 5A/l^2$, we obtain

$$E_2 \cong 2n_2 f \mu_0 l^2 H_x \quad (4.61)$$

For a ring core of the diameter d, the demagnetizing factor is given approximately as $N_d \approx 20A^{0.8}d^{1.6}$ (Marshall 1967), and we obtain

$$E_2 \cong 0.5n_2 f A^{0.2} \mu_0 d^{1.6} H_x \quad (4.62)$$

From relationships (4.61) and (4.62), we can see that the sensitivity of the fluxgate sensor depends on

- Number of turns n_2 of the pick-up coil
- Frequency f of the exciting magnetic field
- Length of the core l (or diameter d)

The most effective is to increase of the core length because the sensitivity increases with the square of the length l (or $d^{1.6}$). Indeed one of the highest sensitivity 10 mV/nT was demonstrated for a fluxgate sensor with core length 0.5 m (and $N_d = 2 \times 10^{-6}$, $n_2 = 4000$, $f = 10$ kHz) (Saito 1980).

A typical fluxgate sensor (Förster Institute (Förster 1955)) has the following parameters: $n_1 = n_2 = 1000$, $f = 3$ kHz, $l = 2 \times 60$ mm, $A = 3 \times 0.1$ mm and exhibits sensitivity of 10 μV/nT. Similar typical values of the sensitivity (between 3 and 20 μV/nT) are reported by other authors (Moldovanu et al. 1997).

A typical measuring range of a fluxgate sensor is from resolution of around 10 pT to upper linear limit of about 1 μT. The upper limit is usually extended to 100 μT by applying a feedback—for instance, to make possible measurement of Earth's magnetic field with 0.1 nT resolution.

The resolution of a magnetic sensor is strongly influenced by magnetic (Barkhausen) noise. Therefore, quality of the core material is very important. It should exhibit large permeability, small hysteresis, negligible magnetostriction, low eddy-current loss (high resistivity), small magnetic noise, and high Curie temperature. Fortunately, the core needs small amount of magnetic material and therefore, expensive materials can be tolerated. In the last century, commonly as the core material, low-magnetostrictive permalloy was used, for example, 81.5Ni6Mo (Gordon et al. 1968). Recently, commonly cobalt-based amorphous materials are more popular (Mohri 1984, Shirae 1984, Narod 1985, Tejedor et al. 1995, Moldovanu et al. 1997).

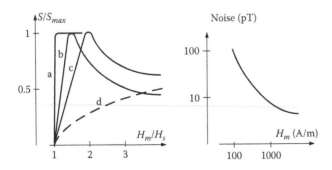

FIGURE 4.55
Sensitivity (theoretical) and noise (experimental) versus magnetizing field H_m (a, rectangular magnetization; b, sinusoidal magnetization; c, triangular magnetization; d, orthogonal magnetometer). (From Afanasjev, J.B., *Fluxgate devices* (in Russian, Energoatomizdat), 1986.)

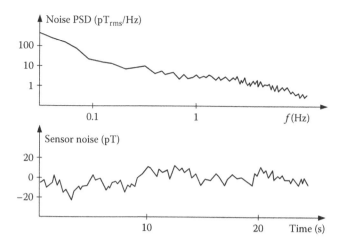

FIGURE 4.56

An example of noise characteristics of low noise fluxgate sensor. (From Ripka, P., *Sens. Actuat.*, A106, 8, 2003.)

There are many papers analyzing the fluxgate sensor noise. Not always it is clear if the analysis also includes the influence of the electronic circuit—generally such circuit should improve the SNR. Figure 4.2 presents a typical noise spectrum of fluxgate sensors. Similar results are reported by other authors (Scouten 1972, Dyal and Gordon 1973, Snare and McPherron 1973, Acuna 1974, Primdhal 1979, Shirae 1984, Primdhal et al. 1989, Ripka 1996a, b, 2003). One of the best case was reported by Ripka (Ripka 1992, Ripka et al. 2001)—in the sensor prepared from $Co_{67}Fe_4Cr_7Si_8B_{14}$ amorphous ribbon frequency range 60 mHz–10 Hz it was 8.8 pT rms (3.8 pT/$\sqrt{Hz}$ at 1 Hz)—as shown in Figure 4.56. In theoretical single-domain, rod-core fluxgate sensor, noise was estimated as 1.4 pT/$\sqrt{Hz}$ at 1 Hz (Koch and Rozen 2001). Sensor prepared from an HTS superconducting material in liquid nitrogen exhibited resolution of 0.17 nT (Gershenson 1991). Attainable lowest noise level at 10–70 pT rms at 0.01–10 Hz bandwidth can be assumed as realistic (Primdhal 1979).

The resolution of the sensor is also limited by offset (including temperature zero drift) and long-time stability. Primdahl (Primdahl 1979) studied both of these parameters. Offset depends on the core material, depth of the drive saturation, initial sensor demagnetization, magnetostriction. Investigations at the temperature range −40°C to +60°C showed that a typical offset stability is at the level of 0.1–0.4 nT. Long-term stability investigated over a period of 1 month and longer was at the level of several nT. Thus, we can see that the best range of operation of a fluxgate sensor is 1 nT–1 µT.

The fluxgate sensors are usually used for measurements of DC and low-frequency magnetic fields. Typical excitation frequency is 1–10 kHz and we assume that the modulated frequency should be no larger than 1/5 of the carrier frequency. Using modern amorphous materials, it is possible to extend the measured field bandwidth to several kHz (Ripka et al. 1995). In higher frequency range, the search coil sensors can better support this area of application.

4.3.3 Design of Fluxgate Sensors

Although the first paper on a fluxgate sensor of Aschenbrenner (Aschenbrenner and Goubau 1936) presented a ring-core fluxgate sensor earlier dominated designs with open cores: single (Figure 4.57a) (Rabinovici et al. 1989), double with common excitation coil (Figure 4.57b) (Vacquier et al. 1947), or double with separate excitation coils (Figure 4.57c) (Förster 1955).

After papers of Geyger in 1962 (Geyger 1962), the ring-core design became more popular (Figure 4.58) and recently, this design is most often realized due to its simplicity (although less sensitivity due to larger demagnetization coefficient).

The closed-core sensors can have both coils wound directly on the core (Figure 4.58a), but more convenient

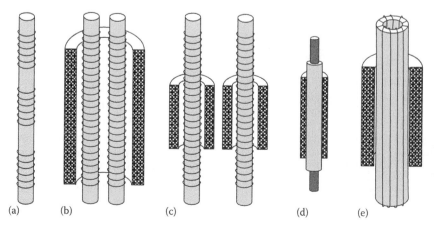

FIGURE 4.57

Various fluxgate sensors with open cores—longitudinal (a, b, c) or orthogonal (d, e).

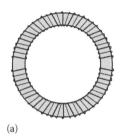 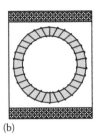 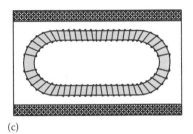

(a)   (b)   (c)

FIGURE 4.58
Various fluxgate sensors with closed cores.

is to use external sense coil (Figure 4.58b). Because a circular ring core has a larger demagnetization factor in comparison with a double-rod design, as a compromise also racetrack (oval) design was proposed (Gordon et al. 1965, 1968, Ripka 2000a,b). One of the advantages of the racetrack design is lower sensitivity to perpendicular fields.

Indeed, the ring-core design (Figure 4.59) is easier to manufacture than the double-rod sensor. But the main advantage of the ring-core sensor is its much smaller power consumption necessary to magnetize the closed sample to saturation, which is important when the sensor is used for outer space investigation (Musmann and Afanassiev 2010). A typical core is wound as a ring with diameter 15–25 mm from 4 to 15 turns of a magnetic tape 1–2 mm wide and 25 μm thick.

The winding of the ring core is usually carried out manually. To obtain perfect properties of the sensor, this coil should be wound very regularly. This is why, recently, many designers tried to substitute hand-made sensors by machine-made ones. For this goal, the printed circuit board (PCB) technology can be useful (Dezauri et al. 1999, 2000, Kubik et al. 2007).

The sensors presented in Figure 4.60 consist of two 100 μm epoxy boards with 35 μm copper lamination. The amorphous Vitrovac 6025 (CiFeMo$_{73}$SiB$_{27}$) foil was used as the ferromagnetic material. The sensitivity

 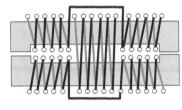

FIGURE 4.60
Examples fluxgate sensors prepared in PCB technology. (From Dezauri, O. et al., *IEEE Trans. Magn.*, 35, 2111, 1999.)

about 60 μV/μT at 30 kHz excitation and five turns of the detection coil was reported. Similar design for a racetrack shape was described by Kubik and coworkers (Kubik et al. 2007). Low-power (4 mW) sensor exhibited sensitivity of about 100 μV/μT.

Further miniaturization is possible by applying planar coil technology. There are many interesting achievements reported in this area (Vencueria et al. 1994, Choi et al. 1996, Gottfried-Gottfied et al. 1996, Chiesi et al. 2000, Ripka et al. 2001, Drljaca et al. 2004). The core of the sensor is usually sputtered thin film permalloy layer. Because planar coils exhibit poor efficiency, it was also proposed to manufacture quasi-solenoid coil prepared by a deposition of two conducting layers (Kawahito et al. 1996, Liakopoulos and Ahn 1999). The main advantage of these technologies (beside small dimensions of the sensor) is possibility of integration with the electronic part in CMOS technology.

The sensor presented in Figure 4.61a (Choi et al. 1996) enables measurement of magnetic field with sensitivity of around 90 μV/μ at 610 kHz excitation field. The whole sensor with dimensions 5 mm × 5 mm includes full electronics—driving part as well as an RF amplifier, phase-sensitive detector, and a low-pass filter.

The sensor presented in Figure 4.61b (Chiesi et al. 2000, Drljaca et al. 2004) enables simultaneous measurement of two components of magnetic field. The sensor with dimensions 2.5 mm × 2.5 mm (including full electronics) exhibited sensitivity of 3.8 mV/μT at 125 kHz excitation, with 26 dB amplification and 12.5 mW power consumption. The 1/f noise (introduced mainly by the

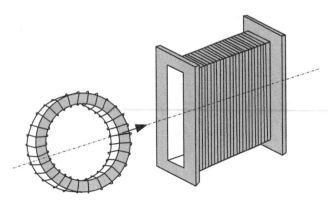

FIGURE 4.59
Assembly of a ring core sensor. (From Acuna, M.H., *IEEE Trans. Magn.*, 10, 519, 1974.)

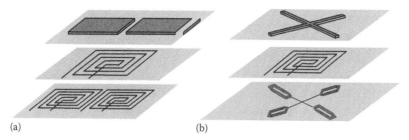

FIGURE 4.61
The planar technology fluxgate sensors. (From Choi, S.O. et al., *Sens. Actuat.*, A55, 121, 1996; Chiesi, L. et al., *Sens. Actuat.*, 82, 174, 2000.)

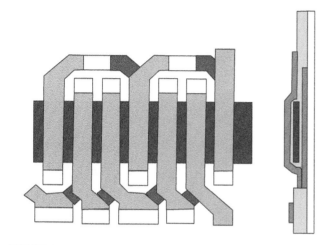

FIGURE 4.62
The planar solenoid coil fluxgate sensors. (From Kawahito, S. et al., *Sens. Actuat.*, A54, 612, 1996.)

electronic part) was around $70\,\text{nT}/\sqrt{\text{Hz}}$ and rms noise in the 0.2–10 Hz range was around 150 nT.

The sensor presented in Figure 4.62 resembles design of a PCB technology (Figure 4.60) but is made in microscale. The sensor (with included electronics) exhibited high sensitivity $2.7\,\mu\text{V/nT}$ at an excitation frequency 3 MHz. Its resolution (noise level), in the frequency range DC to 10 Hz, was around $40\,\text{nT}_{\text{p-p}}$, what was only slightly better than in typical and much simpler AMR sensors.

4.3.4 Multicomponent Sensor Systems (Compass, Gradiometer)

In some applications, for example, Earth's magnetic field measurement, there is a need to measure two- or even three components of magnetic field. Because modern magnetic sensors are relative small and simple, the obvious solution of the compass problem is to use three individual sensors positioned in the three principal axes (Ripka and Kaspar 1998). In such case, two groups of problems should be overcome. Firstly—how to distribute the sensors in order to eliminate mutual influence (magnetic fields generated by sensors and

their feedback). The second problem is the influence of orthogonal (transverse) component on the correct operation of the sensor—the *cross-field effect*. This problem can be significant especially if the orthogonal field is much larger than measured component.

Results of study on the cross-field effect are widely reported (Brauer et al. 1997, Ripka 2000a, b, Ripka and Billingsley 2000b). Fortunately, the error caused by the transverse field is not very significant in fluxgate sensors—it causes errors at the level of a few nT (if the transverse field is not too large—below 60 μT). It was estimated that the shape anisotropy is a factor influencing the cross-field effect—elongated double-rod and racetrack sensors had smaller cross-field effect than ring-core sensors. Transverse field can change the sensitivity and linearity of the sensor—these errors can be suppressed to some degree by using feedback. The most harmful is offset caused by a transverse field because this effect cannot be removed by individual feedback for each axis. If we use a three-axial (spherical) feedback coil (compact spherical coil [CSC]), theoretically we avoid the problem of crossfield because the sensors are in "magnetic vacuum (Figure 4.63)." Unfortunately, such solution (tested in Oersted satellite magnetometer (Primdahl and Jensen 1982, Nielsen et al. 1995, 1997) is rather complex and expensive. Therefore, individually

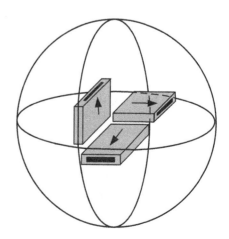

FIGURE 4.63
Three sensors in spherical feedback coil system.

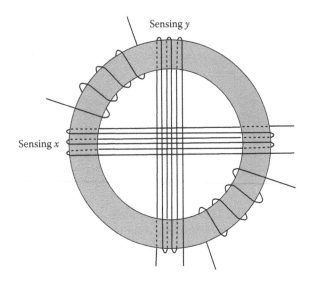

FIGURE 4.64
Two-axis ring core sensor. (From Acuna, M.H. and Pellerinn, C.J. *IEEE Trans. Geosci. Electron.*, 7, 252, 1969.)

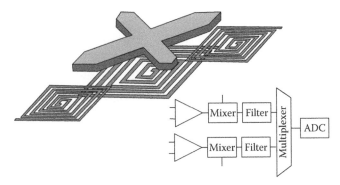

FIGURE 4.65
Two-axis fluxgate microsensor and read-out integrated circuit (only one pair of the pick-up coils is presented for simplicity). (From Baschirotto, A. et al., *IEEE Trans. Instrum. Meas.*, 56, 25, 2007; Baschirotto, A. et al., *Measurement*, 43, 46, 2010.)

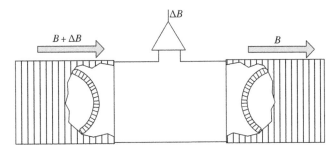

FIGURE 4.66
Two fluxgate sensors connected in gradiometer system.

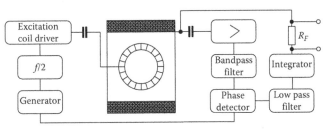

FIGURE 4.67
A block diagram of a second-harmonic fluxgate magnetometer.

compensated three sensors were also used (e.g., in Astrid-2 satellite) with only slight deterioration of immunity to transversal fields (Brauer et al. 2000).

Acuna and Pellerin (1969) demonstrated a possibility to measure two components of magnetic field by using two orthogonal pick-up coils (Figure 4.64). But such neat idea did not find followers. In a ring core, there is a problem with careful positioning of the coils in order to avoid coupling between excitation and sensing coils. Most likely in a two-coil system, this problem is twice as big and it is simply easier to use two independent sensors. Nevertheless, a similar sensor was used as a compass to automobile navigation (Peters 1986).

Mechanical positioning is not a problem in the case of microsensors where every sensor has the same geometry because of the precise manufacturing technology. Therefore, a 2D fluxgate returned in this technology (Chiesi et al. 2000, Kejik et al. 2000, Drljaca et al. 2004). Figure 4.65 presents 2D fluxgate microsensor with a cross-shaped core. Similar sensor with a fully integrated acquisition circuit (including ADC) was developed by Baschirotto (Baschirotto et al. 2007, 2010). The dimensions of the sensor were 1.7 mm × 1.7 mm. Sensitivity is 450 μV/μT 11 LSB/μT) and angular accuracy is 1.5° when the sensor is in Earth's magnetic field.

A system of two distanced sensors known as *gradiometer* (Figure 4.66) is very useful for measurement of small magnetic fields in the presence of large field. The Earth's magnetic field is very homogeneous—the field variation is less than 25 pT/m. Therefore, if source of magnetic field is closer to one sensor, it is possible to detect very small magnetic field (less than nT) in the presence of Earth's magnetic field (which is around 50 μT). Merayo presented a fluxgate gradiometer with

accuracy 0.3 nT/m for two sensors spaced by 20 cm (Merayo et al. 2001).

4.3.5 Electronic Circuits of Fluxgate Sensors

Figure 4.67 presents typical block diagram of a second harmonic fluxgate magnetometer. The frequency of the signal from the generator is halved to $f/2$ frequency drive the excitation coil. The driver circuit forms appropriate wave shape of excitation signal (triangular, sinusoidal, etc.). The output signal from the pick-up coil after amplification and filtration (to remove parasitic signals like odd harmonics) is connected to a phase-sensitive detector (PSD). A signal from the generator is used as a reference signal for the PSD. The output signal after low-pass filtering and integration can be used as a feedback signal. The pick-up coil can be also used as a feedback coil but often another separate coil is used to ensure uniformity of the feedback magnetic field. In feedback system, the fluxgate sensor is used as the zero field detector.

Although the block diagram presented in Figure 4.67 seems to be simple, there are many details significantly influencing performance of magnetometers as discussed in numerous papers:

- Matching between the excitation generator and the excitation coil (wave shape, frequency, value of capacitance of the driving coil) (Primdhal 1979, Ripka 2001). Figure 4.68 presents an example of the influence of parallel capacitor on the driving signal.

- Matching between the pick-up coil and the amplifier (tuned or untuned output, influence of capacitor, etc.) (Russel et al. 1983, Narod et al. 1984, Primdahl et al. 1987, Player 1988, Ripka and Kaspar 1998). The output amplifier is often substituted by a low resistance current-to-voltage transducer (Primdahl et al. 1989, Ripka et al. 2000).

In second-harmonic fluxgate magnetometers, we utilize the feature of PSD known as one of the best tools used for noise rejection (Smith et al. 1992, Tumanski 2006). But also nonselective detection techniques are reported. For example, in a phase-delay method, a difference in time interval of both half-periods of the output signal is measured (see Equation 5.52 and Figure 4.53) (Scarzello and Usher 1977, Heinecke 1978). Figure 4.69 presents an exceptionally simple sensor working in an auto-oscillation mode. If the core is magnetized by an external DC magnetic field, the second-harmonic voltage is generated. A square waveform voltage has different intervals for positive and negative half cycles. The reported resolution was about 0.1 nT.

However, there is also visible tendency to design more complex digital electronic circuits interfaced with the fluxgate sensor. Although simplicity of second-harmonic

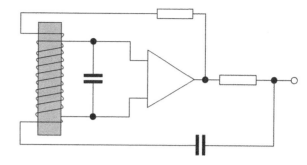

FIGURE 4.69
Resonant-type magnetic sensor. (From Takeuchi, S. and Harada, K., *IEEE Trans. Magn.*, 20, 1723, 1984.)

analog circuit is one of the main advantages of fluxgate sensor, many quite complex digital circuits were proposed (Primdahl et al. 1994, Auster et al. 1995, Piil-Henriksen et al. 1996, Pedersen et al. 1999, Kawahito et al. 1999, 2003, Cerman and Ripka 2003, Cerman et al. 2005, O'Brien et al. 2007, Forslund et al. 2008).

Three main ideas of digitalization of fluxgate magnetometers were proposed:

1. An ADC (most often sigma-delta) included at the output of an analog magnetometer.

2. A digital–analog hybrid in which part of classical analog circuit (integrator and sensor as summation node) serve as an element of sigma-delta modulator, thus the whole magnetometer works as sigma-delta converter of magnetic field into bitstream.

3. Fully digital device in which after analog amplification, the output signal is processed digitally.

All three concepts have been tested and analyzed by Cerman (Cerman et al. 2005). Surprisingly (or not), it was concluded that the best performance was displayed by classical analog magnetometer with ADC connected to the output. However, it was concluded that analog circuits practically reached end of their development whereas digital technology might potentially open completely new possibilities.

One of the main weaknesses of digital instruments is that if we use analog feedback (which is almost always applied), it is necessary to include in return a DAC. This way we introduce a new source or errors (nonlinearity, noise, temperature dependence, etc.). In analog magnetometers, the circuit errors are negligible in comparison with the sensor errors and digital circuit does not improve the sensors features as such. Therefore, the proposed digital magnetometers did not exhibit better performance than best analog devices.

On the other hand, all analog elements can be easily substituted by equivalent or better digital devices

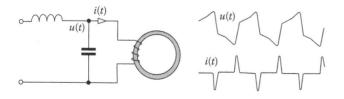

FIGURE 4.68
Pulse excitation current with a capacitor and inductance. (From Acuna, M.H., *IEEE Trans. Magn.*, 10, 519, 1974.)

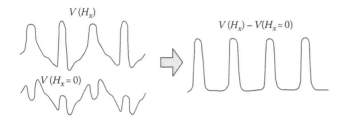

FIGURE 4.70
Cleaning of the output signal by digital subtraction $H_x=0$ signal. (After Primdahl, F. et al., *Meas. Sci. Technol.*, 5, 359, 1994.)

(excluding the initial amplifier)—PSD can be substituted by a digital multiplier (mixer), the same applies to filters. Figure 4.70 illustrates a new possibility resulting from a digital processing, which would be very difficult to realize in an analog way. Theoretically, the output signal for $H_x=0$ should be zero and the output signal for H_x should contain only even harmonics. In the real sensor, the odd harmonics do not compensate completely and due to the coupling between the excitation and the pickup coils, the zero-field signal is very distorted (signal on the left bottom in Figure 4.70). As direct result, also signal for presence of measured field is distorted (signal on the left top in Figure 4.70). By digitally subtracting

the zero-field signal, we can recover the correct signal (signal on the right in Figure 4.70). This operation is difficult to perform with analog circuit but digitally, we can simply remove the parasitic components correlated with prestored zero-field signal.

Also, other operation can be performed digitally, for example, improvement of noisy signal by using a moving average digital filter, remove average value (offset), improvement of linearity, statistical analysis, calibration of the device, change online operating conditions. That is probably why, in the future, digital fluxgate magnetometers will be more commonly used.

Figure 4.71 presents a sigma-delta fluxgate magnetometer and, for comparison, classical sigma-delta modulator. We can see that a part of the circuit is analog and the sensor is a summation node. Unfortunately, initial results of such digital magnetometer are not fully satisfactory (Kawahito et al. 2003, Magnes et al. 2003, Cerman et al. 2005). The application of a sigma-delta modulator to the sensor feedback increases total noise and the whole device is noisier at higher frequency, which strongly limits the frequency bandwidth.

Figure 4.72 presents an example of a fully digital fluxgate magnetometer. The amplified output signal of the sensor converted to digital form by a sigma-delta ADC

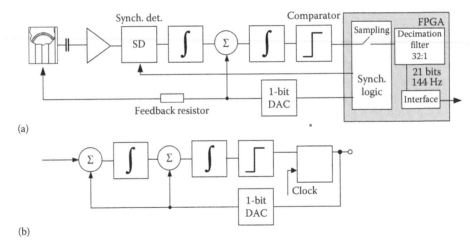

(a)

(b)

FIGURE 4.71
Sigma-delta fluxgate magnetometer (a) and classical sigma-delta modulator (b). (After Magnes et al. 2003.)

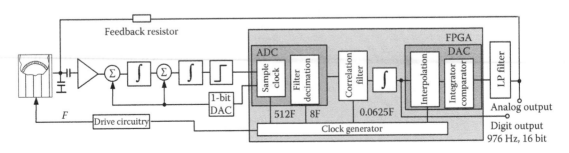

FIGURE 4.72
The example of digital fluxgate magnetometer. (From O'Brien, H. et al., *Meas. Sci. Technol.*, 18, 3645, 2007.)

and the following operations are performed digitally. Noise level was below 30 pT in 0.1–60 Hz range. The FPGA technology was chosen because such circuit is immune to radiation and the whole magnetometer was designed for space investigations.

Fluxgate sensor is an excellent choice for measurement of magnetic field in the range between SQUID and AMR sensors—that is, between 1 nT and 100 μT. It enables measurement of DC and low-frequency magnetic fields with resolution of a few pT. Both the sensor and the associated electronics are relatively simple.

As an example of a fluxgate sensor, we can consider the device designed for Oersted satellite (Nielsen et al. 1995). The ring-core sensor was prepared from Vitrovac 6025 amorphous ribbon 1 mm wide and 24 μm thick. It was wound into ring with inner diameter 15 mm—11 wraps (with the effective cross-sectional area $A = 0.27$ mm²). The excitation coil wound on the ring was 196 turns of 0.2 mm Cu wire. The excitation frequency was 15.625 kHz. The detector coil was wound with 2 × 103 turns of 0.18 Cu wire.

The sensor works a short-circuit output mode with sensitivity 36 μA/μT. The sensor noise was 14 pT$_{rms}$ in the frequency range 0.06–10 Hz corresponding to a noise-power density of 6.2 pT/$\sqrt{Hz}$ at 1 Hz. The power consumption of three-axial sensor was 160 mW and for the electronics including ADC and interface 940 mW. The measurement range was ±65,536 nT, 18-bit resolution corresponding to 0.5 nT, linearity, and accuracy 0.5 nT. Temperature operating range was from –20°C to 40°C.

A promising feature of the fluxgate sensor is it miniature size—it is possible to design devices of dimensions 2 mm × 2 mm including all electronics. Because sensitivity depends on the sensor dimensions, there can be a slight reduction of performance with comparison to normal-size sensors. But sensitivity increases with frequency of excitation and miniature sensors can operate for much higher frequency.

There are many market-available fluxgate sensors, magnetometers, and compasses. For example, Stephan Mayer Instruments offers three-axis fluxgate sensor with dimensions 25 mm × 500 mm, field range 200 μT, and accuracy 1%. The sensitivity is 2 V/70 μT, noise level is less than 0.5 nT$_{rms}$ (0.1–10 Hz) corresponding with 120 pT/$\sqrt{Hz}$ at 1 Hz, zero drift less than 2 nT/K for $T = 15°C–60°C$, and frequency bandwidth DC to 1 kHz.

Another company Billingsley Aerospace & Defense offers several fluxgate magnetometers. Their analog three-axis fluxgate magnetometer TFM100 with the size 3.5 × 3.2 × 8.3 cm measure magnetic field in a range ±100 μT, with sensitivity 100 μV/nT, accuracy 0.75%, linearity 0.015%, noise less than 12 pT/$\sqrt{Hz}$ at 1 Hz, and zero drift 0.6 nT/K. Frequency bandwidth is DC to 500 Hz. The digital version 24-bit resolution three-axial

fluxgate magnetometer DFMG24 with a range ±65 μT provides 20 pT resolution, digital linearity 0.007% of FS, accuracy 0.03% of FS, noise less than 10 pT/$\sqrt{Hz}$ at 1 Hz, and zero offset less than 5 nT. There is a serial digital interface with cable length up to 1 km.

The fluxgate sensors are mainly used in space research (Musmann and Afanassiev 2010), geophysics, underwater magnetic field investigation, mine detection, navigation, and nondestructive testing of materials (Kaluza et al. 2003).

4.4 Magnetoresistive and Magnetoimpedance Sensors

4.4.1 Magnetoresistance—General Remarks

The magnetoresistive (MR) effect (change of resistance caused by magnetic field) exists practically in all metals but it is detectable only at very low temperature and in high magnetic field. However, there are several materials for which the MR effect is sufficiently large at room temperature: ferromagnetic metals, semiconductors, and some minerals as bismuth or lanthanum-based oxides. The MR effect can be large in some artificial multilayer structures like valve or superlattices.

Figure 4.73 presents a comparison of the main MR effects. Historically, the first MR sensor of magnetic field was made from bismuth. A bismuth spiral exhibited a MR coefficient $\Delta\rho/\rho \approx 2\%$ at room temperature but unfortunately in rather large magnetic fields.

After development of semiconductors, the bismuth spiral was substituted by semiconductor devices. The MR sensors made from semiconductor materials like InSb operated on similar physical principles as the

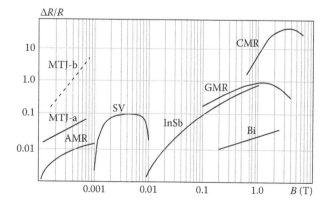

FIGURE 4.73

The main magnetoresistive effects: MTJ, magnetic tunnel junction; AMR, anisotropic magnetoresistance; SV, spin valve; InSb, semiconductor magnetoresistors; GMR, giant multilayer magnetoresistance; Bi, bismuth; CMR, collosal magnetoresistance.

Hall effect—the change of the path of the moving free charge carrier under influence of the Lorentz force. The change of resistance of special geometrical form of the sensors (Carbino disk, Feldplatte (Tumanski 2001)) was relatively large (even up to 20%), but again in rather large magnetic fields (Figure 4.73). Due to several drawbacks, like nonlinearity and influence of temperature, the semiconductor magnetoresistors are currently seldom manufactured—in this range of magnetic field, Hall sensors provide much better performances.

In 1993–1994, new MR effect in lanthanum-based doped manganate perovskites material—called *collosal magnetoresistance* was discovered (von Helmolt et al. 1993, Jin et al. 1994a, b, McCormack et al. 1994). Unfortunately, this effect also occurs only at low temperatures and high magnetic fields. At room temperature, it exhibits only moderate MR of about 30% for high magnetic field (Manoharan et al. 1994).

In 1857, Thomson (Lord Kelvin) discovered that the longitudinal resistivity ρ_l of ferromagnetic materials magnetized parallel to the current direction is larger than transverse resistivity ρ_t of material magnetized perpendicularly to the current direction (Figure 4.75). Although this effect known as *anisotropic magnetoresistances (AMR)* is not significantly large (about 2%), it is commonly used as MFS due to simplicity of the sensor design and relatively large sensitivity of the sensor.

In 1988, the *giant magnetoresistance (GMR)* effect (see Figure 2.42) was discovered (Baibich et al. 1988). In the beginning, it was detected at low temperature and high magnetic fields. However, within just one year demonstrated also at room temperature (Figure 4.76).

The GMR effect is related to the transition between antiferromagnetic and ferromagnetic states between the magnetic layers. In a multilayer structure with relatively thin nonmagnetic layer between the magnetic layers, the antiferromagnetic state is forced by atomic nature of the coupling between the layers. Because this coupling is rather strong, in order to overcome it, also high magnetic field is necessary. Such multilayer

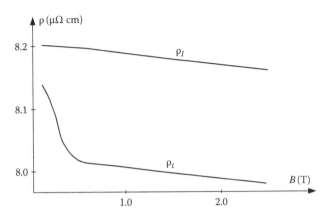

FIGURE 4.75
The anisotropic magnetoresistance—the change of resistivity of NiCo alloy at room temperature versus the external magnetic field. (From McGuire et al. 1975.)

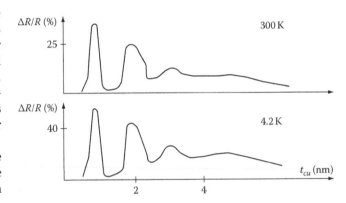

FIGURE 4.76
The GMR magnetoresistance of (CoCu) × 16 multilayer structure versus thickness of Cu layer. (From Parkin, S.S.P. et al., *Phys. Rev. Lett.*, 66, 2152, 1991.)

structures exhibit rather large MR but at high magnetic field. Therefore, such sensors are used when large MR is necessary despite poor sensitivity. More often, another GMR structure known as *spin valve (SV)* is used (Figure 4.77).

In SV structures, the antiferromagnetic state is induced artificially by deposition of additional layer from antiferromagnetic material. In such a case, the distance between magnetic layers can be larger and the sensitivity of the SV sensor is much higher than in the case of classical multilayer GMR structure (but at the expense of smaller change of resistance).

In 1995, a new class of GMR sensors was introduced—*magnetic tunnel junction (MTJ)* in which the conducting layer was substituted by an insulating layer (Moodera et al. 1995, Miyazaki and Tezuka 1995a,b). Such sensors exhibited large change of resistance for much smaller magnetic fields—thus the sensitivity was larger. Especially promising was the substitution of AlO

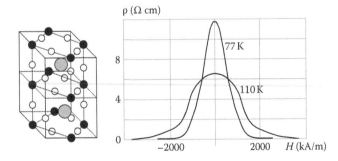

FIGURE 4.74
The crystal structure of manganate perovskites La$_{67}$Ca$_{33}$MnO$_3$ and its change of resistance versus magnetic field. (From McCormack, M. et al., *Appl. Phys. Lett.*, 64, 3045, 1994.)

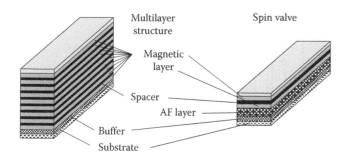

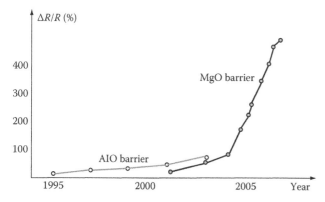

FIGURE 4.78
The progress in magnetoresistivity ratio obtained in magnetic tunnel junction structures. (After Ikeda, S. et al., *IEEE Trans. Electron. Dev.*, 54, 991, 2007.)

insulation barrier by MgO insulator (Figure 4.78). This was the turning point from which a rapid increase of MR in such structures can be observed.

In further part of this chapter, only the recently manufactured MR sensors are described—AMR, TMJ, GMR, and SV sensors.

4.4.2 AMR Magnetic Field Sensors

We can start the analysis of a MR sensor from the known Voigt–Thomson formula describing the change of resistance of a thin film strip of initial resistivity ρ and MR coefficient Δρ/ρ:

$$\rho(\vartheta) = \rho + \Delta\rho \cos^2 \vartheta \tag{4.63}$$

Thus, the resistance of the thin film strip depends on the direction of magnetization with respect to the current direction ϑ and the relative change of the resistance of the sensor can be written as

$$\frac{\Delta R_x}{R_x} = -\frac{\Delta\rho}{\rho} \sin^2 \vartheta \tag{4.64}$$

On the other hand, from the free energy balance (see Equation 2.54), we obtain that the direction of magnetization φ with respect to the anisotropy axis depends on both components of external magnetic field (H_x, H_y) and the anisotropy field H_k:

$$\sin\varphi = \frac{H_x}{H_y + H_k} \tag{4.65}$$

The anisotropy is induced during deposition process of thin film. Thus, in the initial state ($H_x = 0$), the magnetization direction lies along the anisotropy axis (easy axis of anisotropy). We usually direct the magnetic strip also along the anisotropy axis and, therefore, $\varphi = \vartheta$ and

$$\frac{\Delta R_x}{R_x} = -\frac{\Delta\rho}{\rho} \frac{1}{(H_y + H_k)^2} H_x^2 \tag{4.66}$$

The measured magnetic field component H_x is directed perpendicularly to the anisotropy axis and if its value increases, then the direction of magnetization rotates from its initial position along the anisotropy axis to a final state—perpendicular to the anisotropy axis. The component H_y of the external magnetic field should be much smaller than the anisotropy field H_k to minimize its influence on the change of resistance.

Figure 4.79 presents theoretical and experimental characteristics of a single strip MR sensor. The experimental characteristics differ from theoretical due to nonlinear demagnetizing field in the ferromagnetic layer. As can be seen, the dependence $\Delta R/R = f(H_x)$ is nonlinear.

To obtain linear response of the sensor, it is necessary to move its operating point into linear part of characteristics (or the deflection point of real characteristic—e.g., point A in Figure 4.79). The simplest way to achieve this could be biasing of thin film by an additional hard magnetic layer (Figure 4.80). Such method is not used in

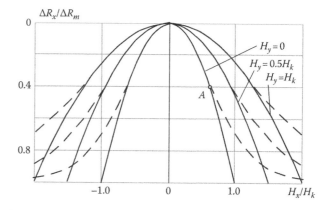

FIGURE 4.79
The transfer characteristics of a single strip magnetoresistive sensor (continuous curves—theory, dashed—experiments).

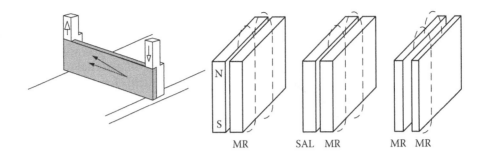

FIGURE 4.80
The single-path sensor and methods of linearization by biasing the thin-film layer: additional hard magnetic layer, additional soft magnetic layer, second thin-film sensor.

practice because the resulting bias is unstable and temperature dependent.

Instead of hard magnetic layer, additional soft magnetic layer (*soft adjacent layer* [*SAL*]) can be used—Figure 4.81. This method was introduced by Beaulieu and Nepala (Beaulieu and Nepala 1975). The current flowing in the MR layer produces a vertical magnetic field in the soft adjacent layer. In the magnetized biasing layer, the magnetic poles are created and these poles generate in return a vertical bias field in the MR layer. Sometimes, additional hard magnetic layers at the ends of the MR layer support the biasing (Figure 4.81).

Instead of using a passive soft magnetic layer, it would be more reasonable to use the second magnetoresistor as SAL (Figure 4.82). In such technique, magnetic field of one element serves as a magnetically bias the other element in a differential sensing mode (Figure 4.82).

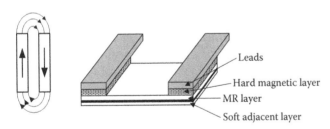

FIGURE 4.81
Biasing of the magnetoresistive film by a soft adjacent layer.

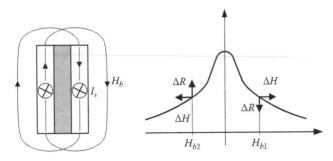

FIGURE 4.82
Biasing of the magnetoresistive film by the second MR layer.

Biasing of the MR element by an additional ferromagnetic layer is convenient when the sensor should be as narrow as possible, for example, in a reading head, sensor of the grain or the domain structure (Mohd Ali and Moses 1989, Nicholson 2000). In the MR sensor of magnetic field, another technique is used. If we incline the thin film strip by an angle ε from the direction of anisotropy axis, we introduce an initial angle shift between the current direction and the magnetization direction (see Figure 2.39) (it is de facto the same as to use a bias field where the initial position of magnetization is shifted).

If we introduce that (see Figure 2.39)

$$\vartheta = \varepsilon - \varphi \qquad (4.67)$$

the expression $\Delta R/R = f(H_x)$ is

$$\frac{\Delta R_x}{R_x} = \frac{\Delta \rho}{\rho} \left[\cos^2 \varepsilon - \cos 2\varepsilon \frac{H_x^2}{\left(H_y + H_k\right)^2} \right.$$
$$\left. + 2 \sin 2\varepsilon \frac{H_x}{H_y + H_k} \sqrt{1 - \left(\frac{H_x}{H_y + H_k}\right)^2} \right] \qquad (4.68)$$

The change of the resistance $\Delta R/R = f(H_x)$ is the sum of the square and quasi-linear components. For $\varepsilon = 0$, we obtain relation (4.66). Especially interesting is the case $\varepsilon = \pm 45°$ for which the expression (4.68) becomes

$$\frac{\Delta R_x}{R_x} = \frac{\Delta \rho}{\rho} \left[\frac{1}{2} + \frac{H_x}{H_y + H_k} \sqrt{1 - \left(\frac{H_x}{H_y + H_k}\right)^2} \right] \qquad (4.69)$$

The transfer characteristics of such sensor described by the expression (4.69) are presented in Figure 4.83. We can see that for a relatively small value of H_x, this dependence is close to linear,

$$\frac{\Delta R_x}{R_x} \cong \frac{\Delta \rho}{\rho} \left(\frac{1}{2} \pm \frac{H_x}{H_y + H_k} \right) \qquad (4.70)$$

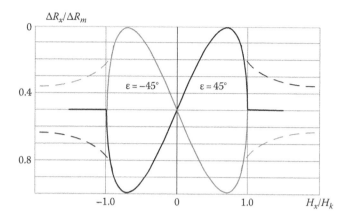

FIGURE 4.83
The transfer characteristics of a magnetoresistive sensor with $\varepsilon = \pm 45°$ (continuous curves—theory, dashed—experiments).

Moreover, the sensor with $\varepsilon = 45°$ is differential with respect to the case of $\varepsilon = \pm -45°$, which is advantageous because for such pair of sensors it is easy to correct the error caused by temperature changes of resistivity.

Initially, such linear sensors were prepared as to create a long meander of magnetic path—by etching appropriate gaps in the film (Figure 4.84). Several examples of such structures are presented in Figure 4.84—the "herring bone" form proposed by Hebbert and Schwee (Hebbert and Schwee 1966, 1968) or designs proposed by Tumanski (Tumanski and Stabrowski 1984).

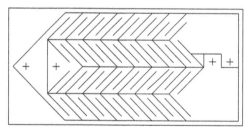

(a)

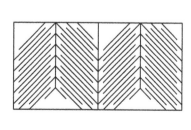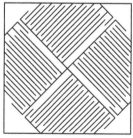

(b)

FIGURE 4.84
The design of linear magnetoresistive sensors: proposed by Hebbert and Schwee (a) and by Tumanski (b). (From Hebberet et al. 1966; Tumanski, S. and Stabrowski M., *IEEE Trans. Magn.*, 20, 963, 1984.)

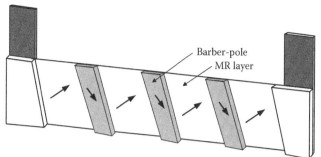

FIGURE 4.85
The design of linear Barber pole magnetoresistive sensors. (From Kuijk, K.E. et al., *IEEE Trans. Magn.*, 11, 1215, 1975; Kuijk, K.E., Magnetoresistive magnetic head, US Patent 4,052,748, 1977.)

The design of a linear sensor presented in Figure 4.84 is not magnetically advantageous because after inclination of the magnetic path from the anisotropy axis, the dispersion of anisotropy increases and the whole sensor is more susceptible to demagnetization. This is why after the invention in 1975 by a Philips team of another design, known as *Barber pole structure*, this type of the AMR sensor is recently manufactured almost exclusively (Kuijk et al. 1975, Kuijk 1977). The example of the Barber pole design is presented in Figure 4.85.

In the Barber pole design, the magnetic path remains along the anisotropy axis, which is advantageous. The inclination of the current direction with respect to the anisotropy axis is forced by special electrodes made from good conducting material (gold or aluminum) deposited on the ferromagnetic layer. These electrodes are inclined by 45° with respect to the anisotropy axis. Of course, if we deposit such electrodes with the angle ±45°, we obtain a pair of differential sensors.

The Barber pole design exhibits also another very important feature. The current flowing in the gold electrodes generates the magnetic field H_{y0}, which additionally biases the magnetic part (Feng et al. 1977, Haudek and Metzdorf 1980, Metzdorf et al. 1980, Tsang et al. 1982, Tumanski and Stabrowski 1985). This magnetic field protects the sensor from demagnetization. This magnetic field generated by the current I in a Barber poles (Figure 4.86) can be approximately determined as

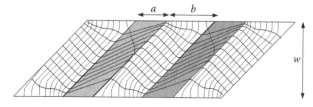

FIGURE 4.86
The calculated equipotential and current lines in the three-segment magnetoresistor. (From Tumanski, S. and Stabrowski, M., *Sens. Actuat.*, 7, 285, 1985.)

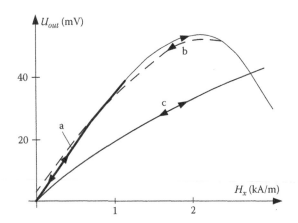

FIGURE 4.87
The return characteristics of the Barber pole sensor: when the measured field does not exceed about half of the range (a), when it exceeds the maximum (b), and with additional stabilizing field H_{y1}(c).

$$H_{y0} \cong \frac{I}{4w}\frac{a}{a+b} \qquad (4.71)$$

For typical values: $a=9\,\mu m$, $b=6\,\mu m$, $w=15\,\mu m$, $t_{Au}=1\,\mu m$, the current of 20 mA generates a magnetic field over 100 A/m, which is usually sufficient to attain the transfer characteristic without hysteresis.

In the real sensors, the internal bias is sufficient if the measured field is not too large—for example, when it does not exceed a half of the maximum value (Figure 4.87 line a). If the measured field exceeds the maximum value, the hysteresis can occur (line b in Figure 4.87). We can return to previous state after application of magnetic field in the y-axis.* To generate this field, we can use an external magnet or a coil wound around the sensor. Some sensors are equipped with an internal planar coil (see Figure 4.9). If we use additional stabilizing field H_{y1}, we extend the hysteresis-free range (Figure 4.87 line c).

Figure 4.88 presents a cross section through the typical structure of a Barber pole sensor. Dibbern and Petersen (Dibbern and Petersen 1983) specified the following steps in the fabrication of the Barber pole sensors of Philips:

- Surface oxidation of a silicon substrate ($1.6 \times 1.63\,mm^2$)
- Sputter deposition of an adhesive layer of titanium (0.1 μm thick)
- Sputter deposition of a permalloy film (30 or 40 nm thick)

---

* It is recommended to use this external field in the same direction as the internal bias. We can recognize such case because the output signal is then proportional to $1/(H_k+H_{y0}+H_{y1})$. If the external field is in opposite direction, we get $1/(H_k+H_{y0}-H_{y1})$, thus the output signal is larger.

- Formation of permalloy stripes using subtractive photolithographic process
- Annealing of the permalloy film in magnetic field (approximately 300°C)
- Sputter deposition of a titanium/tungsten adhesive layer (0.1 μm thick)
- Formation of gold or aluminum barber pole patterns
- Laser trimming of the bridge to give the zero offset voltage at 25°C

When we design the AMR sensor, we should take into account the demagnetizing field H_d because the resulting anisotropy field is

$$H_k = H_{k0} + H_d \cong{\sim} H_{k0} + M\frac{t}{w} \qquad (4.72)$$

Thus, the change of resistance of the single path sensor of the thickness t and width w made from the material with anisotropy field H_{k0} is

$$\frac{\Delta R_x}{R_x} = \frac{\Delta\rho}{\rho}\left(\frac{1}{H_{k0}+M(t/w)}\right)^2 H_x^2 \qquad (4.73)$$

We can change the sensitivity by appropriate choice of the width of the path. Most often as a sensor material is used, a non-magnetostrictive permalloy 81/19 with following parameters:

- Magnetoresistivity coefficient $\Delta\rho/\rho = 2.2\%$
- Resistivity $\rho = 22\,\mu\Omega\,cm$
- Anisotropy field $H_{k0} = 250\,A/m$
- Magnetization $M_s = 8.7 \times 10^{-5}\,A/m$

Moore (Moore et al. 1972) estimated the highest acceptable current density as $10^{10}\,A/m^2$, but usually this value is assumed as $J = (2-5) \times 10^9\,A/m^2$.

Some more complex designs of Barber pole sensors are described in many papers (Casselman et al. 1980, Dibbern 1983, 1984, 1986, 1989, Tumanski and Stabrowski 1985, Eijkel and Fluitman 1990, Pant and Krahn 1991, Pant 1996a,b, Tumanski 2001). The output signal of the sensor of a path length L can be determined from the following expression:

$$V_{out} = 2JL\frac{\Delta\rho}{\rho}\rho\frac{(a/w)}{(a/w)+(b/w)}$$

$$\times\left(0.5+0.14\frac{a}{w}\right)\frac{1}{H_{k0}+Mt/w+H_{y0}} \qquad (4.74)$$

where H_{y0} is the self-bias magnetic field (Equation 4.71).

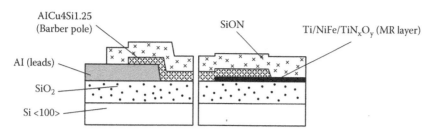

FIGURE 4.88
The typical structure of a Barber pole sensor. (From Linke, S., *Proc. Symp. Dortmund*, 10, 1992.)

We can see that the important parameters are the ratios a/w and b/w (see also Figure 4.86). Influence of these relationships is presented in Figure 4.89.

From the data presented in Figure 4.89, we can summarize following practical design guidelines (Tumanski 2001):

- The Barber electrode width b should be as small as possible

- For assumed width b, there is an optimal value of the magnetic strip width a limited by $(0.5–1.5)w$

- The strip aspect ratio t/w can be recommended as 10^{-3} (width in μm equal to the thickness in nm)

TABLE 4.2

Design Parameters of the Barber Pole Sensors of Philips

Parameter	KMZ10A	KMZ10B	KMZ10C
Sensitivity [mV/(kA/m)] at $U=5$ V	80	20	7.5
Sensitivity [mV/V(kA/m)]	16	4	1.5
Range [kA/m]	0.5	2.0	7.5
R [kΩ]	1.3	1.7	1.4
H_k [kA/m]	1.2	3.8	17.6
t [nm]	33	44	130
w [μm]	30	10	6
b [μm]	4	4	4
a [μm]	10	4	4
Number of paths	14	13	30
Recommended H_{y0} [kA/m]	0.5	3	3

Source: Dibbern, U., Magnetic sensors, in *Magnetoresistive Sensors*, Chapter 9, VCH Publishers, Weinheim, Germany, 1989, pp. 342–378.

Table 4.2 collects the typical parameters of the Barber pole sensors as based on the example of the Philips devices.

There are several manufacturers of AMR sensors. The typical performance of this type of sensors is represented by Philips and Honeywell devices. Figure 4.90 shows the layouts of these sensors and Table 4.3 collects the main data.

From the data presented in Table 4.3, we can see that typical AMR sensors measure the magnetic field with sensitivity 20–200 V/T in the field range 0.2–2.5 mT (0.2–2 kA/m). The linearity error is acceptably small—0.1% if the measured field does not exceed 0.5 FS (for full scale, it increases to 1%). The temperature error can be negligibly small if the sensor is supplied by a current source (for the voltage source, it is 0.3%/K).

Figure 4.91 presents the results of investigations of resolution of the KMZ10A1 sensor performed by Ripka (Ripka 1996a, b). These results confirm that an AMR sensor can measure the magnetic field with the resolution as good as a few nT. But all MR sensors exhibit rather large temperature zero drift at the level of several μV/K.

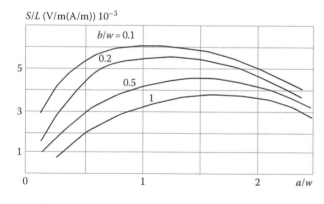

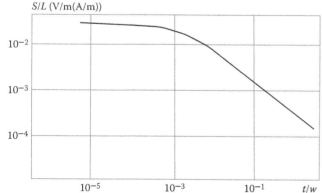

FIGURE 4.89
Sensitivity over length S/L versus geometry of the Barber pole sensor.

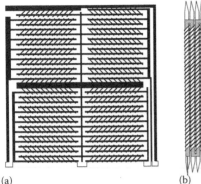
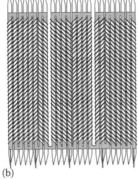

(a)   (b)

FIGURE 4.90
Two examples of Barber pole layouts: design of Philips (a) and design of Honeywell (b). (From Pant, B.B., *Electrochem. Soc. Proc.*, 95, 62, 1996a; Pant, B.B., *J. Appl. Phys.*, 79, 6123, 1996b.)

TABLE 4.3

Typical Data of AMR Sensors

	KMZ10B of Philips	HMC1001 of Honeywell
Sensitivity by $U_0 = 5\,V$ (mV/kA/m)	20	200
Sensitivity by $U_0 = 5\,V$ (V/T)	16	160
Field range (A/m)/μT	2000/2500	160/200
Resistance [Ω]	1600–2600	600–1200
Linearity error 50% FS	0.5	0.1
Temperature error (fixed I_0) [%/K]	−0.1	−0.06
Temperature offset	15 μV/K	15 μV/K
Orthogonal field error		3% FS for $H = 80\,A/m$
Noise density at 1 Hz [nV/Hz$^{-1/2}$]	15 nT$_{pp}$ 0.2–10 Hz	30 nV/Hz$^{-1/2}$ at 1 Hz
Resolution at 10 Hz		3 nT
Frequency bandwidth [MHz]	0–1	0–5

The effective way to suppress this drift is to use alternating bias as it is presented in Figure 4.6 (Tumanski 1984).

Honeywell applies a set/reset technique to reduce the offset influence in their AMR sensors (Figure 4.92). To measure the magnetic field H_x, the sensor is first activated by a "set" magnetic field pulse along the anisotropy axis. Next, the voltage is measured for the "reset" state. The resultant magnetic field is

$$V_{set} - V_{reset} = (SH_x + V_{offset}) - (-SH_x + V_{offset}) = 2SH_x \quad (4.75)$$

By using the set/reset method, it is possible to reduce the temperature offset error from 0.03%/K to 0.001%/K. Figure 4.93 presents a design of AMR sensor with additional planar coils to perform set/reset mode.

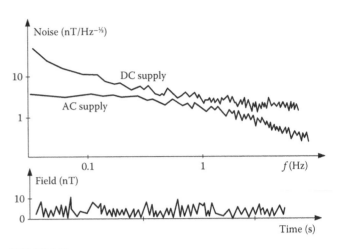

FIGURE 4.91
The resolution of KMZ10A1 sensor of Philips. (From Ripka, P., *J. Appl. Phys.*, 79, 5211, 1996a; Ripka, P., *J. Magn. Magn. Mater.*, 157–158, 424, 1996b.)

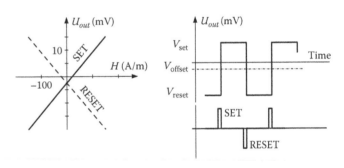

FIGURE 4.92
The SET/RESET technique of reduction of the temperature offset.

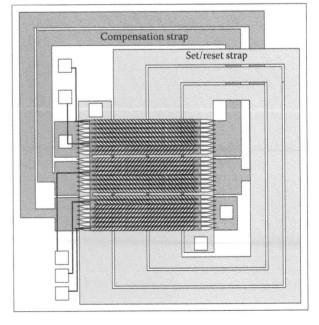

FIGURE 4.93
The HMC1001 sensor of Honeywell with two planar coils—one for feedback the second for set/reset mode.

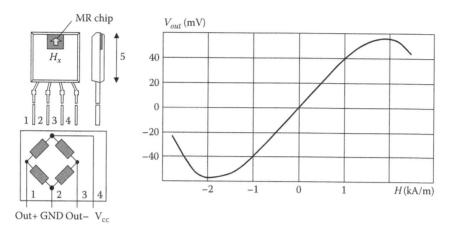

FIGURE 4.94
The bridge sensor output voltage characteristic of KMZ10B sensor of Philips (supply voltage 5 V).

The set/reset mode and feedback enable to decrease the second significant drawback of AMR sensors—the sensitivity to orthogonal component of magnetic field (crossfield sensitivity) (Ripka et al. 2009). The error caused by 80 A/m orthogonal magnetic field equals 3% of FS after using of the feedback and set/reset mode was reduced to about 0.5% FS.

Figure 4.94 presents typical characteristic of an AMR sensor. The linearity of 0.5% FS is better than 0.5% and hysteresis is less than 0.5% of FS. Dimensions of the sensor are 5 mm × 5 mm although the active area of the sensor is 1 mm × 1 mm.

4.4.3 Multilayer GMR Magnetic Field Sensors

The first paper about GMR informed about large change of resistance in special multilayer structure known as *superlattice* (Baibich et al. 1988). The superlattice is the artificially grown structure consisting of periodically alternating single-crystal layers—in this first paper, the described superlattice had the periodically ordered layers of 3 nm Fe separated by 0.9 nm Cr and repeated 40 times (see Figure 2.42). To obtain such structure, the *molecular beam epitaxy* (MBE) technique was used and the effect was observed at 4.2 K for magnetic field 200 kA/m.

Such extremely difficult to obtain and measure conditions were soon repeated by Parkin but in this case, much cheaper sputtering technique was used and the GMR effect as large as 65% was measured at room temperature (Parkin et al. 1991). Since then many other teams reported similar results—typical change of resistance was around 30% (but around 100% at low temperature). Recently, this kind of effect* is rather for-

gotten and is substituted by the spin-valve GMR effect, which is much more sensitive—the sensitivity of the multilayer effect is around 0.2%: kA/m while in SV, it is around 5%: kA/m.

Nowadays, the SV structure practically overshadowed the old classical multilayer effect for sensor applications but the latter still remains interesting. First of all, multilayer sensors exhibit much higher MR—about 30%. SV sensors theoretically can exhibit MR even 25% but practically it seldom exceeds 10%. Therefore, if we would like to have a large signal of the sensor, the multilayer structure is better—the Infineon Company announced their GMR multilayer sensor of angle as the sensor, which does not require an output amplifier.

SV is superior as reading head in comparison with multilayer structure because it exhibits a sharp transition between zero/one state in digital reading of information (Mallinson 2001). But in the case of MFS, advantages of the SV are not so obvious. A comparison of both classes of sensors was presented by Hill (Hill 2000). He concluded that SVs exhibit higher sensitivity and better resolution—10 times smaller noise at 10 Hz. But as a sensor of vector field (immune to orthogonal field component), the GMR multilayer sensor was vastly superior. Also, linearity of multilayer structure is better.

Not all GMR sensor manufacturers inform in detail about the structure of their sensors—but from the data, it can be suspected that the multilayer structure is still in use.

An elementary unit of the multilayer GMR structure consists of a sandwich-type triple-layer: two thin ferromagnetic films separated by a very thin conducting film (Figure 4.95). If this conducting spacer is sufficiently thin, there appears coupling between the two ferromagnetic films causing their antiparallel magnetization. The change of resistance depends on the MR factor ΔR (difference of resistance for ferromagnetic and

---

* This effect does not have its own name—sometimes it is called as multilayer GMR but SV is also multilayer, sometimes it is known as a sandwich structure—but SV is also a sandwich. In this book, this effect is called as multilayer GMR.

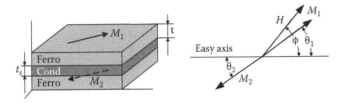

FIGURE 4.95
The sandwich-type multilayer GMR structure.

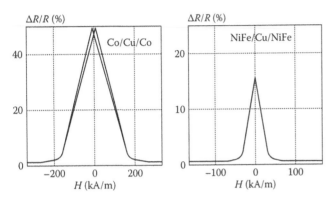

FIGURE 4.96
Typical characteristics of multilayer GMR sensors: Si/Fe5/(Co1/Cu0.9) × 16 and Si/NiFe5/(NiFe1.5/Cu0.8) × 14. (From Parkin, S.S.P. et al., *Phys. Rev. Lett.*, 66, 2152, 1991; Parkin, S.S.P., *Appl. Phys. Lett.*, 61, 1358, 1992.)

antiferromagnetic state) and the value of the coupling field H_s:

$$R(H) = R + \Delta R \left[1 - \frac{H_x^2}{\left(H_k + H_s\right)^2} \right] \quad (4.76)$$

Most frequently, cobalt is used as ferromagnetic films and a copper as the spacer. Sometimes cobalt is substituted by permalloy with smaller MR but higher sensitivity (Figure 4.96).

Table 4.4 presents typical data for various multilayer GMR sensors. The thickness of the spacer is chosen to

obtain antiferromagnetic coupling—for example, in the case of Co/Cu, it can be 0.8 or 2 nm (Figure 4.76). The ferromagnetic layer should be as thin as possible but if this layer is too thin, then the MR vanishes (Sato et al. 1993). On the other hand, the more bilayers the higher MR—from Table 4.4, we can see that more than 50 repetitions are possible. There is a limitation resulting from the thickness of the whole structure—for above around 100 nm, we do not observe further increase of MR (Sato et al. 1993). Often, the whole structure is supplemented by a buffer layer. Sometimes additional thin layer of Co or Co-based layer is added for improvement of the GMR effect (Parkin 1992, Hosoe et al. 1992) (Figure 4.97).

An important drawback of GMR sensor in comparison with AMR is the difficulty of obtaining the differential pair necessary to temperature error compensation (in the AMR case, it is sufficient to form the current path inclined by +45° and −45° with respect to the anisotropy axis). One of the methods of preparation of differential sensors is to bias the sensors by additional magnetic field as it is presented in Figure 4.98.

Hill obtained the differentially biased magnetoresistors by using an additional hard magnetic layers (as it is presented in Figure 4.99). The sensor consists of six bilayers NiFeCo2.6/Cu2.3. The constructed sensor exhibited sensitivity of 20 mV/V kA/m (150 mV/kA/m) for the field range of ±800 A/m. The determined noise level was 0.28×10^{-6} A/m √Hz for the bandwidth from 10 Hz to 10 kHz. The sensor output as a function of applied field is presented in Figure 4.99.

Technologically biasing of every magnetoresistor with different magnetic field is rather complex, hence, more often, a pair composed of active and passive magnetoresistors is used. An example of such sensors developed by Nonvolatile Electronics Corporation (NVE) is presented in Figure 4.100. One pair (passive) of the sensors is covered by a thin film permalloy layer that serves as a shield. The same permalloy layers act as flux concentrator for active pair.

TABLE 4.4

Selected Parameters of Multilayer GMR Magnetoresistors ($T = 300$ K)

Structure	$\Delta R/R$ (%)	H_s (kA/m)	$\Delta R/R$: H_s (%:kA/m)	Reference
(Fe0.5/Cr1.2) × 50	42	6400	0.06	Schad et al. (1994)
(Co1.0/Cu0.8) × 30	38	450	0.08	Grundy et al. (1993)
(Co0.8/Cu0.8) × 0	70	880	0.08	Parkin et al. (1991)
(FeCo1.5/Cu0.9) × 30	32	160	0.20	Jimbo et al. (1993)
(NiFe1.5/Cu0.8) × 14	16	56	0.28	Parkin (1992)
(NiFe2.0/Cu1.0) × 30	18	56	0.32	Sato et al. (1993)
(NiFe3.0/Au1.2) × 12	10	32	0.31	Parkin (1994)
AMR sensor	*1.6*	*1*	*1.6*	

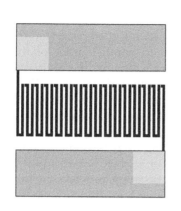

Ta Capping
4 nm NiFeCo
1.5 nm CoFe
1.5 nm Cu
1.5 nm CoFe
2 nm NiFeCo
1.5 nm CoFe
1.5 nm Cu
1.5 nm CoFe
2 nm NiFeCo
1.5 nm CoFe
1.5 nm Cu
1.5 nm CoFe
4 nm NiFeCo
Silicon nitride
Silicon substrate

FIGURE 4.97
The example of multilayer GMR structure reported by Nonvolatile Electronics Inc. (NVE). (From Daughton, J., *Proc. IEEE*, 91, 681, 2003.)

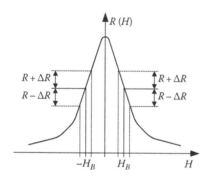 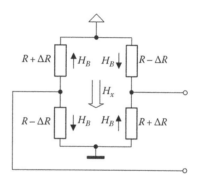

FIGURE 4.98
The differential pair of the GMR sensors obtained by additional biasing.

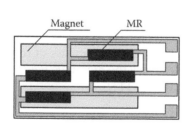 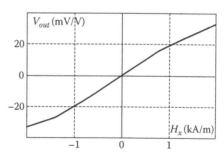

FIGURE 4.99
The SV sensor with differentially biased magnetoresistors by the magnet layers. (From Hill, E.W. et al., *Sens. Actuat.*, A59, 30, 1997.)

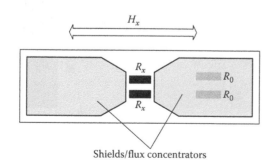 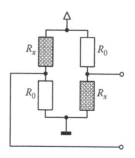

FIGURE 4.100
The GMR bridge sensor with shielding and flux concentrators. (From Daughton, J.M. et al., *IEEE Trans. Magn.*, 30, 4608, 1994.)

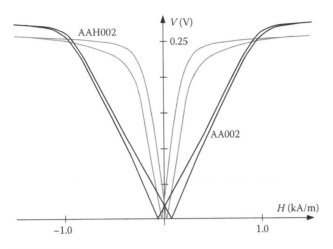

FIGURE 4.101
The bridge sensor output voltage characteristics of typical GMR sensors. Supply voltage 5 V. (NVE catalogue.)

The first commercial GMR sensors were introduced in 1995 by NVE (Daughton et al. 1993, 1994, Daughton 1999, 2003). An example of the bridge sensor output voltage characteristics of two NVE sensors is presented in Figure 4.101. In comparison with KMZ10 AMR sensors, the range of magnetic field is comparable (although GMR sensors of NVE use the flux concentrator of about 10×). The output signal is almost five times larger. The transfer characteristic is with hysteresis—4% for AA sensors and 15% for AAH sensors. The GMR sensors are much more immune to orthogonal field in comparison with the AMR sensors.

The sensor AA002 has resistance 5 kΩ, saturation field $H_s = 1.2$ kA/m, and sensitivity $S = 4.2$ mV/V Oe = 52 mV/(V kA/m). The sensor AAH002 has resistance 2 kΩ, saturation field $H_s = 0.5$ kA/m, and sensitivity $S = 18$ mV/V Oe = 144 mV/(V kA/m). Figure 4.102 presents a typical layout of such sensor.

4.4.4 Spin-Valve GMR Magnetic Field Sensors

The GMR effect appears for transition between antiparallel to parallel order of thin magnetic layers. In multilayer sensors, the antiparallel order is caused by close coupling of two layers. But this coupling field H_s is rather large (see Table 4.4) and therefore the sensitivity is small because large external magnetic field is needed to overcome the coupling field.

Starting from 1990, many authors (Barnas et al. 1990, Dupas et al. 1990, Chaiken et al. 1991) reported GMR effect also in uncoupled structures where the antiparallel order was induced by an artificial way. These sensors are known as *uncoupled GMR sensors* or *pseudo SV*. The principle is presented in Figure 4.103.

In uncoupled GMR structures, the two magnetic layers are separated by a distance larger than in the case of the multilayer coupled structures. Therefore, the antiferromagnetic order is not induced and both layers are magnetized in the same direction. But both ferromagnetic layers are magnetized in another way—one layer, known as *free layer* is magnetized easily while the second, known as *pinned layer*, needs much higher field to change its magnetization. Usually, the free layer exhibits much lower coercivity than the pinned layer. This difference in coercivity can be obtained by using different materials for each layer (e.g., Co for one and NiFe for the other) or the same material but with different thickness or by technologically induced different coercivity.

As it is presented in Figure 4.103, in initial state ($H = 0$), both layers are magnetized in the same direction. Next, if the magnetic field increases, then magnetization of the free layer (of smaller H_{c1}) changes first. Thus, we obtain a transition to an antiferromagnetic order and, at the same time, a large increase of resistance. If we then further increase the magnetic field, then after reaching the coercivity field H_{c2} the second pinned layer also changes

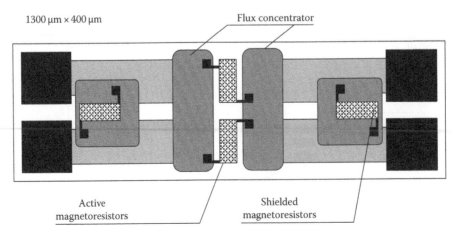

FIGURE 4.102
Typical layout of GMR magnetic field sensor of NVE.

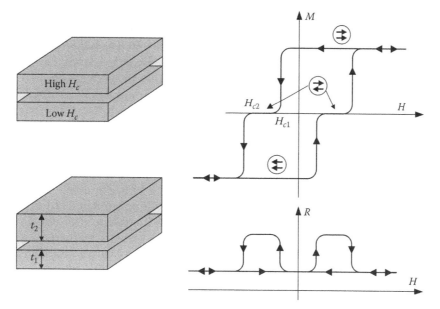

FIGURE 4.103
The principle of operation of uncoupled GMR sensors. (From White, R.L., *IEEE Trans. Magn.*, 28, 2482, 1992.)

the direction of magnetization. Thus, we returned to a ferromagnetic order and the resistance decreases again (Figure 4.103).

Recently, the uncoupled (or pseudo SV) structures are used rather seldom because much better performance are provided by sensors known as *SV sensors*. In 1991, Dieny (Dieny et al. 1991a, b) proposed another method of pinning of one magnetic layer. He used exchange biasing of the ferromagnetic film by adjacent antiferromagnetic layer known as *exchange coupling*.

The exchange coupling was discovered in 1956 by Meiklejohn and Bean (Meiklejohn and Bean 1956). If we connect two layers—ferromagnetic and antiferromagnetic one between these layers, there appears coupling known as *exchange anisotropy*. As a result, the ferromagnetic layer is biased by the exchange field H_{ex} and the hysteresis of ferromagnetic layers is shifted (Figure 4.104). In this way, we obtain stable pinning of one layer

by deposition of the additional layer from antiferromagnetic material onto the first one. Table 4.5 presents properties of typical antiferromagnetic materials used for pinning of the layer in the SV sensors.

FeMn is the most commonly used as the antiferromagnetic materials due to its robustness of deposition process and material design flexibility (Coehoorn et al. 1998). Unfortunately, this material exhibits very low corrosion resistance.

An important parameter of the antiferromagnetic materials is their *blocking temperature*. It is the temperature at which the biasing field disappears. This temperature is rather low in the case of FeMn. High blocking temperature is exhibited by NiMn but this material requires annealing, which can destroy other layers. As an alternative to FeMn, often IrMn is considered.

Figure 4.105 illustrates the principle of operation of a SV sensor. Initially (for $H_x = 0$), both magnetic layers are magnetized parallel. When magnetic field increases, the free layer switches the direction while the pinned layer remains fixed. This way we obtain antiparallel ordering

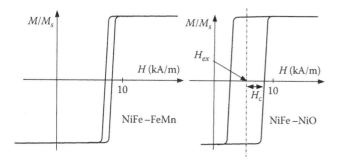

FIGURE 4.104
The hysteresis loops of permalloy layer exchange coupled with antiferromagnetic layer FeMn or NiO. (From Lin, T. et al., *IEEE Trans. Magn.*, 31, 2585, 1995.)

TABLE 4.5

Parameters of Selected Antiferromagnetic Materials

	FeMn	IrMn	NiO	NiMn
Exchange field (kA/m)	6.2	3.6	3.7	16
Coercive field (kA/m)	0.5	2.5	2.8	9.3
Blocking temperature (°C)	150	310	200	450
Neel temperature (°C)	230	420	250	800
Critical thickness (nm)	7	15	35	25

Source: Lin, T. et al., *IEEE Trans. Magn.*, 31, 2585, 1995.

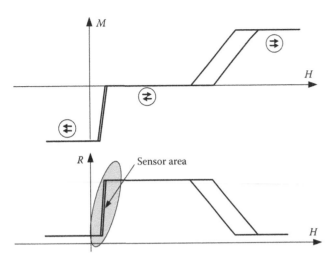

FIGURE 4.105
The principle of operation of the spin-valve sensor.

and such transition causes large, close to linear change of the resistance. This part of characteristic can be used for sensor application. When the external field exceeds the exchange field, the pinned layer changes the direction of magnetization and we return to parallel order. Usually, this part of characteristic exhibits larger coercivity loop. The experimentally determined characteristic of the SV sensor is presented in Figure 4.106.

Comparing the characteristics of multilayer GMR sensor (Figure 4.96) with those presented in Figure 4.106, we can see that the change of resistance is much smaller but also for smaller external magnetic field.

Sometimes the conventional antiferromagnetic material used for pinning of one magnetic layer is substituted by so-called synthetic antiferromagnet. For pinning of one layer, a conventional multilayer sandwich system (typically Co/Ru/Co) is used ensuring the antiparallel magnetization if the distance between both Co layer is very small (Figure 4.107b) (van den Berg et al. 1996, Leal and Kryder 1998, Zhu 1999). The synthetic antiferromagnet can be supported by antiferromagnetic material.

In the SV sensors, it is not possible to increase the change of resistance by increasing the number of layers (as it was in the case for multilayer GMR). It is only possible to double MR by using symmetrical double structure (Figure 4.107c).

Usually, two layers of permalloy separated by a copper layer are used as SV materials. It is possible to obtain larger MR in Co-Cu-Co structure but on the expense of significantly larger hysteresis (Figure 4.108). Nevertheless, Parkin proved that as a compromise in a NiFe-Cu-NiFe structure with additional thin Co layers, it is possible to obtain both: low hysteresis and large MR (Figure 4.108).

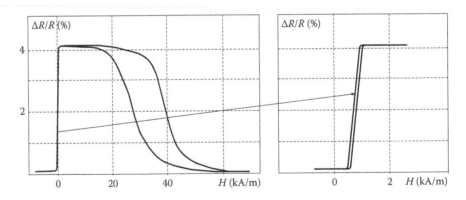

FIGURE 4.106
The characteristic of the spin-valve sensor with a structure NiFe6.2/Cu2.2/NiFe4/FeMn7. (From Dieny, B. et al., *J. Magn. Magn. Mater.*, 93, 101, 1991a; Dieny, B. et al., *J. Appl. Phys.*, 69, 4774, 1991b.)

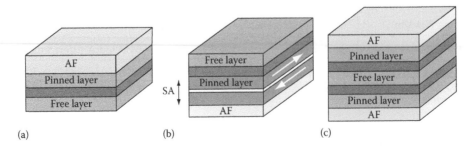

FIGURE 4.107
The conventional spin-valve sensor CSV (a), synthetic antiferromagnet sensor SAF (b), and double spin-valve sensor (c).

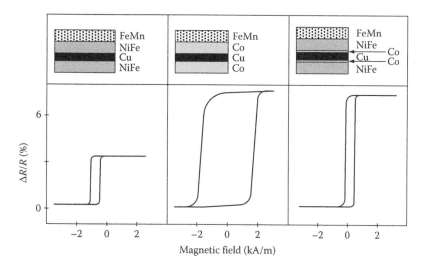

FIGURE 4.108

The typical structures of spin-valve magnetoresistors. (From Parkin, S.S.P., Giant magnetoresistance and oscillatory interlayer coupling in polycrystalline transition metal multilayers, in *Ultrathin Magnetic Structures II*, Springer Verlag, Berlin, Germany, 1994, pp. 148–186.)

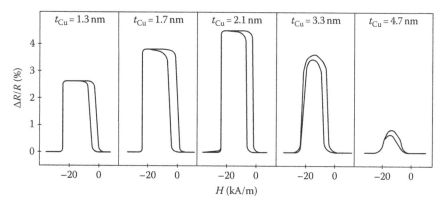

FIGURE 4.109

The influence of a copper layer thickness on the properties of a spin-valve sensor. (From Rijks, Th.G.S.M. et al., *Appl. Phys. Lett.*, 65, 916, 1994a; Rijks, Th.G.S.M. et al., *J. Appl. Phys.*, 76, 1092, 1994b.)

The thickness of the spacer layer (made mostly from Cu) should be chosen carefully because if it is too small we can meet again strong coupling between layers resulting in poor sensitivity. On the other hand spacer which is too thick causes shunting effect. Figure 4.109 presents an example of the influence of the copper layer thickness on the properties of a spin valve sensor.

Table 4.6 presents typical parameters of the SV sensors. In comparison with the multilayer sensors (Table 4.4), the magnetic field range is much smaller but the sensitivity is larger. The SV magnetoresistors have excellent properties for digital reading of information (Heim et al. 1994, Tsang et al. 1994). Figure 4.110 presents examples of the design of SV reading heads.

Sensors of this type do not exhibit such spectacular change of resistance as the multilayer ones. Moreover they suffer from hysteresis and nonlinearity and as in all GMR sensors it is very difficult to obtain differential

pair recommended for correction of temperature error (Figure 4.111).

Spong (Spong et al. 1996) proposed to use an additional current conducting layer to introduce the biasing method. The SV sensor was prepared with the following

TABLE 4.6

Parameters of Selected Spin Valve Structures (at Room Temperature)

Structure	$\Delta R/R$ (%)	ΔH_s (kA/m)	$\Delta R/R$: ΔH_s (%:kA/m)
NiO/Co/Cu/Co/Cu/Co/NiO	24.8	3.3	7.51
Ta/NiFe/Cu/NiFe/FeMn/Ta	4.2	0.8	5.2
Co/Cu/Co/FeMn/Cu	6.9	0.8	8.6
NiFe/Cu/NiFe/FeMn/Cu	2.7	0.4	6.7

Sources: Egelhoff et al. (1995), Dieny, B., *J. Magn. Magn. Mater.*, 136, 335, 1994; Dieny, B. et al., *J. Magn. Magn. Mater.*, 151, 378, 1995.

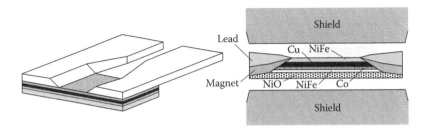

FIGURE 4.110
The design of spin-valve reading heads: unshielded and shielded.

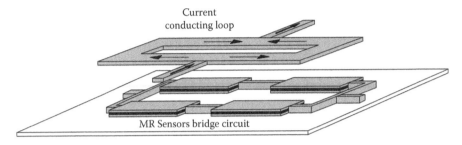

FIGURE 4.111
The design of spin-valve sensor with differentially biased magnetoresistors. (From Spong, J.K. et al., *IEEE Trans. Magn.*, 32, 366, 1996.)

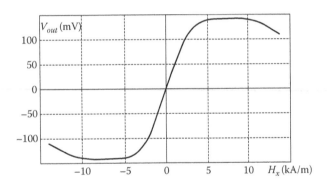

FIGURE 4.112
The characteristic of the spin-valve sensor presented in Figure 4.111. Supply voltage 5 V. (From Spong, J.K. et al., *IEEE Trans. Magn.*, 32, 366, 1996.)

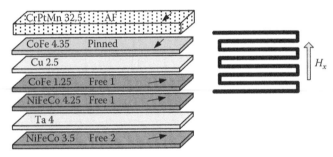

FIGURE 4.113
The linear spin-valve sensor of NVE. (From Qian, Z. et al., *IEEE Trans. Magn.*, 39, 3322, 2003; Qian, Z. et al., *IEEE Trans. Magn.*, 40, 2643, 2004.)

layers: Si/Al$_2$O$_3$100/Ta 5/NiFe 4.2/Co 0.7/Cu 2.5/Co 0.9/ NiFe 3.2/FeMn 15/Ta 0.5. The magnetoresistivity was 5.8%. The resolution of the sensor depending on noise was estimated as 0.26 nT √Hz. The nonlinearity error did not exceed 2% of FS. The characteristic of the sensor is presented in Figure 4.112.

Another realization of a linear SV sensor was reported by a team from NVE (Qian et al. 2003, 2004) (Figure 4.113). They used a pair of active and a second pair of passive (covered by the shield) magnetoresistors —similarly as it is presented in Figure 4.100. In this design, the next additional free layer was used—it improves magnetic properties (decreases hysteresis) but because the spacer exhibits large resistivity, only the first free layer contributes to GMR. Moreover, the pinned layer is magnetized perpendicularly, which also improves linearity (Rijks et al. 1994a, b). By changing the thickness of the additional free layer, it is possible to change the internal bias field to move the operating point to the linear part of characteristic.

Figure 4.114 present an output characteristic of the sensor. The magnetoresistors in form of 4 μm serpentine exhibit the GMR effect of 5.4%. The linearity error was less than 0.05%. Although the magnetoresistors themselves were without hysteresis, the whole sensor exhibits small hysteresis due to the influence of the flux concentrator.

4.4.5 Spin-Dependent Tunneling GMR Magnetic Field Sensors

The spin-dependent tunneling (SDT) sensors (known also as MTJ or *tunnel magnetoresistance* [TMR]) are a good example when the theoretical predictions preceded and advanced a technology. The first investigations of the SDT effect between two ferromagnetic

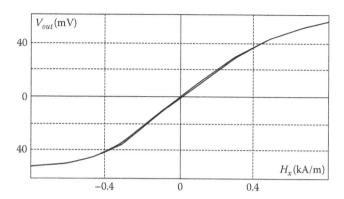

FIGURE 4.114
The response of a bridge spin-valve sensor of NVE. Supply voltage 5 V. (From Qian, Z. et al., *IEEE Trans. Magn.*, 39, 3322, 2003; Qian, Z. et al., *IEEE Trans. Magn.*, 40, 2643, 2004.)

films were reported in 1971 by Tedrow and Meservey (Tedrow and Meservey 1971). Later the SDT was reported in 1975 by Julliere (Juliera 1975) in Fe/Ge/Co junction. These experiments were performed in low temperatures.

Julliere proposed a theoretical model of SDT effect and predicted that in Al-Al$_2$O$_3$-ferro junction, the MR as large as 26% should be observed. Also, Slonczewski in 1989 proposed theoretical explanation of the tunneling effect between two ferromagnetic layers (Slonczewski 1989). But at that time there still was another barrier—difficulties in preparation of very thin and without defects (pin holes) insulation layers.

In 1995, the successful preparation and investigations of room temperature thin ferromagnetic film tunnel junction was reported (Miyazaki and Tezuka 1995a, b, Moodera et al. 1995). Both teams as the insulation barrier used thin aluminum film next oxidized to an aluminum oxide (Al$_2$O$_3$) insulation barrier. Various experiments confirmed the Julliere model. Extended description of MTJ effects were since described in many papers (Gallagher et al. 1997, Daughton 1997, Moodera et al. 1999, Tsymbal et al. 2003).

From Julliere model, it was predicted that the TMR effect as large as 70% at room temperature should be obtained. Indeed the investigations of various tunnel junctions enabled obtaining the tunnel junction effect based on CoFeB amorphous layers as large as 70.4% (Wang et al. 2004). It was the largest reported TMR effect obtained for an aluminum oxide barrier.

The turning point in investigations and applications of TMR effects was the substitution of the amorphous aluminum oxide barrier by a crystallite magnesium oxide (MgO) barrier. In 2001, Butler and Mathon (Butler et al. 2001, Mathon et al. 2001) theoretically predicted that if the barrier exhibits appropriate crystalline texture and the ferromagnetic layers have the texture corresponding with this texture, the coherent tunneling

phenomenon (symmetry filtering (Butler 2008)) should significantly amplify the TMR effect.

In fact, in 2004, two teams reported giant TMR at room temperature with MgO (100) barrier (Parkin et al. 2004, Yuasa et al. 2004) with the TMR ratio as large as 180% and 220%. Since then the research efforts are continued and still higher TMR is reported (see Figures 4.78 and 4.118). The appropriate texture can be obtained by the MBE epitaxial deposition or easier by the appropriate annealing and recrystallization.

The TMR junction is designed in a similar way as the SV structure. Two ferromagnetic layers separated by a thin insulation layer are often supported by an antiferromagnetic layer. But the main difference is that in the tunnel junction we use perpendicular effect, which means that the current is perpendicular to surface of the film (Figure 4.115). Such geometry of the sensor is very useful for memory application.

The Julliere model of the MTJ effect is described by the following expression (Figure 4.116):

$$\frac{\Delta R}{R} = \frac{2P_1P_2}{1+P_1P_2} \tag{4.77}$$

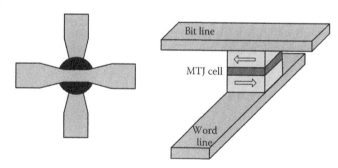

FIGURE 4.115
The design of MTJ sensor (on the right in application for MRAM memory).

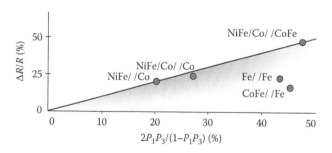

FIGURE 4.116
TMR effect according to the Julliere model (circles, experimental data). (From Inomata, K. and Sakakima, H., Tunnel type GMR devices, in *Giant Magnetoresistance Devices*, Chapter 5, Springer, Berlin, Germany, 2002.)

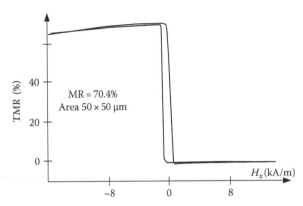
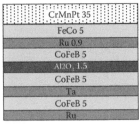

FIGURE 4.117
TMR response of Al-based MTJ sensor. (From Wang, D. et al., *IEEE Trans. Magn.*, 40, 2269, 2004.)

where P is the spin polarization,

$$P = \frac{n\uparrow - n\downarrow}{n\uparrow + n\downarrow} \qquad (4.78)$$

The spin polarization was determined as 44% for iron, 34% for cobalt, and 11% for nickel (Meservey et al. 1994). Figure 4.117 presents the characteristic of 70% Al-based MTJ—from this dependence, the polarization of $Co_{60}Fe_{20}B_{20}$ can be estimated as 61%.

Figure 4.118 presents one of the highest reported (at the time of writing this book) TMR result obtained for pseudo SV tunnel junction (Lee et al. 2007) and Table 4.7 presents the maximum TMR ratio at room temperature of various MTJ structures. We can see that MR effect as large as 500% is reported.

TMR sensors have several important drawbacks. One is that with the increase of supply voltage the TMR ratio rapidly decreases. If we multiply the bias voltage

value and TMR effect, we obtain the optimum value of supplying (bias) voltage (Figure 4.119). We can see that the optimum bias voltage for Al-based MTJ devices is 0.3–0.7 V while for Mg-based devices around 0.5 V. One solution applied to MFS supplied by 5 V is to connect in series many MTJ junctions (Tondra et al. 1998).

Another problem appears if we plan to use the MTJ as the MFS. As it is presented in Figure 4.120, the characteristics are nonlinear and exhibit hysteresis. An improvement is possible and can be achieved by applying additional transverse bias magnetic field generated, for instance, by a hard magnetic layer (Tondra et al. 1998) or an additional planar coil (Daughton 2003).

Also, another problem of using the MTJ effect in sensors is its inherent noise. The SNR of MTJ was estimated as (Freitas et al. 2007)

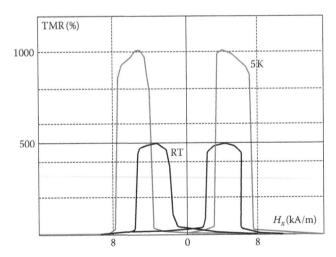

FIGURE 4.118
TMR response of MgO-based MTJ sensor having two 4.0 and 4.3 CoFeB electrodes and 2.1 nm thick MgO layer. (From Lee, Y.M. et al., *Appl. Phys. Lett.*, 90, 212507, 2007.)

TABLE 4.7

Maximum TMR Ratio at Room Temperature of Various MTJ Structures

Fixed Layer	Free Layer	TMR [%]
Al-O based		
CoFe	CoFe	42
CoFeB	CoFeB	70
MgO based		
Co50Fe50	Co40Fe40B20	50
Co90Fe10	Co40Fe40B20	75
Co40Fe40B20	Co90Fe10	130
Co40Fe40B20	Co50Fe50	277
Co40Fe40B20	Co40Fe40B20	355
Pseudo spin valve MgO based		
Co40Fe40B20	Co40Fe40B20	450
Co20Fe60B20	Co20Fe60B20	500

Sources: Parkin, S.S.P. et al., *J. Appl. Phys.*, 85, 5828, 1999; Wang, D. et al., *IEEE Trans. Magn.*, 40, 2269, 2004; Ikeda, S. et al., *IEEE Trans. Electron. Dev.*, 54, 991, 2007.

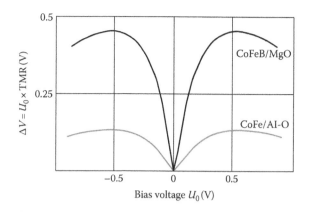

FIGURE 4.119
Bias voltage dependence of the signal voltage of MTJ. (From Ikeda, S. et al., *IEEE Trans. Electron. Dev.*, 54, 991, 2007.)

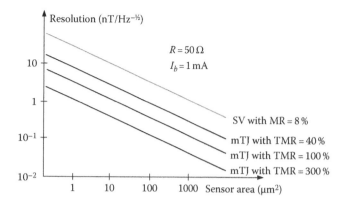

FIGURE 4.121
The resolution calculated for 10 Hz as a function of the sensor area and magnetoresistance. (From Freitas, P.P. et al., *J. Phys. Condens. Matter*, 19, 16221, 2007.)

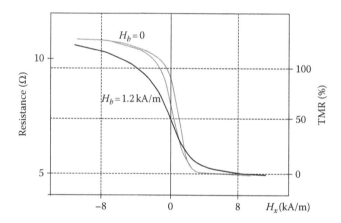

FIGURE 4.120
The transfer characteristic of MTJ sensor with and without the bias field. (From Freitas, P.P. et al., *J. Phys. Condens. Matter*, 19, 16221, 2007.)

$$\text{SNR}_{\text{MTJ}} = \text{TMR}\frac{2V_{1/2} - RI_b}{2V_{1/2}}\sqrt{\frac{RI_b^2}{4k_BT\Delta f + 2eI_bR\Delta f}} \quad (4.79)$$

where $V_{1/2}$ is the voltage bias at which TMR is reduced to a half of its zero-bias value.

For comparison, the same value estimated for SV is

$$\text{SNR}_{\text{SV}} = \text{MR}\sqrt{\frac{RI_b^2}{4k_BT\Delta f}} \quad (4.80)$$

Thus, we obtain the relationship (Freitas et al. 2007),

$$\frac{\text{SNR}_{\text{MTJ}}}{\text{SNR}_{\text{SV}}} = 0.3\frac{\text{TMR}}{\text{MR}} \quad (4.81)$$

Nevertheless, as it is presented in Figure 4.121, the MTJ sensors exhibit better resolution (smaller value) than SV sensors.

Based on the experiences with GMR sensors, NVE developed an SDT device using an Al_2O_3 barrier. The magnetoresistor consists of about 16 MTJ in series connected into bridge as passive (shield)/active (concentrator) pair. First attempts with simple NiFeCo12.5/$Al_2O_3$2.5/CoFe12.5 structure did not give satisfying results because the sensor exhibited hysteresis and non-linearity. But after introduction of the pinning layers including a synthetic antiferromagnetic, the final characteristic is non-hysteric (although only after application of a bias field of about 1.6 kA/m). Figure 4.122 presents the structure and transfer characteristics of the sensor.

The SDT sensor of NVE exhibited sensitivity 37%/(kA/m) or 30%/mT. Other sensors with picotesla resolution based on MgO barrier were also reported (Chaves et al. 2007, 2008, Almeida et al. 2008). The sensor with $Co_{94}Zr_3Nb_4$ flux guides was in a form of a pillar of diameter 14–50 μm.

Figure 4.123 presents a design and transfer characteristics of the MgO-based SDT sensor. The magnetoresistor consists of the following layers: Ta3/CuN30/Ta5/PtMn20/CoFe2.5/Ru0.7/CoFeB3/MgO1.2/CoFeB3/Ta5/TiW(N₂)15. The sensitivity of 870%/mT was estimated. The noise level was 100 pT √Hz at 10 Hz and 2 pT √Hz at 500 kHz.

4.4.6 Giant Magnetoimpedance Sensors

Giant magnetoimpedance (GMI) sensors are an example of such sensors, which are still promising (as, e.g., magnetotransistors), but we still wait for the final commercial product. They have very attractive performances (large change of impedance, even 400% at small magnetic field, below 1 kA/m). There are numerous papers devoted to this subject but there is a poor commercial representation. The sensors of only magnetic field have a rather small area of application and it is necessary to find other important areas to expand their use. For example,

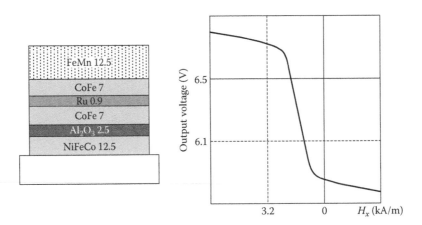

FIGURE 4.122
The structure and the transfer characteristic of an SDT sensor of NVE. (From Tondra, M. et al., *J. Appl. Phys.*, 83, 6688, 1998; Daughton, J., *Proc. IEEE*, 91, 681, 2003.).

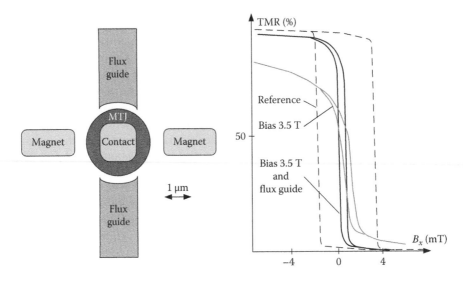

FIGURE 4.123
The structure and the transfer characteristics of a SDT sensor based on MgO barrier. (From Chaves, R.C. et al., *Appl. Phys. Lett.*, 91, 102504, 2007; Chaves, R.C. et al., *J. Appl. Phys.*, 103, 07E931, 2008.)

MR sensors are used as reading heads and MRAMs, fluxgate sensors are important for military purposes, SQUID sensors are important for biomagnetic investigations, Hall sensors are widely applied as current and displacement sensors. But GMI sensors are still waiting for some application—not necessary in direct magnetic field measurement. A good area would be profit on their sensitivity to traffic detection systems or better as competition to optic bar code reading (supermarkets without cash due to remote reading of the magnetic code). At last in October 2001, Aichi Steel Corporation (Japan) introduced magnetoimpedance (MI) amorphous wire MI sensors based on the design of the sensors developed by the Mohri team (Kanno et al. 1997, Kawajiri et al. 1999, Mohri et al. 2002a,b, Honkura 2002).

The MI effect is developed in a thin wire, ribbon, or thin film (Panina and Mohri 1994, Panina et al. 1994, Beach and Berkowitz 1994). In the wire, the high-frequency current generates circular magnetic field $H_\varphi = Ir/2\pi a^2$, where r is a radial coordinate. Thus, the magnetic wire should exhibit circular anisotropy (i.e., following perimeter of a cylinder around the wire as shown in Figure 4.124) and large circular permeability μ_φ.

The amorphous wire is a good candidate for such purpose because it has a circular domain structure. A circular anisotropy can be induced by appropriate annealing of the sample. This high-frequency, circular permeability strongly depends on a DC or low-frequency axial magnetic field as it is presented in Figures 4.125 and 4.126.

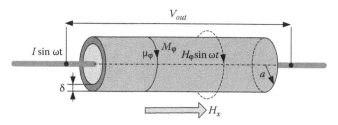

FIGURE 4.124
The wire magnetoimpedance sensor (the current is flowing in the layer indicated as δ due to a skin effect).

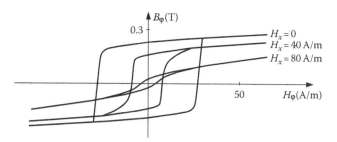

FIGURE 4.125
Hysteresis loop measured for circular magnetization in the presence of longitudinal field. (From Panina, L.V. et al., *IEEE Trans. Magn.*, 31, 1249, 1995.)

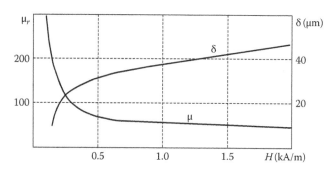

FIGURE 4.126
The changes of permeability μ_r and skin effect depth δ as a function of external DC magnetic field. (From Vazquez, M. et al., *Sens. Actuat.*, A59, 20, 1997.)

In thin ferromagnetic wire, the impedance changes with external field due to the change of permeability and the *skin effect* (current flowing only in thin outer layer δ of the wire—Figure 4.124). The skin effect also depends on the change of permeability (and resistivity ρ) because

$$\delta = \sqrt{\frac{\rho}{\pi f \mu}} \tag{4.82}$$

The resulting impedance can be derived from classical electrodynamics Landau–Lifschitz equation in a form

$$Z = R_{dc} ka \frac{J_0(ka)}{2J_1(ka)} \tag{4.83}$$

where J_0, J_1 are the Bessel functions, $k = (1-j)/\delta$. In a simplified form, this relationship can be expressed as

$$Z = \frac{l\rho}{a} \frac{1}{2\pi\delta} + j\omega \frac{l\mu}{8\pi} \frac{2\delta}{a} \tag{4.84}$$

At low frequency, when the skin effect is weak, the real part of the impedance changes only a little with the external magnetic field and the change of impedance is caused mainly by the change of inductance $L = l\mu/8\pi$. This effect is known as the *magnetoinductance effect* (Mohri et al. 1992, 1993) and can be detected as the voltage pulses induced in the coil wound on the wire.

At high frequency of the current, the skin effect is dominating and both real and imaginary parts of impedance change significantly—this effect is known as the *giant magnetoimpedance (GMI) effect*. It should be noted that in high-frequency operation, the complex permeability should be considered $\mu_\varphi = \mu'_\varphi - j\mu''_\varphi$. The change of both components of permeability for a wire in external magnetic field is presented in Figure 4.127.

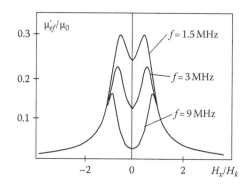

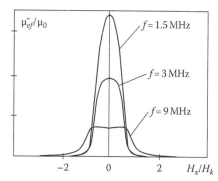

FIGURE 4.127
The effective permeability of a wire versus the external magnetic field. (From Panina, L.V. et al., *IEEE Trans. Magn.*, 31, 1249, 1995.)

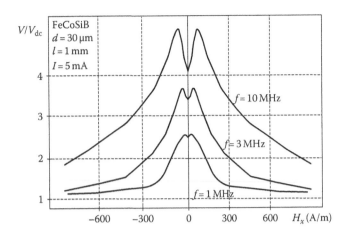

FIGURE 4.128
Typical transfer characteristics of the magnetoimpedance wire. (From Mahdi, A.E. et al., *Sens. Actuat.*, A105, 271, 2003.)

The characteristics of impedance versus the magnetic field determined for the wire exhibit reflection of the change of permeability presented in Figure 4.127. The example of typical characteristics of the MI wire is presented in Figure 4.128.

The transfer characteristic of a typical MI sensor (Figure 4.128) is not convenient due to its shape near the zero field. That is why often asymmetry is introduced to this characteristic as shown in Figure 4.129. Such asymmetry can be achieved in a number of ways—most of them introduce helical anisotropy, for example, by additional current bias and twisting of the wire, by the annealing of a twisted wire, or by pulse excitation (Kitoh et al. 1995, Mohri et al. 1997, 2002, Panina et al. 1999).

Instead of a thin wire, a thin film can be used with transverse permeability. The relation (4.80) in this case is

$$Z = R_{dc} k \frac{t}{2} \cot\left(k \frac{t}{2}\right) \qquad (4.85)$$

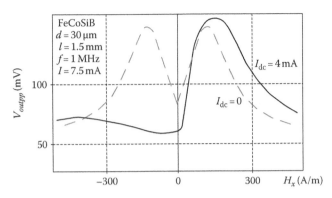

FIGURE 4.129
Introduction of asymmetry of the transfer characteristic by twisting of the wire and supplying by a DC current. (From Kitoh, T. et al., *IEEE Trans. Magn.*, 31, 3137, 1995.)

In comparison to the thin wire, a thin film enables decrease of the dimensions of the sensor and increase of current frequency. To obtain GMI effect in thin film structures, the significant skin effect is not required. The main contribution to impedance is introduced by the inductance of the coil, which in turn depends on permeability. An example of such sensor and its characteristic are presented in Figure 4.130. The magnetic layers of the thickness 50 nm form a closed loop around a copper layer with thickness 100 nm and width 4 μm. The impedance of a sandwich structure with length *l* and dimensions d_m, d_c, *b* as defined in Figure 4.130 can be described with the expression (Hika et al. 1996),

$$Z = R - j\omega \frac{d_m}{2b} l\mu(H) = \frac{l}{2bd_c\sigma_c} - j\omega \frac{d_m}{2b} l\mu(H) \quad (4.86)$$

Morikawa (Morikawa et al. 1996) proposed further enhancement of layered structure by introducing an additional insulator separation (Figure 4.131). The driving current was flowing only in the Cu layer (due to SiO₂ separation) and the GMI effect $(\Delta Z/Z)_{max} = 700\%$ for $H_{max} = 0.9$ kA/m and $f = 20$ MHz was obtained.

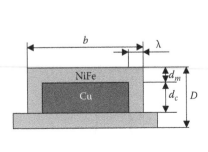

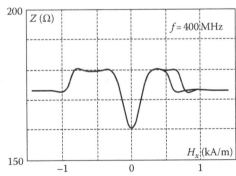

FIGURE 4.130
Thin-film sandwich magnetoimpedance sensor and its transfer characteristic. (From Hika, K. et al., *IEEE Trans. Magn.*, 32, 4594, 1996.)

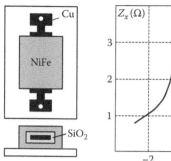
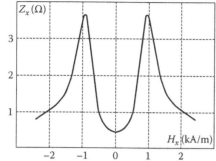

FIGURE 4.131
The design and the transfer characteristic of a sandwich CoSiB/SiO$_2$/CoSiB thin-film sensor. (From Morikawa, T. et al., *IEEE Trans. Magn.*, 32, 4965, 1996.)

The material used for the preparation of GMI sensors should exhibit high magnetic permeability, high saturation magnetization, and low resistivity. Co-based amorphous wire is commonly used as the best for this purpose. But other materials as nanocrystalline materials, permalloy, glass-covered microwire, and thin films are also reported (Hauser et al. 2001, Phan and Peng 2008). Usually, as the main parameter of GMI sensors, the MI factor $\eta = [Z(H) - Z(H_{max})] \times 100\%/Z(H_{max})$ is proposed. Table 4.8 presents typical parameters of GMI sensors.

Because the transfer characteristic of GMI sensors is rather complex, usually such sensors are designed including also the electronic circuit. One of the most convenient electronic circuits is the Colpitts oscillator (Bushida et al. 1995, Uchiyama et al. 1995).

In the Colpitts oscillator, the oscillation frequency depends on the inductance of the sensing element $L(H)$:

$$f = \frac{\sqrt{\frac{1}{C_2} + \left(1 + \frac{R(H)}{R}\right)\frac{1}{C_1}}}{2\pi\sqrt{L(H)}} \tag{4.87}$$

where R is the input resistance of a transistor. In the circuit presented in Figure 4.132, the amplitude of the oscillation voltage is used as the output signal.

The sensor presented in Figure 4.132 was a simple thin-film ribbon. Similar oscillating circuit was used for an amorphous wire sensor (Bushida et al. 1995). The sensor was connected to a feedback circuit. The linearity was better than 0.1% of FS, cutoff frequency was 300 kHz, and resolution 1 nT was reported. Because GMI sensors are very small, it is possible to design the gradiometer with two sensors at some distance from each other (Bushida et al. 1996, Mohri et al. 1997).

By using two asymmetrical GMI elements (see Figure 4.129), it is possible to design the linear MFS as it is illustrated in Figure 4.133.

In 2001, Aichi Micro Intelligent Corporation started production of GMI sensors. The sensor is composed of

TABLE 4.8

Typical Performances of Giant Magnetoimpedance Sensors

		η (%)	S (%/(A/m))	f (MHz)	Reference
CoFeSiB	Wire	600	4	15	Pirota et al. (2000)
CoFeSiB	Wire	125	0.6	3	Vazquez et al. (2000)
Vitrovac 6025	Ribbon	420	0.7	3.5	Sanchez et al. (2003)
FeSiBCuNb	Ribbon	400	0.5	4.5	Guo et al. (2001)
Mumetal	Ribbon	310	0.3	0.6	Nie et al. (1999)
CoSiB	Thin film	700	3.8	20	Morikawa et al. (1996)
NiFe	Thin film	200	1.2	51	Hika et al. (1996)

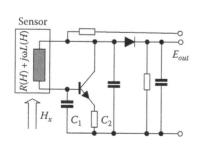
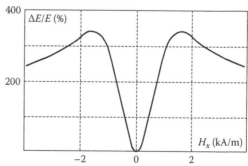

FIGURE 4.132
The magnetoimpedance sensor in the Colpitts oscillator circuit. (From Uchiyama, T. et al., *IEEE Trans. Magn.*, 31, 3182, 1995.)

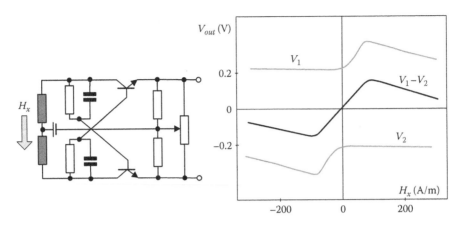

FIGURE 4.133
The linear multivibrator-type GMI sensor with two asymmetrical GMI elements. (From Kitoh, T. et al., *IEEE Trans. Magn.*, 31, 3137, 1995.)

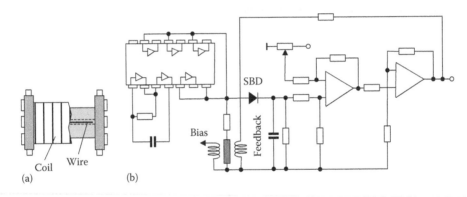

FIGURE 4.134
The MI-CB sensor design of Aichi Micro Intelligent Corporation (a) and the electronic circuit (b). (From Honkura, Y., *J. Magn. Magn. Mater.*, 249, 375, 2002.)

an amorphous wire with diameter 20 μm and length 2 mm and two coils—for bias and for feedback (Figure 4.134a). The sensor is fed by a pulse generator. The output voltage is detected via a Schottky barrier diode and the output signal is send to the feedback coil. The response characteristic is only 1% nonlinear over a full range of 2 μT. The sensitivity is 1 V/μT and the noise level 1 nT in 0.1–10 Hz bandwidth.

There are many measuring devices based on these sensors: three-axis electronic compass AMI204, magnetic field detector MI-CB-1DS, or gyrometer sensor AMI602. There is also commercially available digital magnetic field meter MGM-1DS with the range 200 μT and resolution 10 nT.

The MR and MI sensors cover large area of possible applications. They have very simple design and can be manufactured in large volumes for modest price.

The AMR sensors are recently in the shadow of more trendy GMR sensors. But AMR sensors have many advantages: very simple design and low price, linear characteristic, high sensitivity with nT resolution. It should be noted that commercially available GMR sensors exhibit similar sensitivity but with additional flux concentrator.

Also advantageous is that it is easy to obtain differential pair of magnetoresistors necessary for temperature error compensation. The significant drawback of these sensors is the sensitivity to orthogonal field component (crossfield effect), which requires some a priori knowledge about the direction measured field.

Although the multilayer GMR sensors are recently less popular, it should be noted that these sensors exhibit very large MR without such limitation of the supply voltage as it is in the case of tunnel junction sensors. Therefore, it is possible to obtain large output signal—unfortunately only for rather large magnetic fields.

Recently, the main efforts are concentrated on the SDT sensors. Indeed attainable TMR coefficient between 500% and 1000% at room temperature makes possible extreme miniaturization of the sensors—to the dimensions of μm. It should be possible to design the dense magnetic sensors arrays for the on-line detection of magnetic field distribution (e.g., for NDT or DNA analysis).

The GMI sensors are very good alternative for fluxgate sensors—similar sensitivity but for smaller dimensions of the sensor. Certain problems arise from the necessity of using a high-frequency excitation in such sensors.

4.5 Hall-Effect Sensors

4.5.1 Physical Principles of the Hall Effect

The Hall-effect sensor* was discovered in 1879 by a student Edwin Hall (Hall 1879). As it was described in Section 2.9.4, the Lorentz force acts on the particle with electric charge q moving with velocity v in EM field described by E and B values:

$$F = q(E + v \times B) \qquad (4.88)$$

In the absence of a magnetic field, the particles (electrons or holes), described by their mobility μ_p and density N, are moving with velocity v_p as the current with density J_p in straight lines between the supplying electrodes as the electric current (Figure 4.135a):

$$v_p = \mu_p E, \quad J = q\mu_p N E \qquad (4.89)$$

Under influence of the magnetic field, the moving particles are deflected in the direction perpendicular to the magnetic field vector B (Figure 4.135b), and the second component E_H of electric field that counterbalances this action appears as

$$E_H = -(v \times B) = -\mu_p (E \times B) \qquad (4.90)$$

The direction of the current is deflected by the angle θ_H known as the Hall angle (Figure 4.136):

$$\tan \theta_H = \frac{|E_H|}{|E|} \qquad (4.91)$$

Taking into account the expression (4.90), we can describe the Hall electric field as

$$E_H = -\frac{1}{qN}(J \times B) = -R_H (J \times B) \qquad (4.92)$$

where R_H is the Hall coefficient.

Usually, voltage between the sensing electrodes of the plate of width w is used as the output signal of the Hall sensor:

$$V_H = \mu_p w E_x B_y = R_H w J B \qquad (4.93)$$

By substituting the current density J in the plate of the thickness t by I/wt, we obtain the most commonly used relation describing the Hall sensor:

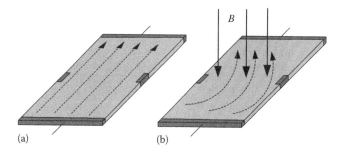

FIGURE 4.135
Current lines without (a) and with magnetic field (b).

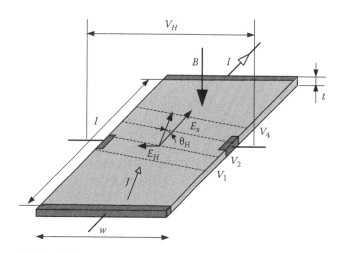

FIGURE 4.136
The Hall sensor (V_1, V_2, V_4—the equipotential lines).

$$V_H = \frac{R_H}{t} I \cdot B \qquad (4.94)$$

Thus, for a given bias current I, the Hall sensor measures directly the magnetic field value (represented by the flux density B) directed perpendicularly to the field plate. The transfer function of this sensor depends on the physical dimensions (width w and thickness t) and the material properties represented by the Hall coefficient R_H.

The relationship (4.94) is slightly misleading because it does not take into account the power dissipation limit and suggests that the larger Hall coefficient R_H (i.e., as smaller carrier density) the better the sensor. However, the most important factor affecting the sensitivity is the carrier mobility μ_p. We can rearrange the relationship (4.94) as dependence on the supplying voltage V. Because resistivity depends on the carrier density N and carrier mobility μ_p ($\rho = qN\mu_p$), we obtain that for the plate of the length l, the output voltage is

$$V_H = \mu_H \frac{w}{l} VB \qquad (4.95)$$

---

* Henceforth, called for simplicity as Hall sensor.

TABLE 4.9

Performances of Typical Materials Used for Hall Sensors

	μ_p (cm²/Vs) Electrons	μ_p (cm²/Vs) Holes	E_g [eV]	R_H (cm³/As)	α_T (%/K)
Si	1,400	1,200	1.12	3,000	−0.4
GaAs	8,500	400	1.42	60	−0.2
InAs	33,000	460	0.36	100	−0.17
InSb	80,000	1,250	0.17	380	−0.75

Approximate values because they depend strongly on the doping and preparation condition.

Thus, to obtain high sensitivity and high output signal, we should use materials with high carrier mobility. The output voltage of the sensor taking into account the dissipated power P is (Popovic 2004)

$$V_H \approx \sqrt{\frac{\mu_p}{Nt}}\sqrt{P}B \qquad (4.96)$$

The relationship (4.96) confirms that in order to obtain large sensitivity of the Hall sensor, the materials should exhibit as large as possible carrier mobility μ_p. Table 4.9 presents data on performance of typical materials used for Hall sensors.

From the data presented in Table 4.9, we can see that indium antimonide (InSb) is a semiconductor material exhibiting the largest electron mobility. Therefore, it is commonly used as the material for Hall sensors. The electrons have much larger mobility than holes, so n-type semiconductors are privileged. But InSb has some drawbacks. As it is presented in Table 4.9, it has the largest coefficient of temperature error α_T. Moreover, indium-based semiconductors have small energy band gap E_g, which indirectly increases the carrier density (Popovic 2004). That is why, also other materials are used, for example: Bell uses InAs and GaAs, Asashi-Kasei uses GaAs while Sentron uses Si.

4.5.2 Design of Hall Sensors

It should be noted that there are two classes of Hall sensors. There are so-called signal Hall sensors used mainly as magnetic field detectors in brushless motors, proximity sensors, etc. These sensors are usually very cheap (less than one dollar) and their main feature is a large output signal. The second class of Hall sensors are so-called measuring Hall sensors. In this case, the main efforts are in obtaining good performances: linearity, small temperature errors, small level of noises, etc. even at the cost of the sensitivity. Such sensors can be very expensive.

The material selection was discussed partially in the previous section. All data have been presented for

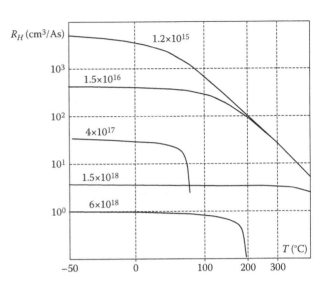

FIGURE 4.137
The Hall coefficient of InAs for various doping densities (indicated electron concentration in cm⁻³). (After Popovic, R.S., *Hall Effect Devices*, IOP Publishing, Boston, MA, 2004; Folberth, O.G. et al., *Z. Naturforsch.*, 9a, 954.)

pure semiconductors although their performances can be significantly altered when they are doped. Figure 4.137 presents influence of the doping densities on the temperature changes of R_H. There exists certain doping density at which this temperature influence is negligible. Similar results have been reported for n-InSb where for electron concentration 1.6×10^{18}, the temperature error was compensated (Oszwaldowski 1998, Berus et al. 2004). Similarly by changing of the impurity concentration, it is possibly to improve the linearity (Popovic 2004).

The Hall sensors are commonly prepared by thin-film deposition, often as epitaxial growth. Recently, the new technology known as *2DEG* (2D electron gas) or *quantum well* seems to be very promising. In this structure, the semiconductor layer is very thin (of the order of 10 nm—smaller than the mean free path of electrons) and the mobility is possible only within a 2D area (so electrons move similarly billiard balls). Such structures exhibit large mobility of electrons and are commonly used in the digital *quantum Hall devices* used as a standard of resistance (see Figure 2.70). They are also often used for preparation of very small (smaller than 1 μm) micro-Hall devices (Boero et al. 2003). Figure 4.138 presents an example of the micro-Hall device with dimension $0.8 \times 0.8\,\mu m$ used in a scanning Hall microscope (Sandhu et al. 2001). Similar heterostructure is used as a quantum Hall resistance standard (Hartland 1992).

The expressions (4.94) and (4.95) were derived under the assumption of an infinitely long Hall plate. For real Hall plates, additional geometrical factor $G = V_H/V_{H\infty}$ is introduced and

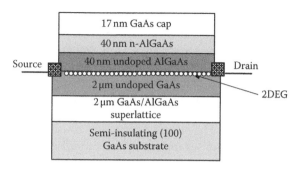

FIGURE 4.138
The heterostructure of micro-Hall 2DEG device. (From Sandhu, A. et al., *J. Cryst. Growth*, 227–228, 899, 2005.)

$$V_H = G \frac{R_H}{t} I \cdot B \qquad (4.97)$$

The geometrical factor G is between 0.7 and 0.9. Figure 4.139 presents various shapes of the Hall plates.

Although all semiconductor Hall elements can be easily incorporated into IC, the most natural solution for this idea is to use the silicon. The majority of ICs is made in silicon technology; therefore, recently there are available quite complex integrated silicon-based Hall circuits, including bias source, amplification, offset reduction circuit, temperature compensation, etc. (Kordic 1986, Schott et al. 1997). Figure 4.140 presents the simple example of a silicon Hall plate integrated with two transistors as a part of differential amplifier (Takamiya and Fujikawa 1972, Huang et al. 1984).

The design of a Hall sensor in a form of the plate presented in Figure 4.135 seems to be obvious as clear illustration of the Lorentz force. In such design, the magnetic field is directed perpendicular to the plate. In certain applications, it can be more useful to detect the field directed in a vertical direction. The design of such sensor is presented in Figure 4.141.

Apart from measurement of different direction of the magnetic field, the design presented in Figure 4.141 revealed also some other advantages: low noise, long-term stability, small offset, and better response to nonuniform magnetic field (Popovic et al. 2001). Very

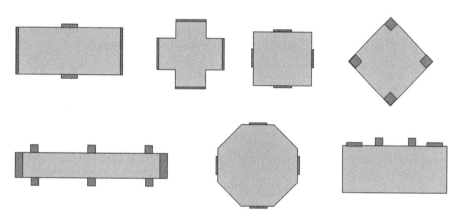

FIGURE 4.139
Various shapes of the Hall plates.

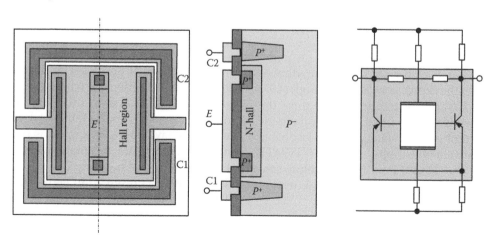

FIGURE 4.140
An example of the Hall sensor with two transistors as the part of differential amplifier. (From Huang, R.M. et al., *IEEE Trans., Electron. Dev.*, 31, 1001, 1984.)

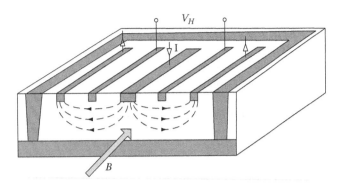

FIGURE 4.141
Design of vertical Hall sensor.

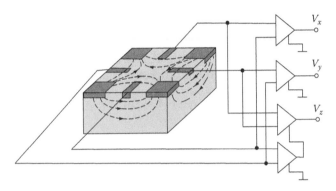

FIGURE 4.142
Three-axis Hall sensor with amplifiers for extracting three voltage signals. (From Schott, C. et al., *Sens. Actuat.*, 82, 167, 2000.)

high-accuracy Hall probes based on vertical, silicon structure were developed. The accuracy better than 0.004% was reported (Schott et al. 1997, 1999).

Other important feature of the vertical design is possibility to develop two- or three-axis Hall sensors. Figure 4.142 presents the example of a three-axis Hall device made as a single silicon chip (Schott et al. 2000).

4.5.3 Performance of the Hall Sensors

The sensitivity of the Hall sensors depends on many factors such as material of the sensor, sensor dimensions—according to the expressions (4.94) through (4.97) but also on the actual sensor technology. Because the output signal depends on two input signals (magnetic field and bias current or bias voltage), it is reasonable to consider the current sensitivity* $S_i = V_H/IB$ [V/AT], voltage sensitivity[†] $S_v = V_H/VB$ [V/VT], or absolute sensitivity $S_0 = V_H/B$ [V/T] determined for nominal (recommended by the manufacturer) bias conditions.

---
* According to expression (4.94), the current sensitivity S_i corresponds with R_H/t.
[†] According to expression (4.95), the voltage sensitivity S_V corresponds with $\mu_H w/l$.

Table 4.10 presents the sensitivity of the market-available sensors. It should be noted that the sensitivity is proportional to the sensor area. Thus, the most sensitive commercially available sensor B-850 of F.W. Bell with sensitivity 180 V/T (bulk indium) has the length 225 mm and width 11.5 mm (bias current 200 mA).

The resolution of Hall sensors is limited mainly by the offset and noise. The offset voltage ΔV is caused by upset of symmetry—for example, by asymmetrical (imprecise) configuration of the electrodes or the local defect. It is possible to reduce the asymmetry by the laser trimming (at the manufacturing stage) or by the change of external resistor (Figure 4.143a). Table 4.11 presents typical errors of commercially available Hall sensors.

The offset voltage can be recalculated to offset-equivalent magnetic field $\Delta B = \Delta V \cdot S_0$ (Table 4.11). Thus, the resolution of typical Hall sensors is at the level of several mT. This offset voltage can be partially reduced by differential connection of two or four Hall elements (Figure 4.143b). Such method is effective only for the case of uniform distribution of asymmetry, for example, mechanical stress and linear gradient of temperature. More general offset reduction is possible by using electronic auto-zero circuits applied in integrated Hall circuits (described later).

Figure 4.144 presents the results of investigations of the silicon Hall sensor. The 12 h noise and offset changes were at the level of 30 μV, which for the sensitivity of 1 V/T gives the noise-equivalent magnetic field on the level of tens of μT. This noise is mainly $1/f$ type and depends on the sensor dimensions. Generally, it can be assumed that the main limitations of the resolution are the temperature offset changes.

From the expression (4.97), results the linear dependence between the output voltage and measured magnetic field. But the Hall coefficient R_H and geometrical factor G depend on the value of measured induction:

$$V_H = R_{H0}\left(1 - \alpha\mu_H^2 B^2\right)G_0\left(1 + \beta\mu_H^2 B^2\right)\frac{1}{t}I \cdot B \quad (4.98)$$

Thus, the error of nonlinearity depends on the measured value of flux density whose range is usually limited to about 0.1 T, and in this range the nonlinearity error is reasonably small (Table 4.11). Because α and β coefficients have opposite signs, it is possible in certain range to compensate the material and the geometrical nonlinearities. An example of nonlinearity error of BH-200 sensor is presented in Figure 4.145.

The other limitation of application of Hall sensor is the temperature error. From the data presented in Table 4.11, results that InSb-based sensors have large temperature error mainly due to significant change of resistance—1.8%/°C. This error can be eliminated by biasing the sensor from the constant voltage source as it

TABLE 4.10

Sensitivity of the Commercially Available Sensors

	S_0 (V/T)	I_n (mA)	S_i (V/AT)	S_V (V/VT)	Dimensions l-w-t (mm)	
BH-200	0.15	150	1	0.4	5.2-1.75-0.47	InAs bulk
FH-301	0.1	25	4	0.2	2-1-0.5	InAs
GH-600	0.5	5	100	0.22	4 diameter	GaAs
SH-400	2.9–11.2	5	580–2240	2.4–4.0	0.3 diameter	InSb
HW-101	10.4	5	2080	5.2		InSb
HQ 0111	2.6	5	650	0.87		InAs 2DEG

Sensors: BH, FH, GH—F.W.Bell/Sypris; HW, HQ—Asashi Kasei.

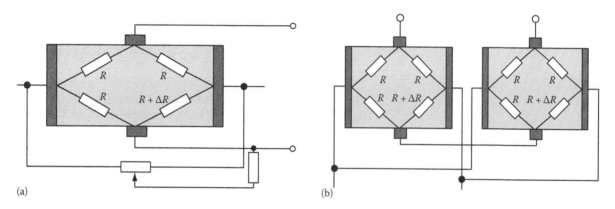

(a)   (b)

FIGURE 4.143
The manual symmetrization of Hall sensor (a) and differential connection of two Hall elements (b).

TABLE 4.11

Errors of the Commercially Available Sensors

	ΔV (mV)	ΔB (mT)	$\Delta V/T$ (μV/K)	$\Delta R/T$ (%/K)	δS (%/K)	Linearity
BH-200	0.1	0.6	1	0.15	0.08	1% (0–0.1 T)
FH-301	2	20	10	0.1	0.1	
GH-600	14	28	1	0.15	0.07	2% (0–0.1 T)
SH-400	20	6		1.8	1.8	
HW-101	7	0.7		1.8	1.8	

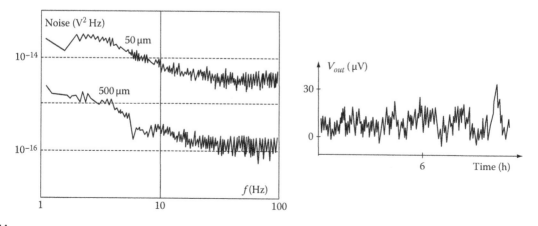

FIGURE 4.144
The example of noise in silicon Hall sensors of different dimensions. (From Schott, C. et al., *IEEE Trans. Instrum. Meas.*, 46, 613, 1997; Schott, C. et al., *Sens. Actuat.*, 82, 167, 2000.)

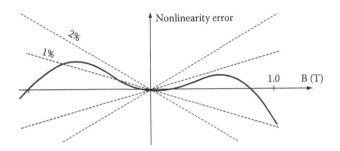

FIGURE 4.145
The example of nonlinear error of BH-200 sensor (F.W. Bell data).

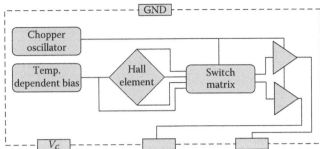

FIGURE 4.147
The linear IC Hall sensor—model HAL-401. (Micronas Data Sheet.)

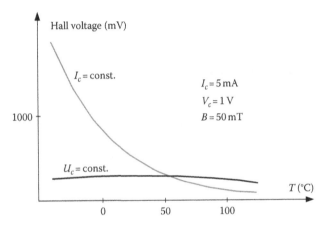

FIGURE 4.146
The example of temperature dependence of the output InSb Hall sensor signal—HW-101 sensor. (Asashi Kasei catalogue.)

is presented in Figure 4.146. In the case of low mobility sensors (GaAs, InAs or Si) the constant supplying current is more often used.

The Hall sensors usually have excellent frequency properties—theoretically, even up to THz values (Mittleman et al. 1997). In practice, the parasitic capacitances and the inductive loop created by pick-up wires limits this frequency bandwidth to MHz level (sometimes even to hundreds of kHz).

4.5.4 Integrated Circuit Hall Sensors

After development of silicon Hall elements in CMOS technology, more and more IC Hall sensors appeared on the market. Recently, IC Hall sensors are dominating in the market. These sensors can be divided into two kinds: linear IC Hall sensors used for magnetic field measurement and switch IC Hall sensors used as the proximity sensors (switches and latches).

Figure 4.147 presents the example of the linear IC Hall sensor equipped with a temperature measurement device, temperature error correction, and the offset voltage auto-zero elimination. The principle of the auto-zero function is presented in Figure 4.148. According to the clock frequency, the bias diagonal of Hall element

is changed. Thus, the Hall voltage is also periodically changed and in the first period is $V_H + \Delta V$ and in the next period is $-V_H + \Delta V$. Next after demodulation, the offset component is removed.

Figure 4.149 presents another IC Hall sensor—a digital one. All data of the sensor such as transfer characteristic or temperature changes are saved in EPROM memory. Therefore, it is possible in digital signal processing (DSP) module to correct linearity or temperature changes because the sensor is equipped with internal sensor of temperature. The sensor works in *ratiometric mode*—it means that the output signal can depend on both input signals: magnetic field value and voltage. This way, the sensor can be used as an MFS (fixed value of the bias voltage) or as a multiplier of both signals. It is interesting that the sensor has only three pins—programming is performed by the modulation of the supply voltage. Table 4.12 presents the typical data of selected commercially available linear IC Hall sensors.

Often, it is necessary to detect the situation when the magnetic field exceeds a given threshold. Such device is known as a switch (or latch if the state is memorized when the magnetic field is removed). An example of IC switch device is presented in Figure 4.150. Practically all manufacturers of Hall sensors (Allegro Microsystems, Asashi Kasei, Infineon, Melexis, Micronas) offer large variety of switching Hall devices.

4.5.5 Hall-Effect-Based Semiconductor Magnetoresistors

The Hall effect is related to the change of the current lines of moving particles due to the Lorentz force, so it is obvious that also their resistance depends on the magnetic field. This effect is known as *semiconductor magnetoresistance* and practically always accompanies the Hall effect.

Similarly, as in the case of Hall sensors, the MR effect depends mainly on the carrier mobility μ_H:

$$R(B) = R_0 \frac{\Delta \rho}{\rho} \left(1 + \mu_H^2 C B^2\right) \qquad (4.99)$$

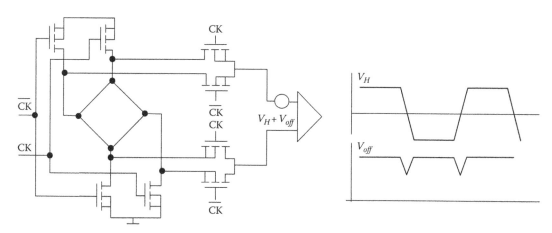

FIGURE 4.148
The auto-zero operation switch circuit. (From Bilotti, A. et al., *IEEE J. Solid State Circuits*, 32, 829, 1997.)

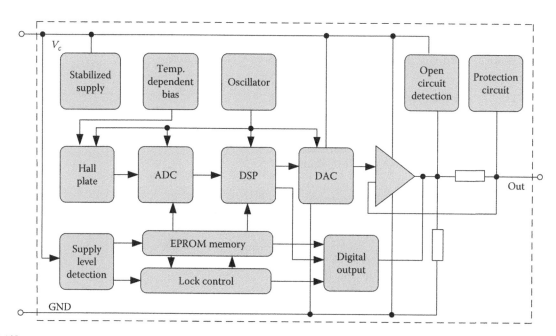

FIGURE 4.149
The linear digital programmable IC Hall sensor—model HAL-805. (Micronas Catalogue.)

TABLE 4.12

Properties of the Commercially Available IC Hall Sensors

	HAL 401 Micronas	A1321 Allegro Microsystems	MLX90242 Melexis
Sensor area [mm × mm]	0.37×0.17		0.15×0.15
Sensitivity [mV/mT]	48	50	39
Range FS	±50 mT		
$\Delta V/T$	25 μT/K	10 mV at OT	25 mV at OT
$\Delta S/T$	4% at OT	4% at OT	0.07%/°C
Nonlinearity	0.5% for FS	1.5	0.5% of FS
Noise	10 μT at BD	40 mV for FS	5 mV for FS
Bandwidth BD	0–10 kHz	0–30 kHz	
Operating temp. OT (°C)	−40 to 150	−40 to 150	−40 to 150

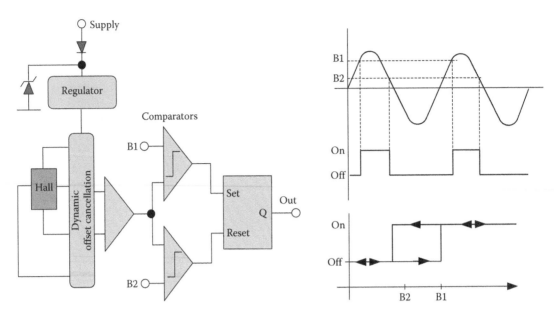

FIGURE 4.150
The example of the switch IC Hall device. (From Ramsden, E., *Hall Effect Sensors*, Newnes, 2006.)

The resistance R_0 is the resistance of the magneto-resistor in absence of external magnetic field and C is the factor depending on the geometry of the sensor. The geometry factor was calculated by Lippmann and Kuhrt (Lippmann and Kuhrt 1958) for length-to-width ratio $l/w < 0.35$ as

$$C = 1 - 0.54 \frac{l}{w} \qquad (4.100)$$

In a special geometrical form proposed by Corbino (Corbino 1911) and known as the *Corbino disc* (Figure 4.151b), it is possible to obtain a pure MR effect (without the Hall effect). The Corbino disc is equivalent to the plate with $l/w \approx 0$ (the current lines are curved along logarithmic spirals).

Figure 4.152 presents a transfer characteristic of the commercially available MR sensor of Asashi Kasei. The dimensions of the sensor are 2×3 mm. The MR ratio

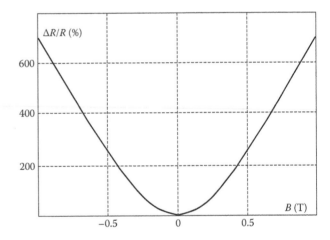

FIGURE 4.152
The example of the transfer characteristic of the semiconductor magnetoresistor type MS-0080 of Asashi Kasei.

$\Delta\rho/\rho$ is about 200% for $\Delta B = 0$–45 mT. The characteristic is nonlinear because resistance is proportional to B^2. But in some applications, this nonlinearity is less important than the actual amplitude of the output signal, which is quite large. Figure 4.153 presents the typical application of the semiconductor magnetoresistor.

In comparison with Hall sensor, the MR sensor is physically bigger and its characteristic is less linear. Unfortunately, semiconductor MR sensor exhibits rather large temperature error, due to the large change of resistivity with the temperature. Figure 4.154 presents the influence of ambient temperature on the sensitivity of semiconductor magnetoresistor.

The Hall-effect sensor is one of the oldest but still frequently used MFS. Its main advantages are small

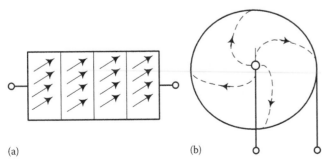

(a)   (b)

FIGURE 4.151
The semiconductor magnetoresistors: thin film element (a) and Corbino disc (b).

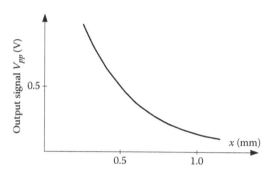
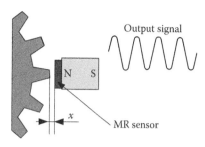

FIGURE 4.153

The output signal of the MR sensor used for detection of rotation of the gear-tooth wheel—supply voltage 5 V. (From Asashi Kasei catalogue.)

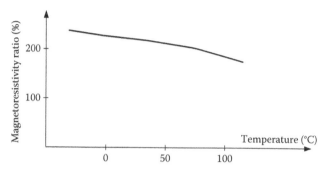

FIGURE 4.154

The change of a magnetoresistivity ratio with temperature for a semiconductor MR sensor. (From Asashi Kasei catalogue.)

dimensions, large output signal, linearity, good frequency properties, output signal directly depending on the magnetic field value. An advantage is also the potential ease of incorporation to an IC.

Its applications are mainly limited by poor sensitivity. In the best case, it is about 10 V/T for InSb (without amplifier). Therefore, typical magnetic field range of this sensor is 1–50 mT although the upper limit is practically unrestricted.

Recently, the market is somewhat dominated by the IC sensors based on silicon Hall sensor. Such sensor enables the elimination of one of the main drawbacks of the Hall sensor—the temperature zero drift can be corrected with an auto-zero function. The sensitivity of these sensors (with amplifier) is typically 50 V/T.

4.6 SQUID Sensors

4.6.1 Operating Principle of SQUID Sensors

SQUID joins two effects: quantization of magnetic flux and tunneling by a weak link (Josephson effect). These devices were mentioned in Section 2.8 and are described in many books or review papers (Swithenby 1980,

Clarke 1980, 1989, Koch 1989, Ryhänen and Seppä 1989, Wikswo 1995, Weinstock 1996, Jenks et al. 1997, Koelle et al. 1999, Fagaly 2001, 2006, Pizella et al. 2001, Clarke and Braginski 2004, Sternickel and Braginski 2006).

In the closed superconducting ring surrounding the area with the flux Φ_{ex} (Figure 4.155), can be induced flux Φ_{in}. Due to condition of continuity of the wave function representing superconducting current, the magnetic flux inside the ring has to be a multiple of the flux quantum $\Phi_0 = h/2e$ and

$$\Phi_{in} = n\frac{h}{2e} = n\Phi_0 = \Phi_{ex} - Li_s \qquad (4.101)$$

where i_s the current in the ring of inductance L.

According to the Meissner effect, the current i_s flows only in a very thin boundary area called London penetration depth λ_L—for niobium, this thickness is about 40 nm. This current acts as a perfect screen and therefore, the superconducting material is perfectly diamagnetic so that it does not to allow the external magnetic flux to penetrate the inside the material. Thus, the internal flux Φ_{in} is trapped and to change it when the external flux Φ_{ex} is increased would be necessary to realize transition from superconducting state to normal state.

The solution of this problem is to introduce very thin tunnel junction proposed by Josephson and known as a Josephson junction (Josephson 1962). The Josephson barrier should be thinner than the coherence distance between electrons in the pair, that is 0.1–2 nm. In circle with Josephson barrier are "open" the superconducting ring for external flux penetration without breaking the superconducting circuit for the internal flux.

Indeed next year after Josephson proposal, Rowell experimentally confirmed that the current in the Josephson junction changes periodically with the external magnetic field according to the flux quantum (Figure 4.156) (Rowell 1963). In the Rowell experiment, the external magnetic field penetrated the junction area and adjacent part of superconductors and the wave function representing the current corresponded

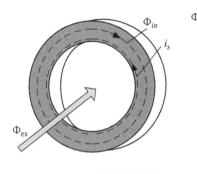

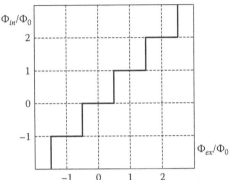

FIGURE 4.155
Quantization of magnetic flux in superconducting ring.

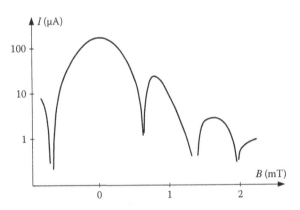

FIGURE 4.156
Experiment of Rowell—the current in the Josephson junction. (From Rowell, J.W., *Phys. Rev. Lett.*, 11, 200, 1963.)

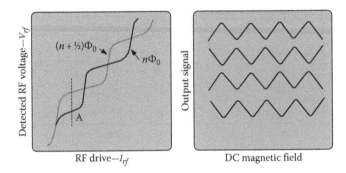

FIGURE 4.157
Experiment of Zimmerman and coworkers—the oscilloscope pictures of RF SQUID signals. (From Zimmerman, J.E. et al., *Phys. Rev.*, 141, 367, 1966; Zimmerman, J.E. et al., *J. Appl. Phys.*, 41, 1572, 1970.)

with the $n\pi$ value. Thus, the superconducting current is described by the following relationship (Kleiner and Koelle 2004):

$$I(B) = I(0)\frac{\sin(\pi\Phi_{in}/\Phi_0)}{\Phi_{in}/\Phi_0} \qquad (4.102)$$

This expression and Figure 4.156 resemble interference diffraction relation known from classical optics.

If the Josephson junction is included into a closed-ring sample discussed earlier, the current i_s is (according to the first Josephson equation (2.83)) $I_c\sin\theta$ and the relation (4.102) takes a form

$$\Phi_{in} = \Phi_{ex} - LI_c\sin\theta = \Phi_{ex} - LI_c\sin\frac{2\pi\Phi_{in}}{\Phi_0} \qquad (4.103)$$

This relation describes a ring device with one Josephson junction known as RF SQUID. We can see that the internal flux depends on the external magnetic flux. Figure 4.157 presents the example of the signals obtained in RF SQUID device.

Two years after the Josephson invention, Jaklevic and coworkers (Jaklevic et al. 1964, 1965) demonstrated a quantum interference effect in two parallel-connected Josephson junctions (Figure 4.158). In such device, we have interference between two different currents in two branches of the ring $I_b - I_{circ}$ and $I_b + I_{circ}$ where I_b is the half of the bias current $2I_b$ and the circular current is the current induced by external magnetic flux $I_{circ} = -\Phi_{ex}/L$. As a result, the current in such device is (Chesca et al. 2004)

$$I \approx 2I_0\left|\cos\frac{\pi\Phi_{ex}}{\Phi_0}\right| \qquad (4.104)$$

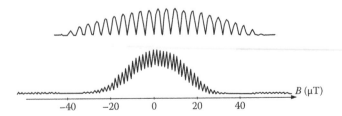

FIGURE 4.158
Experiment of Jaklevic and coworkers—the current in the double Josephson junctions (DC SQUID). (From Jaklevic, R.C. et al., *Phys. Rev.*, 140, A1628, 1965.)

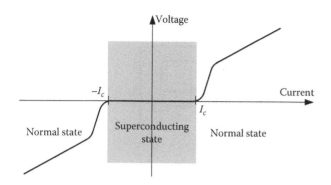

FIGURE 4.159
Voltage versus current of a shunted Josephson junction.

$$I = I_c \sin\theta + \frac{h}{2e}\frac{1}{R}\frac{d\theta}{dt} + \frac{h}{2e}\frac{d^2\theta}{dt^2} + I_n \qquad (4.105)$$

where I_n is the noise source. According to RSJ model, the SQUID devices can be presented using schematic diagrams presented in Figure 4.160.

4.6.2 Design and Properties of SQUID Sensors

Initially, SQUID devices were designed with RF technology because such solution needs only one junction. Recently, as the technology of junction became more established, the DC SQUIDs are more frequently prepared. In the first RF SQUIDs, to obtain weak link, a screw made from niobium was used (Figure 4.161a) (Silver and Zimmerman 1967, Zimmerman et al. 1970). Such technology is sensitive to shocks and vibrations and today is rather avoided.

First DC SQUIDs were prepared in thin-film technology where two thin tin films were separated by an oxide layer (Figure 4.161b) (Jaklevic et al. 1965). Another weak link was created by introducing a narrowing—a bridge known as *Dayem bridge* (Figure 4.161c).

Recently, the SQUID sensors are prepared using thin-film deposition and photolithography, enabling to design even very complex shapes (Wikswo et al. 1998, Cantor and Ludwig 2004). In LT SQUIDs (low-temperature SQUIDs), niobium is used almost exclusively as the superconducting material. The niobium layers are separated by niobium oxide or aluminum oxide film.

This relation describes the ring device with two Josephson junctions known as DC SQUID. We can see that the current flowing through the DC SQUID depends on the external magnetic flux.

The Josephson junction is usually shunted by the resistor R and the capacitor C (known as RSJ model—resistivity shunted junction). By appropriate choice of the R and C values, it is possible to obtain hysteresis-free characteristic I–V (Figure 4.159)—when the McCumber factor $\beta_C = 2\pi I_c R^2 C/\Phi_0 < 1$.

From the characteristic of a shunted Josephson junction (Figure 4.159), we can see that when the current exceeds the critical value I_c, a sharp transition occurs between superconductive state and normal state.

For the RSJ model, we can describe the Josephson junction by the following relationship (McCumber 1968):

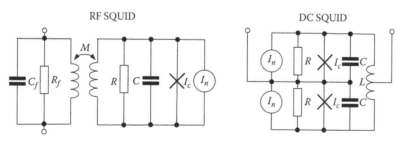

FIGURE 4.160
Schematic diagrams of RF SQUID and DC SQUID devices.

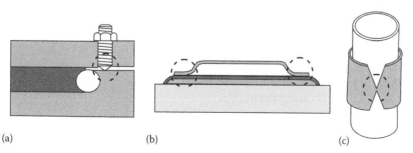

FIGURE 4.161
The main method of weak link (Josephson junction) preparation.

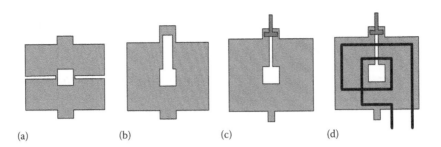

FIGURE 4.162
The examples of the planar thin-film LT SQUIDs design. (From Cantor, R. and Ludwig, F., SQUID fabrication technology, in *The SQUID Handbook*, Clarke J. and Braginski A.I. (Eds.), Chapter 3, Wiley-VCH Verlag, Weinheim, Germany, 2004; Koelle, D. et al., *Tev. Modern Phys.*, 71, 631, 1999.)

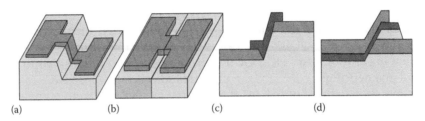

FIGURE 4.163
The examples of the thin-film HT SQUIDs design. (From Koelle, D. et al., *Tev. Modern Phys.*, 71, 631, 1999.)

Figure 4.162 presents various examples of the planar thin film SQUIDs—with a microbridge (Figure 4.162a and b) or a superconductor–oxide–superconductor junction (Figure 4.162c). Often, the sensor is prepared with the coil used as a flux transformer (Figure 4.162d). Usually, sensor is manufactured with the shunt thin-film resistor prepared, for example, from molybdenum.

Preparation of the high-temperature SQUIDS (HT SQUIDs) is more difficult. $YBa_2Cu_3O_{7-x}$ known as YBCO is often used as the base material. This material is not very suitable for deposition because it is ceramics. Figure 4.163 presents some examples of the HT SQUIDs joint. Two technologies are used: GB grain boundary barrier (Figure 4.163a and b) or SNS—superconductor-normal-superconductor material (Figure 4.163c and d).

Usually the SQUID device is not directly contacted with the magnetic source. For this purpose special flux transformer is used, often in form of gradiometer. Frequently the second coil is coupled directly with SQUID structure in form of a spiral as it is presented in Figure 4.162d. Both the flux transformer and the SQUID device one are in special vacuum flask known as Dewar. Such cryostat is filled with liquid helium in the case of LT SQUID (boiling point 4.2 K) or with liquid nitrogen in the case of HT SQUID (boiling point 77 K) (Figure 4.164).

The detected flux density depends on the area of the sensing coil. Because SQUID is usually very small (less than 1 mm²), larger pick-up coil in flux transformer helps in improvement of sensitivity. The mutual

inductance between transformer coil and SQUID M_{SQ} and the inductance of the coil of flux transformer L_i should be matched to the inductance of the SQUID L_{SQ} (Koch 1989):

$$M_{SQ} = nL_{SQ} \quad \text{and} \quad L_i = n^2 L_{SQ} \qquad (4.106)$$

where n is the number of turns of spiral coil.

The conditions (4.106) are rather difficult to fulfill because usually inductance of the spiral coil is of several nH while the inductance of the SQUID is of several pH.

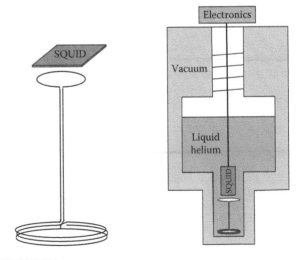

FIGURE 4.164
The flux transformer and the Dewar flask used for SQUID devices.

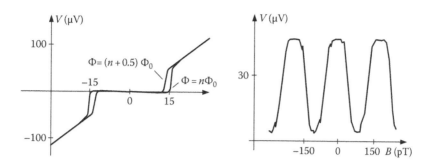

FIGURE 4.165
The characteristic and output signal of typical DC SQUID device. (After Koch, H., SQUID sensors, in *Magnetic Sensors*, Boll R. and Overshott K.J. (Eds.), Chapter 10, VCH Verlag, Weinheim, Germany, 1989.)

TABLE 4.13

Parameters of Typical DC SQUID

SQUID inductance L_{SQ}	620 pH	Critical current I_c	15 μA
Input coil turns n	19	Shunt resistance $R/2$	3.6 Ω
Pickup loop inductance L_p	60 nH	Sensitivity	0.25 nT/Φ_0
Input coil inductance L_i	110 nH	Noise $f > 1$ Hz	$2 \cdot 10^{-5}$ Φ_0/√Hz
Mutual inductance M_{SQ}	6 nH		5 fT/√Hz
Junction capacitance	0.7 pF		

Source: Koch, H., SQUID sensors, in *Magnetic Sensors*, Boll R. and Overshott K.J. (Eds.), Chapter 10, VCH Verlag, Weinheim, Germany, 1989.

Additionally, in order to obtain non-hysteric operation and large output signal, the *Mc-Cumberland factors* should be fulfilled:

$$\beta_C = \frac{2\pi I_c R^2 C}{\Phi_0} < 1 \quad \text{and} \quad \beta_L = 2 L_{SQ} I_c \Phi_0 \approx 1 \quad (4.107)$$

Moreover, the noise energy E of a SQUID loop is (Koch 1989)

$$E = 16 k_B T \sqrt{LC} \quad (4.108)$$

and therefore, the SQUID inductance L_{SQ} and junction capacitance C should be as small as possible. Figure 4.165 and Table 4.13 present the current/voltage characteristics, the output signal, and parameters of a typical DC SQUID device.

From Figure 4.165, we can see that the typical output signal of the SQUID device is rather small—usually it does not exceed 100 μV/Φ although sensitivity of about 300 μV/Φ_0 is reported (Drung et al. 1991). In special design known as double relaxation oscillation SQUID (DROS), it is possible to increase this sensitivity to the level of 1–10 mV/Φ_0 (Pizella et al. 2001).

The noise level is a crucial parameter for the performance of the SQUID device. Figure 4.166 presents typical frequency spectrum of magnetic noise of a DC SQUID. For low frequency, the 1/f type noise dominates but above 0.1 Hz, the resolution of about 5 fT/Hz$^{-1/2}$ is

attainable in DC LT SQUIDs. The resolution as small as 2.5 fT is reported (with sensitivity 370 μV/Φ_0 and 0.47 nT/Φ_0 (Drung et al. 1990, 1991, 2007). In the case of HT SQUIDs, this resolution is about 10 times larger although the resolution of about 30 fT is reported (Cantor et al. 1995, Beyer et al. 1998, Faley et al. 2001) (Figure 4.167).

Sensitivity depends on the input inductance of the device. Hence, in order to compare various SQUID devices, instead of magnetic noise S_B or S_Φ, the energy sensitivity E_N is considered:

$$E_N(f) = \frac{S_\Phi(f)}{2L} \quad [\text{J/Hz}] \quad (4.109)$$

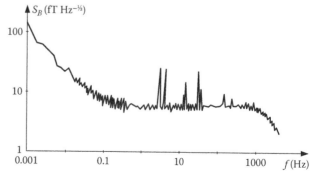

FIGURE 4.166
Typical magnetic noise spectrum of the DC SQUID. (From Koch, H., SQUID sensors, in *Magnetic Sensors*, Boll R. and Overshott K.J. (Eds.), Chapter 10, VCH Verlag, Weinheim, Germany, 1989.)

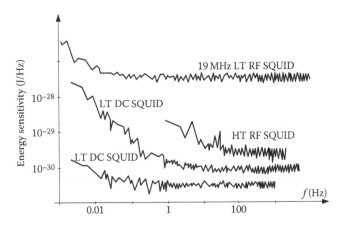

FIGURE 4.167
Energy sensitivity of various SQUID sensors. (From Fagaly, R.L., *Rev. Sci. Instrum.*, 77, 101101, 2006.)

This energy sensitivity is often expressed in terms of Planck's constant $h = 6.6 \times 10^{-34}$ J/Hz. Typical commercially available RF SQUIDs exhibit sensitivity $E_n < 10^{-29}$ J/Hz while for DC SQUIDs this parameter is $E_n < 10^{-31}$ J/Hz (thus, around $100h$). It is assumed that theoretical limit of sensitivity is $h/2\pi = 1.1 \times 10^{-34}$ J/Hz. The energy sensitivity as small as $3h$ at 4.2 K and $1.6h$ at 1.5 K has been reported (Van Harlingen et al. 1982, Wakai and Van Harlingen 1988). Such small resolution was obtained for high-frequency field (3 MHz) when the $1/f$ noise was negligible and only white noise was considered.

4.6.3 SQUID Magnetometers

Usually, the SQUID device works in a *flux locked loop* (FLL) configuration (Figure 4.168) where the output current is used as a feedback. This way the SQUID device operates as zero magnetic field detector.

Because the output signal is very small and noisy usually is modulated by alternating magnetic flux Φ_{osc} and synchronous detector (lock-in amplifier) helps to remove the noise. To take in a feedback, the SQUID sensor is often equipped with additional feedback coil as it is illustrated in Figure 4.169. This feedback coil can also operate as the modulation coil.

Figure 4.170 presents the basic circuit of an RF SQUID magnetometer. The SQUID device with one Josephson link is connected to the resonant tank circuit of an RF oscillator (typically 19 MHz) via some mutual inductance. In the SQUID ring, the current induced by external measured magnetic flux is flowing as well as the RF current induced from tank circuit. If this sum of currents periodically exceeds the critical value, we obtain periodical transition between superconducting and normal state enabling to absorb every new flux quantum. If the operating point (current in the loop) is chosen on the plateau of V_{fr} (I_{rf}) characteristic (e.g., point A in Figure 4.157), we obtain triangular changes of V_{rf} with external magnetic field (see Figure 4.157).

Figure 4.171 presents the basic circuit of the DC SQUID magnetometer. The SQUID device with two Josephson links is supplied by DC current. The voltage drop across the SQUID (the output voltage) changes as $\cos \Phi$ with the period Φ_0—see Figure 4.165. The operating point can be chosen on linear part of sinusoid or on the maximum point (see Figure 4.168). The modulating signal is then amplified and synchronously rectified.

Figure 4.172 presents a DC SQUID magnetometer in which the negative feedback is supported by a positive feedback—feedback signal is taken from the output of the SQUID device. Such setup is known as *additional positive feedback AFP* device (Drung et al. 1990, 1991, Drung and Mück 2004). Additional positive feedback introduces asymmetry into transfer characteristic—one slope is significantly steeper and therefore, the sensitivity for the point A (Figure 4.172) is better in comparison with conventional DC SQUID.

By applying the AFP, it is possible to increase the SQUID sensitivity from around 100 μV/Φ_0 to 300 μV/Φ_0. But more significant increase of the SQUID sensitivity is possible with a different operating principle. Usually, the weak link is shunted by the resistor and capacitor to avoid hysteresis of $V(I)$ dependence. But if the shunt resistor is appropriately chosen and the SQUID is operating in the hysteric mode, it is possible to observe the relaxation oscillation. These oscillations known as *relaxation oscillation of SQUID* (ROS) can be used as a measure

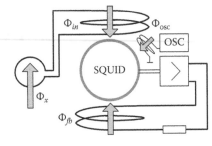
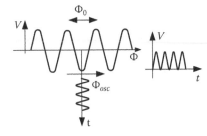

FIGURE 4.168
FLL mode of operation of SQUID magnetometer.

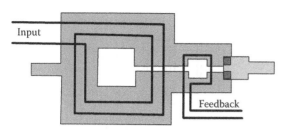

FIGURE 4.169
The SQUID layout with input and feedback planar coils. (From Cantor, R. and Ludwig, F., SQUID fabrication technology, in *The SQUID Handbook*, Clarke J. and Braginski A.I. (Eds.), Chapter 3, Wiley-VCH Verlag, Weinheim, Germany, 2004.)

of external magnetic field. Both the frequency and the average value as the pulses depend on the measured magnetic flux. Therefore, such principle was used as the flux-to-frequency converter (Mück et al. 1988, Mück and Heiden 1989).

By using the ROS principle, it is possible to obtain sensitivity as high as $4\,\text{mV}/\Phi_0$. But by connecting the SQUID device in series with a second reference (Figure 4.173) device and by appropriate choice of the R and L values, it is possible to develop the SQUID device known as DROS with relaxation frequency around 80 MHz, energy sensitivity 160 h, and flux-to-voltage transfer $80\,\text{mV}/\Phi_0$ (Gugoshnikov et al. 1989, 1991, Adelerhof et al. 1994, 1995).

SQUID devices are recently indispensable tool for measurement of magnetic field in the range fT to pT with typical resolution of about 5 fT. Such performance is attainable with LT DC SQUID. About 10 times worse performance (mainly the resolution) is typical for HT (77 K) SQUID devices for which sensitivity is around $100\,\mu\text{V}/\Phi_0$, what requires to use high-quality preamplifier (Drung 1997). The SQUID devices are used most frequently for measurement of DC and low-frequency

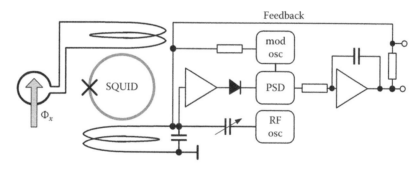

FIGURE 4.170
The RF SQUID magnetometer.

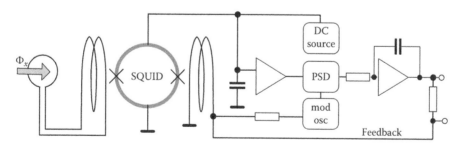

FIGURE 4.171
The DC SQUID magnetometer.

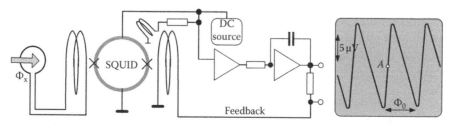

FIGURE 4.172
The AFP DC SQUID magnetometer.

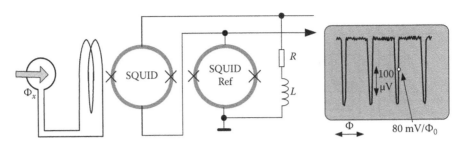

FIGURE 4.173
The DROS DC SQUID magnetometer.

magnetic fields but the range of operation can be extended even to MHz. The SQUID magnetometers measure the vector of magnetic field.

By using standard Deware cryostat, it is possible to develop mobile and multisensor SQUID devices at reasonable price. Although the main areas of application of the SQUID magnetometers are biomagnetic investigations (Romani et al. 1982, Zeng et al. 1998, Zhang et al. 2000, Pizella et al. 2001, Koch 2001, Sternickel and Braginski 2006) recently still increases the importance of application of the SQUIDs to nondestructive testing (Weinstock 1991, Vrba et al. 1993, Wikswo 1995, Jenks et al. 1997, Braginski and Krause 2000, Chatraphorn et al. 2000, Krause and Kreutzbruck 2002) and in geophysical exploration (Zhang et al. 1995, Bick et al. 1999).

4.7 Resonance Sensors and Magnetometers

4.7.1 General Remarks

The nuclear magnetic resonance (NMR) is commonly known as a *magnetic resonance imaging* (MRI) tool used in medicine diagnostics as tomography device. This subject is widely described in many books and papers (e.g., Kuperman 2000, Vlaardingerbroek and den Boer 2003, Guy and Ffytche 2005). This application of NMR is described later.* In the shadow of the MRI, the nuclear resonance is used for other purposes, among others:

- NMR spectroscopy as a method of chemical analysis
- NMR magnetometers for measurements strong magnetic field with high precision
- NMR free-precession magnetometers used for measurements of weak magnetic fields

In all these instruments, we apply the magnetic properties of atomic nucleus. The nuclei exhibit magnetic moment

resulting from their rotation of the charge—thus, in the external magnetic field, they act as elementary magnets. But in contrary to the compass needle, they do not align toward the magnetic field but rather rotate in selected quantum directions with the resonance frequency f_0 strictly dependent on the value of magnetic field B_0.

In the presence of external magnetic field, a nucleus can take two distinct energy states depending on the spin quantum number (the spin-aligned state has lower energy than the unaligned state—Figure 4.174). This difference of energy states ΔE depends linearly on the value of external field B_0:

$$\Delta E = \hbar \gamma B_0 \qquad (4.110)$$

where γ is the *gyromagnetic ratio*.

When we apply alternating magnetic field perpendicular to B_0, we can cause the transition between two energy states and this transition is the largest at resonance frequency f_0:

$$f_0 = \gamma_p B_0 \qquad (4.111)$$

and

$$\gamma = 2.67515255 \times 10^8 \, \text{rad/sT} = 42.5760812 \, \text{MHz/T}$$
$$\text{for protons } {}^1\text{H}$$

$$\gamma = 0.4104415 \times 10^8 \, \text{rad/sT} = 6.535692 \, \text{MHz/T}$$
$$\text{for deuterons } {}^2\text{H}$$

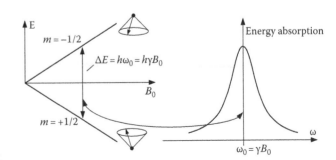

FIGURE 4.174
The nuclear resonance principle.

---

* Section 6.3.3 is devoted to magnetism in medicine.

In the resonance state, the transition between energy states results in the absorption of energy (Figure 4.174). Thus, we can observe the resonance in two ways: by changing the frequency $f_0 \pm \Delta f$ of alternating magnetic field for fixed value of B_x or by changing the additional magnetic field ΔB (thus, by changing the measured field $B_x \pm \Delta B$) for fixed frequency of alternating field.

The important advantage of NMR magnetometers is their exceptionally high accuracy. The value of the gyromagnetic ratio is a physical constant and its value is known with high precision. We are able to measure the frequency also with high accuracy; therefore, it is possible to measure magnetic field with uncertainty at the level of 1 ppm (10^{-6}). Because of this high precision, these magnetometers are often used as a standard device to test and scale, for example, Hall magnetic field meters (Schott et al. 1999, Weyand 1999).

The NMR magnetometers with continuous wave have limited resolution—below the field of about 50 mT the energy split is rather small and such value is often the smallest in commercial magnetometers. For measurements of much smaller magnetic fields in the range 1 pT–100 µT, mainly for geophysics applications, different approach is used. Such NMR magnetometers work by measurement of frequency of free precession.

In space magnetic field investigations, the resolution offered by NMR magnetometers is too coarse due to small gyromagnetic ratio. For example, Earth magnetic field 50 µT corresponds with resonance frequency of only 2100 Hz. Therefore, for smaller magnetic field, instead of NMR methods are used magnetometers utilizing *electron spin resonance* (ESR) (known also as *electron paramagnetic resonance*). Because the gyromagnetic ratio depends on the particle mass, therefore, for Bohr magneton representing free electron, we obtain

$$\gamma = 28.02468 \text{ GHz / T} \quad \text{for } {}^4\text{He}$$

which is about 660 times larger than gyromagnetic ratio for nuclear magnetic resonance.

In the case of ESR, the magnetic field causes the energy split known as *the Zeeman effect*. The resonant transition of electrons is related to the frequency corresponding with the light wave. Therefore, *magnetometers with optical pumping* are used as ESR instruments for magnetic field measurements.

All resonance magnetometers (especially free-precession devices) need certain time to determine the frequency. This poses a problem with continuous magnetic field measurement. Moreover, optical pumping magnetometers consume significant power for the lamp, which is an obstacle in the space research. Solution of these two problems can be found in the Overhauser magnetometer, which joins properties of two resonances NMR

and ESR (therefore, also known as double-resonance magnetometer).

The other drawback of practically all resonance magnetometers is that they generally do not measure the vector value of magnetic field.

4.7.2 NMR Magnetometers for Measurements of Strong Magnetic Fields

Most of NMR measuring instruments apply the theory and principle of operation proposed in 1946 by Bloch (Bloch 1946, Bloch et al. 1946). The fundamental Bloch equation describes the changes of nuclear magnetization M caused by external magnetic field B:

$$\frac{dM}{dt} = \gamma (M \times B) - i \frac{M_x}{T_2} - j \frac{M_y}{T_2} - k \frac{M_z - M_0}{T_1} \quad (4.112)$$

where

M_0 is the magnetization in equilibrium conditions in external magnetic field B_0

T_1 is the spin-lattice relaxation time, while T_2 is the spin-spin relaxation time

We can use a sample filled with proton-reach substance (e.g., water) as the sensor. Such sample is magnetized by the modulation alternating field B_{RF} perpendicular to measured magnetic field B_0. The sample is placed inside a pick-up coil connected to the resonance tank circuit. The state of resonance is detected as the increase of absorption of energy (absorption method) or as increase of magnetization when the resonance circuit is tuned.

Solving the Bloch equation, we can determine magnetization M in the sample of susceptibility $\chi_0 = M_0/B_0$ magnetized by additional oscillating magnetic field B_{RF} (Fiorillo 2004):

$$M = \sqrt{M_x^2 + M_y^2} = \frac{\mu_0 \chi_0 \omega_0 T_2}{\sqrt{1 + (\omega_0 - \omega)^2 T_2^2}} B_{RF} \quad (4.113)$$

Equation 4.113 is the so-called Lorenzian curve describing resonance as a function of frequency ω for a fixed value of B_0. Or we can determine the absorption of energy as a function of magnetic field—for a fixed frequency (Weyand 1989)

$$P(B) = \frac{\chi}{2\mu_0} \frac{T_2}{1 - \gamma^2 (B - B_0)^2 T_2^2 + 0.5 \gamma^2 B_{RF}^2 T_1 T_2} \gamma B_0^2 B_{RF}^2 \quad (4.114)$$

It is also the relation describing the resonance curve (Figure 4.175). Its characteristic parameter ΔB—*full width at half maximum* (FWHM) depends on the relaxation time

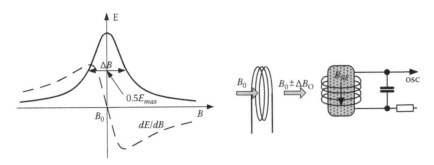

FIGURE 4.175
The NMR sensor utilizing the detection of absorption.

$$\Delta B = \frac{2}{\gamma T_2} \qquad (4.115)$$

The appropriate choice of alternating magnetic field and relaxation times is necessary to ensure that in a thermal equilibrium, the protons continuously absorb energy from alternating field at the resonance frequency. If the relaxation time is too short, the ΔB value is too broad to correctly detect the resonance. If the relaxation time is long, the resonance curve is narrow but its intensity and absorption signal are small. For pure water at room temperature, the T_2 is equal to about 2.3 s and $T_1 \approx 1.5 T_2$. But by addition of small amount of paramagnetic salts as $CuSO_4$, $NiSO_4$ or FeNO, we can decrease this time to milliseconds.

Usually, the magnetizing coil wound on the sample is connected to the LC resonance circuit that is supplied from an oscillator via a resistor. This is helpful because when absorption increases at the resonance, the voltage on the LC circuit is changing. To detect the resonance condition, the derivative dE/dB instead of resonance curve can be used (Figure 4.175)—in such solution, the zero-crossing point is easy detectable.

Figure 4.176 presents the first NMR magnetometer developed by Bloch (Bloch et al. 1946). The oscillating magnetic field B_{RF} is perpendicular to the measured

field B_0. The receiver coil is wound directly on the spherical water-filled sample perpendicular to both measured DC magnetic field and magnetizing B_{RF} field (of the frequency close to expected resonance frequency). Both coils (magnetizing and receiver one) are placed at right angle to each other to avoid the RF component induced in the receiver coil. To help the detection of resonance, the measured DC magnetic field is modulated by small 60 Hz magnetic field parallel to measured field—generated by additional coils. The state of resonance was detected for the largest second harmonic of modulated field (Figure 4.176). After tuning the oscillator to resonance state, the magnetic field was calculated from measured resonance frequency.

Recently, absorption of energy is commonly used as a method of detection of the resonance. Most of NMR magnetometers operate in field-tracking system—it means that the output frequency follows the changes of measured magnetic field due to auto-tuning (Coles 1972, Kubiak et al. 1979, Weyand 1989). To detect the resonance, additional magnetic field ΔB_0 along the measured field is often used. A typical magnetometer is presented in Figure 4.177. The measured magnetic field B_0 is modulated by triangular magnetic field. Every time when the resonance appears (twice per one cycle of swept field), the output signal of the sensor after amplification is changed using the trigger device to rectangular wave.

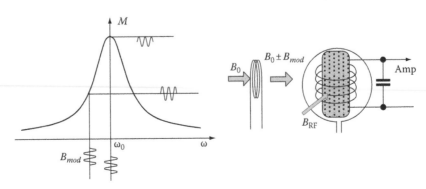

FIGURE 4.176
The NMR Bloch sensor utilizing the detection of induced voltage.

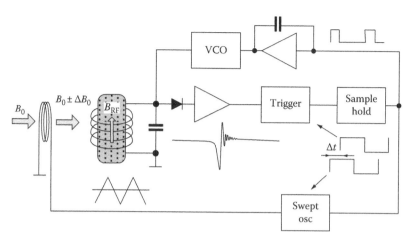

FIGURE 4.177
The principle of operation of typical NMR magnetometer.

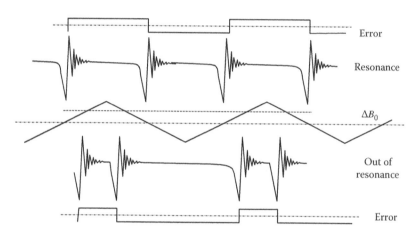

FIGURE 4.178
The output signals of the NMR sensor when the oscillator frequency is the same as the resonance frequency and when the oscillator is out of tune.

This wave is compared with the signal used to control the swept oscillator where every step corresponds with zero-crossing point. In the resonance, both signals should be the same—because it means that the resonance appears exactly for zero of triangular magnetic field (see Figure 4.178). The error voltage representing the difference between both signals is used for tuning of the frequency via voltage-controlled oscillator (VCO) (e.g., by using varicaps).

Figure 4.178 presents an example of simple creation of the error signal. In the state of resonance, the width of positive and negative pulse is the same and the average value of the signal is zero. If the oscillator is out of tune, the average value of the error signal is different from zero and can be used to control the VCO device.

The sensors (probes) of commercially available NMR magnetometers (for instance, PT2026 model of Metrolab) have dimensions $16 \times 12 \times 230$ mm. The active part has diameter only 4 mm and length 4.5 mm. Much better

spatial resolution can be obtained from an integrated chip CMOS sensor developed by Boero and coworkers (Boero et al. 2001). The sensor is presented in Figure 4.179. It consists of two planar coils, preamplifier, mixer, and amplifier. The detection coil is composed of two identical coils wrapped in opposite direction to avoid coupling with the excitation coil. Above the detection coils, solid sample of *cis*-polyisoprene of volume $1\,\text{mm}^3$ is placed. The exciting coil is supplied by 40 μs pulses from external RF oscillator. The response is analyzed with the Fourier transform by a computer. The device can measure the magnetic field in range 0.7–7 T with resolution 1 ppm.

The market-available NMR magnetometer PT2026 model of Metrolab enables measurements of the magnetic field in range 0.2–20 T (in 10 subranges) with accuracy 5 ppm and resolution 0.01 ppm. Measurements can be taken up to 60 measurements per second. In the range of 0.04–2 T, a water doped with $NiSO_4$ is used as a sample

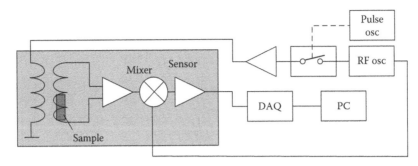

FIGURE 4.179
Block diagram of NMR magnetometer integrated in CMOS technology. (From Boero, G. et al., *Rev. Sci. Instrum.*, 72, 2764, 2001.)

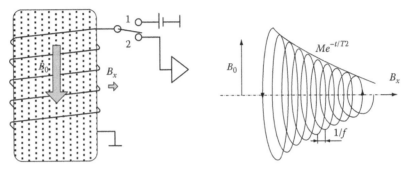

FIGURE 4.180
Principle of operation of free proton precession magnetometer.

material, for higher range the heavy water doped with $GdCl_3$. Unfortunately, as is the case of all resonance magnetometers, the sensors require homogeneity of measured field—required homogeneity is about 1200 ppm/cm (for large magnetic field about 300 ppm/cm).

4.7.3 NMR Magnetometers for Measurements of Weak Magnetic Fields

The signal of above-described NMR sensors for very weak magnetic field is covered by noise. Therefore, in this range, quite a different principle is used (Figure 4.180). In first step, the sample of much larger volume (typical 1 L) is polarized using strong DC magnetic field B_0 perpendicular to the measured field B_x. In the second step, the polarizing field is switched-off and the magnetic moment of protons begins to align parallel to the external measured field. This process is performed by decay precession movement.

The frequency of this precession movement depends on the measured magnetic field with coefficient equal to the gyromagnetic ratio. Thus, to measure the magnetic field value is sufficient to observe the periodic signal of the sensor and to measure the frequency of induced voltage.

The principle of operation of free-precession proton magnetometer seems to be rather simple. No wonder that it is possible to find in the Internet various

information about this type of magnetometers, for example, "Practical guidelines for building a magnetometer by hobbyists" by Willy Bayot.* But amateurs ready to build the proton magnetometer can find many obstacles:

- The output signal of typical sensor is very small (at the level of a few μV), often with noisy and with industrial frequency interferences. Therefore, high-quality amplifier is required.
- The time of precession is relatively short—several seconds (see Figure 2.51). Moreover, frequency of precession is not very high—Earth's magnetic field 50 μT corresponds with frequency of only 2130 Hz. Therefore, accurate measurement of the output frequency is quite difficult.
- To observe the precession, it is necessary to ensure homogeneity of measured magnetic field. Also, if the measured magnetic field is not exactly perpendicular to the excited field H_0, the precession cannot be detected.

---

* In the Internet, it is possible to find a very useful paper by J.A. Koehler, "Proton precession magnetometers." Unfortunately, address of this page is changing and the paper does not seem to be published in any printed source.

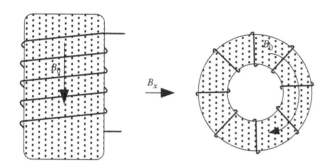

FIGURE 4.181
Two main shapes of proton precession magnetometer sensors.

- The switch-off should be sufficiently fast—the polarizing field should be zero before starting the precession. It means that it should be much shorter than the period of measured signal. Usually, it is time shorter than 50 μs. On the other hand, the coil has large inductivity and fast switching-off can create electric arching and large overvoltage (or sometimes oscillation). Therefore, amplifier should be connected with a small delay. Also, circuits with the Zener diode are usually connected to the output for prevention of the overvoltage.

A typical proton precession sensor is designed in form of a solenoid with glass core filled with proton-reach liquid—water or benzene. Usually, the same coil is used for polarization and detection. To obtain sufficient sensitivity, this sample should be physically rather large—typically with volume of 0.5 or 1 L. The solenoid sensor has several drawbacks. It is not immune to external AC EM fields and operates correctly only if the measured magnetic field is perpendicular to the coil axis. This is why Acker proposed to substitute the cylinder or a spherical shape by the toroid (Figure 4.181 right side) (Acker 1971, Primdahl et al. 2005). In such sensor, measured field is perpendicular to the polarizing field.

The output signal of the cylinder shape sensor can be determined as (Faini and Svelto 1962, Koehler 2004)

$$e \cong \frac{\chi \omega \mu_0}{\sqrt{2}} \frac{n^2 I}{b^2} v \sin \alpha \qquad (4.116)$$

while the same signal for a toroidal sample can be determined as (Acker 1971, Koehler 2004)

$$e = \frac{\chi \omega \mu_0}{2\sqrt{2}} n^2 I \left(R - \sqrt{R^2 - r^2} \right) \left(2 - \sin^2 \alpha \right) \qquad (4.117)$$

where
  n is a number of turns
  b is the length of solenoid
  v is the volume
  I is a polarizing current
  R and r are the external and internal radii of the toroid
  α is an angle between measured field and the axis of a toroid

We can see that in order to obtain large amplitude of the signal, the sample dimensions, the polarizing current, and number of turns should be large. On the other hand, because such magnetometers are often supplied by a battery, the current cannot be too large—usually it is designed only to ensure the polarizing field of about 50 mT. Similarly limited is the number of turns. The output signal is very small; therefore, resistance of the coil should be small to limit the thermal noise.

Bayot (Bayot 2008) described parameters of the following sensor: plastic 100 mL bottle with length 100 mm and diameter 60 mm filled with water. On the bottle there are 1130 turns wound with 0.8 mm wire (AWG 20)—DC resistance 8.3 Ω, inductance 57 mH. For the polarization, 12 V polarizing current 1.46 A was used. The output signal was 1.6 μV rms while SNR was 45 dB.

The block diagram of a typical proton precession magnetometer is presented in Figure 4.182. Due to small frequency of the signal, measurement of frequency is substituted by the measurement of the period. The gate is open for 2^N pulses (typically 2048) and the magnetic field is calculated as $H_x = 2^N f / C\gamma$ where f is the frequency

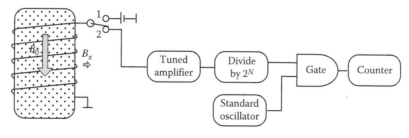

FIGURE 4.182
Two block diagram of typical proton magnetometer. (From Stuart, W.F., *Rep. Prog. Phys.*, 35, 803, 1972.)

of standard oscillator and *C* is the state of pulse counter. For such parameters, time of measurement of Earth's magnetic field was about 1 s.

In modern magnetometers, the classical frequency or period measurement is substituted by DSP, for example, by FFT calculation. The market-available magnetometer model G-856 of Geometrics exhibits the following parameters: resolution 0.1 nT, accuracy 0.5 nT, ranges 20 μT–90 μT, acceptable gradient (nonuniformity) 1 μT/m, sensor dimensions 9 × 13 cm. Another instrument GSM-19 model of GEM Systems exhibits resolution 0.01 nT, accuracy 0.2 nT, ranges 20 μT–120 μT, acceptable gradient 7 μT/m, sensor 7 × 17 cm.

An important problem of the measurement of free proton precession is the relative long time of measurement. For polarization, it is necessary to wait a time of several T_1, similarly the actual reading is also a time of several T_2. Therefore, single measurement can take about 10 s for water and by definition such measurement is discontinuous.

The solution of such drawbacks can be a spatial separation of the polarizing part and detection part with flowing water (Figure 4.183). This type of magnetometer was proposed by Sherman (Sherman 1959) and later improved by other designers (Pendlebury et al. 1979, Kim et al. 1993, Woo et al. 1997). In the systems with flowing water, the water is firstly polarized by permanent magnet and then is transmitted to the detection sensor. In such design, the water can be much better polarized and the continuous measurement of the output signal is possible.

In the system proposed by Sherman (Figure 4.184), the water is polarized by permanent magnet in the same direction as the measured field. The exciting coil is generating the oscillating field perpendicular to measured field with frequency close to expected measured precession frequency. By using additional modulating coils, it is possible to precisely detect the resonance condition (see signal in Figure 4.184).

Many other flowing-water magnetometers have been proposed, including pure free precession where water was polarized perpendicular to the measured field and the free precession was detected (Stuart 1972). On the other hand, the quality of the detected signal in flowing-water systems was so promising that such systems have been adapted to classical NMR resonance—for example, with detection of absorbed energy. Indeed although such systems seem to be rather complex and heavy to exploitation, the commercially flowing-water magnetometers are available.

The FW101 model of Virginia Scientific Instruments enables measurement of the magnetic field in the range 1.4 μT–2.16 T (in extended version FW201 0.7 T–23.4 T) with accuracy 5×10^{-6}. The flowing-water magnetometer consists of five devices: a high magnetic field polarizer, a flowing-water type probe, a synthesized frequency source (50 Hz–100 MHz with resolution 0.01 Hz), a proton magnetic polarization analyzer, and a water pump system. Beside the flexibility (one probe for an extremely broad range of fields), advantages in flowing-water system are small dimensions of the

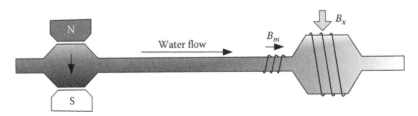

FIGURE 4.183
Proton precession magnetometer with flowing water.

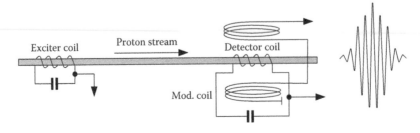

FIGURE 4.184
Proton precession magnetometer with flowing water proposed by Sherman. (From Sherman, C., *Rev. Sci. Instrum.*, 30, 568, 1959.)

sensor (2 mm × 3 mm)—thus higher field nonuniformity can be accepted.

4.7.4 Optically Pumped Sensors and Magnetometers

Relatively poor resolution of free-precession magnetometers (on the level of nT) theoretically can be improved using the electron resonance because the gyromagnetic ratio of electron resonance is almost 1000 times larger than the same ratio of proton resonance. To detect the electron resonance as the sample should be used the material with free electrons. Such material in form of various free radicals was tested with modest achievements (Duret et al. 1991). Recently free radicals are commonly used as a sample for electron resonance in Overhauser magnetometers (see Section 4.7.5) while for detection of *ESR* are *used optically pumped magnetometers*.

The optically pumped magnetometers are based on the three main physical effects:

1. Zeeman effect discovered by Zeeman (Zeeman 1897)
2. Optical pumping effect developed by Kastler (Kastler 1950)
3. Optical detection of magnetic resonance in alkali vapor magnetometers introduced by Bell and Bloom (Bell and Bloom 1957)

The Zeeman effect (described in Section 2.6.3) results in the splitting of energy level of atom under the influence of magnetic field. This energy splitting ΔE is proportional to magnetic field, as

$$\Delta E = \frac{\hbar \gamma}{\mu_0} B = \frac{\hbar}{\mu_o} f_0 \qquad (4.118)$$

and can be detected as the emission or absorption of light during excitation or decay of electrons between the sublevels. Because the energy split corresponds to the resonant frequency f_0, the largest (resonant) absorption/emission occurs when the sample is under the influence of oscillating magnetic field of resonant frequency. Thus, by changing the frequency of magnetic field, we can detect the resonance and next according to the relation (4.115), we can determine the magnetic field B. For many materials, the Zeeman splitting can be very complex resulting in many resonant frequencies, but we can select most efficient and least ambiguous frequency to measure the magnetic field.

As the sample material, the best are alkali metal vapors because such atoms have a single electron on the outer shell and the electron spins are unpaired. The first magnetometers used rubidium (^{85}Rb or ^{87}Rb) as the sample material—recently, commonly cesium (^{135}Cs) and potassium (^{39}K) are employed. Alternatively, also helium isotope ^{4}He can be used.

Unfortunately, even if we excite the electron, the transition is difficult to detect because in the thermal equilibrium, the sublevels are almost equally populated and the resonance signal is small. By using the optical pumping, we can introduce the disequilibrium of the electron population and this way we can obtain sufficiently large resonance signal.

The principle of the optical pumping is illustrated in Figure 4.185 (and described earlier in Section 2.6.3). If we deliver circularly polarized light beam of an appropriate length, we can transfer the angular momentum from resonant polarized radiation to the atom. But due to the *selection rule*, only electrons with the spin corresponding with the direction of the polarization can be transferred to the higher energy level. Thus, in the first step (Figure 4.185), only the electrons from the ground G_1 state (in our example, four electrons) are transferred to the H level. Next, these electrons decay uniformly to both ground states and under influence of the light, only two electrons are excited. After several steps, remains only electrons on the level from that the excitation is forbidden. Such state is detected as the transparency of the vapor because none of electrons can be absorbed. Thus, the optical pumping resembles pumping of the

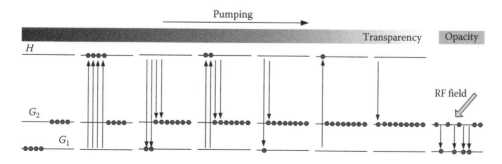

FIGURE 4.185
Two principle of optical pumping (explanation in the text). (From Lowrie, W., *Fundamentals of Geophysics*, Cambridge University Press, New York, 2007.)

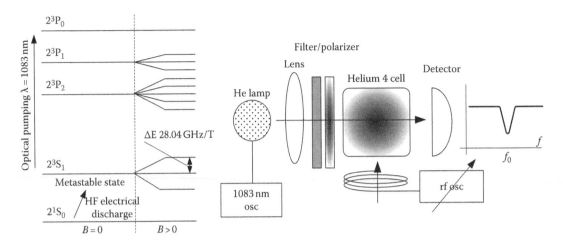

FIGURE 4.186
The helium optical pumped magnetometer.

water from the valley to a high-altitude lake (Grivet and Malna 1967).

If we now apply additional oscillating magnetic field of resonant frequency (corresponding to energy splitting) and perpendicular to the polarized light beam, we force the electron transfer between sublevels, which is indicated by the change of transparency (opacity due to the light emission). Thus, the polarized light beam has two roles—introduction of disequilibrium of the energy levels and detection of the resonance. Combination of two effects—optical pumping and resonance light emission causes that the modulated light has the smallest intensity at the resonance.

The principle of optically pumped magnetometers was developed in the second half of the last century (Bell and Bloom 1957) and is still developed and enhanced (Alexandrov 2003). Recently, the rubidium vapor is substituted by cesium (Kim and Lee 1998, Groeger et al. 2005), potassium (Beverini et al. 1998, Alexandrov 2003), or helium (Guttin et al. 1994, Gilles et al. 2001, Leger et al. 2009) and the discharge lamp as light source is replaced by the laser beam. The principle of operation of a helium optically pumped magnetometer is presented in Figure 4.186.

In the magnetometer presented in Figure 4.186, the helium (^{4}He) filled lamp is emitting light of the length $1.083\,\mu m$ corresponding with transition between 2^3S_1 and 2^3P_0 levels. This light beam is next filtered and circularly polarized. The detector measures the light intensity. The RF oscillator is tuned to the Larmor resonance frequency detected by the smallest light intensity. This frequency corresponds with measured field according to the gyromagnetic ratio, which for ^{4}He is $28.05\,GHz/T$.

Figures 4.187 and 4.188 present examples of laser-pumped magnetometers. In the first magnetometer, the output signal after amplification controls the VCO, which is automatically tuned to the resonance frequency.

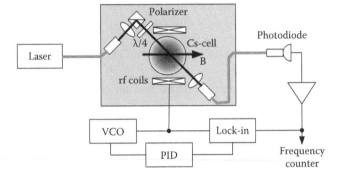

FIGURE 4.187
The laser pumped cesium magnetometer. (From Groeger, S. et al., *Sens. Actuat.*, A129, 1, 2006.)

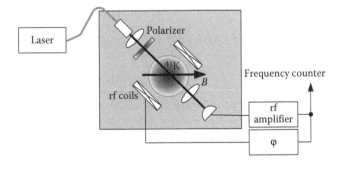

FIGURE 4.188
The laser pumped self-oscillating potassium magnetometer. (From Beverini, N. et al., *Ann. Geofis.*, 41, 427, 1998.)

This magnetometer known as *locked* is usually tuned to selected resonance line and therefore needs perfectly separated resonance frequencies. Such kind of magnetometers are commonly used for potassium magnetometers where the frequency lines are very narrow and do not overlap. On the contrary, in the second magnetometer known as *self-oscillating*, the output signal is transmitted in appropriate phase as a positive feedback to the

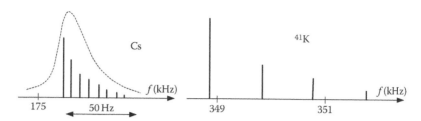

FIGURE 4.189

The spectral lines of cesium and potassium at a magnetic field 50 μT. (From Alexandrov, E.B., *Phys. Scr.*, T105, 27, 2003.)

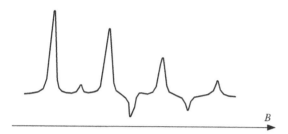

FIGURE 4.190

Experimentally determined spectral lines of potassium for fixed frequency 82 kHz and swept magnetic field in vicinity of 12 μT (only the main resonances are shown). (From Alexandrov, E.B. et al., *Laser Phys.*, 6, 244, 1996.)

RF coils and the whole system is oscillating according to the resonance frequency. Such approach is usually used for cesium magnetometer, where spectral lines are strong, but also densely spaced, very wide and overlapping. In the cesium magnetometers, the spectral lines are often lumped together—Figure 4.189.

Although the optically pumped magnetometer principally measures the scalar value of magnetic field,* it is sensitive to position of the cell with respect to the measured magnetic field. This *heading error* can manifest itself as an asymmetry of spectral lines and sometimes the resonance cannot be detected (dead zone), especially when the resonance frequencies are close to each other and overlapping. This is the main drawback of the cesium magnetometers although also in potassium magnetometers, the spectral lines cannot be precisely resolved.

Figure 4.190 presents the experimentally determined spectrum for a fixed frequency and swept magnetic field. The heading error is smallest if the polarization beam is inclined at about 45° with respect to measured magnetic field. Guttin (Guttin et al. 1994) proposed to decrease the heading error by means of small nonmagnetic piezoelectric motor for correction of the sensor position.

The helium resonance has the largest known gyromagnetic ratio equal to 28 Hz/nT. But the resolution

---

* It is also possible to design the vector version of optically pumped magnetometers.

TABLE 4.14

Properties of the Main Materials Used in Optical Magnetometers

Cesium	Potassium	4 Helium
$\gamma = 3.498572\,Hz/nT$	$\gamma = 7.00533\,Hz/nT$	$\gamma = 28.024954\,Hz/nT$
$\Delta\omega = 20\,nT$	$\Delta\omega = 0.1 - 1\,nT$	$\Delta\omega = 70\,nT$
$t = 23°C$	$t = 63°C$	–

depends also on the other factors as resonance linewidth $\Delta\omega_0$ and signal to noise S/N:

$$\Delta H_{min} = \frac{k\Delta\omega_0}{\gamma(S/N)} \qquad (4.119)$$

Table 4.14 present the properties of three main materials used in optically pumped magnetometers.

From Table 4.14 results that potassium allows achieving the best resolution and fast response. But the helium magnetometer can be an alternative for alkali metal magnetometers. Both cesium and potassium requires heating to obtain vapor, and stability of the temperature influences the accuracy. Helium is gaseous, so heating is not required, which improves stability. Scalar and vector magnetometers based on helium are developed by Polatomic Company (Slocum and Ryan 2002, 2003, Slocum et al. 2002).

In the helium magnetometer presented in Figure 4.191a, the helium is first excited by weak radiation to the metastable 2^3S_1 level (see Figure 4.186) and then is pumped by a tunable diode laser beam with $\lambda = 1082.91$ nm modulated by the Larmor frequency. This way it is not necessary to use additional RF perpendicular coils. The second harmonic from the phase demodulator is used as the control signal. It is also possible to design vector version of such magnetometer with feedback 3D Helmholtz coils (Figure 4.191b) (Slocum and Reilly 1963, Slocum and Ryan 2002, 2003). Figure 4.192 presents typical noise spectrum of the helium magnetometer.

Table 4.15 presents the performances of the commercially available optically pumped magnetometers. These magnetometers are capable of measuring the weak magnetic field with resolution better than part

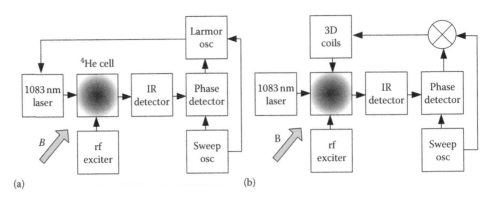

FIGURE 4.191

Laser pumped helium magnetometers principles: scalar (a) and vector (b). (From Slocum, R.E. and Ryan, L., Design and operation of miniature vector laser magnetometer, in *ESTC Conference Paper* B1P8, Polatomic Inc., Richardson, TX, 2003.)

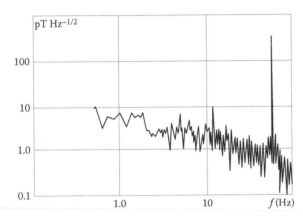

FIGURE 4.192

The noise spectrum of the helium magnetometers. (From Slocum, R.E. et al., Polatomic advances in magnetic detection, in *OCEANS Conference Paper*, Polatomic Inc., Richardson, TX, Vol. 2, 2002, pp. 945–951.)

of fT. An important advantage of optically pumped magnetometer (in comparison with free-precession principle) is that they can perform almost continuous measurements.

All three main materials used for optically pumped magnetometers have certain specific advantages. Efforts to join all these advantages in one type of magnetometer are reported. In so-called tandem magnetometers, the cell is filled with two substances, for example, mixture

of cesium and potassium (Pulz et al. 1999, Alexandrov 2003). Unfortunately, two light sources need to be employed in such instruments.

Yet another principle, which was proposed in 1974, namely nuclear free-precession magnetometer of optically polarized ^{3}He gas (Slocum and Marton 1974) was considered (Moreau et al. 1997, Gilles et al. 2001, Alexandrov 2003). Such magnetometers exhibit resolution as high as 10 pT $\sqrt{Hz}$ but every nuclear precession magnetometer suffers from relatively slow response.

4.7.5 Overhauser Magnetometers

The Overhauser type (or *dynamic nuclear polarization* [DNP]) magnetometers (see Section 2.7.5) offer performance of the best optically pumped instruments without their drawbacks: need of discharge lamp or laser, lens system and polarizer, preheating in order to obtain a vapor, heading error, large power consumption. Indeed commercially available magnetometer/gradiometer GSM-19 of GEM Systems has the following parameters: sensitivity: 22 pT $\sqrt{Hz}$, resolution: 0.01 nT, absolute accuracy: 0.1 nT, range: 20–120 µT, gradient tolerance: <10 µT/m, 5 samples/s. Moreover the Overhauser magnetometers consume extremely small power, which is advisable for the space research or design of portable instruments.

TABLE 4.15

Performance of Commercially Available Optically Pumped Magnetometers

Manufacturer	Geometrics	Scintrex	GEM Systems	Polatomic
Sample	Cesium	Cesium	Potassium	4 helium
Principle	Self-oscillating	Self-oscillating	Locked	Modulated laser
Range	20–100 µT	15–100 µT	20–100 µT	100 µT
Resolution	0.5 fT	0.6 fT	0.1 fT	5 pT
Accuracy	3 nT	2.5 nT	0.1 nT	1 nT
Heading error	<0.15 nT	<0.25 nT	<0.05 nT	

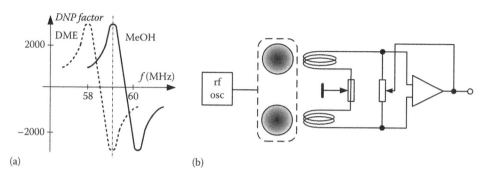

FIGURE 4.193
The DNP spectrum of the sensor for various solvents (a) and the DNP-NMR Overhauser magnetometer diagram (b). (From Kernevez, N. and Glenat H., *IEEE Trans. Magn.*, 27, 5402, 1991; Kernevez, N. et al., *IEEE Trans. Magn.*, 28, 3054, 1992.)

Overhauser magnetometers utilize two resonances—ESR and NMR. The ESR is used for polarization of NMR sensor instead of being conventionally used in NMR sensor large magnetic field. An advantage of such solution is small power and excitation in high-frequency range not disturbing NMR operation (this way, practically, continuous NMR operation is possible—5 samples/s in GSM-19 instrument). Moreover, the polarization by ESR resonance is much stronger than conventional polarization, which leads to high sensitivity and resolution.

To obtain both types of resonances, the sensor should contain the material with free electrons—usually nitroxide free radical diluted in proton-reach solvent, for example, methanol. The electrons are excited by using a high-frequency field and transfer the energy to the hydrogen nuclei. Such type of energy transfer enables to increase the polarization even by several thousand times more than in a conventional way.

It is possible to change the polarization frequency by appropriate selection of solvent. Figure 4.193a presents spectrum characteristics for two solvents—methanol (MeOH) and dimethoxymethane (DME). If two sensors with such solvents are polarized in parallel and antiparallel way, the differential sensor is obtained (Kernevez and Glenat 1991, Kernevez et al. 1992, Duret et al. 1995).

In the GEM Systems magnetometers short 90° submillisecond pulse are used for deflecting the proton magnetization to the plane of precession, after which the frequency of precession is measured. At the same time, the new deflection by Overhauser coupling is created. This way, it is possible to obtain close to continuous operation of the instrument (Hrvoic 1989).

Figure 4.193 presents the differential type DNP–NMR magnetometer (Kernevez and Glenat 1991, Kernevez et al. 1992, Duret et al. 1995). The sensor is composed of two cells with different solvents. Two symmetrical coils induce the resonance and operate as pick-up coils. The NMR signals of opposite signs are amplified by differential amplifier and the amplifier rejects the excitation voltage. Figure 4.194 presents the resonance signal and noise spectral density of this instrument.

The magnetometer developed by Kernevez and coworkers (Figures 4.193 and 4.194) exhibits following performances: field range 20–70 μT, output frequency −1 to 3 kHz, sensitivity 10 pT Hz$^{-1/2}$, power consumption 2 W, ESR excitation 68.9 MHz, SNR at Larmor frequency 57 dB.

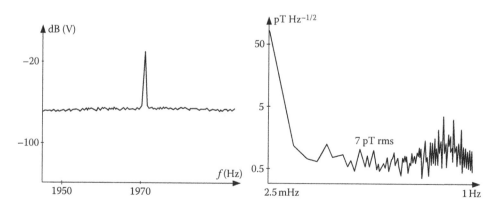

FIGURE 4.194
The resonance signal and noise spectrum of DNP-NMR magnetometer. (From Kernevez, N. and Glenat H., *IEEE Trans. Magn.*, 27, 5402, 1991; Kernevez, N. et al., *IEEE Trans. Magn.*, 28, 3054, 1992.)

4.8 Other Magnetic Sensors

4.8.1 Magnetoelastic and Wiegand-Effect Sensors

By an appropriate choice of technology and chemical composition, it is possible to create the magnetic materials relatively insensitive to stress (zero magnetostriction materials), for example, permalloy 81/19 NiFe, 6.5% SiFe, or amorphous Metglas 2705. But normally, all magnetic materials are sensitive to stress and although the magnetostriction coefficient is rather small ($\lambda_s = 0$–30×10^{-6}), often it is sufficient to design the magnetoelastic sensors. Exist also materials with exceptionally large magnetostriction, for example, Terfenol—see also Table 2.8.

In the magnetoelastic sensors, the measured value is changing the dimensions or stresses of the magnetic material and such changes are detected as the change of magnetic properties. The imagination of inventors is unlimited—such exotic values as acidity pH (Cai and Grimes 2000), humidity and fluid-flow (Grimes et al. 2000), ammonia (Cai et al. 2001), blood coagulation (Puckett et al. 2003), liquid density and viscosity (Grimes and Kouzodis 2000), ricin concentration (Shankar et al. 2005), or even Salmonella bacteria (Lakshmanan et al. 2007). In most of these sensors, the concentration of measured component is influencing the resonance frequency of vibrating magnetoelastic material (e.g., by the change of the mass of the material coating the magnetoelastic element).

Table 4.16 presents properties of the main materials used for magnetoelastic sensors. The magnetostriction coefficient λ_S is not the only important parameter. Also, other magnetic properties like saturation polarization J_s or permeability are important, as well as mechanical parameters: Vickers harness HV or Young's modulus E. Useful for assessment of the magnetoelastic material is magnetomechanical coupling coefficient k_{33} defined as the ratio of the magnetoelastic energy E_{mel} to elastic E_{el} and magnetic E_m energies $k = E_{mel} \cdot (E_{el} \cdot E_m)^{-1/2}$ (du Tremolete de Lacheisserie 2005a, b). From Table 4.16

results that the best material for sensors is amorphous Metglas 2605SC with magnetomechanical coupling close to one.

For sensors application, various magnetoelastic effects can be used:

- *Magnetostriction (Joule effect)*: The change of dimensions under influence of magnetic field
- *Villari effect (inverse magnetostriction)*: The change of permeability under influence of the change of dimensions
- *Wiedemann effect*: Twisting of the sample under influence of helical magnetic field
- *Inverse Wiedemann effect*: The change of permeability of the twisted sample
- *Matteucci effect*: The generation of the voltage at the ends of the sample under torsion
- *Wiegand effect*: The switching of polarization (with accompanying pulse) of the specially prepared wire, under influence of external magnetic field
- *Piezomagnetic effect*: Generation of a magnetic field by application of a mechanical stress to certain crystals

The magnetostriction and Wiedemann effects are usually used for actuators, the inverse effects can be obviously used for the stress, pressure, force, torsion detection. The change of permeability versus mechanical stress σ or strain ε (Villari effect) can be described as (Hinz and Voigt 1989)

$$\mu = \frac{J_s^2}{3\mu_0\lambda_s}\frac{1}{\sigma} = \frac{J_s^2}{3\mu_0\lambda_s E}\frac{1}{\varepsilon} \quad (4.120)$$

Figure 4.195 presents the change of magnetization curve of Co-based amorphous material under stress.

A large number of various force, pressure, and torque magnetoelastic sensors are reported. In 2004, ABB

TABLE 4.16

Comparison of Some Materials Used for Magnetoelastic Sensors

Material	$\lambda \times 10^{-6}$	J_s [T]	HV	E [kN/mm²]	K_{33}
Nickel	−36	0.63	75	210	0.31
SiFe3%	+9	2.0	180	150	
Permendur (49Fe,49Co,2V)	+70	2.35	200	230	0.26
Permalloy (65Fe)	+27	1.55	110	140	0.17
Metglas 2605SC (Fe81,B13,Si3.5C2)	+30	1.61	880	200	0.97
Terfenol D (Tb0.3Dy0.7Fe2)	+2000	1		50	0.75

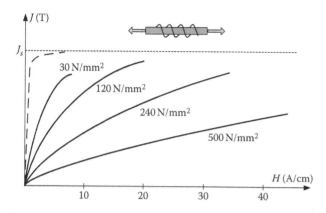

FIGURE 4.195
The variation of the magnetization curve of Co-based amorphous material under the applied mechanical stress. (From Hinz, G. and Voigt, H., Magnetoelastic sensors, in *Magnetic Sensors*, Boll E. and Overshott K.J. (Eds.), Chapter 4, VCH Verlagesellschaft, Weinheim, Germany, 1982, pp. 97–152.)

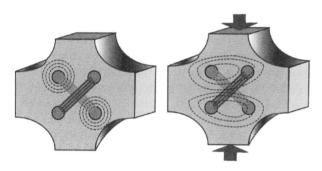

FIGURE 4.196
The pressductor magnetoelastic sensor.

Company celebrated 50 years anniversary of the patent describing one of the most recognized force magneto-elastic sensors known as *pressductor*. This sensor is presented in Figure 4.196.

In the pressductor sensor, two windings, primary and secondary, are orthogonal, and without stress the mutually coupling is zero. After application of some stress, which induces anisotropy in the secondary winding, there appears voltage proportional to the stress. This relatively simple and robust sensor enables the measurements of the force with accuracy of 0.1%, linearity of 0.1%, repeatability error less than 0.1%, and overload up to 300%.

Practically, all magnetoelastic effects have been also used in MFS (Mohri et al. 1981, Mohri and Takeuchi 1982, Chiriac et al. 2000, Kraus et al. 2008). Especially useful for stress detection are the magnetic wires prepared from amorphous material (Mohri 1984, Barandarian and Gutierrez 1997, Vazquez and Hernando 1996, Vazquez et al. 2007, Mohri et al. 2009). The most spectacular results have been reported in application of fiber-optic magnetostrictive sensors for detection of magnetic fields with 70 fT resolution (Bucholtz et al. 1989) (Figure 4.197).

In the fiber-optic magnetometer presented in Figure 4.197, the laser beam is transmitted by two glass fibers—connected to magnetostrictive and reference lines, respectively. Although the difference in length of these two fibers is very small (10^{-13} m), the interference method of measurement of the length exhibits exceptionally good resolution. The *Mach–Zender interferometer* is used to detect the phase shift between two light beams. Such type of magnetometer with the resolution comparable to SQUID magnetometers can be used to detect the undersea ships traffic investigations (Figure 4.197) (Koo et al. 1989, Bucholtz et al. 1995). The sensor with 5 cm long magnetostrictive strip exhibited resolution of 30 pT $\sqrt{\text{Hz}}$. Similar sensors with magnetostrictive ceramic jacket 1 m long and resolution 1 nT is reported (Sedlar et al. 2000).

The *Wiegand sensor* can be very useful in the detection of moving magnetic elements (Wiegand and Velinsky 1974, Rauscher and Radeloff 1989, Dlugas 1998). Its main advantage is capability to generate voltage pulses as large as several volts without a power supply and

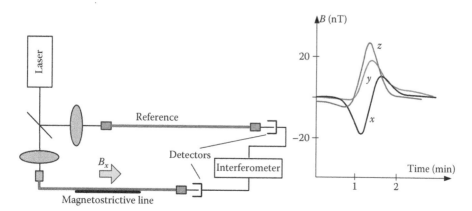

FIGURE 4.197
The fiber-optic magnetometer and the typical detected magnetic signature of a ship. (From Bucholtz, F. et al., *IEEE Trans. Magn.*, 31, 3194, 1995.)

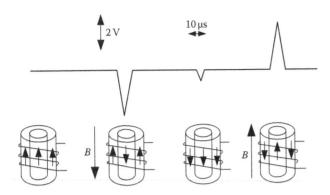

FIGURE 4.198
The operating principle of the Wiegand sensor. (From Dlugos, D., *Sensors*, 15, 32, 1998.)

independently of the rate of change of magnetic field. Wiegand patented in 1974 special composition of twisted wire that generated voltage pulses when the magnetic field exceeds the threshold value. The effect is similar to Barkhausen pulses but concentrated for one value of magnetizing field.

The Wiegand sensor is prepared in a form of a wire (e.g., 0.25 mm of diameter) prepared from special material known as Vicalloy (alloy of 40Fe-50Co-10V) (Rauscher and Radeloff 1989). After twisting the wire and appropriate heat treatment, the wire comprises a magnetically hard core (H_c about 500 A/m) and magnetically soft magnetic shell (H_c about 150 A/m). Initially, both parts of the wire are magnetized in the same direction. The switching pulse of about 10 μs appears when the core switches polarity (Figure 4.198). It is also possible to prepare the Wiegand-type sensor composed of two different coercivity NiFe and CoFe thin films (Takemura and Yamada 2006).

4.8.2 Other and New Principles

As it was demonstrated earlier, the classical old methods are continuously improved, mainly by using new materials and new technologies, for example, by the miniaturization. Also new materials and principles are reported (Jiles and Lo 2003, Lenz and Edelstein 2006). It is hard to say that these new sensors bring revolutionary changes into state of the art, but in some cases, interesting new possibilities are demonstrated.

Very promising are hybrids of various effects, as for example magnetooptical, magnetoacoustic, or magnetoelectric effects. The link between optical interference method and magnetostriction was presented in the previous section (see Figure 4.197). In the *magnetoelectric effect*, the magnetization can be induced by an electric field or reversely electric polarization can be caused by the magnetic field (Fiebig 2005). It is possible to design the composite material in which the magnetostriction is supported by piezoelectric effects (Marmelstein 1992). Figure 4.199 presents two examples of the magnetoelectric sensors.

The sensor presented in Figure 4.199a is a composite lamination of three layers—two external magnetoresistive layers prepared from Terfenol-D (also amorphous material can be used (Dong et al. 2007)) and one internal piezoelectric Pb(Zr,Ti)O$_3$ layer (Dong et al. 2003, 2004a, 2006, 2007). The coil is used to bias the sensor with high-frequency magnetic field with frequency close to resonance (instead of coil AC current can be used (Prieto et al. 2000)). The change of dimensions caused by the measured magnetic field results in output voltage of piezoelectric element. The resolution of this sensor was determined as 10^{-11} T·√Hz. Instead of ribbon the amorphous magnetostrictive disc can be used to detect two components of magnetic field (Prieto et al. 1995, Dong et al. 2004b).

In the second sensor presented in Figure 4.199b, the piezoelectric element detects vibrations of the magnetostrictive element excited by the high-frequency current in the coil. Measured low-frequency magnetic field causes that piezoelectric voltage in the input of the PSD is modulated by the measured signal. Output signal of the PSD is used as a feedback current. The measured magnetic field can be determined from the value of this current. The resolution of this sensor was determined as 10^{-7} T·√Hz (Pantinakis and Jackson 1986).

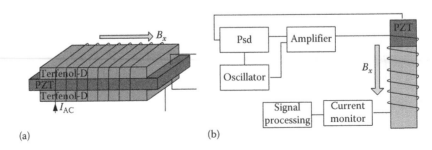

(a)   (b)

FIGURE 4.199
Two examples of magnetoelectric sensors. (From Dong, S. et al., *Appl. Phys. Lett.*, 88, 082907, 2006; Pantinakis, A. and Jackson, D.A., *Electron. Lett.*, 22, 737, 1986.)

Magnetoacoustic sensors mainly based on the ΔE effect—dependence of Young's modulus of amorphous ribbon on the magnetic field (Anderson 1982, Meeks and Hill 1983). This way the velocity of acoustic (ultrasonic) shear wave depends on the magnetic field. It is sufficient to use the amorphous ribbon and two piezoelectric elements (transmitter and receiver) to design the magnetoacoustic shear wave MFS (Squire and Gibbs 1988a, b). The sensitivity of order 2.5 nT in the frequency range 0–1.6 Hz was determined.

Also, propagation of surface acoustic wave (SAW) depends on the magnetic field. This effect was implemented in SAW MFS. As the information about measured static magnetic field, the phase velocity of SAW was used. Thus, the converter of static magnetic field into oscillator frequency was developed (Hanna 1987). As the SAW element, the garnet film of the composition $(Y_{1.5}Lu_{0.3}Sm_{0.3}Ca_{0.9})(Ge_{0.9}Fe_{4.1})O_{12}$ was used. Resulting sensitivity of the sensor was determined as 70 Hz/100 μT, which enabled to detect the magnetic field with resolution 1 μT.

Although the magnetoacoustic sensors exhibited rather modest performances, interesting application of the SAW device was proposed by Seindl and coworkers (Steindl et al. 1999, Hauser et al. 2000, 2001) (Figure 4.200). The reflectivity and acoustic emission of the SAW device depends on the termination impedances; therefore, SAW device can be used as a wireless transponder connected to the impedance sensor. The GMI sensor is used as a reflective device and the difference of delay of pulse acoustic wave between GMI load and the reference electrode can be used to wireless detect the magnetic field.

The magnetooptical Kerr effect is commonly used for measurement of the local value of flux density (see Kerr effect-based hysteresis loop tracker—Figure 2.46) and for magnetic domain imaging (discussed later). All magnetooptical effects: Faraday, Voigt, Cotton–Mouton, and Kerr (see Section 2.6.3) can be used for magnetic sensors.

The most frequently it is used the Faraday effect with garnet crystal as the active part. The polarized light beam after passing the garnet cell of the length

l changes the angle of polarization and this angle β is proportional to the measured magnetic field with the Verdet constant *v*:

$$\beta = vl \cdot B \tag{4.121}$$

As the sample materials commonly is used yttrium-iron-garnet (YIG) crystal (Zvezdin 1997). Better performances exhibit bismuth-substituted rare-earth iron garnet (BIG). Comparison of properties of both materials is presented in Table 4.17.

The magnetooptical Faraday effect is widely used in current sensors or galvanic insulators (Ning et al. 1995). The main advantage of such sensors is galvanic separation of input and output. Such current sensors are often used in high-voltage devices. Figure 4.201 presents the basic circuit of magnetooptical magnetometer. Polarized light is delivered via fiber link to the sensor, for example, garnet crystal. Under influence of magnetic field, the direction of polarization is rotated by the angle β. After Wollaston prism, we detect both main axes of the ellipse (see Figure 2.45) and the same the angle of inclination. Similar circuit can be used for Faraday effect (we detect transmitted light) as well as Kerr effect (we detect the reflected light).

Although the performances (sensitivity, resolution) of magnetooptical magnetometers are modest (Deeter et al. 1993, 1994, Itoh et al. 1995, Kamada 1996, Rochford et al. 1996), their important advantage is gigahertz response.

New possibilities appeared with the development of *micromechanical systems* (MEMS). Figure 4.202 presents two examples of such devices. In the first one, designers returned to the old principle of compass. Small balance beam is deflected under the influence of magnetic field. This deflection can be detected by capacitance method (Yang et al. 2002). The second example realizes principle of a xylophone resonator. The aluminum bar is vibrating

TABLE 4.17

Performances of Main Garnet Crystal Used for Magnetooptical Sensors

	YIG	BIG
Verdet constant for 1300 nm [min/cm T]	10.5×10^4	-800×10^4
For 1550 nm	9.2×10^4	-600×10^4
Saturated magnetooptic rotation for 1300 nm [degree/mm]	20.6	−136
For 1500 nm	18.5	−94
Thickness for 45° rotation for 1300 nm [mm]	2.14	0.33
For 1550 nm	2.43	0.48

Source: Young, D. and Pu, Y., Magnetooptics, in *The Electrical Engineering Handbook*, Chapter 57, CRC Press, Boca Raton, FL, 2000.

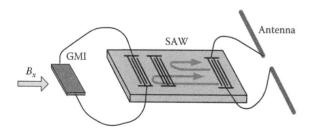

FIGURE 4.200
Giant magnetoimpedance sensor with SAW wireless transponder. (From Hauser, H. et al., *IEEE Instrum. Meas. Mag.*, 4, 28, 2001.)

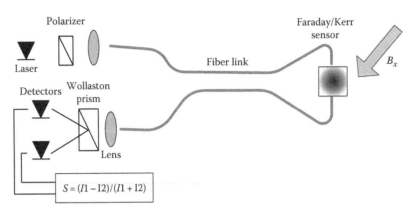

FIGURE 4.201
The basic circuit of magnetooptical magnetometer.

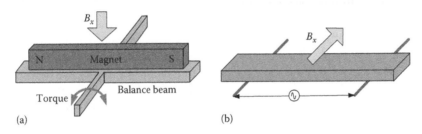

FIGURE 4.202
Two examples of the MEMS magnetic sensors. (From Yang, H.H. et al., *Sens. Actuat.*, A 97–98, 88, 2002; Givens, R.B. et al., *Appl. Phys. Lett.*, 69, 2755, 1996.)

excited by the alternating current. The damping of vibration (its amplitude) depends on the external magnetic field. This amplitude is measured using the diode laser beam reflected from one end of the bar (Givens et al. 1996).

References

Acker F.E., 1971, Calculation of the signal voltage induced in a toroidal proton precession magnetometer sensor, *IEEE Trans. Geosci. Electron.*, **9**, 98–103.

Acuna M.H., 1974, Fluxgate magnetometers for outer planets exploration, *IEEE Trans. Magn.*, **10**, 519–523.

Acuna M.H., Pellerinn C.J., 1969, A miniature two-axis fluxgate magnetometer, *IEEE Trans. Geosci. Electron.*, **7**, 252–260.

Adelerhof D.J., Kawai J., Uehara G., Kado H., 1995, High sensitivity double relaxation oscillation superconducting quantum interference devices with large transfer from flux to voltage, *Rev. Sc. Instrum.*, **66**, 2631–2637.

Adelerhof D.J., Nijstad H., Flokstra J., Rogalla H., 1994, Double relaxation oscillation SQUIDs with high flux-to-voltage transfer: Simulation and experiments, *J. Appl. Phys.*, **76**, 3875–3886.

Afanasjev J.B., 1986, *Fluxgate devices* (in Russian, Energoatomizdat).

Alexandrov E.B., 2003, Recent progress in optically pumped magnetometers, *Phys. Scr.*, **T105**, 27–30.

Alexandrov E.B., Balabas M.V., Pasgalev A.S., Vershovskii A.K., Yakobson N.N., 1996, Double resonance atomic magnetometers: From gas discharge to laser pumping, *Laser Phys.*, **6**, 244–251.

Alldredge L.R., 1951, Magnetometer, US Patent 2,856,581.

Almeida J.M., Wisniowski P., Freitas P.P., 2008, Field detection in single and double barrier MgO magnetic tunnel junction sensors, *J. Appl. Phys.*, **103**, 07E922.

Anderson P.M., 1982, Magnetomechanical coupling, ΔE effect and permeability in FeSiB and FeNiMoB alloys, *J.Appl. Phys.*, **53**, 8101–8103.

Anderson P.W., Rowell J.M., 1963, Probable observation of the Josephson superconducting tunneling effect, *Phys. Rev. Lett.*, **10**, 230–232.

Aschenbrenner H., Goubau G., 1936, Eine Anordnung zur Registrierung rascher magnetischer Störungen, *Hochfrequenztechnik und Elektroakustik*, **47**, 177–181.

Auster H.U., Lichopoj A., Rustenbach J., Bitterlicht H., Fornacon K.H., Hillemnmeier O., Krause R., Schenk H.J., Auster V., 1995, Concept and first results of a digital fluxgate magnetometer, *Meas. Sci. Technol.*, **6**, 477–481.

Baibich M.N., Broto J.M., Fert A., van Dau F.N., Petroff F., Eitene P., Creuzet G., Friederich A., Chazelas J., 1988, Giant magnetoresistance of (001)Fe/(001)Cr magnetic superlattices, *Phys. Rev. Lett.*, **61**, 2472–2475.

Barandarian J.M., Gutierrez J., 1997, Magnetoelastic sensors based on soft amorphous magnetic alloys, *Sens. Actuat. A*, **A59**, 38–42.

Barnas J., Fuss A., Camley R.E., Grünberg P., Zinn W., 1990, Novel magnetoresistance effect in layered magnetic structures: Theory and experiment, *Phys. Rev.*, **B42**, 8110–8119.

Baschirotto A., Dallago E., Ferri M., Malcovati P., Rossini A., Venchi G., 2010, A 2D microfluxgate earth magnetic field measurement systems with fully automated acquisition setup, *Measurement*, **43**, 46–53.

Baschirotto A., Dallago E., Malcovati P., Marchesi M., Venchi G., 2007, A fluxgate magnetic sensor: From PCB to micro-integrated technology, *IEEE Trans. Instrum. Meas.*, **56**, 25–31.

Bayot W., 2008, Practical guidelines for building a magnetometer by hobbyists, http://perso.infonie.be/j.g.delannoy/BAT/PPMGuidelines.htm

Beach R.S., Berkowitz A.E., 1994, Giant magnetic field dependent impedance of amorphous FeCoSiB wire, *Appl. Phys. Lett.*, **64**, 3652–3654.

Beaulieu T.J., Nepala D.A., 1975, Induced bias magnetoresistive read transducer, US Patent 3,864,751.

Bell W.E., Bloom A.L., 1957, Optical detection of magnetic resonance in alkali metal vapor, *Phys. Rev.*, **107**, 1559–1565.

Bellm H., Küchler A., Herold J., Schwab A., 1985, Rogowski-Spulen und Magnetfeldsensoren zur Messung transierter Ströme im Nanosekunbereich, *Arch. Elektr.*, **68**, 63–74.

Berus T., Oszwaldowski M., Grabowski J., 2004, High quality Hall sensors made of heavily n-InSb epitaxial films, *Sens. Actuat.*, **A116**, 75–78.

Beverini N. et al., 1998, A project for a new alkali vapor magnetometer, optically pumped by a diode laser, *Ann. Geofis.*, **41**, 427–432.

Beyer J., Drung D., Ludwig F., 1998, Low noise YBa2Cu3O7-x single layer dc superconducting quantum interference device magnetometer based on bicrystal junctions with 30 misorientation angle, *Appl. Phys. Lett.*, **72**, 203–205.

Bick M., Panaitov G., Wolters N., Zhang Y., Bousack H., Braginski A.I., Kalberkamp U., Burkhardt H., Matzander U., 1999, A HTS SQUID vector magnetometer for geophysical exploration, *IEEE Trans. Appl. Supercond.*, **9**, 3780–3785.

Bilotti A., Monreal G., Vig R., 1997, Monolithic magnetic Hall sensor using dynamic quadrature offset cancellation, *IEEE J. Solid State Circuits*, **32**, 829–836.

Black C., McMichaele I., Riggs L., 2005, Investigation of an EMI sensor for detection of large metallic objects in the presence of metallic clutter, *Proc. SPIE*, **5794**, 320–327.

Bloch F., 1946, Nuclear induction, *Phys. Rev.*, **70**, 460–474.

Bloch F., Hansen W.H., Packard M., 1946, The nuclear induction experiment, *Phys. Rev.*, **70**, 474–485.

Boero G., De Mierre M., Besse P.A., Popovic R.S., 2003, Micro-Hall devices: Performance, technologies and applications, *Sens. Actuat.*, **A106**, 314–320.

Boero G., Frounchi J., Furrer B., Besse P.A., Popovic R.S., 2001, Fully integrated probe for proton nuclear magnetic resonance magnetometry, *Rev. Sci. Instrum.*, **72**, 2764–2768.

Boll R., Overshott K.J. (Eds.), 1989, *Sensors: Magnetic Sensors*, Vol. 5, VCH, Weinheim, Germany.

Bornhöfft W., Trenkler G., 1989, Fluxgate sensors, in *Magnetic Sensors*, Chapter 5, VCH Verlagesellschaft, Weinheim, Germany.

Braginski A.I., Krause H.J., 2000, Nondestructive evaluation using high temperature SQUIDs, *Physica C*, **335**, 179–183.

Brauer P., Merayo J.M.G., Nielsen O.V., Primdahl F., Petersen J.R., 1997, Transverse field effect in fluxgate sensors, *Sens. Actuat.*, **A59**, 70–74.

Brauer P., Risbo T., Merayo J.M.G., Nielsen O.V., 2000, Fluxgate sensor for the vector magnetometer onboard the Astrid-2 satellite, *Sens. Actuat.*, **81**, 184–188.

Brown F.W., Sweer J.H., 1945, The flux ball—A test coil for point measurements of inhomogeneous magnetic field, *Rev. Sci. Instrum.*, **16**, 276–279.

Bucholtz F., Dagenais D.M., Koo K.P., 1989, High frequency fibre optic magnetometer with 70fT Hz$^{-1/2}$ resolution, *Electron. Lett.*, **25**, 1719–1720.

Bucholtz F. et al., 1995, Demonstration of a fiber optic array to three-axis magnetometers for undersea applications, *IEEE Trans. Magn.*, **31**, 3194–3196.

Bushida K., Mohri K., Kanno T., Katoh D., Kobayashi A., 1996, Amorphous wire MI micro magnetic sensor for gradient field detection, *IEEE Trans. Magn.*, **32**, 4944–4946.

Bushida K., Mohri K., Uchiyama T., 1995, Sensitive and quick response micro magnetic sensor using amorphous wire MI element Colpitts oscillator, *IEEE Trans. Magn.*, **31**, 3134–3136.

Butler W.H., 2008, Tunneling magnetoresistance from a symmetry filtering effect, *Sci. Technol. Adv. Mater.*, **9**, 014106.

Butler W.H., Zhang X.G., Schulthess T.C., 2001, Spin dependent tunneling conductance of Fe/MgO/Fe sandwiches, *Phys. Rev.*, **B63**, 054416.

Cai Q.Y., Grimes C.A., 2000, A remote query magnetoelastic pH sensor, *Sens. Actuat.*, **B71**, 112–117.

Cai Q.Y., Jain M.K., Grimes C.A., 2001, A wireless remote query ammonia sensor, *Sens. Actuat.*, **B77**, 614–619.

Campbell W.H., 1969, Induction loop antennas for geomagnetic field variation measurements, ESSA Technical Report, ERL123-ESL6.

Cantor R., Lee L.P., Teepe M., Vinetkiy V., Longo J., 1995, Low noise single layer YBa2Cu3O7-x DC SQUID magnetometers at 77K, *IEEE Trans. Appl. Supercond.*, **5**, 2927–2930.

Cantor R., Ludwig F., 2004, SQUID fabrication technology, in *The SQUID Handbook*, Clarke J. and Braginski A.I. (Eds.), Chapter 3, Wiley-VCH Verlag, Weinheim, Germany.

Casselman T.H., Hanka S.A., 1980, Calculation of the performance of a magnetoresistive permalloy magnetic field sensor, *IEEE Trans. Magn.*, **16**, 461–464.

Cavoit C., 2006, Closed loop applied to magnetic measurements in the range of 0.1–50MHz, *Rev. Sci. Instrum.*, **77**, 064703.

Cerman A., Kuna A., Ripka P., Merayo J.M.G., 2005, Digitalization of highly precise fluxgate magnetometers, *Sens. Actuat.*, **A121**, 421–429.

Cerman A., Ripka P., 2003, Towards fully digital magnetometer, *Sens. Actuat.*, **A106**, 34–37.

Chaiken A., Lubitz P., Krebs J.J., Prinz G.A., Hardorf M.Z., 1991, Low-field spin-valve magnetoresistance in Fe-Cu-Co sandwiches, *Appl. Phys. Lett.*, **59**, 240–242.

Chatraphorn S., Fleet E.F., Wellstood F.C., Knauss L.A., Eiles T.M., 2000, Scanning SQUID microscopy for integrated circuits, *Appl. Phys. Lett.*, **76**, 2304–2306.

Chattock A.P., 1887, On a magnetic potentiometer, *Phil. Mag.*, **24**, 94–96.

Chaves R.C., Freitas P.P., Ocker B., Maass W., 2007, Low frequency picotesla field detection using hybrid MgO based tunnel sensors, *Appl. Phys. Lett.*, **91**, 102504.

Chaves R.C., Freitas P.P., Ocker B., Maass W., 2008, MgO based picotesla field sensors, *J. Appl. Phys.*, **103**, 07E931.

Chesca B., Kleiner R., Koelle D., 2004, SQUID theory, in *The SQUID Handbook*, Clarke J. and Braginski A.I. (Eds.), Chapter 2, Wiley-VCH Verlag, Weinheim, Germany.

Chiesi L., Keijk P., Janossy B., Popovic R.S., 2000, CMOS planar 2D microfluxgate sensor, *Sens. Actuat.*, **82**, 174–180.

Chiriac H., Hristoforou E., Neagu M., Darie I., Hison C., 2000, Torsion and magnetic field measurements using inverse Wiedemann effect in glass-covered amorphous wires, *Sens. Actuat.*, **85**, 217–220.

Choi S.O., Kawahito S., Matsumotoi Y., Ishida M., Tadokoro Y., 1996, An integrated microfluxgate magnetic sensor, *Sens. Actuat.*, **A55**, 121–126.

Clarke J., 1980, Advances in SQUID magnetometers, *IEEE Trans. Electron. Dev.*, **ED-27**, 1896–1908.

Clarke J., 1989, Principles and applications of SQUIDs, *Proc. IEEE*, **77**, 1208–1223.

Clarke D.B., 1999, Demagnetization factors of ring cores, *IEEE Trans. Magn.*, **35**, 4440–4444.

Clarke J., Braginski A.I. (Eds.), 2004, *The SQUID Handbook*, Wiley-VCH Verlag, Weinheim, Germany.

Clem T.R., 1997, Advances in the magnetic detection and classification of sea mines and unexploded ordnance, *Naval Res. Rev.*, **49**, 29–45.

Clerc G., Gilbert D., 1964, La contre-reaction de flux appliquée aux bobines a noyau magnétique utilisee pour l'enregistrement des variations rapides du champ magnetique, *Ann. Geophys.*, **20**, 499–502.

Coehoorn R., Kools J.C.S., Rijks Th. G.S.M., Lenssen K.M.H., 1998, Giant magnetoresistance materials for read heads, *Philips J. Res.* **51**, 93–124.

Coles B.A., 1972, A fast response field tracking proton magnetometer for field programming and plotting, *J. Phys. E.*, **5**, 287–290.

Corbino O.M., 1911, Elektromagnetische Effekte die von den Verzerung herrüren, welche ein Feld und der Bahn der Ionen in Metallen hervorbringt, *Phys. Z.*, **12**, 561.

D'Antona G., Carminati E., Lazzaroni M., Ottoboni R., Svelto C., 2002, AC current measurements via digital processing of Rogowski coils signal, *IEEE Instrum. Meas. Technol. Conf., Anchorage*, 693–698.

D'Antona G., Lazzaroni M., Ottoboni R., Svelto C., 2003, AC current-to-voltage transducer for industrial application, *IEEE Instrum. Meas. Technol. Conf., Vail*, 1185–1190.

Daughton J.M., 1999, GMR applications, *J. Magn. Magn. Mater.*, **192**, 334–342.

Daughton J., 2003, Spin-dependent sensors, *Proc. IEEE*, **91**, 681–686.

Daughton J.M., Brown J., Chen E., Beech R., Pohm A., Kude W., 1994, Magnetic field sensor using GMR multilayer, *IEEE Trans. Magn.*, **30**, 4608–4610.

Daughton J.M., Chen Y.J., 1993, GMR materials for low field applications, *IEEE Trans. Magn.*, **29**, 2705–2710.

Deeter M.N., Bon S.M., Day G.W., Diercks G., Samuelson S., 1994, Novel bulk garnets magnetooptic magnetic field sensing, *IEEE Trans. Magn.*, **30**, 4464–4466.

Deeter M.N., Day G.W., Beahn T.J., Manheimer M., 1993, Magnetooptic magnetic field sensor with 1.4 pTHz$^{-1/2}$ minimum detectable field at 1 kHz, *Electron. Lett.*, **29**, 993–994.

Dehmel G., 1989, Magnetic field sensors: Induction coil (search coil) sensors, in *Sensors—A Comprehensive Survey*, Vol. 5, Chapter 6, VCH Publishers, New York, pp. 205–254.

Dezauri O., Belloy E., Gilbert S.E., Gijs M.A.M., 1999, New hybrid technology for planar fluxgate sensor fabrication, *IEEE Trans. Magn.*, **35**, 2111–2117.

Dezauri O., Belloy E., Gilbert S.E., Gijs M.A.M., 2000, Printed circuit board integrated fluxgate sensor, *Sens. Actuat.*, **81**, 200–203.

Dibbern U., 1983, Sensors based on the magnetoresistive effect, *Sens. Actuat.*, **4**, 221–227.

Dibbern U., 1984, The amount of linearization by barber-pole, *IEEE Trans. Magn.*, **20**, 954–956.

Dibbern U., 1986, Magnetic field sensors using the magnetoresistive effect, *Sens. Actuat.*, **10**, 127–140.

Dibbern U., 1989, Magnetic sensors, in *Magnetoresistive Sensors*, Chapter 9, VCH Publishers, Weinheim, Germany, pp. 342–378.

Dibbern U., Petersen A., 1983, The magnetoresistive sensor— A sensitive device for detecting magnetic field variation, *Electron. Comp. Appl.*, **5**, 148–153.

Dieny B., 1994, Giant magnetoresistance in spin-valve multilayers, *J. Magn. Magn. Mater.*, **136**, 335–359.

Dieny B., Gronovsky A., Vedyayev A., Ryzhanova N., Cowache C., Pereira L.G. 1995, Recent results on the giant magnetoresistance in magnetic multilayers (anisotropy, thermal variation and CCP-GMR), *J. Magn. Magn. Mater.*, **151**, 378–387.

Dieny B., Speriosu V.S., Gurney B.A., Parkin S.S.P., Wilhoit D.R., Roche K.P., Metin S., Peterson D.T., Nadimi S., 1991a, Spin-valve effect in soft ferromagnetic sandwiches, *J. Magn. Magn. Mater.*, **93**, 101–104.

Dieny B., Speriosu V.S., Metin S., Parkin S.S.P., Gurney B.A., Baumgart P., Wilhoit D.R., 1991b, Magnetotransport properties of magnetically spin valve structures, *J. Appl. Phys.*, **69**, 4774–4779.

Dlugos D., 1998, Wiegand effect sensors: Theory and applications, *Sensors*, **15**, 32–34.

Dong S., Li J.F., Viehland D., 2003, Ultrahigh magnetic field sensitivity in laminates of Terfenol-D and Pb(Mg1/3Nb2/3) O3-PtTiO3 crystals, *Appl. Phys. Lett.*, **83**, 2265–2267.

Dong S., Li J.F., Viehland D., 2004b, Vortex magnetic field sensor based on ring type magnetoelectric laminate, *Appl. Phys. Lett.*, **85**, 2307–2309.

Dong S., Zhai J., Li J., Viehland D., Cheng J., Cross L.E., 2004a, A strong magnetoelectric voltage gain effect in magnetostrictive-piezoelectric composite, *Appl. Phys. Lett.*, 85, 3534–3536.

Dong S., Zhai J., Xing Z., Li J., Viehland D., 2006, Small dc magnetic field response of magnetoelectric laminate composites, *Appl. Phys. Lett.*, **88**, 082907.

Dong S., Zhai J., Xing Z., Li J., Viehland D., 2007, Giant magnetoelectric effect (under a dc magnetic bias of 2 Oe) in laminate composites of FeBSiC alloy ribbons and Pb(Zn1/2,Nb2/3))3–7%PbTiO3, *Appl. Phys. Lett.*, **91**, 022915.

Drljaca P.M., Keijk P., Vincent F., Piquet D., Gueissaz F., Popovic R., 2004, Single core fully integrated CMOS microfluxgate magnetometer, *Sens. Actuat.*, **A110**, 236–241.

Drung D., 1997, Improved DC SQUID read-out electronics with low 1/f noise preamplifier, *Rev. Sci. Instrum.*, **68**, 4066–4074.

Drung D., Assmann C., Beyer J., Kirste A., Peters M., Ruede F., Schurig T., 2007, Highly sensitive and east to use SQUIDs sensors, *IEEE Trans. Appl. Supercond.*, **17**, 699–704.

Drung D., Cantor R., Peters M., Ryhänen T, Scheer H.J., Koch H., 1991, Integrated dc squid magnetometer with high dV/dB, *IEEE Trans. Magn.*, **27**, 3001–3004.

Drung D., Cantor R., Peters M., Scheer H.J., Koch H., 1990, Low noise high speed dc superconducting quantum interference device magnetometer with simplified feedback electronics, *Appl. Phys. Lett.*, **57**, 406–408.

Drung D., Mück M., 2004, SQUID electronics, in *The SQUID Handbook*, Clarke J. and Braginski A.I. (Eds.), Chapter 4, Wiley-VCH Verlag, Weinheim, Germany.

Dupas C., Beauvillain P., Chappert C., Renard J.P., Trigui F., Veillet P., Velu E., Tenard D., 1990, Very large magnetoresistance effects induced by antiparallel magnetization in two ultrathin cobalt films, *J. Appl. Phys.*, **67**, 5680–5682.

Duret D., Bonzom J., Brochier M., Frances M., Leger J.M., Odru R., Salvi C., Thomas T., 1995, Overhauser magnetometer for the Danish Oersted Satellite, *IEEE Trans. Magn.*, **31**, 3197–3199.

Duret D., Moussavi M., Beranger M., 1991, Use of high performance electron spin resonance materials for the design of scalar and vectorial magnetometers, *IEEE Trans. Magn.*, **27**, 5405–5407.

Dyal P., Gordon D.I., 1973, Lunar surface magnetometers, *IEEE Trans. Magn.*, **9**, 226–231.

Eijkel K.J.M., Fluitman J.H.J., 1990, Optimization of the response of magnetoresistive elements, *IEEE Trans. Magn.*, **26**, 311–321.

Eriksson T., Blomgren J., Winkler D., 2002, An HTS SQUID picovoltmeter used as preamplifier for Rogowski coil sensor, *Physica C*, **368**, 130–133.

Estola K.P., Malmivuo J., 1982, Air-core induction coil magnetometer design, *J. Phys. E.*, **15**, 1110–1113.

Fagaly R.L., 2001, Superconducting quantum interference devices, in *Magnetic Sensors and Magnetometers*, Chapter 8, Artech House, Boston, MA, pp. 305–347.

Fagaly R.L., 2006, Superconducting quantum interference device instruments and applications, *Rev. Sci. Instrum.*, **77**, 101101.

Faini G., Svelto O., 1962, Signal to noise consideration in a nuclear magnetometer, *Nuovo Cimento*, **23**, 55–65.

Faley M.I., Poppe U., Urban K., Paulson D.N., Starr T.N., Fagaly R.L., 2001, Low noise HTS DC SQUID flip chip magnetometers and gradiometers, *IEEE Trans. Appl. Supercond.*, 11, 1383–1386.

Fan J., Li X.P., Ripka P., 2006, Low power orthogonal fluxgate sensor with electroplated Ni80Fe20/Cu wire, *J. Appl. Phys.*, **99**, 08B311.

Feng J.S.Y., Romankiw L.T., Thompson D.A., 1977, Magnetic self-bias in the Barber pole MR structure, *IEEE Trans. Magn.*, **13**, 1466–1468.

Fenoglio M.A., 1995, Magnetic and electric fields associated with changes in high pore pressure in fault Jones: Application to the Loma Prieta ULF emissions, *J. Geophys. Res.*, **100**, 951–958.

Fiebig M., 2005, Revival of the magnetoelectric effect, *J. Phys. D.*, **38**, R123–R152.

Folberth O.G., Madelung O., Weiss H., 1954, Die elektrischen Eigenschaften von Indiumaresenid, *Z. Naturforsch.*, **9a**, 954.

Fontana R.E., 1995, Process complexity of magnetoresistive sensors—A review, *IEEE Trans. Magn.*, **31**, 2579–2583.

Forslund A., Belyayev S., Ivchenko N., Olsson G., Edberg T., Marusenkov A., 2008, Miniaturized digital fluxgate magnetometer for small spacecraft applications, *Meas. Sci. Technol.*, **19**, 015202.

Förster F., 1955, Ein Verfahren zur Messung von Magnetischen Gleichfeldern, *Zeitschr. Matallkunde*, **46**, 358–370.

Frandsen A.M.A., Holzer R.E., Smith E.J., 1969, OGO search coil magnetometer experiments, *IEEE Trans. Geosci. Electron.*, **7**, 61–74.

Fraser-Smith A.C., Bernardi A., McGill P.R., Ladd M.E., Helliwell R.A., Villard Jr. O.G., 1990., Low-frequency magnetic field measurements near the epicenter of the M_s 7.1 Loma Prieta earthquake, *Geophys. Res. Lett.*, **17**, 1465–1468.

Freitas P.P., Ferreira R., Cardoso S., Cardoso F., 2007, Magnetoresistive sensors, *J. Phys. Condens. Matter*, **19**, 16221.

Gallagher W.J. et al., 1997, Microstructured magnetic tunnel junctions, *J. Appl. Phys.*, **81**, 3741–3746.

Gatzen H.H., Andreeva E., Iswahjudi H., 2002, Eddy-current microsensors based on thin-film technology, *IEEE Trans. Magn.*, **38**, 3368–3370.

Gershenson M., 1991, High temperature superconductive fluxgate magnetometer, *IEEE Trans. Magn.*, **27**, 3055–3057.

Gilles H., Hamel J., Cheron B., 2001, Laser pumped He4 magnetometer, *Rev. Sci. Instrum.*, **72**, 2253–2260.

Gise P.E., Yarbrlough R.B., 1977, An improved cylindrical magnetometer sensor, *IEEE Trans. Magn.*, **12**, 1104–1106.

Givens R.B., Murphy J.C., Osiander R., Kistenmacher T.J., Wickenden D.K., 1996, A high sensitivity, wide dynamic range magnetometer designed on a xylophone resonator, *Appl. Phys. Lett.*, **69**, 2755–2757.

Gordon D.I., Brown R.E., 1972, Recent advances in fluxgate magnetometry, *IEEE Trans. Magn.*, **8**, 76–82.

Gordon D.I., Lundsten R.H., Chiarodo R.A., 1965, Factors affecting the sensitivity of gamma level ring core magnetometers, *IEEE Trans. Magn.*, **1**, 330–337.

Gordon D.I., Lundsten R.H., Chiarodo R.A., Helms H.H., 1968, A fluxgate sensor of high stability for low field magnetometry, *IEEE Trans. Magn.*, **4**, 397–401.

Gotffried-Gottfied R., Budde W., Jähne R., Kück H., Sauer B., Ulbricht S., Wende U., 1996, A miniaturized magnetic field sensor system consisting of a planar fluxgate sensor and a CMOS readout circuitry, *Sens. Actuat.*, **A54**, 443–447.

Grimes C.A., Kouzodis D., 2000, Remote query measurements of pressure, fluid flow velocity and humidity using magnetoelastic thick film sensors, *Sens. Actuat.*, **84**, 205–221.

Grimes C.A., Kouzodis D., Mungle C., 2000, Simultaneous measurement of liquid density and viscosity using remote query magnetoelastic sensors, *Rev. Sci Instrum.*, **71**, 3822–3824.

Grivet P.A., Malna L., 1967, Measurement of weak magnetic fields by magnetic resonance, *Adv. Electron. Electron Phys.*, **23**, 39–151.

Groeger S., Bison G., Knowles P.E., Wynands R., Weis A., 2006, Laser pumped cesium magnetometer fir high resolution medical and fundamental research, *Sens. Actuat.*, **A129**, 1–5.

Groeger S., Bison G., Weis A., 2005, Design and performance of laser pumped Cs magnetometers for the planned UCN EDM experiment at PSI, *J. Res. NIST*, **110**, 179–183.

Groszkowski J., 1937, The vibration magnetometer, *J. Sci. Instrum.*, **14**, 335–339.

Grudzinski E., Rozwalka K., 2004, A wideband magnetic field measurements in environment protection—state of the art and new trends (in Polish), *Przegl. Elektr.*, **80**, 81–88.

Grundy P.J., Pollard R.J., Tomlinson M.E., 1993, Giant magnetoresistance in Co/Cu multilayer thin films, *J. Magn. Magn. Mater.*, **126**, 516–518.

Gugoshnikov S.A., Kaplunenko O.V., Maslennikov Y.V., Snigirev O.V., 1991, Noise in relaxation oscillation driven DC SQUIDs, *IEEE Trans. Magn.*, **27**, 2439–2441.

Gugoshnikov S.A., Maslennikov Y.V., Semenov V.K., Snigirev O.V., Vasiliev A.V., 1989, Relaxation oscillation driven DC SQUIDs, *IEEE Trans. Magn.*, **25**, 1178–1181.

Guidelines on Limiting to Non-Ionizing Radiation, 1999, ICNIRP (International Commission on Non-Ionizing Radiation Protection) and WHO (World Health Organization), ISBN 3-9804789-6-3.

Guo H.Q., Kronmuller H., Dragon T., Cheng Z.H., Sheng B.G., 2001, Influence of nanocrystallization on the evolution of domain patterns and magnetoimpedance effect in FeCuNbSiB ribbons, *J. Appl. Phys.*, **89**, 514–516.

Guttin C., Leger J.M., Stoeckel F., 1994, An isotropic earth field scalar magnetometer using optically pumped helium 4, *J. Phys.*, **4**, 655–659.

Guy C., Ffytche D., 2005, *The Principles of Medical Imaging*, Imperial College Press, London.

Hagedorn A., Mende H.H., 1976, A method for inductive measurement of magnetic flux density with high geometrical resolution, *J. Phys. E*, **9**, 44–46.

Hall E.H., 1879, On the new action of the magnet on the electric currents, *Am. J. Math.*, **2**, 287–292.

Hanna S.M., 1987, Magnetic field sensors based on SAW propagation in magnetic films, *IEEE Trans. Ultrason. Ferroelectr. Freq. Control*, **34**, 191–194.

Haudek H., Metzdorf W., 1980, Magnetisierungvorgänge in Magnetowiderstands-sensoren vom Barber-pole type, *NTG-Fachberichte*, **76**, 61–67.

Hauser H., Kraus L., Ripka P., 2001, Giant magnetoimpedance sensors, *IEEE Instrum. Meas. Mag.*, **4**, 28–32.

Hauser H., Steindl R., Hausleitner C., Pohl A., Nicolics J., 2000, Wirelessly interrogable magnetic field sensor utilizing giant magnetoimpedance effect and surface acoustic wave devices, *IEEE Trans. Instrum. Meas.*, **49**, 648–652.

Hayashi K., Oguti T., Watanabe T., Zambresky L.F., 1978, Absolute sensitivity of a high-μ metal core solenoid as a magnetic sensor, *J. Geomagn. Geoelectr.*, **30**, 619–630.

Hebbert R.S., 1968, Thin film magnetoresistance magnetometer having a current path etched at an angle to the axes of magnetization, US Patent 3,405,355.

Hebbert R.S., Schwee L.J., 1966, Thin film magnetoresistance magnetometer, *Rev. Sci. Instrum.*, **37**, 1321–1323.

Heim D.E., Fontana R.E., Tsang C., Speriosu V.S., Gurney B.A., Williams M.L., 1994, Design and operation of spin valve sensors, *IEEE Trans. Magn.*, **30**, 316–321.

Heinecke W., 1978, Fluxgate magnetometer with time coded output signal of the sensor, *IEEE Trans. Instrum. Meas.*, **27**, 402–405.

von Helmolt R., Wecker J., Holzapfel B., Schulz L., Samwer K., 1993, Giant negative magnetoresistance in perovskite like $La_{2/3}Ba_{1/3}MnOx$ ferromagnetic films, *Phys. Rev. Lett.*, **71**, 2331–2333.

Heremans J., 1993, Solid state magnetic field sensors and applications, *J. Phys. D.*, **26**, 1149–1168.

Herzog R.F.K., Tischler O., 1953, Measurement of inhomogeneous magnetic fields, *Rev. Sci. Instrum.*, **24**, 1000–1001.

Hetrick R.E., 1989, A vibrating cantilever magnetic field sensor, *Sens. Actuat.*, **16**, 197–207.

Hika K., Panina L.V., Mohri K., 1996, Magnetoimpedance in sandwich film for magnetic sensor heads, *IEEE Trans. Magn.*, **32**, 4594–4596.

Hill E.W., 2000, A comparison of GMR multilayer and spin valve sensors for vector field sensing, *IEEE Trans. Magn.*, **36**, 2785–2787.

Hill E.W., Nor A.F., Birtwistle J.K., Parker M.R., 1997, A giant magnetoresistive magnetometer, *Sens. Actuat.*, **A59** (1997), 30–37.

Hinz G., Voigt H., 1989, Magnetoelastic sensors, in *Magnetic Sensors*, Boll E. and Overshott K.J. (Eds.), Chapter 4, VCH Verlagesellschaft, Weinheim, Germany, pp. 97–152.

Hirota T., Siraiwa T., Hiramoto K., Ishihara M., 1993, Development of micro-coil sensor for measuring magnetic field leakage, *Jpn. J. Appl. Phys.*, **32**, 3328–3329.

Honkura Y., 2002, Development of amorphous wire type MI sensor for automotive use, *J. Magn. Magn. Mater.*, **249**, 375–381.

Hortland A., 1992, The quantum Hall effect and resistance standards, *Metrologia*, **29**, 175–190.

Hosoe Y., Hoshino K., Tsunashima S., Uchiyama S., Imura R., 1992, Control of interlayer magnetic coupling and magnetoresistance in magnetic multilayers by insertion of very thin magnetic layer, *IEEE Trans. Magn.*, **28**, 2665–2667.

How H., Sun L., Vittoria C., 1997, Demagnetizing factor of a ring core fluxgate magnetometer, *IEEE Trans. Magn.*, **33**, 3397–3399.

Hrvoic I., 1989, Overhauser magnetometers for measurement of the Earth's magnetic field, in *Magnetic field Workshop on Magnetic Observatory Instrumentation*, Espoo, Finland.

Huang R.M., Yeh F.S., Huang R.S., 1984, Double diffusion amplification magnetic sensor, *IEEE Trans., Electron. Dev.*, **31**, 1001–1004.

Ikeda S., Hayakawa J., Lee Y.M., Matsukura F., Ohno Y., Hanyu T., Ohno H., 2007, Magnetic tunnel junction for spintronic memories and beyond, *IEEE Trans. Electron. Dev.*, **54**, 991–1002.

Inomata K., Sakakima H., 2002, Tunnel type GMR devices, in *Giant Magnetoresistance Devices*, Chapter 5, Springer, Berlin, Germany.

Ioannindis G., 1977, Identification of a ship or submarine from its magnetic signature, *IEEE Trans. Aerospace Electron. Syst.*, **13**, 327–329.

Itoh N., Minemoto H., Ishiko D., Ishizuka S., 1995, Optical magnetic field sensors with high linearity using Bi-substituted rare earth iron garnets, *IEEE Trans. Magn.*, **31**, 3191–3193.

Jaklevic R.C., Lambe J., Silver A.H., Mercereau J.E., 1964, Quantum interference effects in Josephson tunneling, *Phys. Rev. Lett.*, **12**, 159–160.

Jaklevic R.C., Lambe J., Silver A.H., Mercereau J.E., 1965, Macroscopic quantum interference in superconductors, *Phys. Rev.*, **140**, A1628–A1637.

Jenks W.G., Sadeghi S.S.H., Wikswo J.P., 1997, SQUIDs for nondestructive evaluation, *J. Phys. D.*, **30**, 293–323.

Jiles D., 1998, *Magnetism and Magnetic Materials*, Chapman & Hall, London.

Jiles D., Lo C.C.H., 2003, The role of new materials in the development of magnetic sensors and actuators, *Sens. Actuat.*, **A106**, 3–7.

Jimbo M., Kanda T., Goto S., Tsunashima S., Uchiyama S., 1993, Giant magnetoresistance in soft magnetic NiFeCo/Cu multilayers with various buffer layers, *J. Magn. Magn. Mater.*, **126**, 422–424.

Jin S., McCormack M., Tiefel T.H., Ramesh R., 1994a, Colossal magnetoresistance in La-Ca-Mn-O ferromagnetic thin films, *J. Appl. Phys.*, **76**, 6929–6933.

Jin S., Tiefel T.H., McCormack M., Fastnacht R.A., Ramesh R., Chen L.H., 1994b, Thousandfold change in resistivity in magnetoresistive La-Ca-Mn-O films, *Science*, **264**, 413–415.

Josephson B.D., 1962, Possible new effect in superconductive tunneling, *Phys. Lett.*, **1**, 251–253.

Juliera M., 1975, Tunneling between ferromagnetic films, *Phys. Lett.*, **54A**, 225–226.

Kaluza F., Grüger A., Grüer H., 2003, New and future applications of fluxgate sensors, *Sens. Actuat.*, **A106**, 48–51.

Kamada O., 1996, Magnetooptical properties of BiGdY iron garnets for optical magnetic field sensors, *J. Appl. Phys.*, **79**, 5976–5978.

Kanno T., Mohri K., Yagi T., Uchiyama T., Shen L.P., 1997, Amorphous wire MI micro sensor using C-MOS IC multivibrator, *IEEE Trans. Magn.*, **33**, 3358–3360.

Karakelian D., Klemperer S.L., Fraser-Smith A.C., Thompson G.A., 2002, Ultra-low frequency electromagnetic measurements associated with the 1998 M_w 5.1 San Juan Bautista earthquake and implications for mechanisms of electromagnetic earthquake precursors, *Tectonophysics*, **359**, 65–79.

Karp P., Duret D., 1980, Unidirectional magnetic gradiometers, *J. Appl. Phys.*, **51**, 1267–1272.

Kastler A., 1950, Quelques suggestions concernant la production optique et la detection optique d'une inegalite de population des Niveaux de quantifications spatiale des atoms. Application de l'experience de Stern et Gerlach et la resonance magnetique, *J. de Phys et le Radium*, **11**, 255.

Kawahito S., Cerman A., Aramaki K., Tadokoro Y., 2003, A weak magnetic field measurement system using microfluxgate sensors and delta-sigma interface, *IEEE Trans. Instrum. Meas.*, **52**, 103–110.

Kawahito S., Maier C., Schneider M., Zimmermann M., Baltes H., 1999, A 2D CMOS microfluxgate sensor system for digital detection of weak magnetic fields, *IEEE J. Solid-State Circ.*, **34**, 1843–1851.

Kawahito S., Satoh H., Dutoh M., Tadokoro Y., 1996, High resolution microfluxgate sensing element using closely coupled coil structures, *Sens. Actuat.*, **A54**, 612–617.

Kawajiri N., Nakabayashi M., Cai C.M., Mohri K., Uchiyama T., 1999, Highly stable MI sensor using C-MOS IC multivibrator with synchronous rectification, *IEEE Trans. Magn.*, **35**, 3667–3669.

Kejik P., Chiesi L., Janossy B., Popovic R., 2000, A new compact 2D planar fluxgate sensor with amorphous metal core, *Sens. Actuat.*, **81**, 180–183.

Kernevez N., Duret D., Moussavi M., Leger J.M., 1992, Weak field NMR and RESR spectrometers and magnetometers, *IEEE Trans. Magn.*, **28**, 3054–3058.

Kernevez N., Glenat H., 1991, Description of a high sensitivity CW scalar DNP-NMR magnetometer, *IEEE Trans. Magn.*, **27**, 5402–5404.

Khanlou A., Moses A.J., Meydan T., Beckley P., 1995, A computerized on-line power loss testing system for the steel industry, based on the Rogowski Chattock potentiometer compensation technique, *IEEE Trans. Magn.*, **31**, 3385–3387.

Kim C.G., Lee H.S., 1998, Optical pumping magnetic resonance in Cs atoms for use in precise low-field magnetometry, *Rev. Sci. Instrum.*, **69**, 4152–4155.

Kim C.G., Ryu K.S., Woo B.C., Kim C.S., 1993, Low magnetic field measurement by NMR using polarized flowing water, *IEEE Trans. Magn.*, **29**, 3198–3200.

Kitoh T., Mohri K., Uchiyama T., 1995, Asymmetrical magnetoimpedance effect in twisted amorphous wires for sensitive magnetic sensors, *IEEE Trans. Magn.*, **31**, 3137–3139.

Kleiner R., Koelle D., 2004, Basic properties of superconductivity, in *The SQUID Handbook*, Clarke J. and Braginski A.I. (Eds.), Wiley-VCH Verlag, Weinheim, Germany.

Koch H., 1989, SQUID sensors, in *Magnetic Sensors*, Boll R. and Overshott K.J. (Eds.), Chapter 10, VCH Verlag, Weinheim, Germany.

Koch H., 2001, SQUID magnetocardiography: Status and perspectives, *IEEE Trans. Appl. Supercond.*, **11**, 49–59.

Koch R.H., Rozen J.R., 2001, Low noise fluxgate magnetic field sensors using ring- and rod-core geometries, *Appl. Phys. Lett.*, **78**, 1897–1899.

Koehler J.A., 2004, Proton precession magnetometers.

Koelle D., Kleiner R., Ludwig F., Dantsker E., Clarke J., 1999, High-transition-temperature superconducting quantum interference devices, *Tev. Modern Phys.*, **71**, 631–686.

Koo K.P., Bucholtz F., Dagenais D.M., Dandridge A., 1989, A compact fiber optic magnetometer employing an amorphous metal wire transducer, *IEEE Photon. Technol. Lett.*, **1**, 464–466.

Kordic S., 1986, Integrated silicon magnetic field sensors, *Sens. Actuat.*, **10**, 347–378.

Korepanow V.E., 2003, The modern trends in space electromagnetic instrumentation, *Adv. Space. Res.*, **32**, 401–406.

Korepanov V., Berkman R., 2001, Advanced field magnetometers comparative study, *Measurement*, **29**, 137–146.

Kraus L., Malatek M., Dvorak M., 2008, Magnetic field sensor based on asymmetric inverse Wiedemann effect, *Sens. Actuat.*, **A142**, 468–473.

Krause H.J., Kreutzbruck M., 2002, Recent developments in SQUID NDE, *Physica C*, **368**, 70–79.

Krismanic G., Pfützner H., Baumgartinger N., 2000, A handheld sensor for analyses of local distribution of magnetic fields and losses, *J. Magn. Magn. Mater.*, **215–216**, 720–722.

Kubiak J., Ostafin M., Klenitz G., 1979, A new field tracking NMR magnetometer system, *J. Phys. E*, **12**, 640–643.

Kubik J., Pavel L., Ripka P., Kaspar P., 2007, Low power printed circuit board fluxgate sensor, *IEEE Sensors J.*, **7**, 179–183.

Kuijk K.E., 1977, Magnetoresistive magnetic head, US Patent 4,052,748.

Kuijk K.E., van Gestel W.J., Gorter F.W., 1975, The Barber pole, a linear magnetoresistive heads, *IEEE Trans. Magn.*, **11**, 1215–1217.

Kuperman V., 2000, *Magnetic Resonance Imaging*, Academic Press, San Diego, CA.

Lakshmanan R.S., Guntupali R., Hu J., Petrenko V.A., Barbaree J.M., Chin B.A., 2007, Detection of *Salmonella typhimurium* in fat using a phage immobilized magnetoelastic sensor, *Sens. Actuat.*, **B126**, 544–550.

Leal J.L., Kryder M.H., 1998, Spin valves exchange biased by Co/Ru/Co synthetic antiferromagnets, *J. Appl. Phys.*, **83**, 3720–3723.

Lee Y.M., Hayakawa J., Ikeda S., Matsukura F., Ohno H., 2007, Effect of electrode composition on the tunnel magnetoresistance of pseudo spin valve magnetic tunnel junction with MgO tunnel barrier, *Appl. Phys. Lett.*, **90**, 212507.

Leger J.M., Bertrand F., Jager T., Le Prado M., Fratter I., Lalaurie J.C., 2009, Swarm absolute scalar and vector magnetometer based on helium 4 optical pumping, *Proc. Chem.*, **1**, 634–637.

Lenz J., 1990, A review of magnetic sensors, *Proc. IEEE*, **78**, 973–989.

Lenz J., Edelstein A.S., 2006, Magnetic sensors and their applications, *IEEE Sensors J.*, **6**, 631–649.

Liakopoulos T.M., Ahn C.H., 1999, A microfluxgate magnetic sensor using micromachined planar solenoid coils, *Sens. Actuat.*, **77**, 66–72.

Lin T., Tsang C., Fontana R.E., Howard J.K., 1995, Exchange-coupled NiFe/FeMn, NiFe/NiMN and NiO/NiFe films for stabilization of magnetoresistive sensors, *IEEE Trans. Magn.*, **31**, 2585–2590.

Linke S., 1992, Technologie magnetoresistiver Sensoren, Magnetoresistive Sensoren, *Proc. Symp. Dortmund*, 10–28.

Lippmann H.J., Kuhrt F., 1958, Der Geometrieinfluss auf den Widerstandsänderung von InSb, *Z. Naturforsch.*, **13a**, 462–474.

Loisos G., Moses A.J., 2001, Critical evaluation and limitations of localized flux density measurements in electrical steels, *IEEE Trans. Magn.*, **37**, 2755–2757.

Lowrie W., 2007, *Fundamentals of Geophysics*, Cambridge University Press, New York.

Macintyre S.A., 1980, A portable low noise low frequency three-axis search coil magnetometer, *IEEE Trans. Magn.*, **16**, 761–763.

Mahdi A.E., Panina L., Mapps D., 2003, Some new horizonts in magnetic sensing: High Tc SQUIDs, GMR and GMI materials, *Sens. Actuat.*, **A105**, 271–285.

Manoharan S.S., Vasanthacharya N.Y., Hegde M.S., Satyalaksmi K.M., Prasad V., Subramanyam S.V., 1994, Ferromagnetic $La_{0.6}Pb_{0.4}MnO_3$ thin films with giant magnetoresistance at 300 K, *J. Appl. Phys.*, **76**, 3923–3925.

Marmelstein M.D., 1992, A magnetoelastic metallic glass low-frequency magnetometer, *IEEE Trans. Magn.*, **28**, 36–56.

Marshall S.V., 1967, An analytical model for the fluxgate magnetometer, *IEEE Trans. Magn.*, **3**, 459–463.

Mathin J., Umerski A., 2001, Theory of tunneling magnetoresistance of an epitaxial Fe/MgO/Fe (001) junction, *Phys. Rev.*, **B63**, 220403.

McCormack M., Jin S., Tiefel T.H., Fleming R.M., Phillips J.M., Ramesh R., 1994, Very large magnetoresistance in perovskite-like La-Ca-Mn-O thin film, *Appl. Phys. Lett.*, **64**, 3045–3047.

McCumber D.E., 1968, Effect of ac impedance on dc voltage current characteristics of superconductor weak link junctions, *J. Appl. Phys.*, **39**, 3113–3118.

McGuiere T.R., Potter R.I., 1975, Anisotropic magnetoresistance in ferromagnetic 3d alloys, *IEEE Trans. Magn.*, **11**, 1018–1038.

Meeks S.W., Hill J.C., 1983, Piezomagnetic and elastic properties of metallic glass alloys Fe87Co18B14Si1 and Fe81B13.5Si3.5C2, *J. Appl. Phys.*, **54**, 6584–6593.

Meiklejohn W.H., Bean C.P., 1956, New magnetic anisotropy, *Phys. Rev.*, **102**, 1413–1420.

Metzdorf W., Boehner M., Haudek H., 1980, Möglichkeiten and Grenzen magnetoresistiver Leseköpfe, *NTG-Fachberichte*, **76**, 69–75.

Metzdorf W., Boehner M., Haudek H., 1982, The design of magnetoresistive multitrack read heads for magnetic tapes, *IEEE Trans. Magn.*, **18**, 763–768.

Micheel H.J., 1987, Induktionsspulen mit induktiver Gegenkopplung als hochauflösende Magnetfeldsonden, *NTZ Archiv*, **9**, 97–102.

Mittleman D.M., Cunningham J., Nuss M.C., Geva M., 1997, Noncontact semiconductor wafer characterization with the terahertz Hall effect, *Appl. Phys. Lett.*, **71**, 16–18.

Miyazaki T., Tezuka N., 1995a, Giant magnetic tunneling effect in Ge/Al_2O_3/Fe junction, *J. Magn. Magn. Mater.*, **139**, L231–234.

Miyazaki T., Tezeuka N., 1995b, Spin polarized tunneling in ferromagnet/insulator/ferromagnet junctions, *J. Magn. Magn. Mater.*, **151**, 403–410.

Mohd Ali B.B., Moses A.J., 1989, A grain detection system for grain oriented electrical steels, *IEEE Trans. Magn.*, **25**, 4421–4426.

Mohri K., 1984, Review of recent advances in the field of amorphous sensors and transducers, *IEEE Trans. Magn.*, **20**, 942–947.

Mohri K., Humphrey F.B., Panina L.V., Honkura Y., Yamasaki J., Uchiyama T., Hirami M., 2009, Advances of amorphous wire magnetics over 27 years, *Phys. Stat. Sol.*, **A206**, 601–607.

Mohri K., Kawashima K., Kohzawa T., Yoshida H., 1993, Magnetoinductive element, *IEEE Trans. Magn.*, **29**, 1245–1248.

Mohri K., Kohzawa T., Kawashima K., Yoshida H., Panina L.V., 1992, Magnetoinductive effect in amorphous wires, *IEEE Trans. Magn.*, **28**, 3150–3152.

Mohri K., Takeuchi S., 1982, Sensitive bistable magnetic sensors using twisted amorphous magnetostrictive ribbons due to Matteucci effect, *J. Appl. Phys.*, **53**, 8386–8388.

Mohri K., Takeuchi S., Fujimoto T., 1981, Sensitive magnetic sensors using amorphous Wiegand-type ribbons, *IEEE Trans. Magn.*, **17**, 3370–3372.

Mohri K., Uchiyama T., Panina L.V., 1997, Recent advances in micro magnetic sensors and sensing application, *Sens. Actuat.*, **A59**, 1–8.

Mohri K., Uchiyama T., Shen L.P., Cai C.M., Panina L.V., 2002a, Amorphous wire and CMOS IC-based sensitive micromagnetic sensor for intelligent measurements and control, *J. Magn. Magn. Mater.*, **249**, 351–356.

Mohri K., Uchiyama T., Shen L.P., Cai C.M., Panina L.V., Honkura Y., Yamamoto M., 2002b, Amorphous wire and CMOS IC-based sensitive micromagnetic sensor utilizing magnetoimpedance and stress-impedance effects, *IEEE Trans. Magn.*, **38**, 3063–3068.

Moldovanu A., Chiriac H., Macoviciuc M, Diaconu E., Ioan C., Moldovanu E., Tomut M., 1997, Functional study of fluxgate sensors with amorphous magnetic materials core, *Sens. Actuat.*, **A59**, 105–108.

Moodera J.S., Kinder L.R., Wong T.M., Meservery R., 1995, Large magnetoresistance at room temperature in ferromagnetic thin film tunnel junctions, *Phys. Rev. Lett.*, **74**, 3273–3276.

Moodera J.S., Mathon G., 1999, Spin polarized tunnelling in ferromagnetic junctions, *J. Magn. Magn. Mater.*, **200**, 248–273.

Moodera J.S., Nassar J., Mathon G., 1999, Spin tunnelling ferromagnetic junctions, *Annu. Rev. Mater. Sci.*, **29**, 381–432.

Moore G.E., Turner P.A., Thai K.L., 1972, Current density limitations in permalloy magnetic detectors, *AIP Conf. Proc.*, **10**, 217–221.

Moreau O., Cheron B., Gilles H., Hamel H., Noel E., 1997, Magnetometre a 3He pompe par diode laser, *J. Phys. III*, **7**, 99–115.

Morikawa T., Nishibe Y., Yamadera H., Nonomura Y., Takeuchi M., Sakata J., 1996, Enhancement of giant magnetoimpedance in layered film by insulator separation, *IEEE Trans. Magn.*, **32**, 4965–4967.

Mück M., Heiden C., 1989, Simple dc SQUID system based on frequency modulated relaxation oscillator, *IEEE Trans. Magn.*, **25**, 1151–1153.

Mück M., Rogalla H., Heiden C., 1988, A frequency modulated read out system for dc SQUIDS, *Appl. Phys.*, **A47**, 285–289.

Mukhopadhyay S.C, Yamada S., Iwahara M., 2002, Experimental determination of optimum coil pitch for a planar mesh-type micromagnetic sensor, *IEEE Trans. Magn.*, **38**, 3380–3382.

Murgatroyd P.N., 1992, Making and using Rogowski coil, in *ICWA Conference*, Cincinnati, OH, pp. 267–274.

Murgatroyd P., 1996, Progress with Rogowski coils, in *EMCWA Conference*, Chicago, IL, pp. 369–374.

Murgatroyd P.N., Chu A.K.Y., Richardson G.K., West D., Yearley G.A., Spencer A.J., 1991, Making Rogowski coils, *Meas. Sci. Technol.*, **2**, 1218–1219.

Murgatroyd P.N., Woodland D.N., 1994, Geometrical properties of Rogowski sensors, in *IEE Colloquium on Low Frequency Power Measurement and Analysis*, London, pp. 901–910.

Murray I.B., McAuly A.D., 2004, Magnetic detection and localization using multichannel Levinson-Durbin algorithm, *Proc. SPIE*, **5429**, 561–566.

Musmann G., Afanassiev Y., 2010, *Fluxgate Magnetometers for Space Research*, Books on Demand GmbH.

Nafalski A., Moses A.J., Meydan T., Abousetta M.M., 1989, Loss measurements on amorphous materials using a field-compensated single strip tester, *IEEE Trans. Magn.*, **25**, 4287–4291.

Nakata T., Ishihara Y., Nakaji M., Todaka T., 2000, Comparison between the H-coil method and the magnetizing current method for the single sheet tester, *J. Magn. Magn. Mater.*, **215–216**, 607–610.

Nakata T., Kawase Y., Nakano M., 1987, Improvement of measuring accuracy of magnetic field strength in single sheet testers by using two H-coils, *IEEE Trans. Magn.*, **23**, 2596–2598.

Narod B.B., 1985, An evaluation of noise performance of Fe, Co, Si and B amorphous alloys in ring-core fluxgate magnetometers, *Can. J. Phys.*, **63**, 1468–1472.

Nassisi V., Luches A., 1979, Rogowski coils: Theory and experimental results, *Rev. Sci. Instrum.*, **50**, 900–902.

Ness N.F., 1970, Magnetometers for space research, *Space Sci. Rev.*, **11**, 459–554.

Nicholson P.I., So M.H., Meydan T., Moses A.J., 1966, Nondestructive surface inspection system for steel and other ferromagnetic materials using magnetoresistive sensors, *J. Magn. Magn. Mater.*, **160**, 162–164.

Nie H.B., Pakhomov A.B., Yan X, Zhang X.X., Knobel M., 1999, Giant magnetoimpedance in crystalline mumetal, *Solid State Commun.*, **112**, 285–289.

Nielsen O.V., Brauer P., Primdahl F., Risbo T., Jorgensen J.L., Boe C., Deyerler M., Bauereisen S., 1997, A high-precision triaxial fluxgate sensor for space applications: Layout and choice of material, *Sens. Actuat.*, **A59**, 168–176.

Nielsen O.V., Petersen J.R., Primdahl F., Brauer P., Hernando B., Fernandes A., Merayo J.M., Ripka P., 1995, Development, construction and analysis of the Oested fluxgate magnetometer, *Meas. Sci. Technol.*, **6**, 1099–1115.

Ning Y.N., Wang Z.P., Palmer A.W., Gratten K.T.V., Jackson D.A., 1995, Recent progress in optical current sensing techniques, *Rev. Sci. Instrum.*, **66**, 3097–3111.

O'Brien H., Brown P., Beek T., Carr C., Cupido E., Oddy T., 2007, A radiation tolerant digital fluxgate magnetometer, *Meas. Sci. Technol.*, **18**, 3645–3650.

Oszwaldowski M., 1998, Hall sensors based on heavily doped n-InSb thin films, *Sens. Actuat.*, **A68**, 234–237.

Panina L.V., Mohri K., 1994, Magnetoimpedance effect in amorphous wires, *Appl. Phys. Lett.*, **65**, 1189–1191.

Panina L.V., Mohri K., Bushida K., Noda M., 1994, Giant magnetoimpedance and magnetoinductive effects in amorphous alloys, *J. Appl. Phys.*, **76**, 6198–6203.

Panina L.V., Mohri K., Makhnovskiy D.P., 1999, Mechanism of asymmetrical magnetoimpedance in amorphous wires, *J. Appl. Phys.*, **85**, 5444–5446.

Panina L.V., Mohri K., Uchiyama T., Noda M., 1995, Giant magnetoimpedance in Co-rich amorphous wires and films, *IEEE Trans. Magn.*, **31**, 1249–1260.

Pant B.B., 1996a, Design tradeoff for high sensitivity magnetoresistive transducers, *Electrochem. Soc. Proc.*, **95**, 62–76.

Pant B.B., 1996b, Effect of interstrip gap on the sensitivity of high sensitivity magnetoresistive transducer, *J. Appl. Phys.*, **79**, 6123–6125.

Pant B.B., Krahn D.R., 1991, High sensitivity magnetoresistive transducers, *J. Appl. Phys.*, **69**, 5936–5938.

Pantinakis A., Jackson D.A., 1986, High sensitivity low frequency magnetometer using magnetostrictive primary sensing and piezoelectric signal recovery, *Electron. Lett.*, **22**, 737–738.

Parkin S.S.P., 1992, Dramatic enhancement of interlayer exchange coupling and giant magnetoresistance in Ni81Fe19/Cu multilayers by addition of thin Co interface layer, *Appl. Phys. Lett.*, **61**, 1358–1360.

Parkin S.S.P., 1994, Giant magnetoresistance and oscillatory interlayer coupling in polycrystalline transition metal multilayers, in *Ultrathin Magnetic Structures II*, Springer Verlag, Berlin, Germany, pp. 148–186.

Parkin S.S.P., Bhadra R., Roche K.P., 1991, Oscillatory magnetic exchange coupling through thin copper layer, *Phys. Rev. Lett.*, **66**, 2152–2155.

Parkin S.S.P., Kaiser C., Panchula A., Rice P.M., Hughes B., Samant M., Yang S.H., 2004, Giant tunneling magnetoresistance at room temperature with MgO (100) tunnel barriers, *Nat. Mater.*, **3**, 862–867.

Parkin S.S.P. et al., 1999, Exchange biased magnetic tunnel junctions and application to nonvolatile magnetic random access memory, *J. Appl. Phys.*, **85**, 5828–5833.

Pedersen E.B., Primdahl F., Petersen J.R., Merayo J.M.G., Brauer P., Nielsen O.V., 1999, Digital fluxgate magnetometer for the Astrid-2 satellite, *Meas. Sci. Technol.*, **10**, N124–N129.

Pellinen D.G., di Capua M.S., Sampayan S.E., Gerbracht H., Wang M., 1980, Rogowski coil for measuring fast, high-level pulsed currents, *Rev. Sci. Instrum.*, **51**, 1535–1540.

Pendlebury J.M., Smith K., Unsworth P., Greene G.L., Mampe W., 1979, Precision field averaging NMR magnetometer for low and high fields using flowing water, *Rev. Sci. Instrum.*, **50**, 535–540.

Peters T.J., 1986, Automobile navigation using a magnetic fluxgate compass, *IEEE Trans. Veh. Technol.*, **35**, 41–47.

Pettinga J.A., Siersema J., 1983, A polyphase 500 kA current measuring system with Rogowski coils, *IEEE Proc. B*, **130**, 360–363.

Pfützner H., Krismanic G., 2004, The needle method for induction tests—Sources of error, *IEEE Trans. Magn.*, **40**, 1610–1616.

Pfützner H., Schönhuber P., 1991, On the problem of the field detection for single sheet testers, *IEEE Trans. Magn.*, **27**, 778–785.

Phan M.H., Peng H.X., 2008, Giant magnetoimpedance materials: Fundamentals and applications, *Prog. Mater. Sci.*, **53**, 323–420.

Piil-Henriksen J., Merayo J.M., Nielsen O.V., Petersen H., Raagaard Petersen J., Primdahl F., 1996, Digital detection and feedback fluxgate magnetometer, *Meas. Sci. Technol.*, **7**, 897–903.

Pirota K.R., Kraus L., Chiriac H., Knobel M., 2000, Magnetic properties and giant magnetoimpedance in a CoFeSiB glass-covered microwire, *J. Magn. Magn. Mater.*, **221**, L243–L247.

Pizella V., Della Pena S., Del Gratta C., Romani G.L., 2001, SQUID systems for biomagnetic imaging, *Supercond. Sci. Technol.*, **14**, R79–R114.

Popovic R.S., 2004, *Hall Effect Devices*, IOP Publishing, Boston, MA.

Popovic R.S., Schott C., Shibasaki I., Biard J.R., Foster R.B., 2001, Hall effect magnetic sensors, in *Magnetic Sensors and Magnetometers*, Ripka P. (Ed.), Chapter 5, Artech House, Boston, MA.

Prance R.J., Clark T.D., Prance H., 2000, Ultra low noise induction magnetometer for variable temperature operation, *Sens. Actuat.*, **85**, 361–364.

Prance R.J., Clark T.D., Prance H., 2003, Compact room-temperature induction magnetometer with superconducting quantum interference level field sensitivity, *Rev. Sci. Instrum.*, **74**, 3735–3739.

Prance R.J., Clark T.D., Prance H., 2006, Room temperature induction magnetometers, in *Encyclopedia of Sensors*, Vol. 10, Grimes C.A. and Dickey E.C. (Eds.), American Scientific Publishers.

Prieto J.L., Aroca C., Lopez E., Sanchez M.C., Sanchez P., 1995, New type of two-axis magnetometer, *Electron. Lett.*, **31**, 1072–1073.

Prieto J.L., Aroca C., Lopez E., Sanchez M.C., Sanchez P., 2000, Magnetostrictive-piezoelectric magnetic sensor with current excitation, *J. Magn. Magn. Mater.*, **215–216**, 756–758.

Primdahl F., 1970, The fluxgate mechanism. Part 1: The gating curves of parallel and orthogonal fluxgates, *IEEE Trans. Magn.*, **6**, 376–383.

Primdahl F., 1979, The fluxgate magnetometer, *J. Phys. E.*, **12**, 241–253.

Primdahl F., Hernando B., Nielsen O.V., Petersen J.R., 1989, Demagnetizing factor and noise in the fluxgate ring-core sensor, *J. Phys. E.*, **22**, 1004–1008.

Primdahl F., Hernando B., Petersen J.R., Nielsen O.V., 1994, Digital detection of the flux-gate sensor output signal, *Meas. Sci. Technol.*, **5**, 359–362.

Primdahl F., Jensen P.A., 1982, Compact spherical coil for fluxgate magnetometer vector feedback, *J. Phys. E.*, **15**, 221–226.

Primdahl F., Merayo J.M.G., Brauer P., Laursen I., Risbo T., 2005, Internal field of homogeneously magnetized toroid sensor for proton free precession magnetometer, *Meas. Sci. Technol.*, **16**, 590–593.

Puckett L.G., Barrett G., Kouzoudis D., Grimes C., Bachas L.G., 2003, Monitoring of blood coagulation with magneto-elastic sensors, *Biosens. Bioelectron.*, **18**, 675–681.

Pulz E., Jäckel K.H., Linthe H.J., 1999, A new optically pumped tandem magnetometer: Principles and experiences, *Meas. Sci. Technol.*, **10**, 1025–1031.

Qian Z., Daughton J., Wang D., Tonadra M., 2003, Magnetic design and fabrication of linear spin valve sensors, *IEEE Trans. Magn.*, **39**, 3322–3324.

Qian Z., Wang D., Daughton J., Tonadra M., Nordman C., Popple A., 2004, Linear spin valve bridge sensing device, *IEEE Trans. Magn.*, **40**, 2643–2645.

Ramboz J.D., 1996, Machinable Rogowski coil, design and calibration, *IEEE Trans. Instrum. Meas.*, **45**, 511–515.

Ramsden E., 2006, *Hall Effect Sensors*, Newnes.

Rauscher G., Radeloff C., 1989, Wiegand and pulse-wire sensors, in *Magnetic Sensors*, Boll R. and Overshott K.J. (Eds.), Chapter 8, VCH Verlaggesellschaft, Weinheim, Germany, pp. 315–339.

Richter W., 1979, Induction magnetometer for biomagnetic fields, *Exp. Technik Phys.*, **27**, 235–243.

Riggs L.S., Cash S., Bell T., 2002, Progress toward in electromagnetic induction mine discrimination system, *Proc. SPIE*, **4742**, 736–745.

Rijks Th.G.S.M., Coehoorn R., Daenmen J.T.F., de Jonge W.J.M., 1994b, Interplay between exchange biasing and interlayer exchange coupling in $Ni_{80}Fe_{20}/Cu/Ni_{80}Fe_{20}/Fe_{50}Mn_{50}$ layered systems, *J. Appl. Phys.*, **76**, 1092–1099.

Rijks Th.G.S.M., de Jonge W.J.M., Folkerts W., Kools J.C.S., Coehoorn R., 1994a, Magnetoresistance in $Ni_{80}Fe_{20}/Cu/Ni_{80}Fe_{20}/Fe_{50}Mn_{50}$ spin valves with low coercivity and ultrahigh sensitivity, *Appl. Phys. Lett.*, **65**, 916–918.

Ripka P., 1992, Review of fluxgate sensors, *Sens. Actuat.*, **A33**, 129–141.

Ripka P., 1996a, Alternating current-excited magnetoresistive sensor, *J. Appl. Phys.*, **79**, 5211–5213.

Ripka P., 1996b, Noise and stability of magnetic sensors, *J. Magn. Magn. Mater.*, **157–158**, 424–427.

Ripka P., 2000a, New directions in fluxgate sensors, *J. Magn. Magn. Mater.*, **215–216**, 735–739.

Ripka P., 2000b, Race-track fluxgate with adjustable feedthrough, *Sens. Actuat.*, **85**, 227–231.

Ripka P. (Ed.), 2001, *Magnetic Sensors and Magnetometers*, Artech House, Boston, MA.

Ripka P., 2003, Advances in fluxgate sensors, *Sens. Actuat.*, **A106**, 8–14.

Ripka P., Billingsley S.W., 2000, Crossfield effect at fluxgate, *Sens. Actuat.*, **81**, 176–179.

Ripka P., Janosek M., Butta M., 2009, Crossfield sensitivity in AMR sensors, *IEEE Trans. Magn.*, **45**, 4514–4517.

Ripka P., Kaspar P., 1998, Portable fluxgate magnetometer, *Sens. Actuat.*, **A68**, 286–289.

Ripka P., Kawahito S., Choi S.O., Tipek A., Ishida M., 2001, Microfluxgate sensor with closed core, *Sens. Actuat.*, **A91**, 65–69.

Ripka P., Primdahl F., Nielsen O.V., Petersen J.R., Ranta A., 1995, AC magnetic field measurement using the fluxgate, *Sens. Actuat.*, **A46–47**, 307–311.

Rochford K.B., Rose A.H., Day G.W., 1996, Magnetooptic sensors based on iron garnets, *IEEE Trans. Magn.*, **32**, 4113–4117.

Rogowski W., Steinhaus W., 1912, Die Messung der magnetischen Spannung, *Arch. für Elektrotechnik*, **1**, 141–150.

Romani G.L., Williamson S.J., Kaufman L., 1982, Biomagnetic instrumentation, *Rev. Sci. Instrum.*, 53, 1815–1845.

Rowell J.W., 1963, Magnetic field dependence of the Josephson tunnel current, *Phys. Rev. Lett.*, 11, 200–202.

Ryhänen T., Seppä H., 1989, SQUID magnetometers for low-frequency applications, *J. Low Temp. Phys.*, **76**, 287–386.

Sadler D.J., Ahn C.H., 2001, On-chip eddy current sensor for proximity sensing and crack detection, *Sens. Actuat.*, **A91**, 340–345.

Saito T., 1980, Fluxgate magnetometer with a 0.5 m length two core sensor, *Sci. Rep. Tohoku. Univ.*, **27**, 85–93.

Sanchez M.L., Kurlandskaya G.V., Hernando B., Prida V.M., Santos J.D., Tejedor M., 2003, Very high GMI effect in commercial Vitrovac amorphous ribbon, *Sens. Actuat.*, **A106**, 195–198.

Sandhu A., Masuda H., Oral A., Bending S.J., 2001, Room temperature magnetic imaging of magnetic storage media and garnet epilayers in the presence of external magnetic fields using a submicron GaAs SHPM, *J. Cryst. Growth*, **227–228**, 899–905.

Sato M., Ishio S., Miyazaki T., 1993, Magnetoresistance of NiFe/Cu multilayers formed by sputtering, *J. Magn. Magn. Mater.*, **126**, 460–462.

Scarzello J.F., Usher G.W., 1977, A low power magnetometer for vehicle detection, *IEEE Trans. Magn.*, **13**, 1101–1103.

Schad R., Potter C.D., Belien P., Verbanck G., Moshchalkov V.V., Bruynseraede Y., 1994, Giant magnetoresistance in Fe/Cr superlattices with very thin Fe layer, *Appl. Phys. Lett.*, **64**, 3500–3502.

Schilling M., 2003, Rauscheigenschaften magnetoresistiver sensoren, in *Proceedings of the VII Symposium. "Magnetoresistive Sensoren"*, Sensitec.

Schilstra G., van Hateren J.H., 1998, Using miniature sensor coils for simultaneous measurement of orientation and position of small, fast-moving animals, *J. Neurosci. Methods*, **83**, 125–131.

Scholes R., 1970, Application of operational amplifiers to magnetic measurements, *IEEE Trans. Magn.*, **6**, 289–291.

Schott C., Blanchard H., Popovic R., Racz R., Hrejsa J., 1997, High accuracy analog Hall probe, *IEEE Trans. Instrum. Meas.*, **46**, 613–616.

Schott C., Popovic R.S., Alberti S., Tran M.Q., 1999, High accuracy magnetic field measurements with a Hall probe, *Rev. Sci. Instrum.*, **70**, 2703–2707.

Schott C., Waser J.M., Popovic R.S., 2000, Single chip 3D silicon Hall sensor, *Sens. Actuat.*, **82**, 167–173.

Scouten D.C., 1972, Sensor noise in low level fluxgate magnetometer, *IEEE Trans. Magn.*, **8**, 223–231.

Sedlar M., Matejec V., Paulicka I., 2000, Optical fibre magnetic field sensors using ceramic magnetostrictive jackets, *Sens. Actuat.*, **84**, 297–302.

Senaj V., Guillot G., Darrasse L., 1998, Inductive measurement of magnetic field gradients for magnetic resonance imaging, *Rev. Sci. Instrum.*, **69**, 2400–2405.

Senda K., Ishida M., Sato K., Komatsubara M., Yamaguchi T., 1997, Localized magnetic properties in grain-oriented silicon steel measured by stylus probe method, *Trans. IEEE Japan*, **117-A**, 941–950.

Senda K., Ishida M., Sato K., Komatsubara M., Yamaguchi T., 1999, Localized magnetic properties in grain-oriented electrical steel measured by needle probe method, *Electr. Eng. Japan*, **126**, 942–949.

Seran H.C., Fergeau P., 2005, An optimized low-frequency three-axis search coil magnetometer for space research, *Rev. Sci. Instrum.*, **76**, 1–10.

Shankar K., Zeng K., Ruan C., Grimes C.A., 2005, Quantification of ricin concentration in aqueous media, *Sens. Actuat.*, **B107**, 640–648.

Sherman C., 1959, High precision measurement of the average value of a magnetic field over an extended path in space, *Rev. Sci. Instrum.*, **30**, 568–575.

Shirae K., 1984, Noise in amorphous magnetic materials, *IEEE Trans. Magn.*, **5**, 1299–1301.

Shirkoohi G.H., Kontopoulos A.S., 1994, Computation of magnetic field in Rogowski-Chattock potentiometer compensated magnetic tester, *J. Magn. Magn. Mater.*, **133**, 587–590.

Silver A.H., Zimmerman J.E., 1967, Quantum states and transitions in weakly connected superconducting rings, *Phys. Rev.*, **157**, 317–341.

Slocum R.E., Marton B.I., 1974, A nuclear free precession magnetometer using optically polarized 3 He gas, *IEEE Trans. Magn.*, 10, 528–531.

Slocum R.E., Kuhlman G., Ryan L., King D., 2002, Polatomic advances in magnetic detection, in *OCEANS Conference Paper*, Polatomic Inc., Richardson, TX, Vol. 2, pp. 945–951.

Slocum R.E., Reilly F., 1963, Low field helium magnetometer for space applications, *IEEE Trans. Nucl. Sci.*, 10, 165–171.

Slocum R.E., Ryan L., 2002, Self calibrating vector magnetometer for space, in *ESTC Conference Paper* B3P4, Polatomic Inc., Richardson, TX.

Slocum R.E., Ryan L., 2003, Design and operation of miniature vector laser magnetometer, in *ESTC Conference Paper* B1P8, Polatomic Inc., Richardson, TX.

Slonczewski J.C., 1989, Conductance and exchange coupling of two ferromagnets separated by a tunneling barrier, *Phys. Rev. B: Condens. Matter*, **39**, 6995–7002.

Smith R.S., Annan A.P., 2000, Using an induction coil sensor to indirectly measure the B-field response in the bandwidth of the transient electromagnetic method, *Geophysics*, **65**, 1489–1494.

Smith R.W.M., Freesten I.L., Brown B.H., Sinton A.M., 1992, Design of a phase-sensitive detector to maximize signal-to-noise ratio in the presence of Gaussian wideband noise, *Meas. Sci. Technol.*, **3**, 1054–1062.

Snare R.C., McPherron R.L., 1973, Measurement of instrument noise spectra at frequencies below 1 Hz, *IEEE Trans. Magn.*, **9**, 232–235.

Spong J.K., Speriosu V.S., Fontana R.E., Dovek M.M., Hylton T.L., 1996, Giant magnetoresistive spin valve bridge sensor, *IEEE Trans. Magn.*, **32**, 366–371.

Squire P.T., Gibbs M.R.J., 1988a, Shear-wave magnetometry, *IEEE Trans. Magn.*, **24**, 1755–1757.

Squire P.T., Gibbs M.R.J., 1988b, Ultrasonic shear wave absorption in amorphous magnetic ribbons, *J. Appl. Phys.*, **64**, 5408–5410.

Steindl R., Pohl A., Seifert F., 1999, Impedance loaded SAW sensors offer a wide range of measurement opportunities, *IEEE Trans. Microw. Theory Techniq.*, **47**, 2625–2629.

Sternickel K., Braginski A.I., 2006, Biomagnetism using SQUIDS: Status and perspectives, *Supercond. Sci. Technol.*, **19**, S160–S171.

Stoll R.L., 1975, Method of measuring alternating currents without disturbing the conducting circuit, *IEEE Proc.*, **122**, 1166–1167.

Stuart W.F., 1972, Earth's field magnetometry, *Rep. Prog. Phys.*, **35**, 803–881.

Surkov V.V., Hayakawa M., 2006, ULF geomagnetic perturbations due to seismic noise produced by rock fracture and crack formation treated as a stochastic process, *Phys. Chem. Earth*, **31**, 273–280.

Swithenby S.J., 1980, SQUIDs and their applications in the measurement of weak magnetic fields, *J. Phys. E*, **13**, 801–813.

Takamiya S., Fujikawa K., 1972, Differential amplification magnetic sensor, *IEEE Trans. Electron. Dev.*, **19**, 1085–1090.

Takemura Y., Yamada T., 2006, Output properties of zero speed sensors using FeCoV wire and NiFe /CoFe multilayer thin film, *IEEE Sensors J.*, **6**, 1186–1190.

Takeuchi S., Harada K., 1984, A resonant type amorphous ribbon magnetometer driven by an operational amplifier, *IEEE Trans. Magn.*, **20**, 1723–1725.

Tedrow P.M., Meservey R., 1971, Spin dependent tunneling into ferromagnetic nickel, *Phys. Rev. Lett.*, **26**, 192–195.

Tejedor M., Hernandop B., Sanchez M.L., 1995, Magnetization process in metallic glasses for fluxgate sensors, *J. Magn. Magn. Mater.*, **140–144**, 349–350.

Tondra M., Daughton J., Wang D., Beech R., Fink A., Taylor J.A., 1998, Picotesla field sensor design using spin dependent tunnelling devices, *J. Appl. Phys.*, **83**, 6688–6690.

du Tremolete de Lacheisserie, 2005a, Magnetoelastic effects, in *Magnetism: Fundamentals*, Chapter 12, Springer, Berlin, Germany, pp. 351–396.

du Tremolete de Lacheisserie, 2005b, Magnetostrictive materials, in *Magnetism: Materials and Applications*, Chapter 18, Springer, Berlin, Germany, pp. 214–234.

Tsang C., Fontana R.E., Lin T., Heim D.E., Speriosu V.S., Gurney B.A., Williams M.L., 1994, Design, fabrication, testing of spin valve read heads for high density recording, *IEEE Trans. Magn.*, **30**, 3801–3806.

Tsymbal E.Y., Mryasov O.N., LeClair P.R., 2003, Spin dependent tunnelling in magnetic tunnel junctions, *J. Phys. Condens. Matter*, **15**, R109–R142.

Tumanski S., 1984, A new type of thin film magnetoresistive magnetometer—An analysis of circuit principles, *IEEE Trans. Magn.*, **20**, 1720–1722.

Tumanski S., 1986, Analysis of induction coil performances in application for measurement of weak magnetic field, *Przegl. Elektr.*, **62**, 137–141.

Tumanski S., 2001, *Thin Film Magnetoresistive Sensors*, IOP Publishing, Bristol, U.K.

Tumanski S., 2002, A multi-coil sensor for tangential magnetic field investigations, *J. Magn. Magn. Mater.*, **242–245**, 1153–1156.

Tumanski S., 2004, A review of magnetic sensors, *Przegl. Elektr.*, **80**, 74–80.

Tumanski S., 2006, *Principles of Electrical Measurements*, Taylor & Francis, Boca Raton, FL.

Tumanski S., Stabrowski M., 1984, Optimization of the performance of a thin film permalloy magnetoresistive sensor, *IEEE Trans. Magn.*, **20**, 963–965.

Tumanski S., Stabrowski M., 1985, The optimization and design of magnetoresistive Barber-pole sensors, *Sens. Actuat.*, **7**, 285–295.

Uchiyama T., Mohri K., Panina L.V., Furuno K., 1995, Magnetoimpedance in sputtered amorphous films for micro magnetic sensor, *IEEE Trans. Magn.*, **31**, 3182–3184.

Ueda H., Watanabe T., 1975, Several problems about sensitivity and frequency response of an induction magnetometer, *Sci. Rep. Tohoku Univ., Ser. 5. Geophysics*, **22**, 107–127.

Ueda H., Watanabe T., 1980, Linearity of ferromagnetic core solenoids used as magnetic sensors, *J. Geomagn. Geoelectr.*, **32**, 285–295.

Uesaka M., Hakuta K., Miya K., Aoki K., Takahashi A., 1998, Eddy-current testing by flexible microloop magnetic sensor array, *IEEE Trans. Magn.*, **34**, 2287–2297.

van den Berg H.A.M., Clemens W., Gieres G., Rupp G., Schelter W., Vieth M., 1996, GMR sensors scheme with artificial antiferromagnetic subsystem, *IEEE Trans. Magn.*, **32**, 4624–4626.

Van Harlingen D.J., Koch R.H., Clarke J., 1982, Superconducting quantum interference device with very low magnetic flux noise energy, *Appl. Phys. Lett.*, **41**, 197–199.

Vazquez M., Garcia-Benetyez J.M., Garcia J.M., Sinnecker J.P., Zhukov, A.P. 2000, Giant magnetoimpedance in heterogeneous microwires, *J. Appl. Phys.*, **88**, 6501–6505.

Vazquez M., Hernando A., 1996, A soft magnetic wire for sensor applications, *J. Phys. D*, **29**, 939–949.

Vazquez M., Knoble M., Sanchez M.L., Valenzuela R., Zhukov A.P., 1997, Giant magnetoimpedance effect in soft magnetic wires for sensor applications, *Sens. Actuat.*, **A59**, 20–29.

Vazquez M. et al, 2007, Applications of amorphous microwires in sensing technologies, *Int. J. Appl. Electromagn. Mech.*, **25**, 441–446.

Vencueria I., Tudanca M., Aroca C., Lopez E., Sanchez M.C., Sanchez P., 1994, Fluxgate sensor based on planar technology, *IEEE Trans. Magn.*, **30**, 5042–5045.

Vlaardingerbroek M.T., den Boer J.A., 2004, *Magnetic Resonance Imaging*, Springer, Berlin, Germany.

Vrba J. et al, 1993, Whole cortex 64 channel SQUID biomagnetometer system, *IEEE Trans. Appl. Supercond.*, **3**, 1878–1882.

Wakai R.T., Van Harlingen J., 1988, Signal and white noise properties of edge junction dc SQUIDs, *Appl. Phys. Lett.*, **52**, 1182–1184.

Wang D., Nordman C., Daughton J., Qian Z., Fink J., 2004, 70% TMR at room temperature for SDT sandwich junctions with CoFeB as free and reference layers, *IEEE Trans. Magn.*, **40**, 2269–2271.

Ward D.A., Exon J.L.T., 1993, Using Rogowski coils for transient current measurements, *Eng. Sci. Educ. J.*, 105–113.

Weinstock H., 1991, A review of SQUID magnetometry applied to nondestructive evaluation, *IEEE Trans. Magn.*, **27**, 3231–3236.

Weinstock H. (Ed.), 1996, *SQUID Sensors: Fundamentals, Fabrication and Applications*, Kluwer Academic Publishers, Dordrecht, the Netherlands.

Werner E., 1957, Einrichtung zur Messung magnetischer Eigenschaften von Blechen bei Wechselstrommagnetisierung, Austrian patent. No. 191015.

Weyand K., 1989, An NMR marginal oscillator for measuring magnetic fields below 50 mT, *IEEE Trans. Instrum. Meas.*, **38**, 410–414.

Weyand K., 1999, Magnetometer calibration setup controlled by nuclear magnetic resonance, *IEEE Trans. Instrum. Meas.*, **48**, 668–671.

White R.L., 1992, Giant magnetoresistance: a primer, *IEEE Trans. Magn.*, **28**, 2482–2487.

Wiegand J.R., Velinsky M., 1974, Bistable magnetic device, US Patent 3,820,090.

Wikswo J.P., 1995, SQUID magnetometers for biomagnetism and nondestructive testing: Important questions and initial answers, *IEEE Trans. Appl. Supercond.*, **5**, 74–120.

Woo B.C., Kim C.G., Park P.G., Kim C.S., Shifrin V.Y., 1997, Low magnetic field measurement by a separated NMR detector using flowing water, *IEEE Trans. Magn.*, **33**, 4345–4348.

Yabukami S., Kikuchi, K., Yamaguchi M., Arai K.I., 1997, Magnetic flux sensor principle of microstrip pickup coil, *IEEE Trans. Magn.*, **33**, 4044–4046.

Yamaguchi T., Senda K., Ishida M., Sato K., Honda A., Yamamoto T., 1998, Theoretical analysis of localized magnetic flux measurement by needle probe, *J. Phys. IV*, **8**, 717–720.

Yamaguchi M., Yabukami S., Arai K.I., 2000, Development of multiplayer planar flux sensing coil and its application to 1 MHz–3.5 GHz thin film permeance meter, *Sens. Actuat.*, **81**, 212–215.

Yang H.H., Myung N.V., Yee J., Park D.Y., Yoo B.Y., Schwartz M., Nobe K., Judy J.W., 2002, Ferromagnetic micromechanical magnetometer, *Sens. Actuat.*, **A 97–98**, 88–97.

Young D., Pu Y., 2000, Magnetooptics, in *The Electrical Engineering Handbook*, Chapter 57, CRC Press, Boca Raton, FL.

Yuasa S., Djayaprawira D.D., 2007, Giant tunnel magnetoresistance in magnetic tunnel junctions with a crystalline MgO (001) barrier, *J. Phys. D*, **40**, R337–R354.

Yuasa S., Nagahama T., Fukushima A., Suzuki Y., Ando K., 2004, Giant room temperature magnetoresistance in single crystal Fe/MgO/Fe magnetic tunnel junctions, *Nat. Mater.*, **3**, 868–871.

Zeeman P., 1897, On the influence of magnetism on the nature of the light emitted by a substance, *Phil. Mag.*, **43**, 226.

Zeng X.H., Soltner H., Selbig D., Bode M., Bick M., Rüders F., Schubert J., Zander W., Banzet M., Zhang Y., Bousack H., Braginski A.I., 1998, A high temperature RF SQUID system for magnetocardiography, *Meas. Sci. Technol.*, **9**, 1600–1608.

Zhang Y., Panaitov G., Wang S.G., Wolters N., Otto R., Schubert J., Zander W., Krause H.J., Soltner H., Bousack H., Braginski A.I., 2000, Second order high temperature superconducting gradiometer for magnetocardiography in unshielded environment, *Appl. Phys. Lett.*, **76**, 906–908.

Zhang Y., Tavrin Y., Krause H.J., Bousack H., Braginski A.I., Kalberkamp U., Matzander U., Burghoff M., Trahms L., 1995, Applications of high temperature SQUIDs, *Appl. Supercond.*, **3**, 367–381.

Zhao Z.J., Li X.P., Fan J., Seet H.L., Qian X.B., Ripka P., 2007, Comparative study of the sensing performance of orthogonal fluxgate sensors with different amorphous sensing elements, *Sens. Actuat.*, **A136**, 90–94.

Zhao L., van Wyk J.D., Odendaal W.G., 2004, Planar embedded pick-up coil sensor for integrated power electronic modules, *Appl. Power Electron. Conf.*, **2**, 945–951.

Zhu J.G., 1999, Spin valve and dual spin valve heads with synthetic antiferromagnets, *IEEE Trans. Magn.*, **35**, 655–660.

Zijlstra H., 1967, *Experimental Methods in Magnetism*, North-Holland Publishing, Amsterdam, the Netherlands.

Zimmerman J.E., Silver A.H., 1966, Macroscopic quantum interference effects through superconducting point contacts, *Phys. Rev.*, **141**, 367–375.

Zimmerman J.E., Thiene P., Harding J.T., 1970, Design and operation of stable rf biased superconducting point contact quantum devices and a note on the properties of perfectly clean metal contacts, *J. Appl. Phys.*, **41**, 1572–1580.

Zvezdin A.K., 1997, *Modern Magnetooptics and Magnetooptical Materials*, IOP Publishing, London, U.K.

5

Testing of Magnetic Materials

5.1 AC Testing of Soft Magnetic Materials

5.1.1 What Do We Usually Test?

Practically, there is no such term as absolute "parameters of a magnetic material" because magnetic performance is strongly shape dependent (due to the demagnetizing field that causes nonuniformity of magnetization and the change of parameters during the preparation of the sample*). Therefore, when parameters are given, it is recommended to indicate what kind of the sample was used for the investigations (Epstein frame, ring core, sheet, strip, etc.). In such a case, we usually test the average parameters of the whole sample, which is often advantageous because this way we take into account possible heterogeneity of the material.[†]

It is also possible to determine only the local values of the material parameters (Figure 5.1). So it is also important to take into account the possible heterogeneity of the material. For example, in the case of a grain-oriented steel, the performance can differ significantly from grain to grain. Thus, if we use very small sensor, we can determine local parameters of individual grains. If we are interested in average performance, the sensor should be appropriately large to cover significant number of grains (Overshott and Blundell 1984, Moses and Konadu 2001).

We can imagine a case when the sample and magnetizing circuit are relatively large (e.g., 50×50 cm in the case of a single sheet tester [SST]) and we test the sufficiently large (e.g., 10×10 cm) local area in the central part. In this case, the nonuniformity of the demagnetizing field is negligible and we could say about "the parameters of the material." But we should also take into consideration that most of magnetic materials are anisotropic. Therefore, it is necessary to perform this test for various directions of magnetization[‡] to obtain full knowledge about the anisotropy of the material performance.

Practically all important parameters (power loss, permeability, hysteresis, etc.) of magnetic materials depend on the flux density[§] B and magnetic field strength H. Therefore, a basic problem is how to measure these two values. There are two "schools" of the preferred method of determination of magnetic field strength H: indirect and direct measurement (Tumanski 1988, Pfützner and Schönhuber 1991, Moses 1998, Nakata et al. 2000, Stupakov et al. 2010) (Figure 5.2).

In the indirect method, the magnetic field strength is determined from the magnetizing current I_1 in the primary winding n_1 according to the Ampère law:

$$H = \frac{I_1 n_1}{l} \tag{5.1}$$

An advantage of this indirect method is that the measured signal is relatively large, without interferences and proportional to the magnetic field strength. The main drawback of the indirect method is that we never know precisely the value of the magnetic path length l. Only in the case of toroidal sample of recommended dimensions, we can determine the mean length of the magnetic path. But also in this case, the material is not magnetized uniformly; different values of flux density (and magnetic field strength) are found near the inner diameter other near the outer diameter (Ling et al. 1990).

In the direct method, the magnetic field strength is measured as the tangential field component near the magnetized body (Figure 5.2). If the sensor is very close to the magnetized surface according to the Maxwell law, we can assume that the measured magnetic field is the same as that inside the tested sample. Various magnetic field sensors can be employed in this method: flat coil known as H-coil (or Rogowski–Chattock potentiometer [RCP][¶]) or the magnetoresistive sensor [MR] (sometimes

---

* This last effect can be often significantly reduced by reducing mechanical stress from the sample through annealing.

[†] For this reason in the standard SST, sample is large: 50×50 cm.

[‡] In the case of electrical steel, usually such direction is referred to the rolling direction assumed as $0°$.

[§] More precisely standards recommend measuring the polarization J instead of flux density B. The difference between J and B is as follows: $J = B - \mu_0 H$. In the case of soft magnetic materials magnetic field strength does not exceed 10 kA/m and therefore the second component $\mu_0 H$ is negligibly small and practically the flux density B and polarization J have the same value. Other situation is in a case of hard magnetic materials where both flux density and polarization are determined.

[¶] We can assume that the Rogowski coil is a bent H–coil or that a flat H–coil is a straightened Rogowski coil.

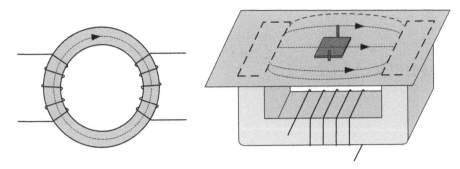

FIGURE 5.1
Testing of the whole sample and testing of the local parameters.

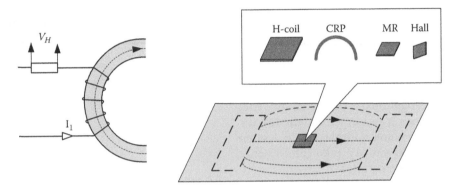

FIGURE 5.2
Indirect and direct measurement of magnetic field strength *H*.

also Hall sensors were used) (Norman and Mende 1980, Pfützner 1980a, Nakata et al. 1982, 1985, 1987, Basak et al. 1985, Flanders 1985, Moses and Jones 1988, Tumanski 1988, Iranmanesh et al. 1992, Pfützner et al. 1992, Shirkoohi and Kontopoulos 1994, Kappel et al. 1997, Tumanski and Winek 1997, Moses 1998, Tumanski 2002b, Lo et al. 2003, Perevertov 2005, 2009, Stupakov et al. 2009, 2010).

The commonly used H-coil has relatively small output signal (in the range of μV) because it should be very thin and the output signal depends on the cross-sectional area of the coil (Figure 5.3). Moreover, the signal can be interferenced by the external magnetic field. Because output signal depends on *dH/dt*, the integrating amplifier is required (Scholes 1970). This amplifier can introduce additional uncertainties mainly due to the zero drift. The Hall and MR sensors measure directly the magnetic field. The examples of output signals of H-coil and MR sensors are presented in Figure 5.3.

The indirect method averages the result for the whole sample. Direct sensors of magnetic field are especially recommended for localized testing of magnetic material, as well as for two-dimensional tests of magnetic materials. Sometimes the small dimension of the sensor can be disadvantageous, but this problem can be solved

by using larger coil sensor or matrix array of MR sensors (Tumanski and Baranowski 2006, 2007).

The flux density is commonly measured by means of the Faraday's law as the voltage *V* induced in secondary winding n_2 (Figure 5.4)

$$\frac{dB}{dt} = -\frac{V}{n_2 A} \tag{5.2}$$

Thus, to determine the value of the flux density *B*, it is necessary to employ an integrating circuit. Moreover, it should be considered that the cross-sectional area *A* can be different in various parts of the sample; therefore, in the case of the Epstein frame, the standard recommends to determine the average cross-sectional area from the weight of the sample.

More complex is the problem of detection of local flux density (Figure 5.4). Such measurements can be realized by means of the Kerr effect but such a method is rather tedious and is used more commonly for domain imaging (Tejedor et al. 1993, Defoug et al. 1996, Hubert and Schäfer 1998, Moses et al. 2005) or for investigations of thin film elements or small particles (Qiu and Bader 2000). To test the local value of flux density, usually small holes are drilled in the sample to wind the

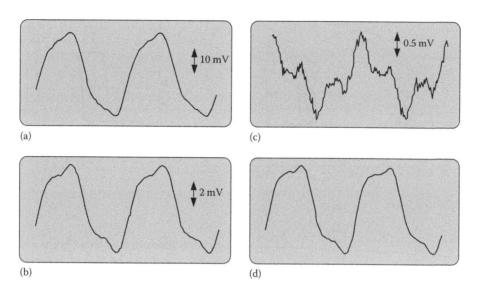

FIGURE 5.3
Output signal of various methods of measurements of magnetic field strength: (a) from the magnetizing current, (b) AMR sensor, (c) H-coil sensor, and (d) H-coil sensor signal after integration and amplification. (From Tumanski, S. and Baranowski, S., *J. Electr. Eng.*, 55, 41, 2004.)

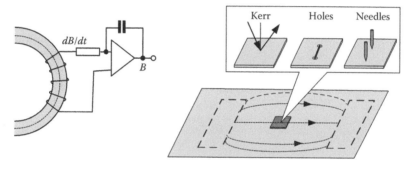

FIGURE 5.4
The measurement of the flux density averaged over whole sample and localized.

micro-coil (Figure 5.4). This method is also troublesome and more important is destructive to the sample under test. That is why for such purposes the needle method was developed recently (Werner 1957, Yamaguchi et al. 1998, Senda et al. 1997, 1999, 2000, Krismanic et al. 2000, Loisos et al. 2001, Pfützner et al. 2004).

Which parameters do we usually determine? The symbolic for magnetic materials is the hysteresis loop $B = f(H)$. Indeed, the "width" of the loop known as a coercive field H_c informs us how magnetically soft the material is. The "high" of the loop known as remanence flux density B_r is important for investigations of hard magnetic materials. For magnetically soft materials, more interesting is the saturation flux density B_s. And generally the area of the hysteresis loop informs us about the magnetic softness of the material and power loss. In the case of AC measurements, we can plot the hysteresis loop relatively easily; an example is presented in Figure 5.5. The horizontal deflection of the oscilloscope is proportional to the magnetizing current, while

the vertical deflection is proportional to the secondary voltage (via integrating circuit).

Similarly in the case of AC measurements, it is relatively easy to derive the second important relation—the magnetization curve $B = f(H)$ (Figure 5.6). It is sufficient to determine the rms or peak value of the magnetizing current and induced voltage (via integrating circuit). The determined magnetization curve is then the connection of the tips of hysteresis loops (see Figure 2.12). Of course, in order to obtain the virgin curve, it is necessary to de-stress and demagnetize the sample.* For special purposes, an anhysteretic magnetization curve (see Figure 2.125) can be determined instead of the normal magnetization curve.

For designers of magnetic devices (especially large transformers and electrical machines), the energy or

---

* In the typical case, it is sufficient to anneal the sample (to de-stress) and demagnetize the sample by applying large AC magnetic field and slowly decrease it to zero (see Figure 2.122).

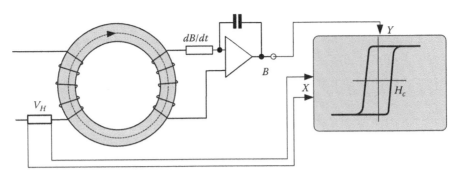

FIGURE 5.5
The principle of a hysteresis loop tracer.

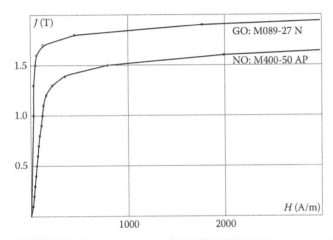

FIGURE 5.6
Typical magnetization curves of electrical steels.

For characterization and categorization of materials, the specific power loss related to the mass m of the sample with density γ is determined:

$$P_\mathrm{w} = \frac{P}{m} = \frac{V}{mT}\int_0^T H\frac{dB}{dt}\,dt = \frac{1}{\gamma T}\int_0^T H\frac{dB}{dt}\,dt \qquad (5.4)$$

The basic circuit for measurement of the power loss is presented in Figure 5.7. Sometimes for local loss measurement, the direct temperature sensor was used as well as the magnetoresistive sensor (Moghaddam and Moses 1993, Krismanic et al. 2003).

It is also possible to determine the specific apparent power in the sample:

$$S = I_{1rms}U_{2rms}\frac{n_1}{n_2}\frac{1}{\gamma l A} \qquad (5.5)$$

From a physical point of view, the magnetic material is best described by permeability μ because it directly links two main magnetic values: magnetic field strength H and flux density B (or polarization J):

$$B = \mu_0\mu_r H \quad \text{and} \quad J = \mu_0 H(\mu_r + 1) \qquad (5.6)$$

power loss is very important. The magnetic loss in volume V also depends on the measured H and dB/dt values:

$$P = \frac{V}{T}\int_0^T H\frac{dB}{dt}\,dt \qquad (5.3)$$

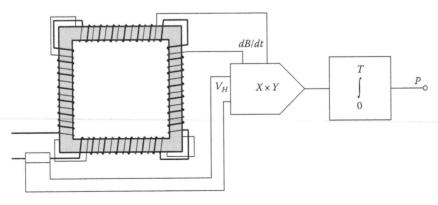

FIGURE 5.7
The basic circuit for power loss measurement.

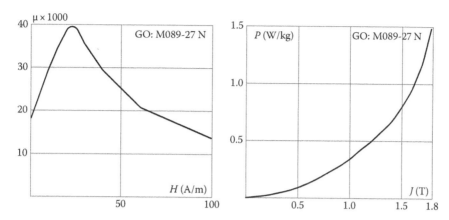

FIGURE 5.8
Typical relations μ(H) and P(B) describing performances of GO electrical steel.

However, it is difficult to use the permeability μ_r as the material parameter because due to nonlinearity of the dependence $B = f(H)$ its value strongly depends on the operating point. Therefore, usually the permeability is determined from the magnetization curve and only certain distinctive values as the maximum value μ_{max} or initial value μ_{in} are given. Often the device is working far from the maximum value of permeability. In some cases, for example, magnetic shielding, the permeability can be a crucial parameter.

In a typical case, the four main relations—$J = f(H)$, hysteresis loop; $J = f(H)$, magnetization curve; $P = f(B)$, and $\mu = f(H)$ (Figures 5.6 and 5.8)—give comprehensive information about properties of the investigated soft magnetic material. Such characteristics are usually presented in manufacturer catalogues. The standards are generally limited to certain typical parameters, for example, $P_{1.5}$ or $P_{1.7}$ (specific power loss at 1.5 or 1.7 T) and J_{800} (polarization at 800 A/m) in the case of grain-oriented electrical steel.*

Of course, some special tests can be performed when required, for example,

- Testing of anisotropy and texture
- Testing of reversible permeability (especially when the material is expected to work under AC + DC magnetization)
- Frequency tests
- Complex permeability for high-frequency applications
- Mechanical and electrical (e.g., resistivity) tests
- Barkhausen noise tests

In physical and geomagnetism investigation, important parameters often tested are susceptibility χ and magnetization M.

Returning to the theoretical concept of "parameters of magnetic material," it should be underlined that there are many other factors influencing the performances of the material. For example, properties of the most magnetic materials strongly depend on the frequency of the magnetizing field (including presence of harmonics). Therefore, for a repeatable test of magnetic properties, all the following testing conditions should be precisely taken into account:

- Geometrical (shape of the sample)
- Operating point (e.g., polarization value) because all parameters as power loss, apparent power, permeability, coercivity, remanence depend on the value of polarization
- Direction of magnetization (usually with respect to the rolling direction in the case of electrical steel)
- Presence of harmonics (usually a pure sinusoidal shape of the polarization waveform is assumed)

5.1.2 Standard Tests of Soft Magnetic Materials

As pointed out in the previous section, results of magnetic tests of materials strongly depend on the measuring conditions: shape of the sample, waveform of polarization, direction of magnetization, etc. To compare the results of tests of various materials, it was necessary to introduce standards describing the measuring conditions in an as detailed way as possible. These standards have been developed especially to test electrical steel because this material dominates the market.

---

* In the case of no-noriented electrical steel, the following parameters are used: J_{2500}, J_{5000}, and $J_{10,000}$.

The Epstein frame (for details, see Section 2.13.3) is the most criticized and at the same time the most frequently used sample (and the testing method) for electrical steel. The reason is that it exhibits exceptionally good repeatability. Because it is described in detail by the international standards (e.g., IEC 60404-2) (the method of preparation of the sample, design of the apparatus, the method of measurement), all investigators use the same equipment and therefore the difference between results of tests of the same sample performed by various laboratories is below 1% (which in testing of magnetic materials is very good uncertainty).

The Epstein frame (Figure 5.9) is criticized because it is arbitrarily assumed that the length of the magnetic path is equal to 0.94 m. Various tests confirm this assumption but only for typical cases (Ahlers et al. 1982, Sievert 1984, Sievert et al. 1990, Wiglasz 1992, Moses et al. 1983, Marketos et al. 2007). For high flux density, for special specimens (e.g., HiB grades, samples cut perpendicularly to the rolling direction) differences can be significant. But because all investigators make the same errors, the conformity of the results is excellent.

There are attempts to analyze numerically the accuracy of the Epstein frame (Cundeva and Arsov 2000, Antonelli et al. 2005) and there are also proposals of the modification of this device (Mthombeni et al. 2007). But the effect on the corners is too complex to obtain satisfying accuracy, and it is reasonable to accept this method with its drawbacks (and some advantages) even if the results are far from correct (from the physical point of view). It is also reasonable to limit this method to typical electrical steel samples for the frequency up to 400 Hz (as recommend by the standard). For testing other materials and other special cases (e.g., samples cut with the angle to the rolling direction), there are other more reliable methods (e.g., single strip tester).

The important drawback of the Epstein method is that it is the off-line method. Although the completion of the frame from prepared strips is not too difficult, the preparation of the samples is time-consuming (cutting of the strips, annealing, etc.). Moreover, the method is practically destructive because further application of the strips of dimensions 3×28 cm is rather limited. That is why the alternative methods are proposed; an example is an SST described by the standards IEC 60404–3.

The SST (see Section 2.13.4 for details) can be advantageous but also brings several drawbacks. The main drawback is that because the mass of the sample of the Epstein frame should be larger than 240 g (to obtain averaging effect of generally heterogeneous material) the comparable area of the sheet is 50×50 cm. Thus, the whole yoke system (that is only for closing of the magnetic circuit) is very heavy (and material consuming). Sometimes special mechanical arrangements including hydraulic lever are necessary.

Nevertheless although the standard also imposes the arbitrary length of the magnetic path equal to 45 cm, the physics of the magnetizing process is not as complex as in the case of the Epstein frame. Therefore, the correctly prepared SST device is a reliable testing instrument. The whole device is not as precisely described by the standards as the Epstein frame; therefore, intercomparison of results of test performed by various laboratories is not as excellent but sufficient (Sievert et al. 1990b, 2000, Sievert 2005). The advantage of the SST device is that the sample can be easily and quickly prepared with a rather large area and hence can be further used. But the large area of the sample practically limits this method to electrical steels.

Testing of the magnetic materials according to standards is practically possible only in professional, laboratory conditions. Preparation of the sample and design of testing apparatus require professional ability. Also the electronic circuit requires appropriate precision. An example of the basic circuit is presented in Figure 5.10.

The power amplifier should exhibit the output resistance as small as possible and it should be able to amplify the signal with the large crest factor. The standards require well-defined waveshape of flux density (seldom magnetic field strength). Typically, the standards require sinusoidal waveform of the secondary voltage, which for sine is form factor $FF = 1.1107\% \pm 1\%$.* Usually to ensure this condition, digital iterative techniques are used (see Section 2.12.4). If there is a demand to test the electrical steel even for polarization equal 1.8 T, both the numerical algorithm as well as power amplifier should be perfectly designed and often this supplying part is the most expensive part of the whole device.

Often a mutual inductance M is in the primary circuit. This part is used for air-flux compensation. The primary winding of the mutual inductance is connected in series with the primary winding of the test transformer while

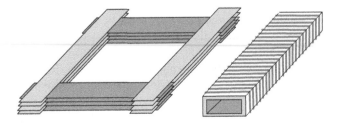

FIGURE 5.9
The Epstein frame and one of the four coils.

---

* Form factor: ratio of rms value to average (rectified) value.

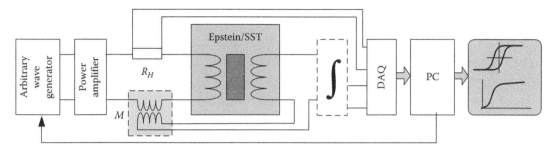

FIGURE 5.10
The basic circuit of Epstein/SST instruments.

the secondary winding is connected to the secondary winding of the test apparatus in series opposition. The adjustment of the mutual inductance is made so that, when passing the nominal current through the primary winding in the absence of the specimen, the output voltage should be compensated and the secondary voltage should be smaller than 1% of the nominal output voltage. The mutual inductance is optional; the air flux can be also compensated numerically.

The choice of the resistor R_H is also important. It should be as small as possible because any additional impedance increases the nonlinearity of the testing transformer. It should be taken into account that the dynamics of the primary current is very large. For the measurement of the properties of typical GO steel at polarization 0.1 T, the sufficient magnetizing current is around 6 mA. For a typical NO steel for polarization 1.7 T, it is necessary to use the magnetizing current around 6 A. Therefore, sometimes it is necessary to change the R_H resistor during the test. This is especially required when even smaller magnetizing currents are tested (e.g., for detection of initial permeability).

The typical output voltage of Epstein apparatus is between 7 and 11 V for GO steel and 13 and 22 V for NO steel for the polarization between 1 and 1.7 T. Because a typical data acquisition board DAQ has the nominal input voltage 5 V, it is necessary to use the a voltage divider. Directly before the DAQ, there can be inserted an analog integrating amplifier. This part is optional because the integration can be performed in the digital part, but the analog integration is sometimes easier to implement.

The data acquisition board DAQ should be of good quality. Because the input current is very distorted, it is necessary to take into account more than 50 harmonics for correct measurement (Sievert et al. 1990). Conventionally used multiplexer for multichannel acquisition is not recommended because delay between channel switching can cause significant error of phase shift. A better solution is to use two AD converters. Fortunately, typical 16 bit DAQ device with simultaneous sampling rate 100 kHz fulfills these requirements. Figure 5.11 presents integrated circuit designed for power and energy measurements. This device has two

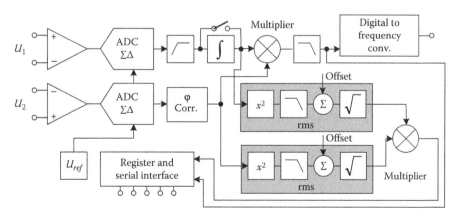

FIGURE 5.11
The integrated power and energy converter: model AD7763 of Analog Devices.

analog-to-digital converters, phase correction, and multiplier. It is interesting that in this device is also included an integrating circuit in one channel to cooperate with the *dB/dt* output.

5.1.3 Single-Sheet and Single-Strip Testers

The standard describes the single-sheet device but it is focused only on 50×50 cm device developed for testing of electrical steel. Such a device is not applicable for many modern materials, for example, amorphous or nanocrystalline materials prepared usually in the form of narrow strips. Fortunately, the recommendation of the standards leave some freedom in design of different devices and therefore various solutions of sensors (H-coils, Rogowski–Chattock potentiometer, and MR sensor) as well as the geometry of such devices have been proposed.

We can differentiate between the sheet testers, when the length and width of the sample are comparable (independently on the sheet area) and the strip testers, where testing of the Epstein strip (width 3 cm) is a natural solution. The main advantage of the strip testers is significantly smaller power necessary to magnetize the sample and dimensions acceptable also in nonprofessional laboratories (including consumers of magnetic materials interesting in assessment of the materials quality). A small disadvantage that the sample is not representative of nonuniform materials can be solved by testing sufficient number of samples (see Figure 2.183) (Tumanski and Baranowski 2007). Of course, it is also possible to use a sheet tester with several strips.

Various shapes and dimensions of alternative apparatus have been considered (Enokizono and Sievert 1990, Nakata et al. 1990a,b, Sievert et al. 1990, Hribernik et al. 1998, Tumanski and Baranowski 2004). Beside the vertical arrangement, horizontal arrangement was also proposed (Nakata et al. 1994). Recommended is a symmetrical double C-type yoke with the magnetizing coil directly on the sample (the yoke is only to close the magnetic circuit). The design used in the standard recommendation is presented in Figure 5.12. As proved by Mikulec et al. (1984) and confirmed by Nakata et al. (1990a,b) in an asymmetrical single C-yoke system, the error caused by the additional eddy currents can be significant. In the double symmetrical system, the clockwise and counterclockwise eddy currents compensate each other.

It is a pity because in a symmetrical double C-yoke and coil on the sample it is not possible to gain access to the specimen to test surface magnetic field. Therefore, the author of this book designed the yoke system with two possibilities: 2C system for SST application and

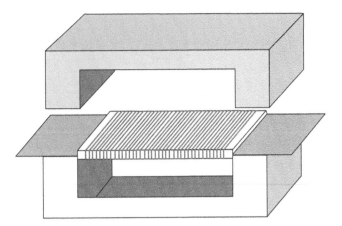

FIGURE 5.12
The single-sheet tester.

FIGURE 5.13
The design and photo of the double C yoke SST device with additional coil on the yoke for surface testing (in an asymmetrical arrangement).

1C with the winding on the yoke for the surface testing (including local measurement; Figure 5.13). Because the length of the sample was significant (28 cm as in the Epstein frame and the height of the yoke was 10 cm), the influence of the eddy currents near the poles was negligible in the central part of the strip.

Figure 5.14 presents the results of computation of the factor affecting the accuracy of a SST by using an H-coil (Nakata et al. 1990b). From these calculations, it was observed that the accuracy of the magnetic field strength measurement is improved by increasing of the length *L* of the yoke. The high *H* and the width *W* of the yoke influence the accuracy less although all these dimensions should be not too small.

Similarly as in the case of a "large" SST device, exist "two schools" of method of measurement of magnetic field strength (Nakata et al. 1982, 1987, 1990a,b, 2000, Enokizono and Sievert 1990). But this time, adherents of direct method (using H-coil) had better arguments. In the case of small device, it is easy to cover the whole specimen with an H-coil. On the other hand, in a small device, the influence of air gaps and yoke permeability

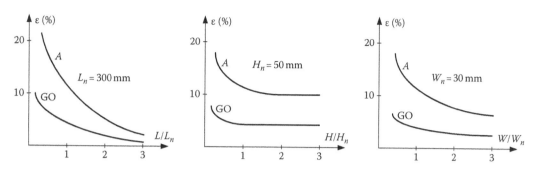

FIGURE 5.14
The error of determination of magnetic field strength by an H-coil for different materials (A, amorphous; GO, grain-oriented steel) and different dimensions of the yoke: length L, high H, and width W. (From Nakata, T. et al., *Ann. Fis.*, B86, 190, 1990a.)

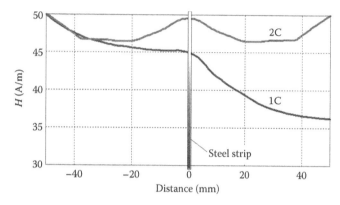

FIGURE 5.15
The tangential magnetic field strength above and under the strip in 1C and 2C yoke system. (From Tumanski, S., *J. Magn. Magn. Mater.*, 242–245, 1153, 2002a.)

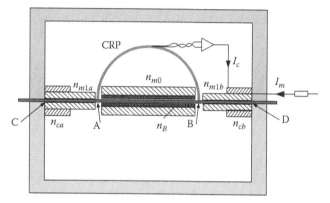

FIGURE 5.16
The compensation RCP single-strip tester.

can cause errors of current method. Figure 5.15 presents the results of computations (confirmed experimentally) of the magnetic field above and under the strip in 1C and 2C yoke systems. Thus, it is possible to easily extrapolate the result of measurement to the sheet surface using two or more sensors distance from the sheet (Nakata et al. 1987, Tumanski 2002a, Perevertev 2005).

Figure 5.14 demonstrates that in the case of conventional GO electrical steel, the measurement of magnetic field strength by means of H-coil ensures satisfying accuracy. But if we test high-permeability amorphous specimen A, the stray external field is very small and the accuracy of measurements using H-coil is worsened. This conclusion was confirmed by other authors (Nafalski et al. 1989). Therefore, recently another design of single-strip tester (especially for high-permeability materials) is developed, known as compensation CRP SST (Mikulec et al. 1984, Moses 1998b, Havlicek and Mikulec 1989, Nafalski et al. 1989, 1990, Shirkoohi and Kontopoulos 1994).

The design and principle of the operation of the compensation SST device is presented in Figure 5.16. The magnetizing coil is divided into three sections: n_{mo},

n_{m1a}, and n_{m1b}. Between the central and external coils is inserted RCP for the measurement of the magnetic field between points A and B.

Without the compensation circuit the magnetic field strength H_{mCD} (between points C and D) measured from the magnetizing current I_m can be determined from the relation,

$$H_{mCD}l_{CD} + 2H_a l_a + H_j l_j = I_m(n_{mo} + n_{m1a} + n_{m1b}) = I_m n_m \quad (5.7)$$

Thus, the measured magnetic field strength H_{mCD} is strongly affected by the unknown magnetic field in the air gaps H_a and magnetic field in the yoke H_j.

With the compensation circuit, the signal from the RCP is connected as a feedback current I_c to the compensation coils n_{c1} and n_{c2}. The condition of the feedback is the magnetic field between points A and B is equal to zero $H_{mAB} = 0$, which means that we compensate magnetic fields in the yoke and air gaps. Therefore, we obtain the following relationship:

$$H_{mAB}l_{AB} = I_m n_m \quad (5.8)$$

and we can determine the magnetic field between points A and B practically without errors. Of course, the length

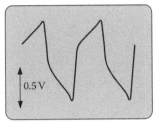

FIGURE 5.17
The AMR sensor array used for SST device and the output signal of the sensors connected in series corresponding with polarization of about 1.5 T.

of magnetic path l_{AB} is known or can be easily measured with an appropriate precision.

Tumanski and Baranowski (2004, 2007) proposed to design the single-strip tester with substituting the H-coil by a thin film magnetoresistive sensor. The advantage of this solution is that the sensitivity and output signal are significantly larger –50 μV/(A/m) for a typical KMZ10B sensor. Moreover, the output signal of the sensor is proportional to the magnetic field and it is not necessary to use the error-introducing integrating circuit. The drawback of this proposal is that the anisotropic magnetoresistive (AMR) sensor is rather small—typical active area is about 1 mm². In the case of grain-oriented steel with very large heterogeneity, such a small area is not acceptable. Therefore, it was proposed to use the sensor array composed of a distributed array of 16 sensors (Figure 5.17) connected in series. In such a design, we obtain exceptionally large sensor signal at the level of about 0.8 mV/(A/m). An example of this signal is presented in Figure 5.17. Moreover by using the sensor array, it is possible to measure the magnetic properties at 16 distinct local points; this feature is described later.

5.1.4 Online Testing Devices

The SST-type devices introduced the possibility of online testing of electrical steel. Two cases can be considered: online control in manufacturing process or testing

of the large final sheets without necessity of cutting it into 50 × 50 cm samples. The first case is more difficult to realization because the device should be in certain distance from the sheet. This results in increasing of the air gaps and increasing the magnetic field uniformity in the magnetized part.

To be immune to the air-gap variations, it is especially recommended to use RCP compensation device because in this system the air gap and yoke magnetic field are compensated. Most of online systems use such principle of measurement (Havlicek and Mikulec 1989, Iranmanesh et al. 1992, Khanlou et al. 1992, 1995, Zemanek et al. 2006, Zemanek 2009). To decrease the influence of stray fields, it is obligatory to use air-flux compensation (flux without the specimen) by means of mutual inductance (see Figure 5.11). Simple and quite effective method of improvement of uniformity of magnetic field is a 2–3 mm thick copper plate inserted between sheet and magnetizing winding. This plate known as Dannatt plate (Dannatt 1933) restraints the AC flux components normal to the sheet plane due to eddy currents generated in the copper plate by the AC flux. Online testing systems are generally used as a comparative quality system with accuracy slightly worse than more quantitative laboratory systems. To simplify the device sometimes, symmetrical yoke is substituted by one-side yoke.

Figure 5.18 presents the testing devices proposed by Beckley (1983). In the case of longitudinal

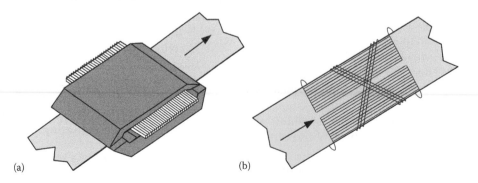

(a)   (b)

FIGURE 5.18
Online testing SST devices: for longitudinal magnetizing of the sample (a) and transversal magnetizing (b) (in the case of transversal magnetization the yoke system is not shown). (From Beckley, 1983.)

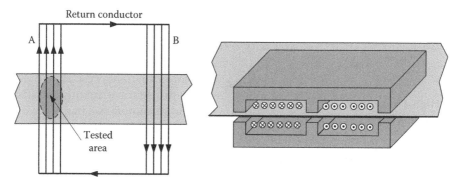

FIGURE 5.19
The non-enwrapping magnetizing systems. (From Beckley, 1983.)

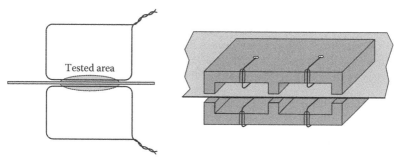

FIGURE 5.20
The non-enwrapping sensing coil systems. (From Beckley, 1983.)

magnetization (the same as rolling direction and the movement direction), the method is relative by easy; the sheet is enwrapped by magnetizing and secondary sensing coils supported by the yoke, which improves the magnetic field distribution (Figure 5.18a). The H-coil was used as the magnetic field strength sensor.

More difficult is testing of the samples magnetized perpendicular to the rolling direction; a proposal for this case is presented in Figure 5.18b. The secondary coils are inclined by ±45° and in this way two components of the polarization are measured (or by appropriate connection of the coils unwanted component can be eliminated).

In an online system, the easy possibility to inserting and removing of the specimen under test is usually required. Therefore, the coil system should not enwrap the specimen. The principle of the non-enwrapping magnetizing system proposed by Beckley is presented in Figure 5.19. If we have the current-carrying rectangular loop, we can create the non-enwrapping system by distancing both opposite sides of the loop (and the same time also the return conductor should be far from the tested area). Figure 5.19 presents also the coil system supported by the yoke.

Figure 5.20 presents a similar system used for the measurement of polarization. Also in this case, the second half of the loop is positioned far from the tested area.

Figure 5.21 presents an example of the computerized online power loss testing system developed by Wolfson

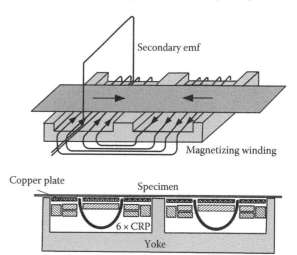

FIGURE 5.21
The example of the online power loss testing system for the steel industry based on the RCP compensation technique. (From Khanlou et al., 1992, 1995.)

Centre for Magnetics Technology at Cardiff University. The results of measurements correlated with the Epstein frame data with deviation smaller than 2.5% and with 99.5% repeatability.

5.1.5 Medium- and High-Frequency Testing of Magnetic Materials

Although principles and rules for medium- and high-frequency testing of magnetic materials are the same as described earlier in this chapter (hysteresis loop, magnetization curve, losses, etc.), there are significant differences in comparison with tests at industrial frequency (50 or 60 Hz). These differences are comprehensively described in the book of Fiorillo (2004).

International standard IEC 60404-10 includes a modified Epstein frame into the methods of testing. It was also proved that the single strip tester is suitable for correct testing of the amorphous ribbons up to 100 kHz (Ahlers et al. 1992). But in medium- and high-frequency tests, the toroid sample (ring core) is dominating. Generally,

it is reasonable to test the magnetic materials in the form similar to which is expected to be used in practice. And most of medium- or high-frequency devices such as inductors, chokes, and transformer use the ring core. Moreover, nanocrystalline materials are delivered in ready-to-use cores, mainly toroids.

The toroid sample has many advantages: almost ideal shape for easy magnetization, the most reliable determined the length of magnetic path (under the condition that appropriate geometrical proportions are fulfilled), and small power necessary to magnetize the sample. An important drawback is the troublesome preparation of the windings. But currently, there exist automated machines correctly winding the toroid sample. Figure 5.22 presents a multipin device for easy "winding" of a ring sample.

Similarly, as in the case of electrical steel, the most frequently recommended is sinusoidal waveform of the flux density. But it is advisable to test the specimen in similar conditions as the future operating conditions. Therefore, the possibility of arbitrary waveform generator controlled by the operator is often suggested. In the case of toroid specimen, it is relatively easy to ensure recommended waveform of excitation, although in the case of rectangular hysteresis loop, it can be a challenge.

Digital forming of arbitrary waveform of flux density or magnetic field strength enables creation various magnetizing conditions. Figure 5.23 presents the picture of major and minor hysteresis loops obtained by appropriate composition of magnetizing signal.

Figure 5.24 presents a typical circuit for testing of soft magnetic materials at medium and high frequencies. It is important to arrange both winding to ensure uniform magnetization (both windings wound uniformly around the whole circuit) and diminish stray inductances and capacitances. Fiorillo (2004) recommends laying both windings of the same number of turns as bifilar single layer. This way the neighboring wires have the same potential and the high-frequency leakage current is limited. Of course, the resistor R_H should be with a flat frequency characteristic. The connecting cables should be short and shielded (e.g., coaxial

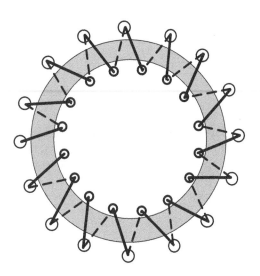

FIGURE 5.22
The device with pins for easy winding of a ring sample. (From Backley, 2002.)

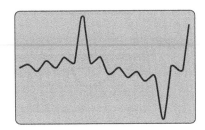 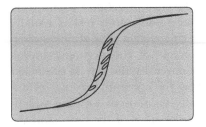

FIGURE 5.23
Minor and major hysteresis loops obtained by appropriate construction of magnetizing signal. (From Disselnköter, R., *J. Appl. Phys.*, 79, 5208, 1996.)

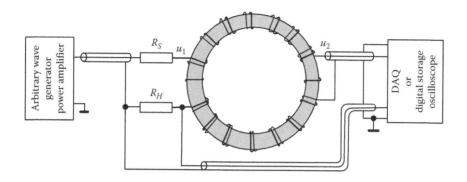

FIGURE 5.24
Typical circuit used for test of soft magnetic materials at medium and high frequencies. (After *Measurement and Characterization of Magnetic Materials*, Fiorillo, F., Copyright (2004), from Elsevier.)

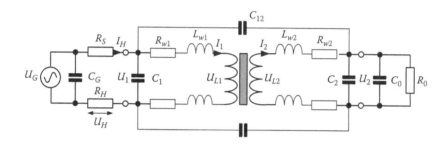

FIGURE 5.25
The equivalent circuit of the measuring setup at higher frequencies (thin line elements that can be neglected in correct prepared sample). (After *Measurement and Characterization of Magnetic Materials*, Fiorillo, F., Copyright (2004), from Elsevier.)

with external shield). The output signals can be connected to a data acquisition board DAQ supported by an appropriate software (Birkfeld and Hempel 1994). Sometimes, a digital storage oscilloscope is used as the output circuit (Thottuvelil et al. 1990, Schmidt and Güldner 1996). Modern digital oscilloscopes are capable of performing signal processing, for example, integrating the secondary voltage. Both DAQ or oscilloscope can be connected to a computer that controls the measurement process.

Figure 5.25 presents an equivalent circuit of the measuring setup. In comparison with low-frequency measurements, interwinding capacitances C_{12}, self-capacitances C_1 and C_2, leakage inductances L_{w1} and L_{w1} as well as generator capacitance C_G, and load capacitance and resistance C_0, R_0 should also be taken into account. The interwinding capacity can be decreased by inserting an electrostatic screen between the windings (for instance, a copper foil). Even if we neglect the leakage inductances L_w and winding resistance R_w, the results of measurement of power P_{meas} and apparent power S_{meas} deviated from the true values P and S (Fiorillo 2004) are

$$P_{meas} = -\frac{1}{m}\frac{n_1}{n_2}\tilde{i}_m\tilde{u}_2\cos\varphi - \frac{1}{m}\frac{\tilde{u}_2^2}{R_0} = P - \Delta P \quad (5.9)$$

$$S_{meas} = -\frac{1}{m}\frac{n_1}{n_2}\tilde{i}_m\tilde{u}_2 - \frac{1}{m}\frac{\tilde{u}_2^2}{R_0}\cos\varphi$$

$$-\frac{1}{m}\left[\left(\frac{n_1}{n_2}\right)^2 C_1 + C_2 + C_0\right]\omega\tilde{u}_2^2\sin\varphi = S - \Delta S$$

$$(5.10)$$

The power loss is not affected by the self-capacitances but can be affected by the power dissipated in the load. The apparent power is influenced by the self-capacitances.

If we cannot neglect the leakage inductances L_w, winding resistances R_w, and interwinding impedance $Z_{12} = R_{12} + jX_{12}$, the expression describing the power loss is much more complex (Fiorillo 2004):

$$P_{meas} \cong P(1 - \omega^2 C_1 L_{w1}) + \frac{1}{m}\frac{\tilde{u}_2^2}{R_0} + \omega C_1 R_{w1}\frac{1}{m}\tilde{i}_m\tilde{u}_m\sin\varphi$$

$$+\frac{1}{m}\left[R_{12}(1 - \omega^2 C_1 L_{w1}) - X_{12}\omega(R_H + R_{w1})C_1\right]\tilde{i}_m^2 \quad (5.11)$$

Thus in the analysis of performance of magnetic materials at medium and higher frequency, it is necessary to take into consideration the complex equivalent circuit of the measuring specimen. Above around 100 kHz, the self-capacitances should be taken into account.

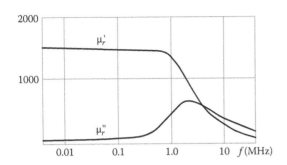

FIGURE 5.26
The real and imaginary components of complex permeability determined for NiZn ferrite. (From Goldfarb et al., 1987.)

Also other high-frequency effects should be considered, for example, skin effect and increase of temperature caused by eddy currents. The skin effect can be limited by using a special type of wire, for example, multiconductor Litz wire. The standards also demand to control the temperature of the specimen by using small thermocouple.

For much higher frequencies, above 1 MHz, the analysis of permeability is also more complex. Figure 5.26 presents the example of investigations of permeability of a NiZn ferrite.

As presented in Figure 5.26, at high frequency, testing both components of the permeability should be considered. If we assume that $H(t) = H_p e^{j\omega t}$ and the flux density $B(t) = B_p e^{j(\omega t - \delta)}$ (H_p, B_p peak values), the complex permeability is

$$\mu = \frac{B_p e^{j(\omega t - \delta)}}{H_p e^{j\omega t}} = \mu' - j\mu'' \qquad (5.12)$$

The loss factor $tg\delta$ is

$$\tan\delta = \frac{\mu''}{\mu'} \qquad (5.13)$$

while the power loss is

$$W = \pi J_p H_p \frac{\tan\delta}{\sqrt{1 + \tan^2\delta}} \qquad (5.14)$$

Testing of magnetic materials at frequencies in the range 20 Hz to 200 kHz is described by the international standard EN 60404-6. The recommended sample is a toroid sample with ratio of external to internal diameter D/d not exceeding 1.4 but the suggested value is smaller than 1.25. For such samples of the height h, the core area A and magnetic path length l_m can be determined as

$$A = \frac{D-d}{2}h, \quad l_m = \pi\frac{D+d}{2} \qquad (5.15)$$

The standard EN 60404-6 recommends two methods of sample analysis: as transformer (as Figure 5.23) or as inductor (impedance). If we test the specimen in form of an inductor, then the permeability can be determined as

$$\mu' = \frac{L l_m}{n_1^2 A \mu_0}, \quad \mu'' = \frac{R l_m}{\omega n_1^2 A \mu_0}, \quad \mu = \frac{l_m}{\omega n_1^2 A \mu_0}\sqrt{R^2 + (\omega L)^2} \qquad (5.16)$$

The impedance can be determined using bridge methods. Recently such measurements are more frequently performed using digital impedance analyzers. The example of an impedance analyzer is presented in Figure 5.27.

The impedance is determined from the Ohm's law by measuring the voltage drop across the impedance and by measuring the supplied current (for this purpose, the current-to-voltage converter is commonly used). Both components of impedance can be determined using phase-sensitive detector (PSD). Other techniques are also proposed, for example, in the impedance to digital

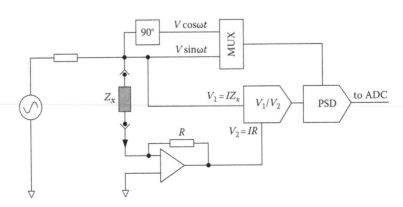

FIGURE 5.27
The basic circuit of an impedance analyzer.

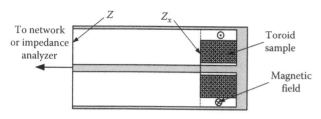

FIGURE 5.28
High-frequency setup for testing of toroid sample. (*Measurement and Characterization of Magnetic Materials*, Fiorillo, F., Copyright (2004), from Elsevier.)

converter AD5933 of Analog Devices real and imaginary parts of impedance is calculated using discrete Fourier transform (DFT).

Neglecting the resistance of the windings, the power loss can be determined from the expression

$$P = \frac{1}{m}\frac{\tilde{u}_2^2}{R} \quad (5.17)$$

Even more complex is the testing of magnetic materials in the radiofrequency range (including GHz range) because we should take into account the wave effects.

At high frequencies, the measuring device (including cabling) should be considered as a transmission line where the signal is a wave traveling through connecting cable and the device under test. Thus, the line is a set of distributed parameters described by the characteristic impedance $Z_0 = \sqrt{L/C}$ and the propagation constant $\beta = \omega\sqrt{LC}$ (where L and C are the inductance and capacitance per unit length). The wave can be reflected from the end of the line if the load impedance is different than the characteristic impedance.

Figure 5.28 presents a setup for testing of toroidal samples at high frequency (Cagan and Guyot 1984, Fiorillo 2004). The sample is placed at the shorted end of coaxial line where the magnetic field is maximum. The height h of the sample should be smaller than the quarter wavelength of the electromagnetic field. The permeability of the sample can be determined by two measurements—with and without the sample. The difference of inductance ΔL of the setup with a toroid with external and internal radii R and r and height h depends on the permeability. Thus, the permeability can be determined from the expression

$$\mu' = \mu_0 + \Delta L \frac{h}{2\pi}\ln\frac{R}{r} \quad (5.18)$$

The second component of permeability can be determined by measurement of the impedance Z_x

$$Z_x = j\omega\frac{\mu_0}{2\pi}h(\mu' - j\mu'')\ln\frac{R}{r} \quad (5.19)$$

If we use the impedance analyzer, the impedance of the transmission coaxial line of the length l, characteristic impedance Z_0, and the propagation constant β should also be taken into account because we measure the whole impedance (Fiorillo 2004):

$$Z = Z_0\frac{Z_x + jZ_0\tan\beta(l-h)}{jZ_x\tan\beta(l-h) + Z_0} \quad (5.20)$$

If we use the network analyzer, it is possible to simulate the line impedance and determine directly the impedance Z_x.

5.1.6 Local Measurements of the Parameters of Magnetic Materials

By "local" measurement, we understand testing of magnetic materials over the area significantly smaller than the whole sample. This area can be as small as several μm^2 but also several cm^2. This kind of measurements can be performed when

- The material is uniform and it is sufficient to test only a selected part
- The whole sample under test is magnetized nonuniformly so that only limited area is subjected to fairly uniform magnetization (Abdallh et al. 2009, Abdallh and Dupre 2010a)
- We would like to test the material uniformity by mapping of the distribution of the materials parameters
- We cannot test the whole sample, as in the case of most of the two-dimensional tests of magnetic material

For local magnetic measurements, we use various sensors. For testing of local values of magnetic field strength, we use H-coil sensors, as well as magnetoresistive, fluxgate, and Hall sensors. For the flux density, we use needle sensors or coils wound through the microholes. For localized power loss, temperature sensors were also used. Knowing the local values of the flux density and magnetic field strength, it is possible to determine local value of permeability, coercivity, or losses.

The flat coil, known also as H-coil, is the most commonly used sensor for testing of the local value of magnetic field strength H. The best would be a one-layer coil placed as close as possible to the tested surface. Therefore, the thickness t of the sensor should be small. But the smaller the thickness, the smaller sensitivity of the sensor. The output signal of the sensor composed of n turns of the wire with diameter d and width w can be calculated using the expression (Tumanski 2002a)

$$V = 8\pi^2 10^{-7} fn(t+d)(w+d)H \qquad (5.21)$$

The H-coil with dimensions 25×25 mm, thickness of the former 0.5 mm wound using 0.1 mm wire (about 150 turns) exhibits the sensitivity only 0.8 µV/(A/m). This corresponds with the output signal of about 20 µV for the typical grain-oriented steel at the flux density of 1 T (see Figure 5.3). Sensitivity can be increased by using thinner wire allowing more turns over the same area. The sensor with the same dimensions but wound with a 0.05 mm wire (about 400 turns) exhibits a sensitivity of about 3 µV/(A/m). Table 5.1 presents parameters of the typical H-coil sensors.

Figure 5.29 presents the experimentally determined magnetic field strength above the magnetized sample. The dependence of the output signal versus the distance from the investigated surface is close to linear; the nonlinear part $0.01x^2$ was smaller than 0.2%. Thus, it is not necessary to use extremely thin sensor; we can use multilayer thicker sensor. Slightly thicker multilayer sensor exhibited sensitivity 22 µV/(A/m).

Moreover, we can use two or more sensors to measure the magnetic field strength and if these sensors are separated from each other by some known distance, we can extrapolate the result to the investigated surface. This idea was proposed by Nakata et al. (1987). In other investigations (Abdallh et al. 2009), the double-coil sensor was used composed of two H-coils

of the area 10×10 mm, thickness 1.5 mm, and 500-turn wire of 0.1 diameter (five layers). The sensors were distant from the investigated surface by 1.5 and 4.6 mm, respectively, and the obtained overall sensitivity was 3 µV/(A/m).

Alternatively, it is possible to use Rogowski–Chattock coil (Figure 5.30) for the measurement of the local value of magnetic field strength (Abdallh and Dupre 2010b). It is de facto a bended H-coil; thus, it is possible to use the coil with large number of turn for small tested area. In every turn of the Rogowski coil of an area A, there is induced voltage $V_i = A\, dB_i/dt$ thus in the whole coil with n turns is induced voltage:

$$V = \sum_{i=1}^{n} A \frac{dB_i}{dt} = A\mu_0 \frac{dH_i}{dt} \qquad (5.22)$$

If we multiply this equation by $\Delta L/\Delta L$ where Δl is the distance between turns (thus, it is $n = L/\Delta L$ where L is the length of the coil) we obtain

$$V = \mu_0 A \frac{n}{L} \sum_{i=1}^{n} \frac{d}{dt} H_i \Delta l = \mu_0 A \frac{n}{L} \frac{d}{dt} H_{AB} l_{AB} = \mu_0 A n \frac{l_{AB}}{L} \frac{dH_{AB}}{dt} \qquad (5.23)$$

Thus, the sensitivity of the Rogowski coil san be expressed as

$$S = \frac{V}{dH_{AB}/dt} = \mu_0 A n \frac{l_{AB}}{L} \qquad (5.24)$$

For $l_{AB} = L$, we obtain the expression for a flat H-coil. If we assume that the turns are wound as close and tight as possible and $n = L/d$ we obtain

$$S = \frac{\mu_0 A l_{AB}}{d} = \frac{\mu_0 wt}{d} l_{AB} \qquad (5.25)$$

We can see therefore that the sensitivity depends on the cross section of the coil $w \times t$ and the length of the tested area l_{AB} and practically does not depend on the length L.

TABLE 5.1

Parameters of Typical H-Coil Sensors

t (mm)	Dimensions (mm)	d (mm)	N	S (µV/[A/m])
0.5	20×20	0.1	150	0.84
0.5	25×25	0.05	400	2.8
0.5 multilayer	25×25	0.05	2000	22

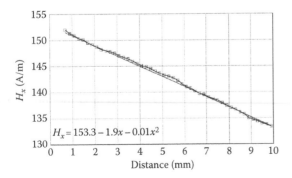

$H_x = 153.3 - 1.9x - 0.01x^2$

FIGURE 5.29
Magnetic field strength above a magnetized steel sample versus the distance of the H-coil from the sample surface. (From Tumanski, S., *J. Magn. Magn. Mater.*, 242–245, 1153, 2002a.)

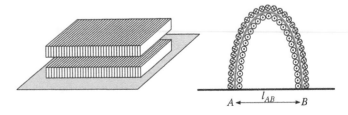

FIGURE 5.30
Double H-coil and Rogowski coil as sensors for local magnetic field strength.

FIGURE 5.31

The distribution of magnetic field strength determined by using AMR sensor for the GO electrical steel of polarization 1 T (tested area 20×60 mm). (From Tumanski, S., *J. Magn. Magn. Mater.*, 75, 266, 1988a.)

These dependencies are very similar to those of H-coil including that the output signal depends on dH/dt, thus the integration is necessary. As an example, Rogowski coil with length 40 cm, number of turns 1308, $A = \pi 0.4^2$ mm², and the tested length $l_{AB} = 0.5$ mm exhibited a sensitivity of only 1.8 nV/(A/m) (Abdallh and Dupre 2010a).

Incomparably, better sensitivity is offered by AMR sensors, typically 50 μV/(A/m). Figure 5.31 presents the map of the distribution of magnetic field strength in typical GO electrical steel determined by using an AMR sensor. We can see that due to the grain structure, the nonuniformity of magnetic field strength is significant, in the area of 2×6 cm, differences are as high as 700%. Thus, the H-coil enables averaging this heterogeneity while the AMR sensor with typical area of 1×1 mm detects very localized values. This is not always disadvantageous. We can solve this problem using the sensor array as presented in Figure 5.17 (Tumanski and Baranowski 2006). Other solution is to test the area at multiple points and to determine the average value. This way we know the average properties for the whole sample but at the same time the dispersion of the particular parameter.

Figure 5.32 presents two examples of the surface distribution of the magnetization curve and anisotropy (magnetic field strength versus the angle of magnetization) determined experimentally by an AMR sensor. Also by using the AMR sensor, it was possible to analyze the grain structure, power loss distribution, and domain structure (Mohd Ali and Moses 1989, Moghaddam and Moses 1992, 1993, So et al. 1995). The scanning microscope with sensor of dimensions 5 μm capable of measuring and imaging the information stored on hard disk is also reported (O'Barr et al. 1996).

The GMR sensors with high magnetoresistivity coefficient can be manufactured with very small dimensions. Figure 5.33 presents an array of GMR sensors developed by Nonvolatile Electronics (NVE). The line

of 2×16 GMR sensors (in half-bridge connection) is 80 μm long with sensors 1.5 μm wide and 6 μm high separated from each other by 5 μm. By "cloning" this arrangement, it is possible to obtain the line composed of 128 sensors, enabling simultaneous investigation of 128 points.

As an alternative, the Hall sensor can be considered. In comparison with the AMR sensors, it has significantly lower sensitivity and moreover it detects normal component of the magnetic field (although there are Hall sensors for tangential components). For example, an application of the array of four vertical sensor with sensitivity 63 μV/(A/m) and area 0.7×0.5 mm is reported (Abdallh and Dupre 2010a) (note that this sensitivity is thousand times lower than in the case of the commercial AMR sensor). Nevertheless, Hall sensors have certain advantages; they are not sensitive to orthogonal component, which is important when very nonuniform magnetic field is tested. Moreover, they can be made with extremely small dimensions (several micrometers), which enables development of the Hall probe microscopy (Chang et al. 1992, Oral et al. 1996). The Hall sensors were used by Pfützner to analyze the grain and domain structures of electrical steels (Pfützner 1980a,b, 1981, Pfützner et al. 1983, 1985, Schönhuber and Pfützner 1989).

In investigation of the local values of the flux density, the obvious method is to drill small holes to wound a microcoil (see Figure 2.175a). But this method means destruction of the material and deterioration of magnetic properties in the area close to holes (Tamaki et al. 2009). Furthermore drilling of very small holes in relatively hard electrical steel is rather difficult. Therefore, the needle probe method is recently more frequently used for localized flux density measurements.

The needle probe method was firstly proposed many years ago (Czeija and Zawischa 1955, Werner 1957). These patents were initially forgotten because the output signal of these sensors was very small. But

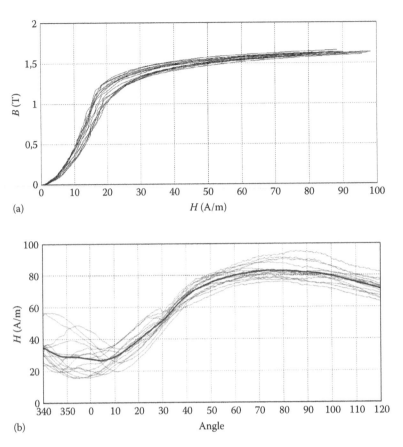

(a)

(b)

FIGURE 5.32
The magnetization curves (a) and magnetic field versus the direction of magnetization (b) determined at 16 points by means of an AMR sensor. (From Tumanski, S. and Winek, T., *J. Magn. Magn. Mater.*, 174, 185, 1997; Tumanski, S., *IEEE Trans. Magn.*, 38, 2808–2810, 2002d.)

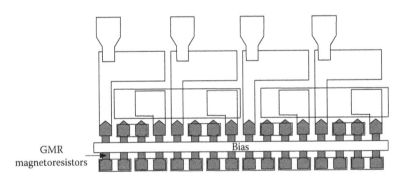

GMR
magnetoresistors   Bias

FIGURE 5.33
The array of GMR sensors developed by NVE. (From Smith, C.H. et al., *J. Appl. Phys.*, 93, 6864, 2003.)

with development of the methods of measurement of small signals. Senda et al. (1997, 1999, 2000) returned to this idea. Also other teams successfully adopted this technique (Yamaguchi et al. 1998, Krismanic et al. 2000, Loisos and Moses 2001, 2003, De Wulf et al. 2003, Pfützner and Krismanic 2004).

In the case of the B-coil method, the relation is relatively simple; the induced voltage depends on the loop area and the rate of the change of flux density:

$$V = ntl_{AB} \frac{dB}{dt} \qquad (5.26)$$

which for one turn, thickness 0.3 mm, frequency 50 Hz, and $l_{AB} = 20$ mm results in the output voltage of about 1.5 mV/T.

In the case of needle probe method, the output voltage depends on the induced voltage caused by flux density

B and voltage caused by two horizontal components of eddy currents i_{ec}. Considering the loop $ABCDA = S_1$, we can write according to the Maxwell's equation (Yamaguchi et al. 1998)

$$\oint_{ABCDA} E dl = -\int_{S_1} \frac{dB}{dt} dS \quad \text{and} \quad \oint_{ABCDA} E dl = 2\int_{AB} \rho i_{ec} dl \quad (5.27)$$

Hence we can write that

$$V = \int_{AB} \rho i_{ec} dl = \frac{1}{2} S_1 \frac{dB}{dt} = \frac{1}{2} t l_{AB} \frac{dB}{dt} \quad (5.28)$$

We should also take into account the voltage induced by the flux density in the air B_a above the sheet (in the loop S_2 on Figure 5.33) and

$$V = \frac{1}{2} t l_{AB} \frac{dB}{dt} - \int_{S_2} \frac{dB_a}{dt} dS \quad (5.29)$$

We can see that the output voltage of the needle probe sensor is only a half of that of the single turn B-coil. The second component should be negligible and therefore the loop S_2 should be as small as possible. For this reason, the connecting wire should be positioned close to the investigated surface (Figure 5.34b). To decrease the influence of the air flux, the leads tightly twisted (Zurek et al. 2008).

The needle sensor method is not as reliable as the B-coil method (Pfützner and Krismanic 2004). It works correctly only under assumption that the magnetic field is uniform in the measured area. For position near the edge of the tested sample, the additional errors due the magnetic field nonuniformity appear. Only a symmetrical yoke system is acceptable because asymmetry will introduce planar eddy currents. The distance between the needles should be ideally much larger than the thickness of the sample.

Most of electrical steel samples are covered by an insulating layer. Therefore, the needles should be sharp and the mechanical force is necessary to maintain electrical contact between the probes and the sample can cause additional errors. The special pyramid needle tip was proposed (Krismanic et al. 2000). The possibility of substituting of the needles with surface capacitors (Figure 5.35b) was also reported.

Various modifications of the needle system were proposed, for example, additional needles on the opposite side of the sheet (Loisos and Moses 2001, 2003). A second pair of the needles was also tested (Figure 5.35a) (Abdallh et al. 2009). In such a system, the value $B_{material} + B_{air}$ is detected between points 1 and 2. Between points 3–4 only B_{air} is detected (because they are shorted). Thus, the difference $V_{12} - V_{34}$ is free from the influence of air flux.

Sometimes, it is necessary to test the local values of a given material parameter (mainly distribution of the flux density) inside large magnetic devices, for example, transformer cores or rotating electric machines. For such investigations, the thin-film sensors deposited directly onto the investigated sheet were employed (Basak et al.

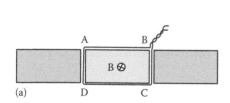
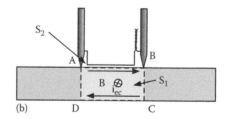

FIGURE 5.34
The hole (B-coil) and the needle probe methods of localized the flux density measurements.

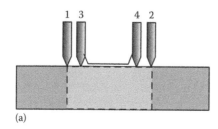
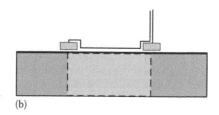

FIGURE 5.35
Modifications of the needle probe method: two pair of needles (a) and capacitors instead of needles (b). (From Abdallh, A. and Dupre, L., *Meas. Sci. Technol.*, 21, 045109, 2010; Zurek, S. and Meydan, T., *Sens. Actuators*, A129, 121, 2006.)

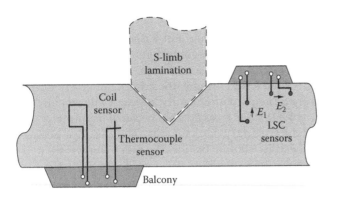

FIGURE 5.36
Thin film sensors for interior analysis of three-phase transformer core. (From Pfützner, H. and Mulasalihovic, E., *Przegl. Elektrotech.*, 85, 39, 2009.)

1995, 1997, Ilo et al. 1998, Pfützner and Mulasalihovic 2009). Figure 5.36 presents an example of the thin film sensors designed for investigation of a transformer core. For power loss, the thermocouple sensor was used. Because it was not possible to drill holes or to use needle sensors, the later ones were substituted by point contact lamination surface contact (LSC) sensors for detecting both components of electric field.

In the sensor fusion presented in Figure 5.36 for the local loss testing, the termocouple sensor is used. Indeed, the local value of power loss can be measured using the temperature sensor, most frequently small thermistor (Ball and Lorch 1965, Pfützner et al. 1982, 1992, Albir and Moses 1990, Derebasi et al. 1992, Krismanic et al. 2003).

The main problem in measurement of the power loss by a temperature sensor is that the power loss does not depend on the temperature itself but rather its rate of change versus time:

$$P = c_p \frac{\partial T}{\partial t} \tag{5.30}$$

where c_p is the material-dependent specific heat capacity constant ($c_p = 486 \, \text{J/°C}$ for silicon-iron and $c_p = 544 \, \text{J/°C}$ for amorphous material) (Derebasi et al. 1992). Thus to measure the local value of the power loss at one point sometimes time of several minutes is necessary (to observe the temperature change and then return to the starting temperature). Figure 5.37 presents the sensor design and the examples of temperature changes at two testing points.

There is also other underappreciated method for local flux density value testing known as "flux injection method." An example of application of this method for local measurement of electrical steel parameters is presented in Figure 5.38.

It can be assumed that if the flux density is measured by using the coils close to the poles of the yoke the flux density in the sheet is only slightly smaller than the flux density in the yoke (due to flux leakage). Indeed investigations performed by Chen et al. (1993) proved that the efficiency coefficient B_{yoke}/B_{sample} was almost the same for various samples (equal to about 80%). Similar investigation was repeated for the samples representing various stage of the production (hot band with permeability 1000, annealed hot band with permeability 990, cold rolled with permeability 367, decarburized with permeability 5,400, and final of permeability 33,000) (Wood et al. 2009). Only final sample with high permeability exhibited high efficiency (about 80% depending on the value of the flux density). Wilkins proposed and tested the special yoke arrangement with adjacent guard yokes, which provided the efficiency of almost 100% for all samples (Wilkins and Drake 1965, 1970).

Figure 5.39 presents the set of results obtained for a selected point by applying the flux injection mode. Since we know the signals proportional to the magnetic field strength $h(t)$ and the flux density $b(t)$, we are able to easily calculate the instantaneous power as

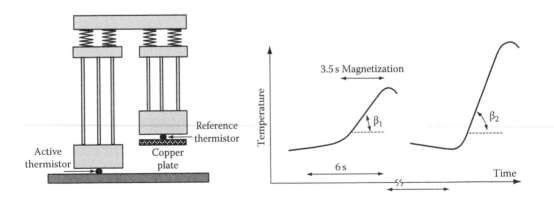

FIGURE 5.37
The thermistor sensor for local loss measurement and the examples of testing results at two points. (From Pfützner, H. et al., Novel nondestructive methods for analysis of crystalline and amorphous soft magnetic materials, in *Proceedings of 2nd International Symposium on Physics of Magnetic Materials*, Beijing, Japan, pp. 427–434, 1992; Krismanic, G. et al., *J. Magn. Magn. Mater.*, 254–255, 60, 2003.)

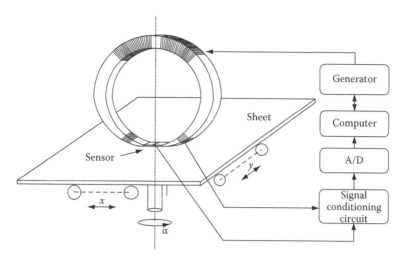

FIGURE 5.38
The measuring system for testing of the localized parameters of electrical steel. (From Tumanski, S. and Winek, T., *J. Magn. Magn. Mater.*, 174, 185, 1997; Tumanski, S. and Fryskowski, B., *J. Phys. IV*, 8, Pr2-669, 1998.)

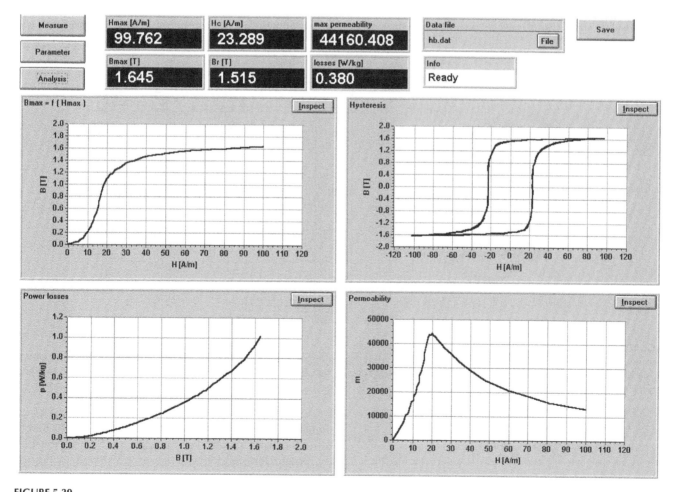

FIGURE 5.39
The example of results of investigations of the steel parameter at selected point. (Tumanski, S. and Winek, T., *J. Magn. Magn. Mater.*, 174, 185, 1997.)

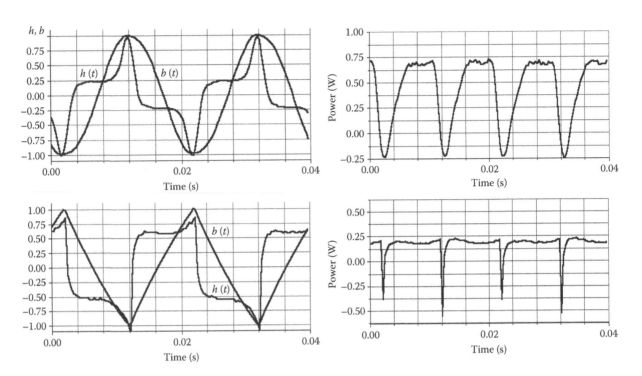

FIGURE 5.40
The example of signals $b(t)$, $h(t)$, and $p(t)$ determined for sinusoidal and triangular polarization. (From Tumanski, S. and Winek, T., *J. Magn. Magn. Mater.*, 174, 185, 1997.)

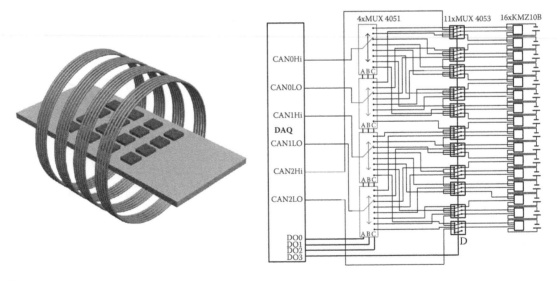

FIGURE 5.41
The measuring setup for determination of local values of electrical steel parameters (magnetizing yoke not shown for clarity). (From Tumanski, S. and Baranowski, S., *J. Electr. Eng.*, 55, 41, 2004; Tumanski, S. and Baranowski, S., *Przegl. Elektrotech.*, 83, 46, 2007; Tumanski, S. and Baranowski, S., *J. Electr. Eng.*, 59, 1, 2008.)

$$p(t) = \frac{m}{\rho} h(t) \frac{db(t)}{d(t)} \qquad (5.31)$$

Figure 5.40 presents an example of signals $b(t)$, $h(t)$, and $p(t)$ determined for sinusoidal and triangular flux

density. Knowing these signals, it is possible to determine localized material parameters such as permeability, power loss, and hysteresis as presented in Figure 5.39.

Figure 5.41 presents another, more reliable method of testing of localized values of parameters of electrical

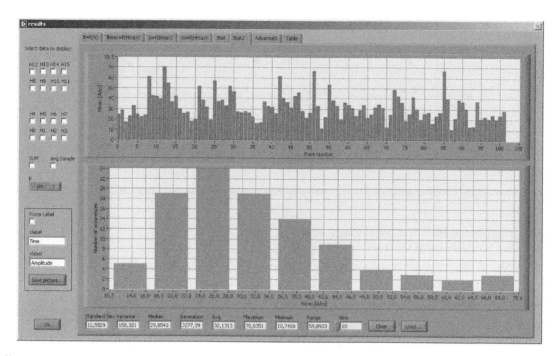

FIGURE 5.42

Magnetic field strength in 100 test locations of the sample (bottom window shows the histogram). (From Tumanski, S. and Baranowski, S., *J. Electr. Eng.*, 55, 41, 2004; Tumanski, S. and Baranowski, S., *Przegl. Elektrotech.*, 83, 46, 2007.)

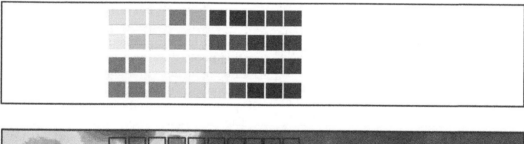

FIGURE 5.43

Distribution of the magnetic field strength measured at 40 points and the map obtained by scanning. (From Tumanski, S. and Baranowski, S., *J. Electr. Eng.*, 59, 1, 2008.)

steel in form of a strip (Tumanski and Baranowski 2004, 2007, 2008). There is a 2D array of 16 magnetoresistive sensors all of which can be connected in series to obtain large signal proportional to the spatially averaged measured value (Figure 5.17). But the same sensor array can be used to test magnetic parameters at 16 separate points by using a multiplexing system.

Within one measurement, iteration magnetic field strength and flux density can be determined simultaneously at 16 points (every row of sensor has B-coil). By

controlled movement of the sample, it is possible to multiply number of the test points. Figure 5.42 presents an example of the magnetic field strength measured at 100 points.

Figure 5.43 presents the results of measurements of the magnetic field strength at 40 points. These results were compared with the map of the magnetic field strength obtained by scanning. Both results are practically the same, but scanning takes about 1 h while multiplexing is carried out in just a few seconds. By knowing the

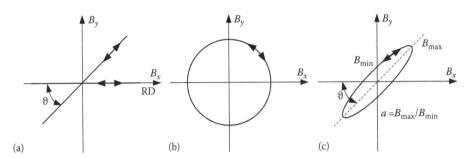

FIGURE 5.44
The axial magnetization along the rolling direction RD or at arbitrary direction (a) and rotational magnetization circular (b) or elliptical (c).

magnetic field strength and flux density at every tested point, it is possible to determine local values of other parameters as well.

5.1.7 Rotational Power Loss

It was assumed in previous considerations that the AC magnetizing field changes its value only with respect to time (Figure 5.44a). This axial magnetization can be directed along the anisotropy axis (usually the rolling direction RD), perpendicular to this axis or in arbitrary 2D direction. However, the magnetizing field can also vary in space as well as in time (Figure 5.44b,c). When the magnitude of the magnetizing field vector is fixed, we can say about circular magnetization (Figure 5.44b). And in the most universal case, the magnetizing field is changing in space and its magnitude is changing with time (Figure 5.44c). Then can be also rotating elliptical magnetization (and its special case the circular magnetization when $a = 1$) present in the cores of rotating machines.

Figure 5.45a presents the loci of the vectors of the flux density and magnetic field strength applied at some angle to the rolling direction (Pfützner 1994). Figure 5.45b presents similarly plotted loci under circular rotating field. In both cases, the magnetization process

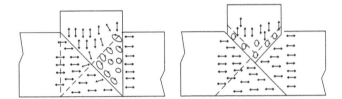

FIGURE 5.46
The rotational magnetization in joint parts of three-phase transformers. (From Moses, A.J. and Thomas, B., *IEEE Trans. Magn.*, 9, 655, 1973.)

is connected with power loss. Unfortunately, the two mechanisms of loss differ and correlation between them is poor. Therefore, it is not possible to predict the level of the rotational loss on the basis on axial experiments for various direction of magnetization (Moses 1994). Also in this book, these two groups of problems known as 2D magnetic measurements are described separately—rotational power loss (this chapter) and anisotropy and 2D performance—in the next chapter.

The rotational magnetization was recognized more than 100 years ago by Weiss and Planer (1908). The first measurements of rotational hysteresis presented Brailsford (1938). But the turning point in thinking about rotation power loss was the paper of Moses and Tomas proving that rotational loss exists not only in rotating machines (which is obvious) but also in certain parts of three-phase transformers (Figure 5.46) (Moses and Thomas 1973). From this time the development of theory and experimental methods concerning rotational power loss commenced. Not only non-oriented steel was taken as the subject but also grain oriented. The Scientific Committee of international conference "Soft Magnetic Materials" decided to organize "Workshop on Two-Dimensional Magnetization Problems."* To date, 10 workshops were organized resulting in hundreds of papers.

The theory of rotational magnetization is rather complex and not all physical phenomena are sufficiently explained or indeed understood. One of proposed

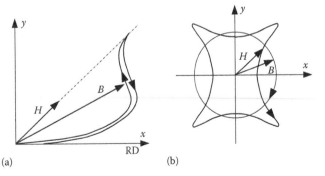

FIGURE 5.45
Loci of the vectors of flux density and magnetic field strength applied at some angle to the rolling direction (a) and similar loci under rotational magnetization (b).

---

* Extended later to one- and three-dimensional problems.

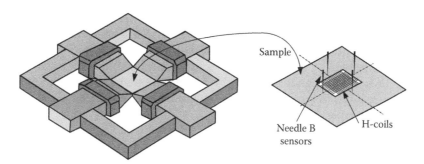

FIGURE 5.47
Typical horizontal-type RSST. (After Enokizono.)

method of describing rotational magnetization is reluctance tensor matrix (Brix and Hempel 1984, Birkfeld and Hempel 1996, Enokizono et al. 1994, 1997, 2003, 2006, Birkfeld and Hempel 1997, Birkfeld 1998, Enokizono and Soda 1999, Enokizono and Urata 2007, Enokizono 2009). For axial magnetization, we can use the reluctance tensor matrix as

$$\begin{bmatrix} H_x \\ H_y \end{bmatrix} = \begin{bmatrix} v_x & 0 \\ 0 & v_y \end{bmatrix} \begin{bmatrix} B_x \\ B_y \end{bmatrix} \tag{5.32}$$

which can be extended to rotational magnetization

$$\begin{bmatrix} H_x \\ H_y \end{bmatrix} = \begin{bmatrix} v_{xx} & v_{xy} \\ v_{yx} & v_{yy} \end{bmatrix} \begin{bmatrix} B_x \\ B_y \end{bmatrix} \tag{5.33}$$

where the reluctance tensor is $v_{xx} = H_x/B_x$, $v_{yy} = H_y/B_y$, etc. This model was further extended by Enokizono and Soda to the following form:

$$\begin{bmatrix} H_x \\ H_y \end{bmatrix} = \begin{bmatrix} v_{xr} & v_{xi} \\ v_{yr} & v_{yi} \end{bmatrix} \begin{bmatrix} \dfrac{\partial B_x}{\partial t} \\ \dfrac{\partial B_y}{\partial t} \end{bmatrix} \tag{5.34}$$

where

v_{xr}, v_{yr} are magnetic reluctance coefficients
v_{xi}, v_{yi} the magnetic hysteresis coefficients obtained from the measurement data

Comprehensive description of the rotational magnetization and losses was presented by Pfützner (1994). Basing on the Poynting's theorem, the following description of the power loss with rotational magnetization is given:

$$P_r = \frac{1}{T\gamma} \int_0^T \left(H_x \frac{dB_x}{dt} + H_y \frac{dB_y}{dt} \right) dt \tag{5.35}$$

This expression is commonly used for determination of the rotational power loss (γ is the specimen density).*

The measuring methods of rotational power loss have a long history. The review is presented in papers of Sievert or Pfützner (Sievert 1990, Sievert et al. 1992, Sievert 1995, 2005, Pfützner 2000, 2002). First experiments were performed by means of torquemetric methods (Brailsford 1938, Kelly 1957, Narita et al 1974, Cechetti et al. 1978, Brix 1982). Also thermometric method was reported (Young and Schenk 1960, Ball and Lorch 1965, Boon and Thompson 1965, Yamaguchi and Narita 1976, Radley and Moses 1981, Fiorillo and Rietto 1988, Albir and Moses 1990). Also recently some advantages of thermometric method were discussed (Fiorillo 2004).

Recently, the most commonly used is the fieldmetric method based on Equation 5.35. In magnetized sample, both components of magnetic field strength and flux density are measured using appropriate sensors. One of the first devices employing such principle known as a *rotational single sheet tester* (RSST) was proposed by Brix et al. (1982, 1984a). To date, it is the most frequently used device copied (with modifications) by other researchers (Enokizono and Sievert 1989, Enokizono et al. 1990a,b, Sievert et al. 1990c, Zhu and Ramsden 1993, 1997, Salz 1994, Xu and Sievert 1997, Makaveev et al. 2000a,b, Shimamura et al. 2000, Maeda et al. 2007, Todaka et al. 2009).

In the device presented in Figure 5.47 (see also Figures 2.176 and 2.189), the sample in form of a square, typically 80×80 mm (or 100×100 mm) is in the center of two orthogonal magnetizing yokes. The poles of the yokes are tapered and between these poles and the sample usually, there is an air gap of about 0.5 mm to improve the uniformity of magnetization. In the middle part of the sample, two-H-coil sensors and two needle probe sensors of flux density are inserted. The tested area is not larger than 30×30 mm to profit uniform magnetization

---

* In some papers this loss is called total power loss and for rotational loss the expression $P_r = \omega/T\gamma\int_0^T(H \times B)dt$ is proposed (Enokizono and Sievert 1989). But this relation is valid only for purely circular magnetization (Moses 1994).

of the center (see Figure 2.190). To obtain better uniformity of magnetization, additional shielding lamination (Makaveev et al. 2000a,b) or guard sheet (Zhu and Ramsden 1993, 1997) were proposed.

Other shapes of the specimen (and magnetizing yoke) were also designed (see Figure 2.191). Team from Vienna proposed to use hexagonal sample and three-phase excitation to obtain rotating field (Hasenzagl et al. 1996, 1997, 1998, Mehnen et al. 2000, Pluta et al. 2003, Pfützner et al. 2007) (Figure 5.48). This testing device is in some way similar to a three-phase electric motor stator core. In such type of magnetizing yoke, it is much easier to obtain uniform and high magnetization than in the case of conventional four-pole device.

The hexagonal sample is also closer to the ideal case of a round sample. The square sample in a four-pole yoke system exhibits better uniformity of magnetization than the round sample (Nencib et al. 1995). But for high flux density and magnetization at the angle of 45° in the corners, there exist large leakage flux. In the round sample and multipole yoke system (see Figure 2.191), the sample is magnetized in the same way for in every direction (Gorican et al. 2000, 2002, Jesenik et al. 2003). As the yoke of round sample rotational loss tester, a typical stator core of electric motor can be used (Matsuo et al. 2007) (Figure 5.49).

Beside horizontal testers also vertical design was proposed (Enokizono et al. 1992, Sievert et al. 1992, Zouzou et al. 1992, Nencib et al. 1994, Enokizono and Tanabe 1997) (Figure 5.50).

Especially interesting is a design with vertical yoke presented in Figure 5.50b because it enables to test the specimen in form of a sheet with nonlimited dimensions (Tumanski 2002a, 2005).

Diversity of rotational loss testers complicates the efforts to accept on a general standard of such

measurements. The possible suggestion is illustrated in Figure 5.51. It seems that the best candidate could be first Brix proposal (Figures 5.47 and 5.52) because it is most commonly used, optimized, and tested. Intercomparison of measurements of the rotational power loss performed by different laboratories gives fairly good agreement for

FIGURE 5.49
Design of round sample RSST. (From Zurek, S., *Przegl. Elektrotech.*, 85(1), 89, 2009.)

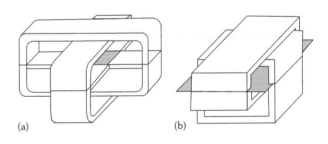

(a)   (b)

FIGURE 5.50
Vertical types of RSST.

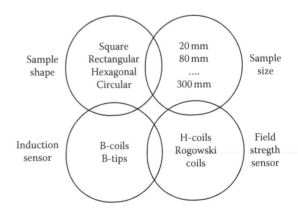

FIGURE 5.51
Variety of different features of rotational loss testers. (From Pfützner, H., Different designs of rotational single sheet testers—Suggestions for procedures of calibration, in *Proceedings of 6th Workshop 2D-Magnetic Measurements*, Badgastein, Austria, pp. 154–162, 2000.)

FIGURE 5.48
The design of hexagonal-type RSST. (From Pluta, W. et al., *Przegl. Elektrotech.*, 79, 151, 2003.)

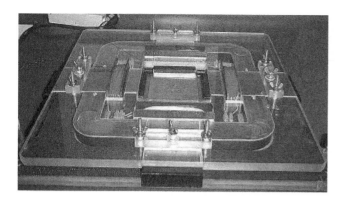

FIGURE 5.52
The design of typical horizontal-type RSST. (Moses, A.J. and Leicht, J., *Przegl. Elektrotech.*, 80(12), 1181, 2004.)

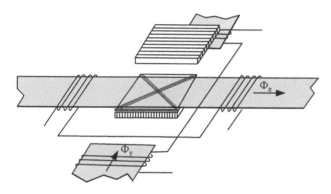

FIGURE 5.53
Strip tester for measurement of rotational loss. (From Sasaki, T. et al., *IEEE Trans. Magn.*, 21, 1918, 1985.)

non-oriented materials (for grain-oriented materials, the results differ considerably) (Sievert et al. 1996).

Completely different testing device was proposed by Sasaki (Sasaki et al. 1985) (Figure 5.53). It is de facto single-strip tester where the Epstein strip is magnetized by magnetizing coil and additional orthogonal magnetizing system. To avoid drilling holes as the B-coils, two diagonal coils are used. For measurement of magnetic field strength, two H-coils are used. Figure 5.54 presents the examples of hysteresis loops determined by this device.

The rotational sheet tester is much more complex than the SST—everything is doubled: number of sensors, two-channel wave control (Makaveev et al. 2001), and so on. By determination of loss often a curious effect is observed; the measured loss values are different for clockwise and counterclockwise rotation (especially for large flux density). There are many papers trying to explain this phenomenon (Zurek and Meydan 2006b, Pfützner et al. 2007, Maeda et al. 2007, Yanase et al. 2007, Todaka 2009). Investigations of the same sample by other method do not detect such asymmetry (Zurek et al. 2009). Therefore, the asymmetry of rotational sheet tester results is often explained as the inaccurate. Indeed even very small angle misalignment of all sensors results in quite large power error. Therefore, the best method to eliminate such asymmetry is determine the results as an average results for two tests: clockwise and counterclockwise (Sievert et al. 1990). Such simple averaging yields quite a good agreement with other methods (e.g., thermometric).

Figure 5.55 presents the typical dependence of the phase shift between B and H vectors. For large flux density, this dependence is irregular, which explains the strange shape of hysteresis loop (Figure 5.54). Figure 5.56 presents the typical dependence of rotational loss versus flux density. Note that in contrary to axial power loss, the rotational loss approaches zero in saturation. The reason is the fundamental difference between both types of the power loss: in the rotating field the sample remains saturated (hence losses are lower) while in the axial field the sample is always demagnetized twice in each cycle, which increases the losses.

Figure 5.57 presents an example of loci of magnetic field strength for rotating circular field. On the basis of this dependence, we can analyze two-dimensional properties of the specimen, which is discussed in next section.

5.1.8 Two-Dimensional Properties

Most of magnetic materials exhibit anisotropy, which means that their properties depend on the direction of

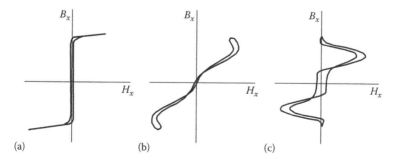

FIGURE 5.54
The examples of hysteresis loop of non-oriented steel (axial) (a), non-oriented steel (rotational) (b), grain-oriented steel (rotational) (c). (From Sasaki, T. et al., *IEEE Trans. Magn.*, 21, 1918, 1985.)

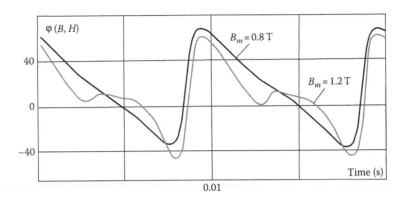

FIGURE 5.55
The example of results of testing of the rotational power loss—the dependence of the phase shift between B and H during the magnetization cycle. Specimen: GO steel M140-27S. (From Anuszczyk, J. and Pluta, W., *Soft Ferromagnetics in Rotating Magnetic Fields* (in Polish), WNT, Warszawa, Poland, 2009.)

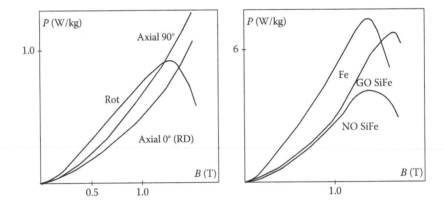

FIGURE 5.56
Two examples of results of testing of the rotational power. (From Zouzou, S. et al., *J. Magn. Magn. Mater.*, 112, 106, 1992; Fiorillo, F. and Rietto, A.M., *IEEE Trans. Magn.*, 24, 1960, 1988.)

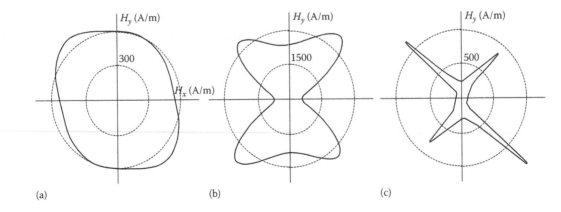

FIGURE 5.57
The example of results of testing of the rotational power loss—the locus of magnetic field strength for circular rotating field: (a) $B = 1.4\,T$ No470-50A steel, (b) $B = 1.4\,T$ M300-35A steel, and (c) $B = 1.2\,T$ M150-30S steel. (From Anuszczyk, J. and Pluta, W., *Soft Ferromagnetics in Rotating Magnetic Fields* (in Polish), WNT, Warszawa, Poland, 2009.)

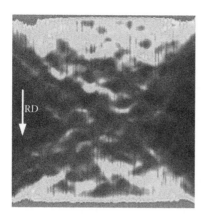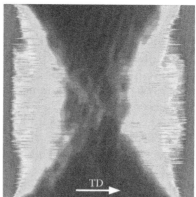

FIGURE 5.58
The distribution of magnetic field strength in the 8 × 8 cm sample of RSST. Sample magnetized along the rolling direction RD and in transversal TD direction. (From Tumanski, S. et al., Analysis of magnetic field distribution in the sample of RSST device, in *Proceedings of Soft Magnetic Materials 16 Conference*, Düsseldorf, Germany, pp. 859–864, 2004.)

magnetization. In some materials, for example, in grain-oriented electrical steel, the anisotropy is significant. But most of manufacturer data sheets inform only about the properties in one preferred direction. Typically, it is the rolling direction or more generally the easy axis of anisotropy. Also most of software used to design magnetic devices do not take into account the anisotropy.

Rarely in some software packages, there are included the magnetization curves for two directions of magnetization and the properties are calculated as intermediate between these two extremes of data. But, for example, in grain-oriented steel, the worse properties are not for 90° but for 55°. Thus, the best solution would be to include in the computer memory the set of characteristics for various directions. Unfortunately, methods of obtaining such data are not straightforward.

Theoretically, the simplest method is to use the RSST described in the previous chapter. To obtain rotation of the flux, both pairs of the yoke should be excited by the signals shifted by 90°. If these signals are in phase, we

have nonrotating field with its direction depending on the ratio of these signals.

Figure 5.58 presents experimentally determined distribution of magnetic field strength in the sample of RSST device. In the case of magnetization in the rolling direction, the uniform magnetized area is about 3 × 3 cm but for transversal direction, it can be much smaller. Additionally, the tested area is limited by the devastated material around the holes (see also Figure 2.190). Thus, the sample should be much larger to obtain reliable results, especially in the case of high oriented steel with large grains. Nevertheless, the application of RSST device to measurement of magnetic characteristics along arbitrary directions is reported (Morino et al. 1992, 1993, Nakata et al. 1993).

Probably a good candidate for such measurements could be a vertical yoke system (as in Figure 5.51b) with the sheet much larger than the yoke. For a very large sheet, the demagnetizing field is far from the tested area and can be neglected. But also in this case, we should be careful. As presented in Figure 5.59a (magnetization

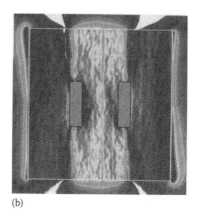

(a)   (b)

FIGURE 5.59
The distribution of magnetic field strength in the RSST device with vertical yoke of the tested area 9 × 9 cm and the specimen in form of the sheet with dimension 30 cm × 30 cm. (a) Rolling direction, tested area 15 × 15 cm and (b) transversal direction, tested area 35 × 35 cm. (From Tumanski, S., *Przegl. Elektr.*, 81(5), 32–34, 2005.)

along rolling direction), the distribution of magnetic field is perfect; exists only in the area between the poles. But for transversal direction (Figure 5.59b), the magnetic flux tries to find better way and practically the whole sheet is magnetized. Nevertheless the RSST devices are actually the best systems for measurement of two-dimensional properties of electrical steel.

Probably also a single-strip tester could be useful for testing of 2D performances. But in this case, the shape anisotropy can strongly influence the results. Test of two samples 300×100 and $300 \times 25\,mm$ cut at the angles $30°$ and $45°$ to the rolling direction gives different results due to shape anisotropy (Layland et al. 1996).

Testing of the whole Epstein frame with samples cut at an angle to rolling direction is more reliable. This method of sample preparation is tedious and a lot of material is wasted. But the method is still reported (Shirkoohi 1994, Shirkoohi and Arikat 1994, Tumanski 2002c). Figures 5.60 and 5.61 present the results of investigation of the hysteresis loop for various angles of magnetization.

In the case of the Epstein frame, the attention is recommended during stacking of the strips. The results obtained for the same order of strips (e.g., all in the direction $45°$) are completely different than in the case of X-stacking order ($45°$ and $-45°$; see Figure 5.62) (Nakata et al. 1984, Fiorillo et al. 2002). On the other hand, in the X-stacked order, we obtain complete flux closure in the plane of the lamination and the emulation of infinitely extended sample. This way we can predict intrinsic properties of the material that does not depend on the sample dimensions.

Figure 5.63 presents another system for measurement of 2D properties of the sample in form of a sheet. The direction of magnetizing field varied by computer-controlled rotation of the yoke. In this case, it is not possible to use generally recommended symmetrical yoke. To diminish the influence of the poles of the yoke and their eddy currents, the sheet is significantly larger than the yoke and the tested area is significantly smaller than the magnetizing area.

The sensors detect magnetic field components with respect to rolling direction (Figure 5.64). The components with respect to the magnetizing direction can be calculated. Figure 5.65 presents an example of measured permeability and power loss versus direction of magnetization with respect to the rolling direction. The dependence of magnetization curves and power loss on the direction of magnetization were presented earlier (Figures 3.18 and 3.24).

Figure 5.66 presents the examples of the dependence of magnetic field strength on the angle of magnetization determined at a fixed value of polarization. Some of materials exhibit bad performances for $55°$, but for others the worse performances were for $90°$. Figure 5.67

presents a similar dependence determined for so-called isotropic non-oriented materials.

Of course, using the needle sensors and H-coils sensors, it is possible to determine local distribution of 2D parameters (Enokizono et al. 1998). Also devices for 3D testing of magnetic materials were proposed (see Figure 2.192) (Zhu et al. 2002a,b, 2003, 2009).

5.1.9 Anisotropy and Texture

Correct modeling of anisotropy plays crucial role in design of magnetic devices (Di Napole et al. 1983, Silvester and Gupta 1991, Moses 1992, Liu and Shirkoohi 1993, Liu et al. 1994, 1996, Pera et al. 1993, Fard and Moses 1994, Shirkoohi and Kontopoulos 1994, Soinski and Moses 1995, Waeckerle et al. 1995, Moses and Soinski 1996, Jiles et al. 1997). Liu tested various models of anisotropic material and concluded that the multicurve model taking into account at least several characteristics representing important directions ($0°$, $55°$, $90°$) was the most effective (Figure 5.68).

The IEC standard 10106 defines only the anisotropy of loss in the case of non-oriented steel taking into account power loss P_0 in the rolling direction and P_{90} in direction perpendicular to rolling direction:

$$\delta P = \frac{P_{90} - P_0}{P_{90} + P_0} \tag{5.36}$$

By analogy, we can define the anisotropy of magnetization curve for fixed polarization:

$$\delta H = \frac{H_{90} - H_0}{H_{90} + H_0} \tag{5.37}$$

But the dependence of magnetic field strength on the direction of magnetization is very irregular and so information only about the border conditions ($0°$ and $90°$) is insufficient. Therefore, it was proposed to extend the definition of anisotropy (Soinski and Moses 1995, Moses and Soinski 1996) to the form representing selected angles α:

$$\Delta B = B_0 - B_\alpha \quad \text{and} \quad \delta B = \frac{\Delta B}{B_0} \tag{5.38}$$

Of course, we can also express the anisotropy energy E_k using the anisotropy constants K in the Akulov equation describing the cubic crystal:

$$E_k = K_0 + K_1(\alpha_1^2\alpha_2^2 + \alpha_2^2\alpha_3^3 + \alpha_3^2\alpha_1^2) + K_2(\alpha_1\alpha_2\alpha_3)^2 \tag{5.39}$$

where α_1, α_2, and α_3 are the directional cosines of the magnetization vector relative to the axes of a crystal.

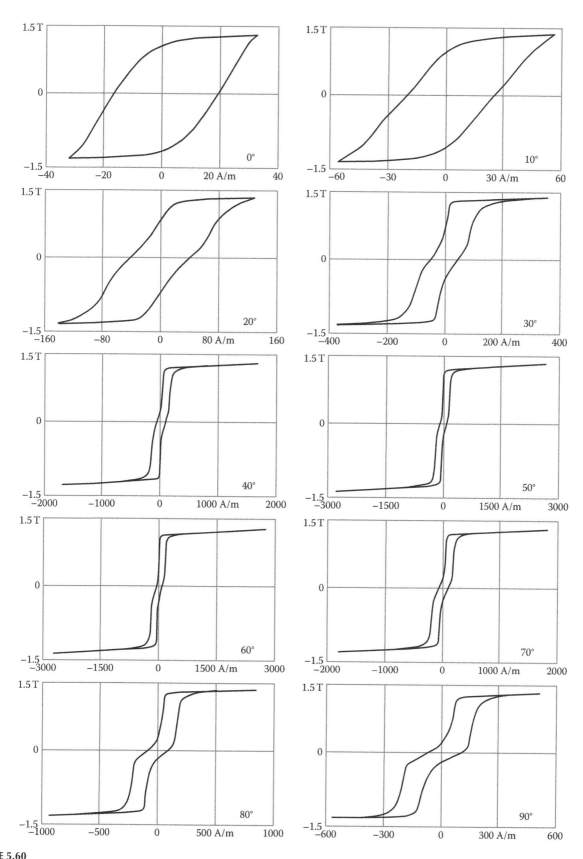

FIGURE 5.60
The set of hysteresis loops determined for grain oriented M089 steel (note various horizontal scales). (From Tumanski, S., *Proceedings of 7th Workshop on 2D Measurements*, Lüdenscheid, Germany, pp. 151–157, 2002.)

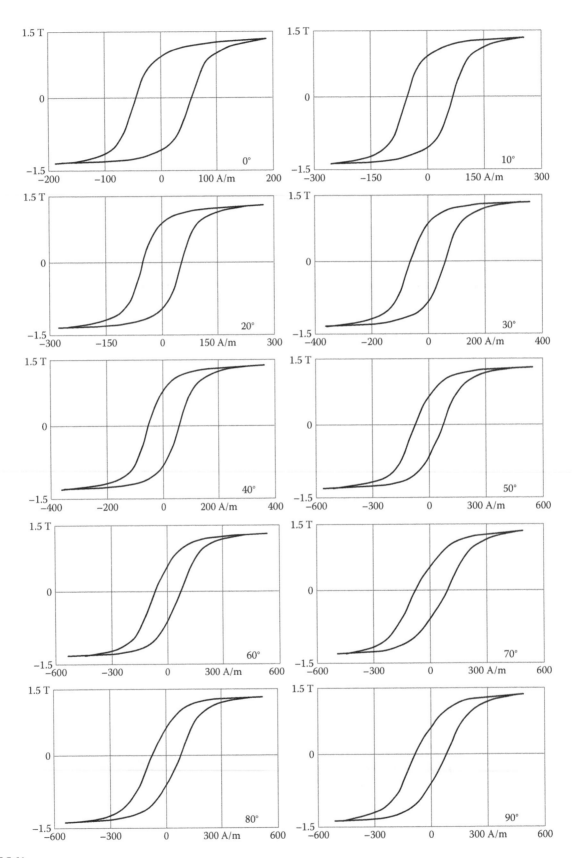

FIGURE 5.61
The set of hysteresis loops determined for non-oriented M330 steel (note various horizontal scales). (From Tumanski, S., *Proceedings of 7th Workshop on 2D Measurements*, Lüdenscheid, Germany, pp. 151–157, 2002.)

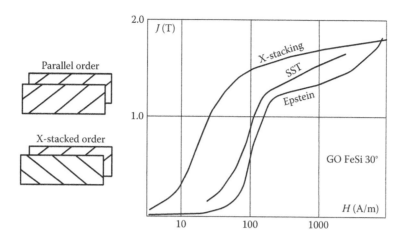

FIGURE 5.62
Experimentally determined magnetization curves in SST and Epstein device (with different stacking of strips). (From Fiorillo, F. et al., *IEEE Trans. Magn.*, 38, 1467, 2002.)

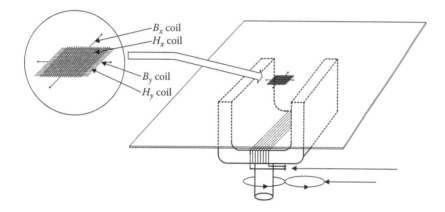

FIGURE 5.63
The testing device for investigation of electrical steel sheet in arbitrary direction. (From Tumanski, S. and Bakon, T., *J. Magn. Magn. Mater.*, 223, 315, 2001; Tumanski S., *IEEE Trans. Magn.*, 38, 2808–2810, 2002.)

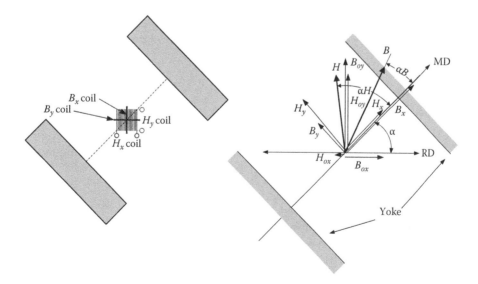

FIGURE 5.64
The sensors and vectors configuration: the rolling direction (RD) and magnetizing direction (MD), respectively; α, angle of magnetization; H_{ox}, H_{oy}, B_{ox}, B_{oy}, field components detected by the sensor; H_x, H_y, B_x, B_y, calculated field components.

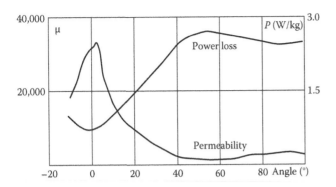

FIGURE 5.65
Permeability and power loss versus the direction of magnetization determined experimentally for GO electrical steel for $B = 1.3$ T.

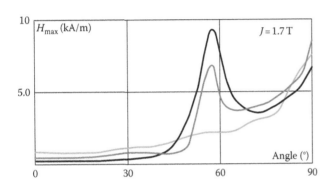

FIGURE 5.66
Magnetic field strength versus the direction of magnetization determined experimentally for three grades of GO electrical steel.

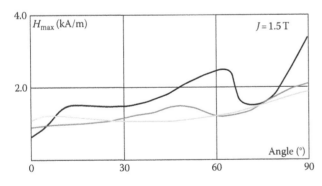

FIGURE 5.67
Magnetic field strength versus the direction of magnetization determined experimentally for three grades of NO electrical steel.

Commonly, numerical design of magnetic devices that takes into account the anisotropy by using finite element method (FEM) is used. The functional to be minimized for the finite element analysis is (Shen et al. 1987)

$$F = \int\int_{\Omega}^{H}_{0} B d\mathbf{H} d\Omega + \int_{\Gamma} B_n \phi d\Gamma \qquad (5.40)$$

where
Ω is the material region
Γ is the boundary
B_n is normal component of flux density

and magnetic scalar potential ϕ is defined as

$$\mathbf{H} = -\text{grad}\phi \qquad (5.41)$$

The unknown potential ϕ is described by the expression

$$\phi(x, y, z) = \sum_{i=1}^{n} \alpha_i(x, y, z)\phi_i \qquad (5.42)$$

where n is the number of nodes.

The example of the final Jacobian matrix is (Shen et al. 1987)

$$\frac{\partial^2 F}{\partial\phi_j\phi_i} = \int_{\Omega} \text{grad}^T\alpha_i \frac{\partial B}{\partial H^T} \text{grad}\alpha_j d\Omega + \int_{\Gamma} \alpha_i \frac{\partial B_n}{\partial\phi}\alpha_j d\Gamma \qquad (5.43)$$

where the anisotropic material is represented by the incremental permeability tensor $[\partial B/\partial H]^T$.

Various models of anisotropy were based on the experimentally determined family of magnetization curves (similar to those presented in Figure 5.68). Di Napoli and Paggi (1983) proposed to draw the loci of vector B with a fixed magnetic field strength (Figure 5.69) and then to substitute it by an ellipse

$$\frac{B_x^2}{a^2} + \frac{B_y^2}{b^2} = H^2 \qquad (5.44)$$

where $a = \mu_x$, $b = \mu_y$, and $\mu_y = r\mu_x$.
The tensor of permeability is

$$\mu = \begin{bmatrix} 1 & 0 & 0 \\ 0 & r & 0 \\ 0 & 0 & r \end{bmatrix}\mu_x \qquad (5.45)$$

and the differential term in Equation 5.43 is (Shen et al. 1987)

$$\frac{\partial B_i}{\partial H_j} = r\mu_x \frac{\partial H_i}{\partial H_j} + r_i H_i \left(\frac{\partial\mu_x}{\partial H^2} \frac{\partial H^2}{\partial H_j} \right) = r_i\mu_x \frac{\partial H_i}{\partial H_j} + 2r_i H_i H_j \frac{\partial\mu_x}{\partial H^2} \qquad (5.46)$$

In the elliptical model, the material is represented by one curve, which is advantageous in comparison with many experimental data in the multicurve model. But the elliptical model can be constructed only if there is no

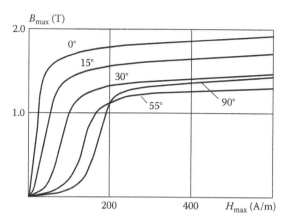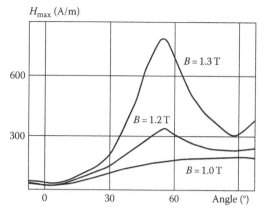

FIGURE 5.68
The distinctive directions of magnetization of grain-oriented steel.

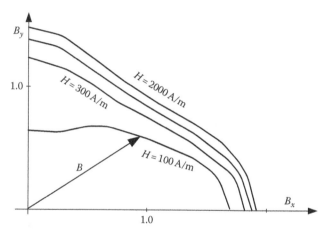

FIGURE 5.69
The elliptical model of anisotropy—locus of the flux density for various directions of magnetization. (From Liu, J. et al., *IEEE Trans. Magn.*, 30, 3391, 1994.)

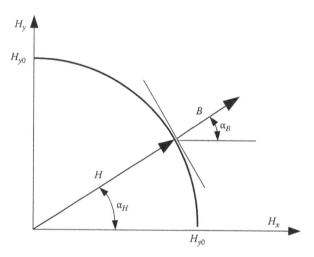

FIGURE 5.70
Determination of *B* direction from the coenergy contour. (From Pera, T. et al., *IEEE Trans. Magn.*, 29, 2425, 1993a.)

$$f(H_x, H_y) = \left(\frac{H_x}{H_{x0}}\right)^n + \left(\frac{H_y}{H_{y0}}\right)^n = 1 \quad (5.49)$$

phase shift between the *B* and *H* vectors. This leads to significant error. Therefore, other model was proposed based on the energy minimization theory (Silvester and Gupta 1991, Pera et al. 1993a,b). According to this theory, the vectorial relation *B*(*H*) is equivalent to a coenergy representation defined as

$$w' = \int_0^H B(H)dH \quad (5.47)$$

where w' is the density of an electromagnetic coenergy stored in the material as a function of applied field intensity vector *H*. The flux density vector *B* can be determined as

$$B(H) = \text{grad}_H w'(H) \quad (5.48)$$

The contour of equal coenergy is described by the relation

Figure 5.70 presents the graphical representation of the expression (5.49) where H_{xo} and H_{yo} represent magnetic fields strength in the rolling and transversal direction, respectively. The exponent n is determined to obtain the condition that maximum energy corresponds with the hard magnetization direction. The direction of flux density is determined knowing that the *B* vector is perpendicular to contour of equal coenergy at an arbitrary point.

The coenergy model is only a simplification because it was derived under assumption that the material is anhysteretic. Thus, the most accurate model of anisotropy of FEM calculation is the multicurve model determined experimentally. It is possible to determine the family of reluctivity curves that can be then used in the

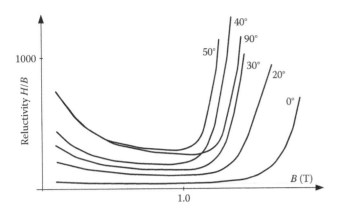

FIGURE 5.71

The anisotropic reluctivity curves for various directions of magnetization. (From Shirkoohi, G.H. and Kontopoulos, A.S., *J. Magn. Magn. Mater.*, 133, 587, 1994.)

FEM calculations (Liu and Shirkoohi 1993, Shirkoohi and Kontopoulos 1994) (Figure 5.71).

Modern 2D testing devices enable us to determine practically all components of the reluctivity tensor. In such case, no modeling of anisotropy is necessary. The problem is that for such computation, huge amount of data in memory is required. Therefore, Moses proposed interactive modeling where FEM package is linked with measuring equipment (Moses 1992).

In the anisotropy measurement, we usually determine the difference of magnetic properties for various directions of magnetization. From physical point of view, it is interesting to determine anisotropy coefficients (according to Equation 5.39) as well as the directions of hard and easy axes of anisotropy. Recently, importance of such measurements is weakened by the development of various RSST and 2D measuring devices, which enable determination of anisotropy (see, e.g., Figures 5.57 and 5.66). But such devices are rather complex and expensive. Moreover, they are not applicable for small samples, for example, thin film elements. Therefore, other specialized methods for testing anisotropy are also developed.

One of the oldest methods is a torsion magnetometer (Kouvel and Graham 1957, Zijlstra 1967, Tejedor et al. 1990, Cornut et al. 2003).* The sample in form of a disk or sphere (to reduce the effect of the shape anisotropy) is suspended in rotating magnetic field (see Figure 2.196). The torque acting on the sample proportional to the derivative of energy $\tau = dE/d\theta$, where θ is an angle of rotation related to suitable axis. Thus for the sample of Goss texture the energy is (see Equation 2.28)

$$E = K_0 + \frac{1}{4}K_1(\sin^4\theta + \sin^2 2\theta) + \frac{1}{4}K_2(\sin^4\theta\cos^2\theta) \quad (5.50)$$

and the torque is (Cullity and Graham 2009)

$$\tau = -\frac{dE}{d\theta} = -\left(\frac{K_1}{4} + \frac{K_2}{64}\right)\sin 2\theta - \left(\frac{3K_1}{8} + \frac{K_2}{16}\right)$$

$$\times \sin 4\theta + \frac{3K_2}{64}\sin 6\theta \quad (5.51)$$

An example of such torque curve is presented in Figure 5.72. From this curve, we can directly indicate the principal axes as well anisotropy coefficients (e.g., by using Fourier analysis).

The rotating magnetic field should be sufficiently large in order to obtain condition that the angle θ (between the magnetization direction and the easy axis) is the same as the angle φ (between direction of magnetic field and the easy axis). That is why sometimes in torque magnetometer Halbach magnet is used (Cornut et al. 2003). If this is not fulfilled, then the appropriate correction should be introduced (Fiorillo 2004)

$$\theta = \varphi - \sin^{-1}\frac{\tau(\varphi)}{\mu_0 M_s H} \quad (5.52)$$

The torque depends on the perpendicular component of polarization. Instead of a torque, the induced voltage in the coil wound on the disk sample can be measured (Figure 5.73). Such principle proposed by Ingerson (Ingerson and Beck 1938) was later improved (Matheisel 1973, Soinski 1984). The sample in form of a disk is rotated in the alternating magnetic field generated by an electromagnet. The obtained angular dependencies are very similar to the torque dependencies.

Figure 5.74 presents the measuring system for testing the anisotropy on the large sheets. Because the yoke is rotating, it must be asymmetrical and the B-coils are wound on the yoke near the poles. Therefore, the

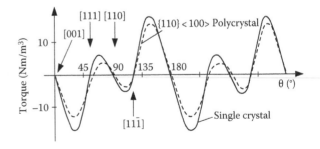

FIGURE 5.72

Torque curves of single crystal and textured polycrystal. (From Graham, C.D., Textured magnetic materials, in *Magnetism and Metallurgy*, Academic Press, New York, Chapter 15, 1969.)

---

* Commonly as magnetometers are assumed devices for measurement of magnetic field. But there are special types of measuring devices, known as magnetometers, for example classical Foner magnetometer (discussed more detailed later).

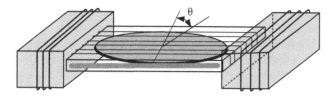

FIGURE 5.73
The induction anisometer. (From Soinski, M., *IEEE Trans. Magn.*, 20, 172, 1984.)

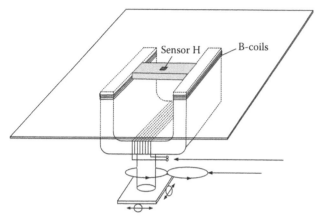

FIGURE 5.74
The device for testing an in-plane distribution of anisotropy. (From Tumanski, S., *IEEE Trans. Magn.*, 38, 2808–2810, 2002.)

accuracy of measurements of flux density is poor, but for comparative measurements it is acceptable. H-coils can be used as the magnetic field strength sensors. In the presented system, small 1×1 mm MR sensor is used, and this way it is possible to test the anisotropy curves grain by grain.

Figure 5.75 presents the examples of the dependence of magnetic field strength on the direction of magnetization as determined for the same steel sheet by using the device presented in Figure 5.74. The sample is polycrystalline consisting of many grains of different texture and direction of anisotropy axis. The Goss texture dominates only in one location in (bottom left); graphs show easy axis for 0° and hard axis for 55°.

Also the torque curve differs for a single crystal and for a polycrystalline sample in which the result is averaged from many grains. The term "texture" is used to description of orientation of crystals. Texture means preferred orientation of majority grains with respect to some coordinate system. For instance, if we say that the sample exhibits the Goss texture, it means that majority of the grains have (110)[100] crystal orientation. But in the polycrystalline sample, there can be crystals of arbitrary orientation as well as with the Goss orientation but different angles with respect to the rolling direction. Texture informs about the number and type of various components of the polycrystalline sample.

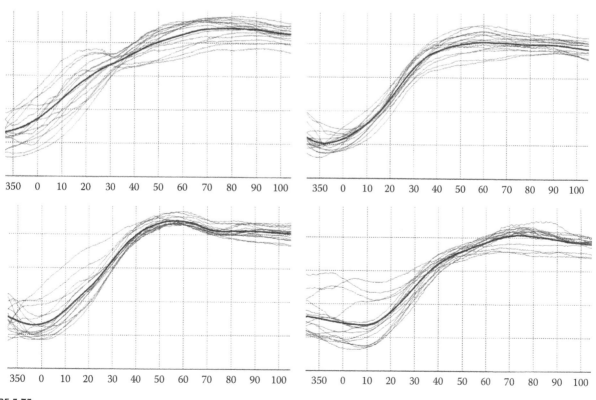

FIGURE 5.75
The examples of local distribution of anisotropy curves $H = f(\theta)$ on the four various areas of the same grain-oriented sheet.

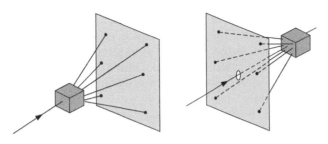

FIGURE 5.76
The transmission or reflection of x-ray on the crystal lattice.

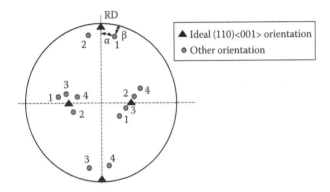

FIGURE 5.77
Stereographic projection showing orientation of typical SiFe grains and the deviation angles. (From McCarthy, M. et al., *J. Appl. Phys.*, 38, 1096, 1967.)

The x-ray or neutron diffraction methods are commonly used for texture analysis (Wecker and Morris 1978, Wenk and Van Houtte 2004). The main method is the Laue method where the beam of x-ray is reflected from the atoms of crystal lattice. Two types of Laue method are used: transmission method and back-reflection method (Figure 5.76). Pole figures attributed to crystal orientation are created as a result of diffraction (see Figure 2.20).

There are a number of sophisticated methods of analysis of texture pattern enabling quantitative determination of the grain size, crystal orientation, and type of texture.

Figure 5.77 presents the pole figures with indicated perfect Goss orientation (triangles) and several misoriented grains. From the position of the "spots," it is possible to determine the misorientation angles.

Figure 5.78 presents the grain structures and corresponding pole figures before and after recrystallization. We can observe the growth of the grains with the correct orientation. Figure 5.79 presents similar figures but with classification of the grain sizes. We can observe that smaller grains exhibit larger deviation from Goss texture.

Instead of point representation, often a kind of map is constructed. Figure 5.80 presents the distribution of the pole normal to {100} plane. After heating (recrystallization), we can observe an improvement of the texture.

5.2 DC Testing of Soft Magnetic Materials

5.2.1 Analog and Digital Fluxmeters

Generally, it is recommended to test the magnetic materials in the conditions similar to expected future application. For this reason, testing of soft magnetic materials in DC magnetic field is often performed although it is usually more difficult in comparison with AC excitation. Moreover, testing under DC conditions brings information about more "virgin" properties of

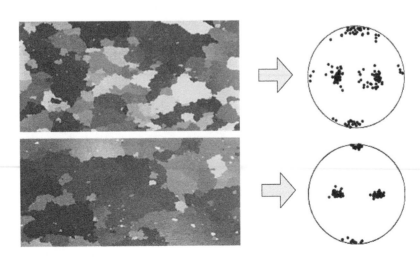

FIGURE 5.78
The grain structure and corresponding diffraction pattern before and after secondary recrystallization. (From Kumano, T. et al., *ISIJ Int.*, 43, 736, 2003.)

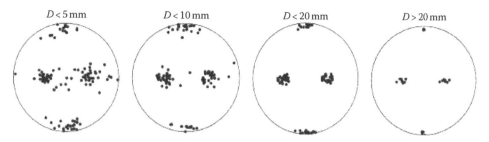

FIGURE 5.79
The pole figures of the sample with classification of the grain size *D*. (From Kumano, T. et al., *ISIJ Int.*, 42, 440, 2002.)

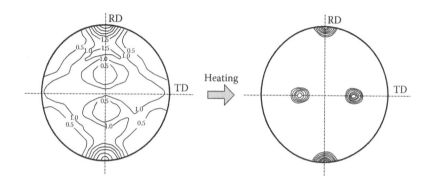

FIGURE 5.80
Recorded {100} pole figures before and after heating of the sample with classification of the grain size *D*. (From Kawasaki, K. et al., Development of new pole figure method for directly projecting grains with high speed and high resolution using synchrotron radiation and its application for observing electrical steel and mild steel sheets at high temperatures, *Nippon Steel Technical Report*, No. 69, 1996.)

the tested material—free from eddy currents and other dynamic influences.

The main obstacle under DC conditions is that coil sensors based on Faraday's law do not work if magnetic flux is not varying. Thus to obtain output signal from the coil, it is necessary to somehow introduce variation of magnetic flux that voltage would be induced:

$$E = -\frac{\Delta \Phi}{\Delta t} \qquad (5.53)$$

This change of magnetic flux can be introduced by switching the flux off* or by removing the sensor from the magnetic field. The use of voltmeter for measurement of the flux is not convenient because the output signal quickly disappears and its value depends on the speed of the change. But if we use the integrating device, we obtain

$$\Delta \Phi = \int E dt \quad \text{or} \quad \Delta B = \frac{1}{nA} \int E dt \qquad (5.54)$$

Earlier for measurements of the flux density, special instruments known as ballistic galvanometer, fluxmeter, or webermeter were used. Such devices had especially

increased moment of inertia or damping and detected the amount of the charge (thus they performing integration of the pulse signal). Recently, these mechanical instruments are old fashioned and are substituted by electronic devices (usually digital known as digital fluxmeters).

Currently, there are various types of commercial electronic integrators. These instruments are commonly used to measure the magnetic field (with induction coil sensor), current (with Rogowski coil sensor), and to DC test of magnetic materials.

There are several methods to build digital fluxmeters:

- Use of analog integrator based on operational amplifier and next analog-to-digital converter
- Conversion of the pulse signal to digital form and performing digital integration
- Conversion of analog signal to frequency and counting pulses

The simple solution to obtain fluxmeter is to use an integrating amplifier as presented in Figure 5.81. The output signal of this circuit is

$$V_{out} = \frac{1}{R_1 C} \int\limits_{t_0}^{t_0 + T} V_{in}(t) dt + V_0 \qquad (5.55)$$

---

* The change of the flux can be $\Delta \Phi$ if we switch off the flux ($\Phi \rightarrow 0$) or $2\Delta \Phi$ if we reverse the polarization (+$\Phi \rightarrow -\Phi$).

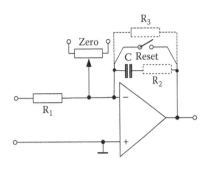
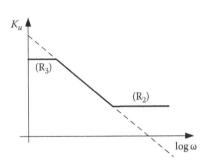

FIGURE 5.81
The integrating amplifier and its frequency characteristic (resistors R_2 and R_3 optionally for changing the characteristic).

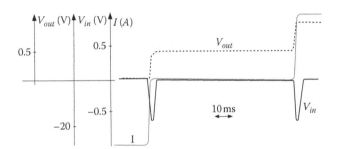

FIGURE 5.82
Signals recorded with an Epstein frame when switching the magnetizing current from −1 to 0 A and then from 0 to 1 A (V_{in}, V_{out}, input and output signals of integrating amplifier. (From Bajorek, J. et al., *Przegl. Elektrotech.*, 80, 161, 2004.)

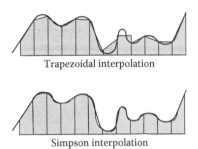

FIGURE 5.83
Two examples of numerical integration algorithms.

An important problem in application of analog integrating amplifier is its zero drift (Stata 1967). Therefore, such fluxmeters have usually possibility of zero correction (short-circuiting the capacitor by a switch "reset"), which is sufficient for short measurements. Sometimes, it is necessary to measure pulses lasting up to thousands of seconds (Spuig et al. 2003). In such a case, more sophisticated amplifiers, including auto-zero option, are recommended.

Figure 5.82 presents typical signals registered in Epstein frame during the testing of non-oriented steel. The resistance R_1 of integrating amplifier was 10 kΩ, which is a typical value (Scholes 1970) and capacitors were 1 or 10 μF (corresponding with the fluxmeter ranges 10 or 100 mWb or time constants RC = 10 or 100 ms). The nonlinearity error for the output signal of 10 V was smaller than 0.2% and zero drift was smaller than 5 μWb/min. The shape of induced pulses strongly depends on the resistance of the primary winding and speed of switching (Bajorek et al. 2004).

When the integrating circuit is a part of a computer measuring system, it is usually better to perform the whole operation numerically. Because the measured signal is a series of pulses, a high sampling rate and number of bits of analog-to-digital converter is required. For example, for testing materials, the 16 bit and rate of

250 kS/s was advisable (Yamamoto et al. 1994, Gozdur and Majocha 2010). The algorithm of numerical integration also plays an important role (Davis et al 2007). Various algorithms of integration have been tested (Figure 5.83) (Papamarkos and Chamzas 1996, Nam Quoc Ngo 2006, Tseng 2006) among other trapezoidal where sampling interval T was interpolated by

$$H_T(z) = \frac{T}{2} \frac{1+z^{-1}}{1-z^{-1}} \qquad (5.56)^*$$

or Simpson algorithm where sampling interval was interpolated by a polynomial:

$$H_T(z) = \frac{T}{3} \frac{1+4z^{-1}+z^{-2}}{1-z^{-2}} \qquad (5.57)$$

Table 5.2 presents the comparison of measurements of a sinusoidal signal using different algorithms of integration. As can be seen, the differences between the algorithms appear negligible and difference between the analog and numerical integration was very small. In these tests, the sampling frequency was chosen to obtain 5000 samples per cycle.

---

* $H(z)$ is a z-transform or more exactly z-transform of impulse response.

TABLE 5.2

Comparison of the Results of Numerical Integration Using Different Algorithms

	Analog	Rectangular	Trapezoidal	Simpson
B_{rms}: 50 Hz	1.203	1.203	1.204	1.203
B_{max}: 50 Hz	1.699	1.701	1.701	1.700
B_{pp}: 50 Hz	3.398	3.399	3.401	3.399
B_{rms}: 1 Hz	1.234	1.237	1.237	1.236
B_{max}: 1 Hz	1.699	1.702	1.700	1.699
B_{pp}: 1 Hz	3.404	3.403	3.403	3.405

Source: Gozdur, R. and Majocha, A., *Przegl. Elektrotech.*, 86(4), 52, 2010.

Other method of digital integration is also reported—by realization of the old idea proposed by de Mott (de Mott 1966, 1970). If we use the voltage-to-frequency converter, then the number of output pulses is proportional to the input voltage × time, thus the time integral of input voltage. Kurihara reported tests of various long-time digital integrators for magnetic measurements utilizing commercially available voltage-to-frequency converters (AD652 of Analog Devices and VFC110 of Burr-Brown) and up-down counters (Kurihara and Kawamata 1998). Obtained performances were very promising. Figure 5.84 presents the voltage to frequency converter model AD562. It enables to convert voltage in the range 0–10 V into frequency up to 1 MHz with linearity error 0.005% FS. This principle is used, for example, in a market available high-precision fluxmeter PDI 5025 of Metrolab.

Before any measurements can be carried out, it is recommended to perform calibration of the fluxmeter. The induced pulse depends on the circuit resistance; therefore, the best would be to calibrate the instrument exactly at the same circuit as the expected application. The resistance of the source is often smaller than R_{in} of the integrator and can be neglected but sometimes it should be taken into account.

The international standard 60404-4 recommends following methods of the calibration of fluxmeter:

- To use standard source of Vs signals
- To use standard source of magnetic field (magnet or Helmholtz coils) and coil sensor
- To use a standard capacitor
- To use a standard mutual inductance

In the first method, it is sufficient to measure the rectangle pulse with precisely determined time and magnitude. The integrator should measure the voltage × time. The Walker Scientific Company (Laboratorio Elettrofisico) offers precise volt-second generator MTC-1 with an accuracy of 0.05% and pulse ranges from 100 μVs to 1 Vs.

In the second method, it is advisable if we plan to use the fluxmeter for magnetic field measurement. It is sufficient to extract the coil from the gap of the magnet or from the Helmholtz coils and determine the fluxmeter constant C according to the relation $N = S_c C \Delta B$ where N is the reading indicated by the fluxmeter (output signal) and S_c is the sensitivity coefficient of the coil sensor. We should also take into account that extraction (or introduction) to "known" fields results actually in the measurement $\Delta B = B - B_0$ where B_0 is the external field (e.g., the Earth's magnetic field). To minimize the influence of the Earth's field, the calibration should be performed along the east–west directions and the tests should be repeated for two polarizations of standard magnetic field.

If we use the capacitor C to calibrate the fluxmeter, we assume that the charge $Q = CV$ of the capacitor previously loaded by the voltage V results the magnetic flux $\Delta\Phi = CVR$ (R is the resistance of the circuit, thus the input resistance of the fluxmeter).

Figure 5.85 presents the commonly used method for calibrating of the fluxmeter in the system for testing of magnetic materials. If we connect the mutual inductance instead of the tested core, the change of the current in the primary winding of mutual inductance results that the flux in the secondary winding is

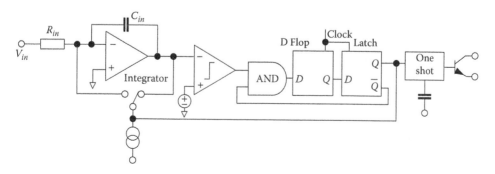

FIGURE 5.84
The voltage-to-frequency converter—model AD652 of Analog Devices.

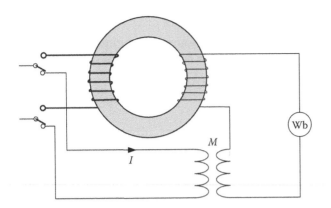

FIGURE 5.85
The mutual inductance M used to calibrate the fluxmeter in the circuit for testing of magnetic materials.

$$\Delta\Phi = M\Delta I \qquad (5.58)$$

To eliminate the influence of the difference of resistance of the windings of the mutual inductance and the tested specimen usually the secondary winding of the, mutual inductance remains in the testing circuit also during the testing of the specimen.

There are many models of digital fluxmeters available on the market, as for instance models: 480 of Lakeshore (Figure 5.86), PDI 5025 of Metrolab, ED5 of Magnet-Pphysik, F10 of Brockhause, F2 of Magsys or CST7 of Xiamen Yuxiang Magnetic Materials. Most of them are equipped with automatic drift compensation, some of them offer many additional functions as control of the coil sensors movement or software to support the testing of magnetic material.

5.2.2 Point-by-Point DC Testing of Soft Magnetic Materials

For DC testing of soft magnetic materials, the same samples can be used as for AC tests including Epstein frame or toroid. The main differences in comparison with the AC testing are as follows:

- To measure the flux density, we should generate the stepwise change of the magnetic field strength.
- In the case of AC measurement, the flux in every cycle is running around a hysteresis loop, thus we measure the distance between the tips of hysteresis. In the case of DC measurement, we are going around the hysteresis loop step by step; therefore, it is always important to go in one direction (to avoid the minor hysteresis loop).

Figure 5.87 presents a typical circuit for DC testing of magnetic materials. The switch S1 allows introduction of the magnetic field difference ΔH. The switch S2 reverses polarity. In this figure, the possibility of manual change of magnetic field is considered. Of course, to introduce the difference of magnetic field strength, computer-controlled bipolar supply unit can be used (enabling also to create the change of magnetic field with change of polarity $+H \rightarrow -H$ without switching the S2). The separate winding is for an additional bias of the core by DC magnetic field U_0. Such bias can be useful, for example, for determination of anhysteretic curve.

In Figure 5.87, the winding supplied by alternating current to demagnetize the sample is also included. Usually, it is the same winding as primary winding for DC magnetization.

Figure 5.88 presents the method of determination of the magnetization curve. The demagnetized sample (after magnetization by AC magnetic field with decreased magnitude) firstly is magnetized by several commutation of DC filed. This way we stabilize the magnetic conditions. Then step by step we introduce changes of magnetic field $0 \rightarrow A, A \rightarrow B, B \rightarrow C$, etc. It is important to make all subsequent changes in one direction (e.g. always increasing the field). In the algorithm presented in Figure 5.88, the changes of magnetic field are selected in such a way as to obtain similar increments of magnetic flux density (otherwise near the saturation the resolution of measurements would be poor).

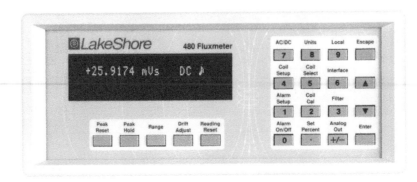

FIGURE 5.86
Picture of the Model 480 fluxmeter. (Courtesy of Lake Shore Cryotronics, Inc.)

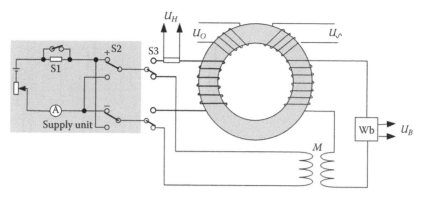

FIGURE 5.87
The typical circuit for DC testing of soft magnetic materials.

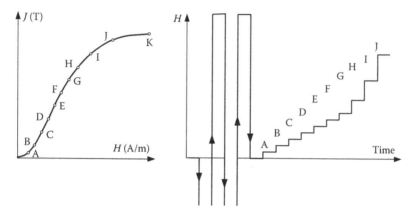

FIGURE 5.88
The algorithm for determination of the magnetization curve.

Figure 5.89 presents an example of algorithm of determination of hysteresis loop. One of the methods is to change the magnetic field step by step, thus: a → A, A → B, B → C, etc. (similarly as shown in Figure 5.88). But such changes of magnetic field strength for a small quant of ΔB the output signal of fluxmeter is small. Figure 5.89 presents different approach. Every time we return to the point "a" and from this point, we create different change of magnetic field strength, thus we have: a → A, a → B, a → C, etc. After arrival to the remanence (point D), we can continue such change of magnetic field but with changed of polarity, thus is: a → E, a → F, etc. In the algorithm presented in Figure 5.89, the second quadrant of the hysteresis loop is determined with a changed starting point, thus the changes are D → E, D → F, etc.

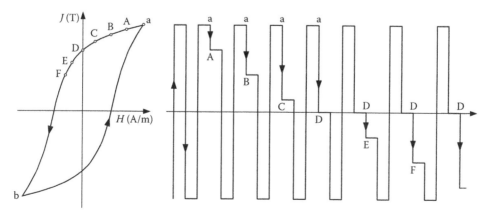

FIGURE 5.89
The example of algorithm for determination of the hysteresis loop.

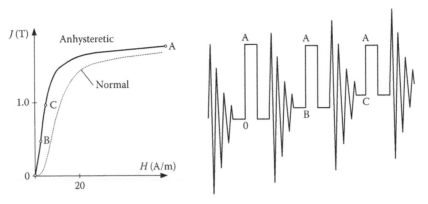

FIGURE 5.90
The example of algorithm for determination of the anhysteretic curve.

Figure 5.90 presents the strategy used for determination of the anhysteretic magnetization curve. Every time we demagnetize the sample with decreasing AC field but with various bias DC magnetic field. For every bias, we can then determine the flux density by changing the flux density ΔB from this point to the fixed point A.

The methods of determination of hysteresis loop or magnetization curve presented above seem to be complex. But with the availability of modern computer-controlled bipolar sources of voltage, such methods are relatively simple and fast.

5.2.3 Continuous Quasi-Static Testing of Soft Magnetic Materials

We can determine static magnetic properties of soft magnetic materials using the classical method of AC testing though performed at very low frequency. The question is, which frequency can be the best for the emulation of

DC conditions: 1, 0.1, or maybe 0.01 Hz? Figure 5.91 presents results of experimentally determined hysteresis loops at low frequency. Between 10 and 1 Hz, the difference is visibly significant. But between 1 and 0.1 Hz, this difference is much harder to detect.

The main differences between the AC and DC hysteresis loops come from the eddy currents. By using classical relation describing the eddy current magnetic field as the dependence on thickness t and conductivity γ

$$H_{\text{eddy}} = \frac{\gamma t^2}{8} \frac{dB}{dt} \tag{5.59}$$

we obtain that for the flux density 1.5 T, $f = 0.1\,\text{Hz}$ and thickness 1 mm the magnetic field strength $H_{\text{eddy}} = 0.75\,\text{A/m}$ (Fiorillo 2004), which is the nonnegligible value. Moreover, Fiorillo (2004) also discussed the dynamics of domain movement depending on the speed of the change of magnetic field. This can result

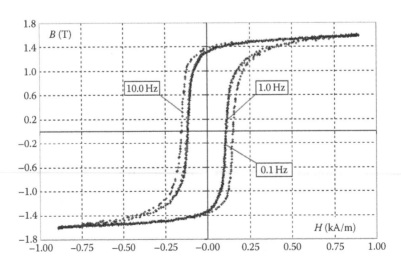

FIGURE 5.91
The quasi-static hysteresis loops determined at low frequency. (From Gozdur, R., *Przegl. Elektrotech.*, 80(2), 147, 2004; Gozdur, R. and Majocha, A., *Przegl. Elektrotech.*, 83(1), 134, 2007.)

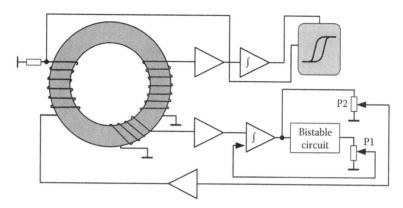

FIGURE 5.92
The measuring system for low frequency and dB/dt constant. (From Mazzetti, P. and Soardo, P., *Rev. Sci. Instrum.*, 37, 548, 1966.)

in differences between fluxmeter and quasi-static performances.

The magnetizing field should be increased and decreased as slow as possible and constant rate of dH/dt is not recommended (Fiorillo 2004). The triangular flux density signal is recommended with $dB/dt = $const. Figure 5.92 presents an example of the hysteresisgraph fulfilling this condition. The difference of a rectangular signal from the bistable circuit and dB/dt signal from additional winding are sent to an integrating amplifier. Such feedback keeps the difference small and the magnetizing current is appropriately formed to ensure linear changes of the flux density with constant dB/dt. The cyclic time is selected experimentally and for some material needs to be as long as 20 min. This value can be selected by the potentiometer P1 while the maximum value by the potentiometer P2.

The continuous magnetization is generally more difficult than the step-by-step method but it enables us to detect the physical phenomena not detectable by the fluxmeter method. An example is presented in Figure 5.93 where Barkhausen jumps are visible on the hysteresis loop.

The advantage of applying the quasi-static testing of soft magnetic materials is that with the same testing equipment we can investigate the sample for medium as well low frequency. An example is a measuring system developed for testing of soft magnetic materials with arbitrary chosen waveshape of the flux density (including pure sinusoidal and triangular) in the frequency range 0.005–400 Hz proposed by Birkelbach et al. (1986).

5.2.4 Permeameters for Testing of Soft Magnetic Materials

Permeameters are recommended by IEC Standard 60404-4 as devices for DC testing of soft magnetic materials. According to the standard 60404-4, DC magnetic permeameters are assumed to be the testing devices

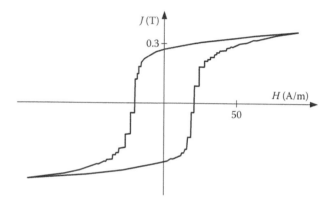

FIGURE 5.93
The example of the hysteresis loop determined for the FeAl sample. The cyclic time was 160 s. (From Mazzetti, P. and Soardo, P., *Rev. Sci. Instrum.*, 37, 548, 1966.)

for characterization of bulk soft magnetic materials. In the case of DC magnetic materials, it is not necessary to prepare laminated sheet because there does not exist the problem of eddy currents. Therefore, these materials, used, for example, for relays or magnetic actuators are manufactured in form of bars, rods, or cylinders of various lengths. To test such specimens, it is necessary to use the magnetic circuit enabling tests of this sample in conventionally closed magnetic circuit. The standard divides permeameters* into two groups: permeameters type A and permeameters type B (Figure 5.94).

In permeameter type A, the magnetizing coil is wound directly onto the specimen. Measuring coils or Hall sensor is positioned near the surface of the specimen. The yoke prepared from good quality magnetic material is used only to close the magnetic circuit. An example of such device is a permeameter proposed by

---

* This terminology is slightly misleading. Sometimes permeameters are considered all measuring instruments for tests of permeability. Often as permeameters are considered high-frequency permeability analyzers and permeameter is also the name of the device for testing of hydraulic conductivity.

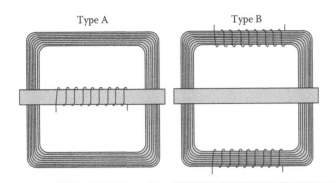

FIGURE 5.94
The design of permeameters type A and B.

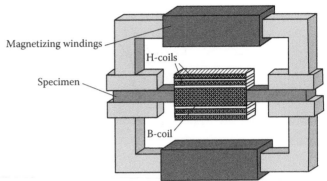

FIGURE 5.96
The design of the Sandford–Bennet permeameter.

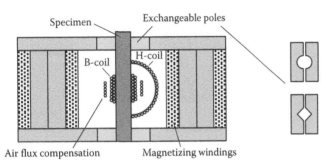

FIGURE 5.95
The design of the permeameter C-500 of Magnet-Physik.

Hopkinson over a century ago (Hopkinson 1885). In permeameter type B, the magnetizing coil is wound on the yoke. Near the specimen there are coil sensors, including Rogowski–Chattock potentiometer. The permeameters proposed by Fahy (1918) or Neumann (1939) are examples of such devices.

In previous century, a large number of various permeameters were proposed (e.g., by Iliivici, Babbit, Sanford, Carr, Burrows, Ewing, Fischer, and more). Recently, for strip samples, the so-called single strip tester is used (it resembles the permeameter type A). This type of device that can be easy adapted for other samples. There seem to be only two or three types of DC permeameters available on the market. One of them is the permeameter Remagraph C-500 proposed by Magnet-Physic (Figure 5.95).

A different approach is proposed by Laboratorio Elettrofisico. It is permeameter based on the Sandford–Bennet permeameter (Figure 5.96) (Sandford and Bennet 1993) with Hall sensor for the measurement of the magnetic field strength. In the original permeameter, two H-coils were used for magnetic field strength and the result of measurements was extrapolated to the surface of the sample.

The permeameters presented in this section can be also used for testing of hard magnetic materials.

5.3 Testing of Hard Magnetic Materials

5.3.1 Testing of the Hard Magnetic Materials in Closed Magnetic Circuit

The hard magnetic materials can be tested using similar circuit as presented in previous chapter devoted to permeameters. Figure 5.97 presents the typical circuit. In comparison with DC testing of soft magnetic materials, the following differences should be considered:

- Much, much higher magnetic field strength should be generated. It is assumed that the generated field strength should be five times larger than the coercive field H_{cj}, which gives about 5×10^6 A/m (Fiorillo 2004). This field should be established sufficiently slowly to avoid problems with eddy currents. Thus, it is necessary to use special a design of electromagnet with tapered poles from FeCo alloy, described more detailed in Section 2.12.3. Figure 5.98 presents the examples of two such electromagnets designed for test of hard magnetic materials.

- It is necessary to determine not only the magnetic field strength H and flux density B but also the polarization J whose value can differ significantly from B (which usually not the case for soft magnetic materials).

- Very large magnetic field strength enables substituting the system of coil sensor and fluxmeter by the Hall sensor.

To ensure uniform magnetization of the sample, the poles of electromagnet should have diameters D_0 more than two times larger than the length of the sample l_m and larger than diameter of the sample D_s (Fiorillo 2004)

$$D_0 \geq 2l_m, \quad D_0 \geq D_s + 1.2l_m \quad (5.60)$$

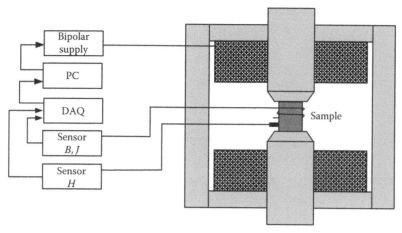

FIGURE 5.97
Typical testing system for hard magnetic materials.

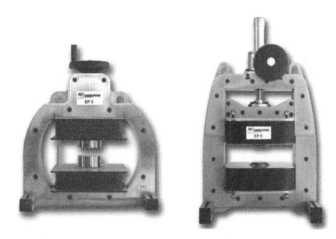

FIGURE 5.98
The electromagnets used for testing of hard magnetic materials in the measuring system of Magnet-Physik. (Courtesy of Magnet-Physik.)

A good contact of the poles with the surfaces of the tested magnet should be established (Figure 5.98).

Often it is not necessary to test the whole hysteresis loop because mostly only the second demagnetization quadrant of the hysteresis is important—with indicated $_BH_c$, $_JH_c$, and eventually energy product $(BH)_{max}$. Figure 5.99 presents an example of the results of hard magnetic material investigations.

The polarization J can be calculated as

$$J = B - \mu_0 H \qquad (5.61)$$

The simple method of determination of both values was proposed by Magnet-Physik (Steingroever et al. 1997). In the pole of electromagnet, two coils are embedded: one under the tested sample for measurement value of B, the second one outside the sample to measure the $\mu_0 H$ component (Figure 5.100).

A relatively simple method of compensation of air flux is recommended by the standard IEC 60404-4. There are two coils as in Figure 5.101 and the area and number of turns of these coils is designed in such a way that the following condition is fulfilled:

$$n_1 A_1 = n_2 A_2 \qquad (5.62)$$

For the sample of area A_s the coils detect the magnetic fluxes:

$$\Phi_{A1} = \mu_0 H n_1 (A_1 - A_s) \quad \text{and} \quad \Phi_{A2} = \mu_0 H n_2 A_2 \qquad (5.63)$$

The detected magnetic flux penetrating the sample is

$$\Phi_s = \mu_0 B n_1 A_s \qquad (5.64)$$

If we connect the coils differentially, we obtain resulting magnetic flux:

$$\Phi = \Phi_{A1} + \Phi_s - \Phi_{A2} = (B_s - \mu_0 H) n_1 A_s = J n_1 A_s \qquad (5.65)$$

In the absence of the specimen, the output signal is zero; thus, the air flux is compensated. With the specimen, we can directly detect the magnetic polarization.

5.3.2 Testing of the Hard Magnetic in Open Magnetic Circuit

The open magnetic circuit can be used to test the quality of the magnet. Which parameters are important for assessment of the quality? Figure 5.102 presents the second quadrant of hysteresis loop commonly used to determine the magnet parameters. The main parameters of magnets are as follows:

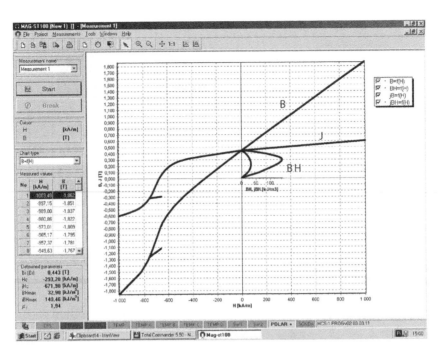

FIGURE 5.99
The example of the protocol of test of hard magnetic material. (Courtesy of R&J Measurement.)

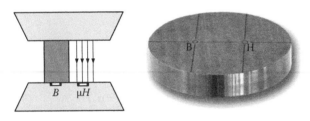

FIGURE 5.100
The pole of electromagnet with two embedded coils. (Courtesy of Magnet-Physik.)

B_r, remanence induction, $_JH_c$, coercivity of J, and $_BH_c$, coercivity of B, $(BH)_{max}$, maximum energy product, μ_r, recoil permeability.

The hysteresis demagnetization curves $B(H)$ and $J(H)$ are different as well as the coercivities $_JH_c$ and $_BH_c$. The real flux density B_d and magnetic field strength H_d are smaller than the remanence B_r and coercivity H_c due to demagnetization of the sample where the working point is determined by load line $(1-N)/N$ where N is the demagnetizing factor.

When the magnet is closed by movable ferromagnetic part, the operating point is moved from the original operating point A by some value x along the demagnetization line, which crosses two points 0, $_BH_c$, and B_r, 0—the slope of this line describes the recoil permeability μ_r. In the modern magnets, the recoil permeability is within the range 1.02–1.15 and in the first approximation can be assumed* as equal to 1.05.

Theoretically, we can determine the flux density of the magnet B_d by means of flux density meter, for example, by Hall sensor meter. But such measurement results in poor accuracy. The standard IEC 60404-14 recommends the use of extraction method, as presented in Figure 5.103. The magnet is placed inside the Helmholtz coils (see Section 2.12.1) in the area of uniform magnetic field (grey area in Figure 5.103). The fluxmeter is reset and the magnet is extracted to the outside of the coils. The measured change of magnetic flux is

$$\Delta\Phi = jK_H \tag{5.66}$$

where
  j is the dipole moment (Wb m)
  K_H is the coefficient of the coil; the relation of the magnetic field to the current is $K_H = H/I$ (for Helmholtz coil with n turns it is 0.7155 n/r)

Next we can determine the polarization J of the sample with volume V:

$$J = \frac{j}{V} \tag{5.67}$$

Although we determine polarization J_d (see Figure 5.102) but because the dependence $J(H)$ is very flat, we can assume that this value is very close to the remanence polarization J_r (and remanence flux density B_r).[†] Instead

---

* For moving armature, the demagnetization line can exhibit slightly other slope and can demonstrate minor hysteresis loop.

[†] The difference is usually about 1%–3%.

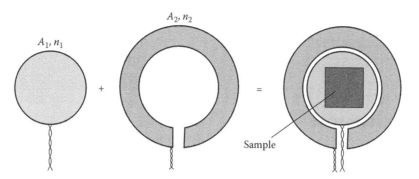

FIGURE 5.101
J-compensated surrounding coil.

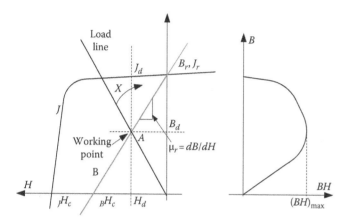

FIGURE 5.102
The main parameters of the permanent magnet.

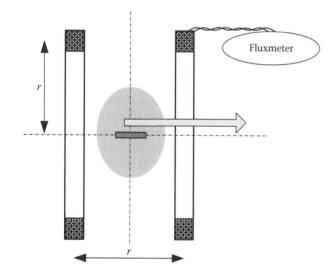

FIGURE 5.103
The method of measurement of magnetic moment of the magnet.

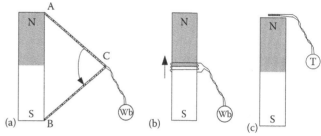

FIGURE 5.104
Fast methods of testing of the magnet quality: by means of potential coil (a), by means of B-coil (b), and by using the Hall teslameter (c).

Figure 5.104a presents a fast method of testing of the magnetic field strength at the working point H_d. For this purpose, the special coil sensor called magnetic potential coil is used (McDonald and Steingroever 1978, Steingroever et al. 1997). This coil is a straightened Rogowski–Chattock potentiometer. The potentiometer detects the magnetic field strength between ends of the coil independently on the shape of the coil. Thus if we measure the magnetic field strength between points A and C and next between B and C, the difference is

$$H_{AC} - H_{BC} = H_{AB} = H_d \qquad (5.68)$$

Knowing H_d and J_d, we can determine the value of B_d as

$$B_d = J_d - \mu_0 H_d \qquad (5.69)$$

We can measure directly the flux density B_d by surrounding a coil on the neutral part of the magnet and next by removing the coil (Figure 5.103b). We can also use the Hall sensor teslameter (Figure 5.104c). If we would like to compare various magnets, we should take into account the volume of the magnet (see also Figure 3.59).

The coercivity can be determined by using the method recommended by IEC Standard 60404-7. The investigated specimen is inserted into a solenoid (Figure 5.105). We increase the magnetic field of the coil to fully

of Helmholtz coil a long solenoid can be used. The described method assumes that the magnet is a bar. It is possible to use this method also for arc-shaped magnets (Trout 1988). Instead extraction of the magnet we can also rotate the magnet by 180°. In this case, we have $\Delta\Phi = j2K_H$.

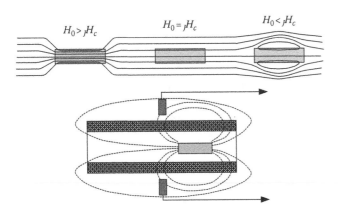

FIGURE 5.105
The method of determination of the coercivity $_jH_c$.

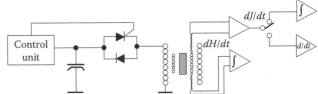

FIGURE 5.106
The example of the pulsed field testing system. (From Dudding, J. et al., *J. Magn. Magn. Mater.*, 142–245, 1402, 2002; Espina-Hernandez, J.H. et al., *Meas. Sci. Technol.*, 18, 893, 2007.)

saturate the sample. Then the polarity is reversed and the magnetic field slowly is increased toward the "negative" saturation. Two differentially connected sensors (e.g., Hall or fluxgate sensors) detect the magnetic field outside the coil; they work as zero field detectors. When the sensors detect state of demagnetization of the sample, we determine the $_jH_c$ value knowing the magnetic field in the coil.

It is also possible to measure the $_BH_c$ in a similar testing device. As a sensor detecting demagnetization the coil wound directly on the sample can be used.

It is recommended to apply the air flux compensation to obtain zero magnetic field in the absence of the specimen. These methods are described in Section 2.13.6 (see Figure 2.195).

To determine the energy product $(BH)_{max}$ it is necessary to know the second quadrant of the hysteresis loop. The optimum magnetic field is not always the same as the H_d value. Assuming that the part of hysteresis $B(H)$ in the second quadrant is linear the maximum energy product can be roughly estimated as

$$(BH)_{max} = \frac{1}{4\mu_0\mu_r}B_r^2 \tag{5.70}$$

5.3.3 Pulsed Field Methods

Modern hard magnetic materials need magnetization with magnetic field as high as 15 T, which is beyond the capabilities of conventional magnetic circuits.* Partially this problem can be solved by using superconducting circuits but such methods are difficult to use in an industrial environment. To obtain such large magnetic fields, pulse magnetization with capacitors bank as a source (see Section 2.12.3) is commonly used. It was

therefore the natural consequence that also measuring methods based on pulse magnetization were proposed (Grössinger et al. 1988, 1993). The pulsed field technique means not only easier to obtain large magnetic field but also much shorter time of testing. Typical pulse duration is at the order of milliseconds and the whole measurements, including postprocessing of the data, need a few seconds.

The example of pulsed field magnetometry (PFM) system is presented in Figure 5.106. The magnetizing system uses the coil system connected to the capacitor battery 8 or 24 mF supplied a DC voltage of 2.5 kV. The parameters of the magnetizing system are presented in Table 5.3.

As presented in Table 5.3, it is possible to generate pulses up to 10 T with a damped sine pulse of 10 ms. The field coil has a bore of 70 mm with homogeneity of 1% over a 50 mm.

As the sensors are used the thin film pickup coils (the Hall sensors were also tested) (Eckert et al. 2001, Espina-Hernandez et al. 2004). The J-coil system is composed of at least two coils: one of compensation of air flux (therefore the differential amplifier in Figure 5.106 is used). The signal proportional to polarization can be conventionally integrated but also can be differentiated. The signal o d^2J/dt^2 is used to analyze of anisotropy by *singular point detection* (SPD). In this method, the maximum of the second derivative of magnetization corresponds with the anisotropy field of polycrystalline sample (Asti and Rinaldi 1974, Grössinger et al. 1998, Espina-Hernandez et al. 2007).

TABLE 5.3

Technical Data of the Pulsed Magnet

Layers		2	4
Inner bore	Mm		70
Turns/layer		24	24
Maximum field	[T] for C = 8 mF	5.9	5.7
	[T] for C = 24 mF	9.6	9.3
Pulse duration	[ms] for C = 8 mF	4.3	9.1
	[ms] for C = 24 mF	7.5	15.7

Source: Grössinger, R., et al., *IEEE Trans. Magn.*, 35, 3971, 1999.

---

* In magnetic field strength units, the modern materials need for magnetization in the magnetic field larger than 20 MA/m.

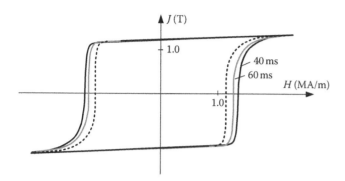

FIGURE 5.107
The hysteresis loop of sintered NdFeB with various times of pulse duration. The dashed line: corrected for eddy-currents result. (From Dudding, J. et al., *J. Magn. Magn. Mater.*, 142–245, 1402, 2002.)

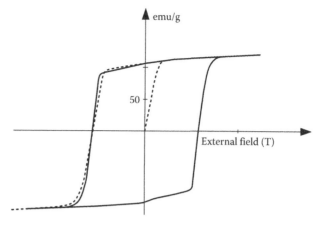

FIGURE 5.108
The hysteresis loop of NdFeB magnet determined with pulse duration 15.7 ms. The dashed line—results of static investigations. (From Springer Science + Business Media: *Springer Handbook of Materials Measurement Methods*, Magnetic characterization in a pulsed field magnetometers, Chapter 10.3, 2006, Grössinger, R.)

In the pulse technique, the large error can appear due to eddy currents and magnetic viscosity. Figure 5.107 presents the result of testing of neodymium magnet for two different pulses. The static value of $_jH_c$ was about 1200 kA/m while for pulse 40 ms pulse the result was 1385 kA/m. This difference can be much larger for very short pulses and for pulses shorter than 1 ms this error can be even 40% (Grössinger et al. 1993). Fortunately, the methods for effective correction of this error were developed (Jewell et al. 1992, Golovanov et al. 2000). The commonly used method known as *f/2f* method consists of two times measurements of hysteresis loop with two different duration of pulses; based on this information, it is possible to calculate the appropriate correction.

The *magnetic viscosity* (known also as the *magnetic aftereffect*) is the effect of delay in magnetization of the sample when the magnetic field strength is changing very rapidly. The reason of this effect is rather complex and depends on the microstructure of the material (Chikazumi 2009). The effect occurs in all magnetic materials. It also exists in hard magnetic materials and depends on the mechanism of coercivity—nucleation of domain walls or pinning. In some materials, it can be quite significant and can influence the coercivity field (Street et al. 1987, Tellez-Blanco et al. 1999, Grössinger et al. 2004). Nevertheless for the pulses of several ms used in pulsed field magnetometry, this effect is negligible (Dudding et al. 2002).

Figure 5.108 presents the comparison of results obtained by pulse field method and by static method. The differences are practically within the line thickness. In the presented example, $\mu_0 H_c$ determined by pulse method was 2.89 T while by static method was 2.85 T.

Hard magnetic materials can be also tested using the magnetometric methods. This subject is discussed in the next chapter.

5.4 Special Methods of Testing of Magnetic Materials

5.4.1 Magnetometric Methods

For many physically small magnetic particles, for example, thin film elements or paleomagnetic samples, it is not possible to create closed magnetic circuit testing device. In such a case, the most efficient is the magnetometer method where the magnetic moment is detected. It is assumed that the sample represents the elementary dipole magnet. To detect very small magnetic fields accompanying the small samples commonly the *vibrating sample magnetometer* (VSM) known as *Foner magnetometer* is used (Foner 1956, 1959, 1996). The principle of the Foner magnetometer is presented in Figure 5.109.

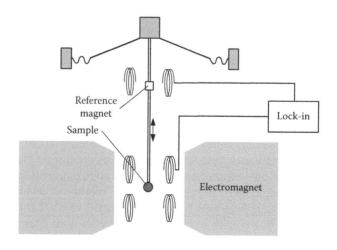

FIGURE 5.109
The principle of a Foner magnetometer.

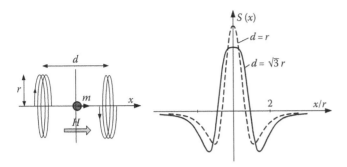

FIGURE 5.110
The pickup coil in the Foner magnetometer.

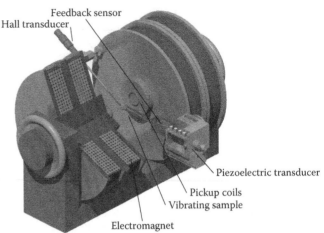

FIGURE 5.111
The example of vibrating sample magnetometer. (From Wrona, J. et al., *J. Magn. Magn. Mater.*, 196–197, 935, 1999; Wrona, J. et al., *J. Magn. Magn. Mater.*, 272–276, 2294, 2004.)

The sample in the form of a sphere or disk is connected to a vibrating device. In the original Foner magnetometer, this was a loudspeaker transducer; recently most often, it is a piezoelectric element. The sample is magnetized by a strong magnetic field generated by external electromagnet. Thus, the magnetized sample operates as a vibrating dipole inducing in the pickup coils the voltage proportional to the magnetic moment. In Figure 5.109, four pickup coils are used, but other arrangements are possible (Zieba and Foner 1982).

Because the induced signal is very small, a reference magnet of the vibrating frequency used in the lock-in amplifier is important. Instead of vibrating magnet other sensors of vibration can be used, for example, capacity sensor (Wrona et al. 1999). By changing the magnetic field of the electromagnet, we can investigate the properties of the sample, including hysteresis loop.

If the magnetic moment is directed along the axis of the coils (Figure 5.110) the EMF induced in the pickup coils depends on the magnetic moment m and the frequency of vibration:

$$e(x,t) = mS(x)\frac{dx}{dt} \tag{5.71}$$

where $S(x)$ is the sensitivity factor depending on the sample geometry, coil arrangement, electromagnet geometry, etc. If we use sinusoidal vibration $x(t) = X_0 \sin\omega t$, the induced voltage is

$$e(x,t) = mS(x)\omega X_0 \cos\omega t \tag{5.72}$$

The sensitivity factor can be analyzed theoretically (Zieba and Foner 1982, Pacyna and Ruebenbauer 1984) but often the magnetometer is calibrated experimentally by using the standard of magnetic moment (Sievert et al. 1990, Eckert and Sievert 1993, Shull et al. 2000). Commonly a sphere of chemically pure Ni with diameter of a few mm is used as a standard. For example, 99.999% pure nickel sphere of diameter 2.283 mm and mass 63.16 mg exhibits in the temperature 298 K

and magnetic field 398 kA/m the magnetic moment* 3.47 ± 0.01 mAm^2, which corresponds to specific magnetization 54.97 mAm^{2}g^{-1}. Of course, it is very important to use the sample of the same geometry as the standard sample used for calibration.

The sensitivity factor depends on the configuration of the pickup coils. Commonly pairs of inverse Helmholtz coils are used as the pickup coils (the coils are connected in opposition). The best sensitivity is for the geometry when the radius of the coil r is the same as the same as distance d (Figure 5.110). But in order to obtain stable sensitivity of the coil in the sample area, this distance is enlarged to $\sqrt{3}r$ at the expense of slightly reduction of sensitivity.

Since the first Foner publication (Foner 1956) various improvements have been reported. An example of a modern VSM system is presented in Figure 5.111. As the vibrating source, commonly the piezoelectric transducer is used. The sample is mounted at the end of the glass rod and the mechanical resonance of the rod is exploited. As the reference vibration sensor, the capacitive sensor is used. To control the magnetizing conditions, the Hall probe is used.

Other improvements include use of superconducting electromagnets (Gerber et al. 1982, Dufeu and Lethuilllier 1999), permanent Halbach magnet (Cugat et al. 1994) or pickup coil modification (Mallinson 1966, Foner 1975).

Most of the investigated samples are anisotropic. The simplest way to obtain two-dimensional information is the mechanical rotation of the sample with respect to the electromagnet axis.

---

* The unit of magnetic moment is Am^2 although old unit emu is also used (1 Am^2 = 10^3 emu).

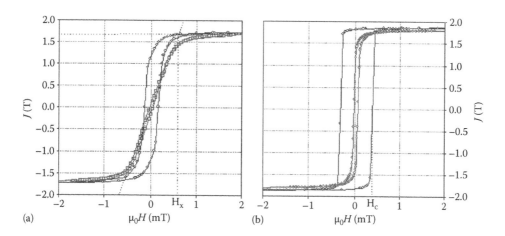

FIGURE 5.112

The example of hysteresis loops determined for direction perpendicular to the anisotropy axis (a) and along the anisotropy axis (b) for two different thin film samples. (From Wrona J., Magnetometers for thin film structures, PhD dissertation, AGH, Cracow, Poland, 2002.)

Figure 5.112 presents the results of investigation of two thin film samples. From these curves, we can estimate the anisotropy field and the coercivity. But the best solution of investigations of 2D properties is to use a biaxial coil system as presented in Figure 5.113. Such system known as vector VSM should include at least two pairs of the coils although 8-coil and 12-coil systems are also reported (Bernards 1993, Samwel et al. 1998).

Especially complex is the investigation of the thin film samples with perpendicular anisotropy. Figure 5.113 presents the hysteresis loop $M_x(H_a)$ determined in a biaxial pickup coil system. Knowing that true internal magnetic field is

$$H_{xcorr} = H_a - H_d \cos\theta \qquad (5.73)$$

it is possible to take into account the correction. This correction was calculated as

$$H_d = 4\pi N_d M_\perp = 4\pi N_d M \cos\phi \qquad (5.74)$$

where N was the demagnetizing factor $N_d = 1$.

After taking into account this relation, it is possible to determine the hysteresis loop $H_{xcorr}(H_a)$ representing true intrinsic material properties.

A limit of resolution of vibrating sample magnetometers is around 10^{-8} to 10^{-9} Am^2. Better resolution is exhibited by different type of magnetometer known as alternating gradient magnetometer (AGM) (also vibrating reed magnetometer). Such magnetometers can be used to detect trace amounts of magnetism—the resolution of 10^{-11} Am^2 enabling to investigate sample with dimension as small as $5\,\mu m$ (Ross et al. 1980).

This type of magnetometer introduced by Zijlstra (1970) was later improved by many other designers (Richter et al. 1988, Flanders 1988, 1990, Frey et al. 1988, Asti and Slozi 1996, Zimmermann et al. 1996). The principle of operation of AGM is presented in Figure 5.114. The sample magnetized by an electromagnet is oscillating under the influence of additional alternating magnetic field generated by a gradient coil system. These oscillations are detected by a vibration sensor; most frequently the piezoelectric bimorph is used. The sensor detects the force

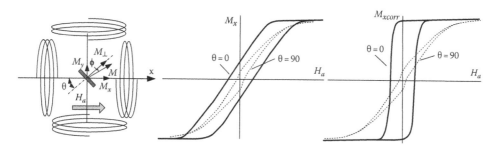

FIGURE 5.113

Biaxial vibrating sample magnetometer and the hysteresis loops: $M_x(H_a)$ determined without correction of demagnetizing field and $M_{xcorr}(H_a)$ determined with correction of the demagnetizing field. (From Ouchi, K. and Iwasaki, S., 1988, *IEEE Trans. Magn.*, 24, 3009, 1988.)

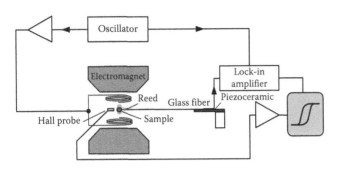

FIGURE 5.114
An example of the alternating gradient magnetometer. (From Ross, W. et al., *Rev. Sci. Instrum.*, 51, 612, 1980.)

$$F_x = \mu_0 m \frac{dH_x}{dx} \qquad (5.75)$$

If the coils are generating sinusoidal magnetic field, the detected signal is

$$e(t) = \mu_0 m S \omega V_0 \cos \omega t \qquad (5.76)$$

where S is the sensitivity factor.

The frequency of the alternating field is chosen as to fit the mechanical resonance of the vibrating sample. For example, in the measuring system presented in Figure 5.114, the reed is a gold wire 10 mm long and 18 μm diameter of the mechanical resonance 62 Hz and quality factor about 70. The vibration of the gold wire is transferred to piezoelectric ceramic bimorph by a glass fiber 70 mm long and 150 μm in diameter.

The AGM magnetometers can be more sensitive and more compact than the classical VSM magnetometers. But they are not as accurate as VSM magnetometers because their mechanical properties depend strongly on the mass of the specimen and therefore are difficult to calibrate. The European intercomparison of the investigations of the same permanent magnet sample (sphere 3 mm in diameter) by VSM method demonstrated excellent conformity (better than for the closed samples) (Sievert et al. 1993). Therefore, VSM technique is commonly used for test of small specimens. VSM professional instruments, for example, Lakeshore VSM model

7400 or MicroSense VSM model EV11, are available on the market.

The magnetic moment and susceptibility can be also measured by magnetic scales. Figure 5.115 presents two main principles of this method. In the Faraday method, the sample is inserted in the gap of an electromagnet with poles of the shape ensuring magnetic field gradient dH/dy. The sample of the volume V is subjected to the force depending on the magnetic dipole moment $j = \mu_0 m$ or susceptibility $\chi = \mu - 1$:

$$F = j \frac{dH}{dy} = V \chi H \frac{dH}{dy} \qquad (5.77)$$

In the Gouy method, only the part of the long bar specimen is in the gap of the electromagnet inside the uniform magnetic field. The force is proportional to the difference of the magnetic field H in the gap and the magnetic field around the second end of the bar H_a:

$$F = \frac{1}{2} A \chi (H^2 - H_a^2) \cong \frac{1}{2} A \chi H^2 \qquad (5.78)$$

The Gouy method is mainly used for paramagnetic and diamagnetic materials whereas the Faraday method is mainly used for ferromagnetic materials.

5.4.2 Magnetovision Method

Various systems designed for scanning of magnetic fields have been reported including market available Magscan system using Hall sensors (Redmag) or scanning microscope using also small Hall sensor (Nano magnetics). The author of this book developed the scanning system using small AMR sensor. This system called Magnetovision (Tumanski 1999, Michalski 2002) primarily designed for testing of the quality of electrical steel (Tumanski and Stabrowski 1998, 2002a, 2007, Tumanski and Fryskowski 1998, Tumanski and Baranowski 2008) was also used in other areas as the tool supporting the design of magnetic devices (Tumanski 2004, Tumanski

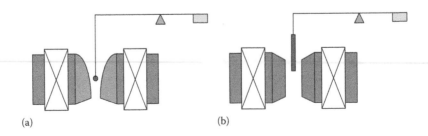

(a)   (b)

FIGURE 5.115
The magnetic scales: Faraday method (a) and Gouy method (b). (From Springer Science + Business Media: *Springer Handbook of Materials Measurement Methods*, Soft and hard magnetic materials: Standard measurement techniques for properties related to B(H) loop, Chapter 10.2, 2006, Ross, G.)

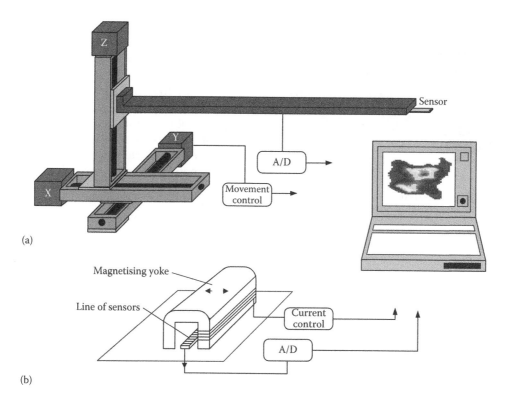

(a)

(b)

FIGURE 5.116
The scanning method in the magnetovision system. (From Tumanski, S. and Liszka, A., *On-Line Evaluation of Electrical Steel Structure and Quality*, INTRERMAG 2002 Paper No. DU09, Amsterdam, the Netherlands, 2002a; Tumanski, S. and Liszka, A., *J. Magn. Magn. Mater.*, 242–245, 1253, 2002b.)

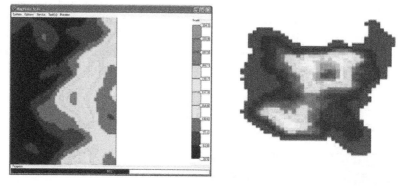

FIGURE 5.117
The example of the map during the scanning and after the scanning.

et al. 2004, Tumanski 2005) or nondestructive testing of materials (Tumanski 2000, Baudouin et al. 2003).

In the magnetovision system, the sensor is moved above the investigated materials by using three stepper motors (Figure 5.116a). As the sensor, the AMR sensor KMZ10B of Philips with sensitivity of about 50 mV/(kA/m) and active area of about 1 mm² was used because its performance suits very well for typical problems of investigation of magnetic materials. Various step sizes of the movement (spatial resolution) can be chosen but taking into account the dimensions of the sensor the best results are obtained for the step

equal to 0.5 mm. It is important that thin film magnetoresistive sensors are sensitive to the magnetic field in their plane so the tangential component of the magnetic field can be determined. Alternatively the "line" of the sensors can be used as handheld device (Figure 5.116b).

It is possible to observe the results during the scanning process but the resolution of the image is poor (Figure 5.117). Much better results are obtained after the postprocessing of the image (Figure 5.118).

Figure 5.119 shows the final result of data processing, where besides the map the value of magnetic field in

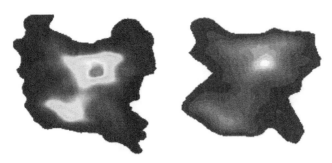

FIGURE 5.118
The examples of the map after postprocessing—with 500-color scale (photo quality) and with gray scale. (Tumanski, S., *Przegl. Elektrotech.*, 83(1), 108, 2007.)

selected line (or point) is also presented. Also a histogram with information about distribution of magnetic field in selected line or area (min/max values, average value, standard deviation, etc.) is very helpful.

Having the possibility of 3D scanning of the magnetic field, the map of the magnetic field in the arbitrary plane can be reconstructed as demonstrated in Figure 5.120 where the magnetic field around the C-yoke system is analyzed.

Figure 5.121 presents the results of magnetic field scanned above five adjacent Epstein strips (3 cm wide). Although steel was produced by a reputed manufacturer, the distribution of magnetic field was nonuniform, which is obvious taking into account the grain structure. But, more importantly the quite large differences between individual strips are visible (which is not detectable using conventional steel testers). For example, strip four is much better than strip two. Thus, we can

say that picture of field distribution can be a graphical signature representing steel grade and quality.

The quality of grain-oriented steel depends on the material heterogeneity. This quality is commonly described by the power loss. Figure 5.122 presents four examples of map of magnetic field determined for the samples with various power loss (determined using the Epstein method). In these maps, the magnetic field strength is represented by colors similarly as in the geographic maps (blue and green, low values; yellow and red, large values). It is clear that the map brings immediately valuable information about steel quality: the more "cold" colors the better the steel.

Figure 5.123 presents the maps of three different strips: typical GO steel (A), steel with various grains (B), and the steel with extremely large grains separated by areas with very small grains (incompletely crystallization process).

Although results of scanning are visually very impressive, it is difficult to recommend this method to steel manufacturers. The main drawbacks of scanning methods are the sophisticated and time-consuming procedure of testing and difficulties in fast and clear interpretation of the testing results. Therefore, statistical methods of assessment of steel quality should be considered.

As presented in Figure 5.124, for further analysis, the map can be interpreted by a histogram showing the average value, standard deviation, and field distribution. The scanning process is rather slow: the investigation of area 10 × 10 cm with resolution of 0.5 mm needs about 1 h. Therefore, the map can be substituted by the diagram describing the magnetic field distribution in

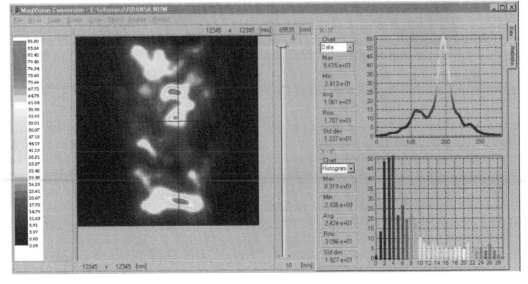

FIGURE 5.119
The example of final result of processing the scanned data.

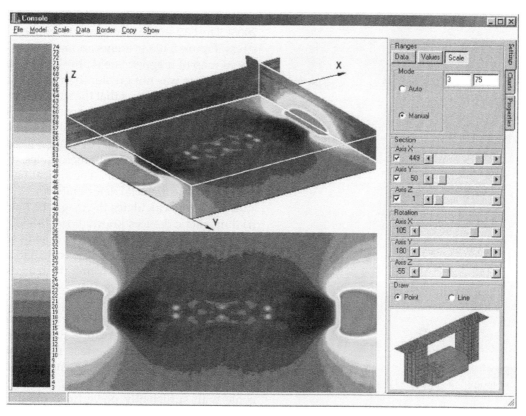

FIGURE 5.120
The tomography of 3D scanning results.

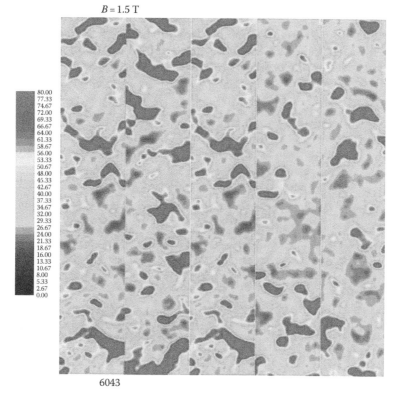

FIGURE 5.121
Example of results of magnetic field scanned above five strips 3 cm wide of grain-oriented steel ($B = 1.5\,$T).

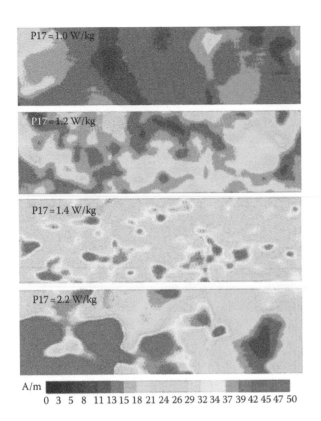

FIGURE 5.122
The relation between average power loss of grain-oriented steel and the map of magnetic field above the sample (loss determined at 1.7 T).

magnetic materials (Tumanski 2000b, Baudouin et al. 2003). The GO steel is especially sensitive to mechanical stress. Figure 5.126 presents an impressive, often reproduced map of magnetic field after writing the letter "A" (the pressure was not greater than as typical writing on a paper). It is interesting that the steel was most sensitive when the line was written diagonally and less sensitive for the line written along the rolling direction. Figure 5.127 presents the changes of magnetic field distribution after drilling two small micro-holes, often used for flux density measurements. The annealing did not completely remove material deterioration.

Anyone who calculates the magnetic fields for industrial magnetic devices knows how difficult is to obtain the results perfectly corresponding with the real data. Only from simple magnetostatic calculations, it is possible to obtain the results fitting the true data quantitatively. Usually only the qualitatively fitting results are appreciated as successful.

There are many reasons causing differences between calculation results and reality. Such differences may arise from not having perfect mathematical model: imperfect representation of geometry, boundary conditions, or mesh. These problems are possible to overcome with sufficiently powerful computers. But there are many unpredictable sources of differences, for example, material data (heterogeneity, anisotropy, coercivity, eddy currents) and the shape data (edge effects, air gaps, etc.). All these obstacles are magnified in the case of 3D calculations. Figure 5.128 presents example of the differences between calculated and experimentally determined distribution of magnetic field in the tested electrical steel sheet.

one line. The examples of such diagrams related to the samples presented in Figure 5.125.

The mapping of magnetic field distribution offers unique possibility for the investigation of the influence of various technological processes on the properties of

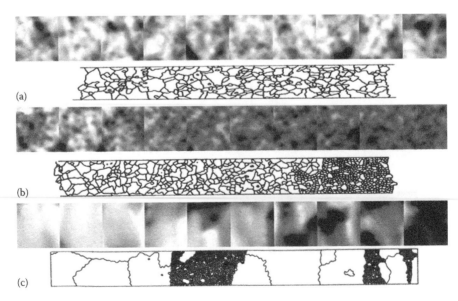

FIGURE 5.123
The magnetovision maps and corresponding grain structure of three different steel samples.

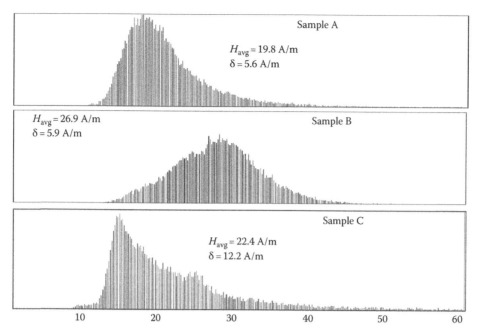

FIGURE 5.124
The histograms of magnetic field strength corresponding to the samples presented in Figure 5.123.

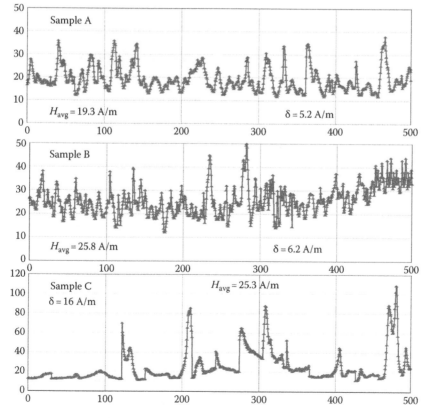

FIGURE 5.125
The $H(x)$ dependence determined for the samples presented in Figure 5.120.

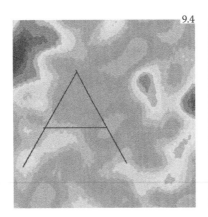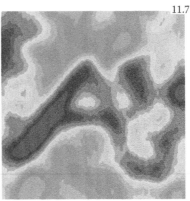

FIGURE 5.126
The magnetic field above the steel sheet before and after writing a letter "A" on the surface.

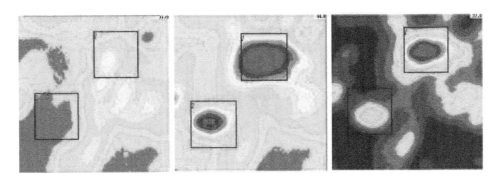

FIGURE 5.127
The magnetic field above the same sample of the GO steel (left) after drilling a small 0.5 mm holes (middle) and next after annealing (right).

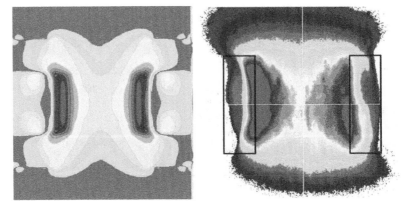

FIGURE 5.128
Comparison of the numerically (left) and experimentally (right) determined distribution of magnetic field in the sheet on the C-yoke.

Thus, the experimental methods can be very useful method for supporting the numerical design methods because they give the possibility to check if the numerical design is correct (Figure 5.129).

Usually the results of scanning are presented in a form of a color map, which expresses the material heterogeneity. For editorial purposes, such maps have been converted to the black-and-white scale. Surprisingly the grain structure was more visible with such

images (Figure 5.130). Especially such structure is well expressed in the case of the steel with large grains, as demonstrated in Figure 5.130b.

5.4.3 Other Scanning Systems

The grain and domain structures are the important information for working on technology of magnetic materials. The most commonly used method for domain

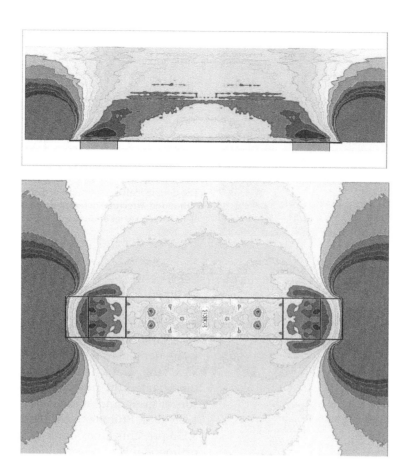

FIGURE 5.129
Experimentally determined distribution of magnetic field above and in the strip placed on the C-yoke system (compare with the results of computation, Figure 2.83).

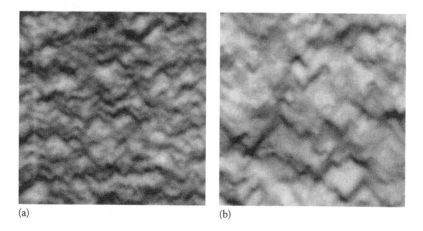

(a)   (b)

FIGURE 5.130
Experimentally determined distribution of magnetic field above the typical GO steel (a) and HiB steel (b).

observation is simple Bitter colloidal technique (Bitter 1931, Shilling et al. 1974, Mohri et al. 1979). In the Bitter technique, the thin layer of suspension of ferrimagnetic powder in liquid is covered on the investigated surface. The colloid particles are attracted to regions of maximum gradient of magnetic field and distribution of this colloid reveals a picture of domain configuration. Although other techniques have been introduced like magnetooptic Kerr method, magnetic force microscopy, and scanning electron microscopy, the colloidal method with many improvements is still used today (see Figure 5.137).

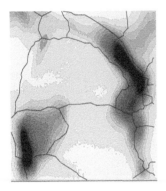

FIGURE 5.131

The example of the grain structure revealed by etching.

FIGURE 5.132

The magnetovision map of an area of GO steel and the same area with grain structure revealed by etching.

Similarly there is a very simple method enabling observation of the grain structure. After removing of coating and appropriate etching (e.g., by sodium hydroxide NaOH or nitric acid HNO_3 + alcohol known as Nital), it is possible to observe the grain structure as illustrated in Figures 3.14 and 5.131.

Above both, domain walls and grain boundaries, there exist stray magnetic fields; thus by using a magnetic field sensor, it should be possible to obtain the picture of grain or domain structure in automatic way by scanning the investigated area. As a good candidate of the magnetic sensors for such testing, we can consider Hall sensors, MR sensors, and Kerr effect devices.

Figure 5.132 presents the magnetovision map of GO electrical steel and possibility of comparison of the grain structure. Although both pictures are not the same, the correspondence between them is obvious. We can even point the "bad" and "good" individual grains.

Magnetovision results presented in Figure 5.132 demonstrate the component of magnetic field strength only in the rolling direction. Moreover, the active

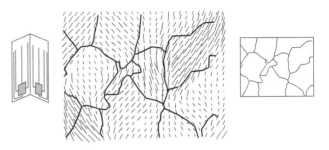

FIGURE 5.133

The double Hall sensor and the distribution of magnetic field (magnitude and direction) detected above the investigated 10×20 mm area of GO steel. (After Pfützner, H., *Z. Elektr. Inform. Energietechnik*, 10, 534, 1980.)

sensor area was relatively large 1×1 mm. It is possible to prepare the Hall sensor of very small dimension (even part of micrometer, Boero et al., 2003) and prepare the map of field distribution with better resolution. First tests using Hall sensors to grain detection were demonstrated by Mohri and Fujimoto (1977). Next the team from Vienna Technical University investigated the possibility of application of Hall sensor for grain and domain structure analysis (Pfützner 1980, 1981, Pfützner et al. 1983, 1984, 1985, 1992). Figure 5.133 presents an example of such investigations. Two orthogonal Hall sensors inclined by an angle 45° were used for testing not only the value of magnetic field but also its direction. The sensors were as small as 30×30 µm.

The investigated magnetic field values were at the level of between tens to hundreds of A/m, which is near the lower limit of sensitivity of Hall sensors. Therefore, the team from Cardiff University (Wolfson Centre for Magnetics Technology) demonstrated the possibility of testing of grain and domain structures by using the more sensitive magnetoresitive sensors (Mohd Ali et al. 1988, 1989, So et al. 1995). Figure 5.134 presents the results of scanning of magnetic field strength above the HiB-coated electrical steel by means of double MR sensor in a chevron configuration.

To investigate of the grain or domain structure, it is more useful to detect not a tangential but rather a normal component of the magnetic field. As depicted in Figure 5.135, stray magnetic fields are generated above the domain walls or grain boundaries. The picture of this magnetic field is rather complex because it is difficult to distinguish the field from the grain boundary H_b, domain wall H_s, and from domain or grain surfaces H_β. The solution for investigations of only the grain structure is to test the field in the saturation state when domain structure does not influence the map of magnetic field.

Figure 5.136 presents the change of normal component of magnetic field during the scanning process by

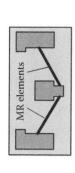

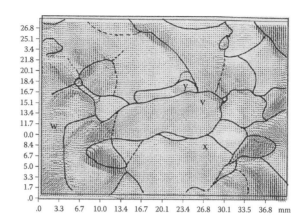

FIGURE 5.134

The double MR sensor and the distribution (magnitude and direction) of magnetic field above the HiB-coated steel. (Map of grain structure; From Mohd Ali, B.B., Computer mapping of grain structures in grain-oriented silicon iron, PhD dissertation, University of Cardiff, Cardiff, U.K., 1985.)

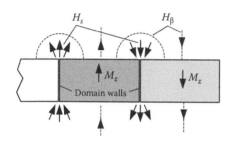

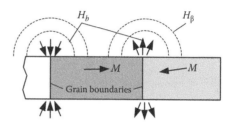

FIGURE 5.135

The magnetic fields: H_s, generated by the domain walls; H_b, generated by the grain boundaries; and stray fields H_β. (From Pfützner, H. et al., *Jpn. J. Appl. Phys.*, 22, 361, 1983.)

means of an MR sensor. Figure 5.137 presents the comparison of the domain structure images obtained by colloid technique and by scanning with an MR sensor. A strong agreement between the images is observed.

Figure 5.138 presents the results of the scanning of the area of amorphous materials using the Kerr device. In

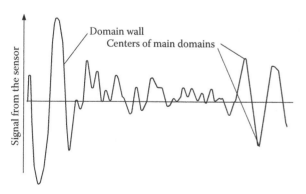

FIGURE 5.136

The magnetic field detected during the scanning process along the line a-b in the sample presented in Figure 1.137. (From Nicholson, P.I., Development of a magnetic domain imaging system for electrical steel using a magnetoresistive sensor based stray field scanning technique, PhD dissertation, Cardiff School of Engineering, Cardiff, U.K., 1999.)

this case, the analysis of domain structure is not distorted by the grain structure.

Comparing the results obtained by simple colloid and etching technique (Figure 5.137, left, and Figure 5.131), with the images obtained by using sophisticated scanning devices, we can see that the simple methods defend themselves with quite satisfying quality of images. Therefore in industrial environment they are still preferred. But for some investigations, scanning techniques cannot be substituted. Figure 5.139 shows one of the examples.

Figure 5.139 presents an example of power loss scanned by means of thermistor. It is detected that there are parts of the transformer core where the local power loss is almost three times higher than average (measured by SST device) power loss.

5.4.4 Investigations of the Magnetostriction and Other Magnetomechanical Effects

The magnetostriction effect causes the acoustic noise and vibration, which in turn result not only in acoustic pollution but also in additional power loss and faster wear and tear of transformer and rotating machines (Weiser et al. 1996, Krell et al. 2000, Belahcen 2006, Snell

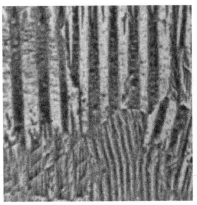
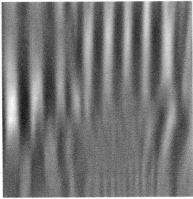

FIGURE 5.137
The domain structure image obtained by Bitter colloid technique (left) and after scanning of magnetic field by means of MR sensor (right). (From Nicholson, P.I., Development of a magnetic domain imaging system for electrical steel using a magnetoresistive sensor based stray field scanning technique, PhD dissertation, Cardiff School of Engineering, Cardiff, U.K., 1999.)

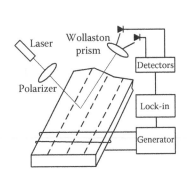
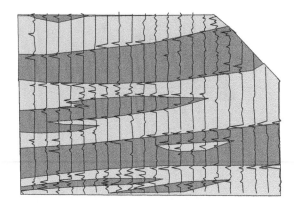

FIGURE 5.138
The example of the domain investigations by scanning of the Kerr reflection beam from the amorphous material. (From Pfützner, H. et al., Novel nondestructive methods for analysis of crystalline and amorphous soft magnetic materials, in *Proceedings of 2nd International Symposium on Physics of Magnetic Materials*, Beijing, Japan, pp. 427–434, 1992.)

2008). For this reason, measurement of magnetostriction is one of the important tests of magnetic materials. No wonder that in IEC TC 68 standards committee is discussed new standard "Methods of measurement of the magnetostriction characteristics by means of single sheet and Epstein test specimen" TR62581 (Stanbury 2010). The review of modern techniques for measurement of magnetostriction was published by Yabumoto (2009).

The measurement of magnetostriction consists of measurement of the change of physical dimensions under magnetization of the sample. The main difficulties in such tests are very small change of dimensions ($\lambda_s = \Delta l/l$ for SiFe is about 10^{-5}) and strong dependence on mechanical stress (Klimczyk et al. 2009). The sample under test should be long enough to obtain sufficient change of dimensions, but on the other hand it is recommended to test the sample similar to those used in the SST or Epstein method. Because magnetostriction

depends on the magnetic flux waveform, the control of this parameter is required (Anderson 2008).

Earlier for the measurement of magnetostriction, resistance strain gauges were commonly used. But because it is only a local measurement, nowadays optical methods are preferred as they can detect changes over the whole length of the sample.

Figure 5.140a presents the measuring system with strain gauges (Hilgert et al. 2005). The sample must deform freely so a small air gap is created, which allowed to deform of the sample freely what results in small difference in comparison with closed magnetizing yoke (Vandevelde et al. 2004). Often the testing focuses on so-called butterfly hysteresis, the dependence of magnetostriction versus the flux density (Figure 5.141).

Figures 5.140b, 5.142, and 5.143 present the magnetostriction measuring devices based on the optical methods. In the system proposed by Ban and Janosi (1996)

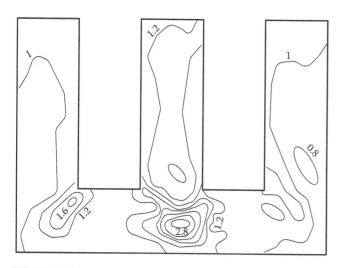

FIGURE 5.139
The example of the power loss investigations by scanning of the rise-of-temperature method (lines indicate the ratio of the power loss to the average SST loss). (From Krismanic, G. et al., *J. Magn. Magn. Mater.*, 254–255, 60, 2003.)

(Figure 5.139b), the change of the dimensions is detected by a commercial laser linear encoder. It is possible to obtain very high accuracy by using the Michelson interferometer (Figure 5.142) with resolution 2×10^{-9}.

Several systems were constructed with the laser Doppler vibrometers (Nakata et al. 1994b, Mogi et al. 1996, Hirano et al. 2003). In the device presented in Figure 5.143, the Doppler frequency depends on the velocity of the moving part. Measured is the difference of velocity with respect to the reference reflector.

The magnetostriction measuring system presented in Figure 5.144 is based on the application of two piezoelectric accelerometers (Anderson et al. 2000, Anderson 2008).

In the measurement system presented in Figure 5.144. the vibrations are detected by two piezoelectric sensors with sensitivity 1 *V/g* (the second sensor is used as a reference). The system was equipped with pneumatic stressing system. The magnetizing part is supplemented by the generator of arbitrary wave of flux density as

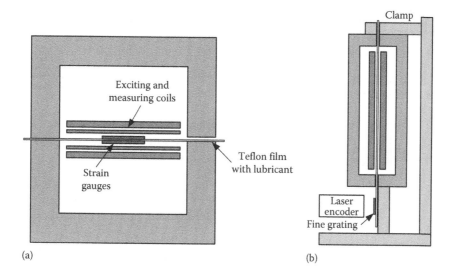

FIGURE 5.140
Two typical measurement systems for investigations of magnetostriction. (From Hilgert, T. et al., *Przegl. Elektrotech.*, 81(5), 87, 2005; Ban, G. and Janosi, F., *J. Magn. Magn. Mater.*, 160, 167–170, 1996.)

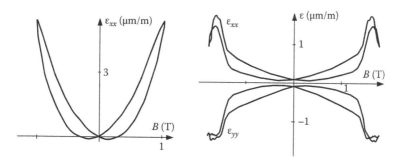

FIGURE 5.141
The examples of test results of measured strain ε for magnetizing field of 1 Hz. (From Hilgert, T. et al., *Przegl. Elektrotech.*, 81(5), 87, 2005.)

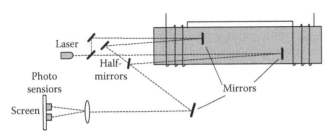

FIGURE 5.142
The measurement system based on Michelson interferometer. (From Sasaki et al., 1996; Yabumoto, M., *Przegl. Elektrotech.*, 85(1), 1, 2009.)

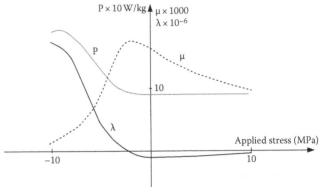

FIGURE 5.145
Typical peak magnetostriction, power loss, and permeability versus applied stress for conventional GO steel at 1.5 T. (From Anderson, P., *J. Magn. Magn. Mater.*, 320, e583, 2008.)

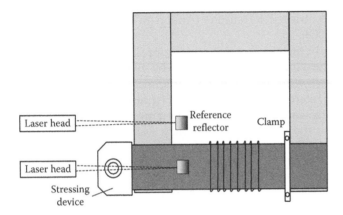

FIGURE 5.143
The measurement system based on the laser Doppler vibrometer. (From Yabumoto, M., *Przegl. Elektrotech.*, 85(1), 1, 2009.)

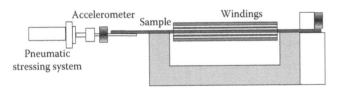

FIGURE 5.144
The magnetostriction measurement system with two piezoelectric accelerometers. (From Anderson, P.I. et al., *J. Magn. Magn. Mater.*, 215–216, 714, 2000.)

well as the measuring system for other parameters like power loss, permeability, etc. Figure 5.145 presents an example of typical measured results.

As we can see, magnetostriction strongly depends on mechanical stress. But Figure 5.145 also demonstrates the second very important problem of the magnetomechanical effects—significant changes of other magnetic parameters under stress. More detailed description of such effects was presented by Permiakov et al. (2003).

The magnetomechanical problems in design of magnetic devices are more complex if we take into account that it is a multidimensional problem. Both magnetic field and stress can vary in arbitrary directions including rotational magnetization and rotational magnetostriction. Therefore, the magnetostriction should be

tested using the RSST devices (Enokizono et al. 1995, Piermiakov et al. 2003, Pulnikov et al. 2003, Krell and Pfützner 2005, Yamaguchi et al. 2008, Somkun et al. 2009, Wakabayashi et al. 2009). Figure 5.146 presents an example of such an approach. For measurement of strain, three strain gauges were used. The magnetostriction in arbitrary direction is

$$\lambda = \frac{\Delta l}{l} = \sum_{i=1}^{2}\sum_{j=1}^{2} e_{ij}\beta_i\beta_j \qquad (5.79)$$

where
β_i and β_j are the components of the unit vector
e_{ij} is the strain tensor

From measurements in three directions (75°, 135°, and 195°), we can determine the strain tensor described by the following matrix (Wakabayashi et al. 2009):

$$\begin{pmatrix} e_{11} \\ e_{12} \\ e_{22} \end{pmatrix} = \begin{bmatrix} (1-\sqrt{3})/3 & 1/3 & (1+\sqrt{3})/3 \\ 1/3 & -2/3 & 1/3 \\ (1+\sqrt{3})/3 & 1/3 & (1-\sqrt{3})/3 \end{bmatrix} \begin{pmatrix} \lambda_{75} \\ \lambda_{135} \\ \lambda_{195} \end{pmatrix} \qquad (5.80)$$

By substituting the strain tensor e_{ij} into Equation 5.79, we can obtain the value of magnetostriction for an arbitrary direction. Other angles of the strain gauges are also possible in such measurements.

Figure 5.147 presents a system for 2D measurements developed at University of Ghent. Specially designed mechanical system enables to manually introduce one-dimensional stress. Two needle sensors and two H-coil sensors are used for measurement of both components of magnetic field strength and flux density. The special nonmagnetic helm and frame with brass grips enables to stress the material (because the yoke is much bigger than the sample the stress in the limbs is negligibly small).

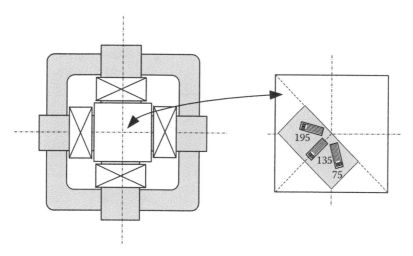

FIGURE 5.146
The system for measurement of the vector magnetostriction. (From Wakabayashi et al., 2009.)

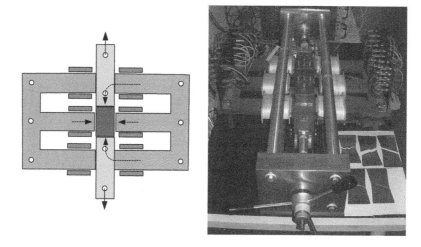

FIGURE 5.147
The system for 2D measurements of the sample under stress. (From Piermiakov, V. et al., *Przegl. Elektrotech.*, 81(5), 68, 2005. With permission.)

In the system presented in Figure 5.147, both alternating and rotational magnetization with arbitrary shape of the waveform are possible. Figure 5.148 presents an example of H-loci under rotational magnetization after applying of the stress.

During the preparation of the magnetic devices, various mechanical stresses and deformations during the cutting, punching, and welding strongly deteriorate magnetic performance of material (Rygal et al. 2000, Schoppa et al. 2000, Kedous-Lebouc et al. 2003, Wilczynski et al. 2004, Kurosaki et al. 2008). Therefore, the best would be the possibility of testing the magnetic material exactly in the working conditions. Figure 5.149 presents the testing method of a stator core by applying an additional testing yoke closing the magnetic circuit through the teeth. As detected by this method, the power loss after manufacturing of the stator increased more than two times.

Similar results were reported by Kedous–Lebouc after direct testing of performances of the teeth area of a stator core (Figure 5.150). As presented in Figure 5.151, the power loss increased almost two-times after teeth punching from the M330-35A fully processed material.

Therefore, the investigations of magnetomechnical properties of the magnetic materials are very important. It is not possible to avoid the deterioration of material performances after different manufacturing steps (only the effect of cutting can be eliminated by appropriate annealing). But importantly the different grades of non-oriented steel have different sensitivity to mechanical treatment (Schoppa et al. 2000). It is possible that the better and more expensive materials after manufacturing can be damaged to such an extent that they will become magnetically worse compared to a material that is cheaper but less sensitive to stress.

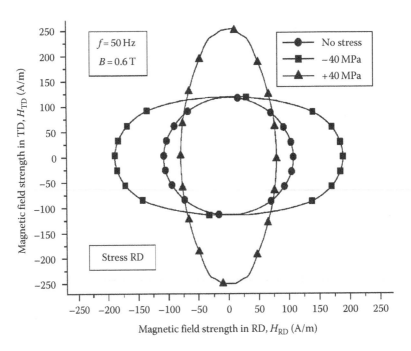

FIGURE 5.148
Results showing the influence of the stress on rotational magnetization. (From Piermiakov, V. et al., *Przegl. Elektrotech.*, 81(5), 68, 2005.)

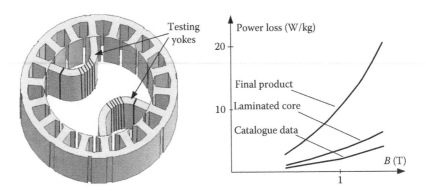

FIGURE 5.149
The method of testing the stator cores during manufacturing process. (From Nakazaki, O. et al., *Przegl. Elektrotech.*, 85(1), 74, 2009.)

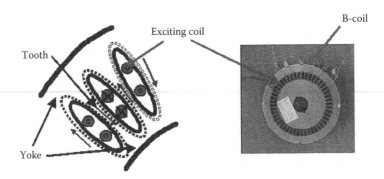

FIGURE 5.150
The direct method of the stator core testing. (From Kedous-Lebouc, A. et al., *Przegl. Elektrotech.*, 83(4), 55, 2007.)

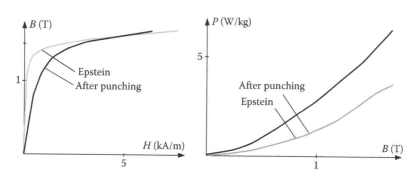

FIGURE 5.151
The degradation of material performance after teeth punching. (From Kedous-Lebouc A. et al., *Przegl. Elektrotech.*, 83(4), 55, 2007.)

References

Abdallh A., Dupre L., 2010a, Local magnetic measurements in magnetic circuits with highly non-uniform electromagnetic fields, *Meas. Sci. Technol.*, **21**, 045109.

Abdallh A., Dupre L., 2010b, A Rogowski–Chattock coil for local magnetic field measurements: Sources of errors, *Meas. Sci. Technol.*, **21**.

Abdallh A., Sergeant P., Crevecoeur G., Vandebossche L., Dupre L., Sablik M., 2009, Magnetic material identification in geometries with non-uniform electromagnetic fields using global and local magnetic measurements, *IEEE Trans. Magn.*, **45**, 4157–4160.

Ahlers H., Nadalski A., Rahf L, Siebert S., Sievert J., Son D., 1992, The measurement of magnetic properties of amorphous strips at higher frequencies using a yoke system, *J. Magn. Magn. Mater.*, **112**, 88–90.

Ahlers H., Sievert J., Qu Q., 1982, Comparison of the single strip tester and Epstein frame measurements, *J. Magn. Magn. Mater.*, **26**, 176–178.

Albir R.S., Moses A.J., 1990, Improved DC bridge method employed to measure local power loss in electrical steels and amorphous materials, *J. Magn. Magn. Mater.*, **83**, 553–554.

Anderson P., 2008, Measurement of the stress sensitivity of magnetostriction in electrical steels under distorted waveform conditions, *J. Magn. Magn. Mater.*, **320**, e583–e588.

Anderson P.I., Moses A.J., Stanbury H.J., 2000, An automated system for the measurement of magnetostriction in electrical steel sheet under applied stress, *J. Magn. Magn. Mater.*, **215–216**, 714–716.

Antonelli E., Cardelli E., Faba A., 2005, Epstein frame: How and when it can be really representative about the magnetic behavior of laminated magnetic steels, *IEEE Trans. Magn.*, **41**, 1516–1519.

Anuszczyk J., Pluta W., 2009, *Soft Ferromagnetics in Rotating Magnetic Fields* (in Polish), WNT, Warszawa, Poland.

Asti G., Rinaldi S., 1974, Singular points in the magnetization curve of a polycrystalline ferromagnet, *J. Appl. Phys.*, **45**, 3600–3610.

Asti G., Slozi M., 1996, Vibrating wire magnetic susceptometer, *Rev. Sci. Instrum.*, **67**, 3543–3552.

Bajorek J., Kalwak A., Kolasa J., 2004, Condition of webermeters operation and calibration, *Przegl. Elektrotech.*, **80**, 161–164

Ball D.A., Lorch H.O., 1965, An improved thermometric method of measuring local power dissipation, *J. Sci. Instrum.*, **42**, 90–92.

Ban G., Janosi F., 1996, Measuring system and evaluation method of DC and AC magnetostriction behaviour to investigate 3.2% SiFe GO electrical steels, *J. Magn. Magn. Mater.*, **160**, 167–170.

Basak A., Rowe D.M., Anayi F.J., 1995, Magnetic flux and loss measurements using thin film sensors, *IEEE Trans. Magn.*, **31**, 3170–3172.

Basak A., Rowe D.M., Anayi F.J., 1997, Thin film senses magnetic flux and loss in rotary electric motors, *IEEE Trans. Magn.*, **33**, 3382–3384.

Baudouin P., Houbaert Y., Tumanski S., 2003, Magnetic local investigations of non-oriented electrical steels after tensile deformation, *J. Magn. Magn. Mater.*, **254–255**, 32–35.

Beckley P., 1983, Continuous power loss measurement with and against the rolling direction of electrical steel strip using nonenwrapping magnetizers, *IEE Proc. A*, **130**, 313–321.

Beckley P., 2002, *Electrical Steels for Rotating Machines*, IEE, London, U.K.

Belahcen A., 2006, Vibrations of rotating electrical machines due to magnetomechanical coupling and magnetostriction, *IEEE Trans. Magn.*, **42**, 971–974.

Bengtsson C., Pfützner H., 1984, Stray field at grain boundaries in demagnetized stress coated HiB sheets, *IEEE Trans. Magn.*, **20**, 1478–1480.

Bernards J.P.C., 1993, Design of a detection coil system for a biaxial vibrating sample magnetometer and some applications, *Rev. Sci. Instrum.*, **64**, 1918–1930.

Birkelbach G., Hempel K.A., Schulte F.J., 1986, Very low frequency magnetic hysteresis measurements with well-defined time dependence of the flux density, *IEEE Trans. Magn.*, **22**, 505–507.

Birkfeld M., 1998, Investigation of the permeability tensor of electrical steel sheet, *IEEE Trans. Magn.*, **34**, 3667–3673.

Birkfeld M., Hempel K.A., 1994, A device for measuring the magnetic properties of ring specimens at high frequencies, *J. Magn. Magn. Mater.*, **133**, 393–395.

Birkfeld M., Hempel K.A., 1996, An extended description of the behaviour of steel sheet under rotational flucx conditions, *J. Magn. Magn. Mater.*, **160**, 17–18.

Birkfeld M., Hempel K.A., 1997, Calculation of the magnetic behaviour of electrical steel sheet under two dimensional excitation by means of the reluctance tensor, *IEEE Trans. Magn.*, **33**, 3757–3759.

Bitter F., 1931, On inhomogenities in the magnetization of ferromagnetic materials, *Phys. Rev.*, **38**, 1903–1905.

Boero G., Demierre M., Besse P.A., Popovic R.S., 2003, Micro-Hall devices: performances, technologies and applications, *Sens. Actuators A*, **106**, 314–320.

Boon C.R., Thompson J.E., 1965, Alternating and rotational power loss at 50 Hz in 3% SiFe sheets, *Proc IEE*, **112**, 2147–2151.

Brailsford F., 1938, Rotational hysteresis loss in electrical sheet steel, *J. IEE.*, **83**, 566.

Brix W., 1982, Measurements of the rotational power loss in 3% silicon-iron at various frequencies using a torque magnetometer, *J. Magn. Magn. Mater.*, **26**, 193–195.

Brix W., Hempel K.A., 1984, Tensorial description of the rotational magnetization process in anisotropic silicon steel, *J. Magn. Magn. Mater.*, **41**, 279–281.

Brix W., Hempel K.A., Schroeder W., 1982, Method for measurement of rotational power loss and related properties in electrical steel sheets, *IEEE Trans. Magn.*, **18**, 1469–1471.

Brix W., Hempel K.A., Schulte F.J., 1984, Improved method for the investigation of the rotational magnetization process in electrical steel sheets, *IEEE Trans. Magn.*, **20**, 1708–1710.

Cagan V., Guyot M., 1984, Fast and convenient technique for broadband measurements of the complex initial permeability of ferrimagnets, *IEEE Trans. Magn.*, **20**, 1732–1734.

Cecchetti A., Ferrari G., Masoli F., Soardo G.P., 1978, Rotational power losses in 3% SiFe as a function of requency, *IEEE Trans. Magn.*, **14**, 356–358.

Chang A.M., Hallen H.D., Harriott L., Hess H.F., Kao H.L., Kwo J., Miller R.E., Wolfe R., van der Ziel J., 1992, Scanning Hall probe microscopy, *Appl. Phys. Lett.*, **61**, 1974–1976.

Chen Z.J., Devine M.K., Jiles D.C., 1993, Measurements of magnetic circuit characteristics for comprehension of intrinsic magnetic properties of materials from surface inspection, *J. Appl. Phys.*, **73**, 5620–5622.

Chikazumi S., 2009, *Physics of Ferromagnetism*, Oxford Scientific Publishing, Oxford, U.K.

Cornelius R. et al., 2002, Pulsed field magnetometer for industrial use, *IEEE Trans. Magn.*, **38**, 2462–2464.

Cornut B., Catellani S., Perrier J.C., Kedous-Lebouc A., Waeckerle T., Fraisse H., 2003, New compact and precise magnetometer, *J. Magn. Magn. Mater.*, **254–255**, 97–99.

Cugat O., Byrne R., McCaulay J., Coey J.M.D., 1994, A compact vibrating sample magnetometer with variable permanent magnet flux source, *Rev. Sci. Instrum.*, **65**, 3570–3573.

Cullity B.D., Graham C.D., 2009, *Introduction to Magnetic Materials*, Wiley, Hoboken, NJ.

Cundeva M., Arsov L., 2000, Modeling and design of Epstein frame using FEM-3D, in *Nonlinear Electromagnetic Systems*, Vol. 18, IOS Press, Amsterdam, the Netherlands, pp. 105–108.

Czeija E., Zawischa R., 1955, Vorrichtung zum Messen des Wechselinduktionsflusses oder der Flussänderung in ferromagnetischen Materialien aus der Induktionsspannung, Austrian Patent No. 180990.

Dannatt C., 1933, Energy loss testing of magnetic materials utilizing a single strip specimen, *J. Sci. Instrum.*, **8**, 276–285.

Davis P.J., Rabinowitz P., 2007, *Methods of Numerical Integration*, Dover Publication Inc., New York.

Defoug S., Kaczmarek R., Rave W., 1996, Measurements of local magnetization by Kerr effect on SiFe nonoriented sheets, *J. Appl. Phys.*, **79**, 6036–6038.

De Mott, 1966, An integrating type of electronic hysteresisgraph, *J. Appl. Phys.*, **37**, 1118–1119.

De Mott, 1970, Integrating fluxmeter with digital readout, *IEEE Trans. Magn.*, **6**, 269–271.

Derebasi N., Meydan T., Goktepe M., So M.H., 1992, Computerised DC bridge method of thermistor measurement of localised power loss in magnetic materials, *IEEE Trans. Magn.*, **28**, 2467–2469.

De Wulf M., Dupre L., Makaveev D., Melkebeek J., 2003, Needle probe techniques for local magnetic flux measurements, *J. Appl. Phys.*, **93**, 8271–8273.

Di Napoli A., Paggi R., 1983, A model of anisotropic grain oriented steel, *IEEE Trans. Magn.*, **19**, 1557–1561.

Disselnköter R., 1996, Automized magnetic hysteresis measurement system, *J. Appl. Phys.*, **79**, 5208–5210.

Dudding J. et al., 2002, A pulsed magnetometer for the quality control of permanent magnets, *J. Magn. Magn. Mater.*, **142–245**, 1402–1404.

Dufeu D., Lethuilllier P., 1999, High sensitivity 2T vibrating sample magnetometer, *Rev. Sci. Instrum.*, **70**, 3035–3039.

Eckert D., Sievert J., 1993, On the calibration of vibrating sample magnetometers with the help of nickel reference samples, *IEEE Trans. Magn.*, **29**, 3001–3003.

Eckert D. et al., 2001, High precision pick-up coils for pulsed field magnetization measurements, *Phys. B.*, **294–295**, 705–708.

Enokizono M., 2009, Vector magnetic property and magnetic characteristic analysis by vector magneto-hysteresis E&S model, *IEEE Trans. Magn.*, **45**, 1148–1153.

Enokizono M., Kanao S., Shirakawa G., 1995, Measurement of arbitrary dynamic magnetostriction under alternating and rotating field, *IEEE Trans. Magn.*, **31**, 3409–3411.

Enokizono M., Kanao S., Yuki K., 1994, Permeability tensot of grain-oriented silicon steel sheet, *J. Magn. Magn. Mater.*, **133**, 209–211.

Enokizono M., Mori S., Benda O., 1997, A treatment of the magnetic reluctivity tensor for rpotating magnetic field, *IEEE Trans. Magn.*, **33**, 1608–1611.

Enokizono M., Sievert J.D., 1989, Magnetic field and loss analysis in an apparatus for the determination of rotational loss, *Phys. Scr.*, **39**, 356–359.

Enokizono M., Sievert J., 1990, Analytical studies of the yoke construction of a single sheet testers, *Ann. Fis.*, **B86**, 123–125

Enokizono M., Sievert J., Ahlers H., 1990a, Optimum yoke construction for rotational loss measurements apparatus, *Ann. Fis.*, **B86**, 320–322.

Enokizono M., Shimoji H., Ikagira A., 2003, Vector Magneto-Hysteric E&SS model for magnetic characteristic analysis, *Przegl. Elektrotech.*, **79**, 671–676.

Enokizono M., Shimoji H., Ikagira A., 2006, Vector Magneto-Hysteric E&SS model for magnetic characteristic analysis, *IEEE Trans. Magn.*, **42**, 915–918.

Enokizono M., Soda N., 1999, Iron loss analysis of transformer core model by FEM considering vector magnetic properties, *IEEE Trans. Magn.*, **35**, 3008–3011.

Enokizono M., Suzuki T., Sievert J., Xu J., 1990b, Rotational power loss of silicon steel sheet, *IEEE Trans. Magn.*, **26**, 2562–2564.

Enokizono M., Tanabe I., 1997, Studies on a new simplified rotational loss trester, *IEEE Trans. Magn.*, **33**, 4020–4022.

Enokizono M., Tanabe I., Kubota T., 1998, Local distribution of magnetic properties in grain-oriented silicon steel sheet, *J. Appl. Phys.*, **83**, 6486–6488.

Enokizono M., Todaka T., Sashikata T., Sievert J., Ahlers H., 1992, Magnetic field analysis of rotational loss tester with vertical yoke, *J. Magn. Magn. Mater.*, **112**, 81–84.

Enokizono M., Urata S., 2007, Dynamic vector magneto-hysteric E&S model taking into account of eddy current effect, *Przegl. Elektrotech.*, **83**, 1–8.

Espina-Hernandez J.H., Grössinger R., Kato S., Hauser H., Estevez-Rams E., 2004, New sensors for measuring M and H in high magnetic fields, *Phys. B.*, **346–347**, 543–547.

Espina-Hernandez J.H., Grössinger R., Martinez-Garcia J.C., Estevez-Rams E., 2007, A new system for local area measurements in a pulsed field magnetometer, *Meas. Sci. Technol.*, **18**, 893–900.

Fahy F.P., 1918, A permeameter for general magnetic analysis, *Chem. Mater. Eng.*, **19**, 339–342.

Fard S.M.B., Moses A.J., 1994, A method of modelling the anisotropy of electrotechnical steels in the finite element method, *J. Magn. Magn. Mater.*, **133**, 536–539.

Fiorillo F., 2004, *Measurement and Characterization of Magnetic Materials*, Elsevier, New York.

Fiorillo F., Dupre L.R., Appino C., Rietto A.M., 2002, Comprehensive model of magnetization cureve, hysteresis loop and losses in any direction in grain-oriented SiFe, *IEEE Trans. Magn.*, **38**, 1467–1475.

Fiorillo F., Rietto A.M., 1988, Extended induction range analysis of rotational losses in soft magnetic materials, *IEEE Trans. Magn.*, **24**, 1960–1962.

Flanders P.J., 1985, A Hall sensing magnetometer for measuring magnetization, anisotropy, rotational loss and time effects, *IEEE Trans. Magn.*, **21**, 1584–1589.

Flanders P.J., 1988, An alternating gradient magnetometer, *J. Appl. Phys.*, **63**, 3940–3945.

Flanders P.J., 1990, A vertical force alternating gradient magnetometer, *Rev. Sci. Instrum.*, **61**, 839–847.

Foner S., 1956, Vibrating sample magnetometer, *Rev. Sci. Instrum.*, **27**, 548.

Foner S., 1959, Versatile and sensitive vibrating sample magnetometer, *Rev. Sci. Instrum.*, **30**, 548–557.

Foner S., 1975, Further improvements in vibrating sample magnetometer sensitivity, *Rev. Sci. Instrum.*, **46**, 1425–1426.

Foner S., 1996, The vibrating sample magnetometer: Experiences of a volonteer, *J. Appl. Phys.*, **79**, 4740–4745.

Frey T., Jantz W., Stibal R., 1988, Compensating vibrating reed magnetometer, *J. Appl. Phys.*, **64**, 6002–6007.

Gerber J.A., Burmester W.L., Sellmyer D.J., 1982, Simple vibrating sample magnetometer, *Rev. Sci. Instrum.*, **53**, 691–693.

Goldfarb R.B., Bussey H.E., 1987, Methods for measuring complex permeability at radio frequencies, *Rev. Sci. Instrum.*, **58**, 624–627.

Golovanov C., Reyne G., Meunier G., Grössinger R., Dudding J., 2002, Finite element modeling of permanent magnets under pulsed field, *IEEE Trans. Magn.*, **36**, 1222–1225.

Gorican V., Hamler A., Hribernik B., Jesenik M., Trlep M., 2000, 2D measurements of magnetic properties using a round RSST, in *Proceedings of 6th Workshop on 2D Measurements*, Badgastein, Austria, pp. 66–75.

Gorican V., Jesenik M., Hamler A., Stumberger B., Trlep M., 2002, Performance of round rotational single sheet tester at higher flux densities in the case of GO material, in *Proceedings of 7th Workshop on 2D Measurements*, Lüdenscheid, Germany, pp.143–149.

Gozdur R., 2004, Determination of quasi-static hysteresis loop of electrical steel, *Przegl. Elektrotech.*, **80** (2), 147–149.

Gozdur R., Majocha A., 2007, Classification criterion of quasi-static magnetic hysteresis loops, *Przegl. Elektrotech.*, **83** (1), 134–137.

Gozdur R., Majocha A., 2010, Digital and analog processing of induction sensor signals, *Przegl. Elektrotech.*, **86** (4), 52–54.

Graham C.D., 1969, Textured magnetic materials, in *Magnetism and Metallurgy*, Academic Press, New York, Chapter 15.

Grössinger R., 2006, Magnetic characterization in a Pulsed Field Magnetometers, in *Springer Handbook of Materials Measurement Methods*, Czichos H. et al. (Eds.), Springer, Berlin, Germany, Chapter 10.3.

Grössinger R., Dahlgren M., 1998, Exchange coupled hard magnetic materials in pulsed high magnetic fields, *Physica B*, **246–247**, 213–218.

Grössinger R., Gigler C., Keresztes A., Fillunger H., 1988, A pulsed field magnetometer for the characterization of hard magnetic materials, *IEEE Trans. Magn.*, **24**, 970–973.

Grössinger R., Jewell G.W., Dudding J., Howe D., 1993, Pulsed field magnetometry, *IEEE Trans. Magn.*, **29**, 2980–2982.

Grössinger R., Turtelli R.S., Tellez-Blanco C., 2004, The influence of magnetic viscosity on pulsed field measurements, *J. Optoelectron. Adv. Mater.*, **6**, 557–563.

Grössinger R., et al., 1999, Large bore field magnetometer for characterizing permanent magnets, *IEEE Trans. Magn.*, **35**, 3971–3973.

Grössinger R. et al., 2002, Eddy currents in pulsed field measurements, *J. Magn. Magn. Mater.*, **242–245**, 911–914.

Hasenzagl A., Pfützner H., Saito A., Okazaki Y., 1997, Status of the Vienna hexagonal single sheet tester, in *Proceedings of 5th Workshop on 2D Measurements*, Grenoble, 33–41.

Hasenzagl A., Pfützner H., Saito A., Okazaki Y., 1998, Field distribution in rotational single sheet testers, *J. Phys. IV*, **8**, 681–684.

Hasenzagl A., Weiser B., Pfützner H., 1996, Novel 3-phase excited single sheet tester for rotational magntization, *J. Magn. Magn. Mater.*, **160**, 180–182.

Havlicek V., Mikulec M., 1989, On-line testing device using the compensation method, *Phys. Scr.*, **39**, 513–515.

Hilgert T., Vandevelde L., Melkebeek J., 2005, Magnetostriction measurement on electrical steels by means of strain gauges and numerical applications, *Przegl. Elektrotech.*, **81**(5), 87–91.

Hirano M., Ishihara Y., Harada mK., Todaka T., 2003, A study on measurement of magnetostriction of silicon steel sheet by laser displacement meter, *J. Magn. Magn. Mater.*, **254–255**, 43–46.

Hopkinson J., 1885, Magnetization of iron, *Trans. R. Soc.*, **A176**, 455–469.

Hribernik B., Hamler A., Kreca B., Trlep M., 1998, Numerical estimation of the field error of small single and double sheet testers, *IEEE Trans. Magn.*, **34**, 3276–3279.

Hubert A., Schäfer R., 1998, *Magnetic Domains*, Springer, Berlin, Germany.

Ilo A., Pfützner H., Bangert H., Eisemnmenger-Sittner Ch., 1998, Sputtered search coil for flux distribution analyses in laminated magnetic cores, *J. Phys. IV*, **8**, Pr2-733–Pr2-736.

Ingerson W.E., Beck F.J., Magnetic anisotropy in sheet steel, *Rev. Sci. Instrum.*, **9**, 31–35.

Iranmanesh H., Tahouri B., Moses A.J., Beckley P., 1992, A computerised Rogowski–Chattock potentiometer compensated on-line power-loss measuring system for use on grain-oriented steel production lines, *J. Magn. Magn. Mater.*, **112**, 99–102.

Jesenik M., Gorican V., Trlep M., Hamler A., Stumberger B., 2003, Calculation of the rotational magnetic fields in the sample of the rotational single sheet testers, *Przegl. Elektrotech.*, **79**, 920–922.

Jewell G., Howe D., Schotzko C., Grössinger R., 1992, A method of assessing eddy current effects in pulsed magnetometry, *IEEE Trans. Magn.*, **28**, 3114–3116.

Jiles D.C., Ramesh A., Shi Y., Fang X., 1997, Application of the anisotropic extension of the theory of hysteresis to the magnetization curves of crystalline and textured magnetic materials, *IEEE Trans. Magn.*, **33**, 3961–3963.

Kappel W., Pascale A., Rusu-Petroaia A., 1997, Power loss measurements in electrical steel using the H-coil method, *Sens. Act.*, **A59**, 320–322.

Kawasaki K., Shimazu T., Sakuma Y., Iwasaki H., Yoshinaga N., Kikuchi T., 1996, Development of new pole figure method for directly projecting grains with high speed and high resolution using synchrotron radiation and its application for observing electrical steel and mild steel sheets at high temperatures, *Nippon Steel Technical Report*, No. 69.

Kedous-Lebouc A., Cornut B., Perrier J.C., Manfe Ph., Chevalier Th., 2003, Punching influence on magnetic properties of the stator teeth of an induction motor, *J. Magn. Magn. Mater.*, **254–255**, 124–126.

Kedous-Lebouc A., Gautreau T., Chevalier T., 2007, Effect of the stato9r punching on the iron loss of a high speed synchronous machine, *Przegl. Elektrotech.*, **83**(4), 55–60.

Kelly J.M., 1957, New techniquwe for measuring rotational hystertesis in ferromagnetic materials, *Rev. Sci. Instrum.*, **28**, 1038–1040.

Khanlou A., Moses A.J., Meydan T., Beckley P., 1992, A non-enwrapping single yoke on-line magnetic testing system for use in the steel industry, *IEEE Trans. Magn.*, **28**, 2479–2481.

Khanlou A., Moses A.J., Meydan T., Beckley P., 1995, Computerized on-line power loss testing system for the steel industry, based on the RCP compensation technique, *IEEE Trans. Magn.*, **31**, 3385–3387.

Klimczyk P., Moses A., Anderson P., Davies M., 2009, Challenges in magnetostriction measurements under stress, *Przegl. Elektrotech.*, **85**(1), 100–102.

Kouvel J.S., Graham C.D., 1957, On the determination of magnetocrystalline anisotropy constants from torque measurements, *J. Appl. Phys.*, **28**, 340–343.

Krell C., Baumgartinger N., Krismanic G., Leiss E., Pfützner H., 2000, Relevance of multidirectional magnetostriction for the noise generation of transformer cores, *J. Magn. Magn. Mater.*, **215–216**, 634–636.

Krell C., Pfützner H., 2005, Magnetostriction of different types of Fe-based soft magnetic alloys under rotational magnetization, *Przegl. Elektrotech.*, **81**(5), 47–51.

Krismanic G., Leiss E., Barsoum S., Pfützner H., 2003, Automatic scanning system for the determination of local loss distribution in magnetic cores, *J. Magn. Magn. Mater.*, **254–255**, 60–63.

Krismanic G., Pfützner H., Baumgartinger N., 2000, A hand-held sensor for analyses of local distribution of magnetic field and losses, *J. Magn. Magn. Mater.*, **215–216**, 720–722.

Kumano T., Haratani T., Ushigami Y., 2002, The relationship between primary and secondary recristallization texture of grain oriented silicon steel, *ISIJ Int.*, **42**, 440–449.

Kumano T., Haratani T., Ushigami Y., 2003, The improvement of primary texture for sharp Goss orientation on grain oriented silicon steel, *ISIJ Int.*, **43**, 736–745.

Kurihara K., Kawamata Y., 1998, Development of a precise long-time digital integrartor for magnetic measurements in a Tokomak, in *17th IEEE/NPSS Symposium on Fussion Engineering*, San Diego, CA, pp. 799–802.

Kurosaki Y., Mogi H., Fujii H., Kubota T., Shiozaki M., 2008, Importance of punching and worability in non-oriented electrical steel sheets, *J. Magn. Magn. Mater.*, **320**, 2474–2480.

Layland N.J., Moses A.J., Takahashi N., Nakata T., 1996, Effect of shape on samples of silicon iron on the directions of magnetic field and flux density, in *Nonlinear Electromagnet Systems*, IOS Press, Amsterdam, the Netherlands, pp. 800–803

Ling P.C.Y., Moses A.J., Grimmond W., 1990, Investigation of magnetic flux distribution in wound toroidal cores taking into account geometrical factors, *Ann. Fis.*, **B86**, 99–101.

Liu J., Basak A., Moses A.J., 1996, The effectiveness of anisotropic magnetic material modelling methods, in *Nonlinear Electromagnetic Systems*, IOS Press, Amsterdam, the Netherlands, pp. 608–611.

Liu J., Basak A., Moses A.J., Shirkoohi G.H., 1994, A method of anisotropic steel modelling using finite element method with confirmation by experimental results, *IEEE Trans. Magn.*, **30**, 3391–3394.

Liu J., Shirkoohi G.H., 1993, Anisotropic magnetic material modelling using finite element method, *IEEE Trans. Magn.*, **29**, 2458–2460.

Lo C.C.H., Paulsen J.A., Jiles D., 2003, A magnetic imaging system for evaluation of material conditions using magnetoresistive devices, *IEEE Trans. Magn.*, **39**, 3453–3455.

Loisos G., Moses A.J., 2001, Critical evaluation and limitations of localized flux density measurements in electrical steels, *IEEE Trans. Magn.*, **37**, 2755–2757.

Loisos G., Moses A.J., 2003, Demonstration of a new method for magnetic flux measurement in the interior of a magnetic material, *Sens. Act.*, **A106**, 104–107.

Maeda Y., Sugimoto S., Shimoji H., Todaka T., Enokizono M., Sievert J., 2007, Study of the counterclockwise/clockwise rotation problem with the measurement of two-dimensional magnetic properties, *Przegl. Elektrotech.*, **83**(4), 18–24.

Makaveev D., de Wulf M., Gyselinck J., Maes J., Dupre L., Melkebeek J., 2000a, Measurement system for 2D magnetic properties of electrical steel sheets: design and performance, in *Proceedings of 6th Workshop on 2D Measurements*, Badgastein, Austria, pp. 48–55.

Makaveev D., Maes J., Melkebeek J., 2001, Controlled circular magnetization of electrical steel in rotational single sheet testers, *IEEE Trans. Magn.*, **37**, 2740–2742.

Makaveev D., von Rauch M., de Wulf M., Melkebeek J., 2000b, Accurate field strength measurement in rotational single sheet testers, *J. Magn. Magn. Mater.*, **215–216**, 673–676.

Mallinson J., 1966, Magnetometer coils and reciprocity, *J. Appl. Phys.*, **37**, 2514–2515.

Matheisel Z., 1973, *Cold Rolling Electrical Steel*, PWN.

Marketos P., Zurek S., Moses A.J., 2007, A method for defining the mean path length of the Epstein frame, *IEEE Trans. Magn.*, **43**, 2755–2757.

Matsuo T., Hirao H., Shimasaki M., 2007, Preliminary study of two-diomensional magnetic propertyy measurement of siolicon steel sheet using stator of induction motor, *Przegl. Elektrotech.*, **83**(4), 67–69.

Mazzetti P., Soardo P., 1966, Electronic hysteresisgraph holds dB/dt constant, *Rev. Sci. Instrum.*, **37**, 548–552.

McCarthy M., Houze G.L., Malagari F.A., 1967, Texture-electrical property correlations in oriented silicon steel, *J. Appl. Phys.*, **38**, 1096–1098.

McDonald D., Steingroever E., 1978, Magnetic potentiometer as an aid in testing and analyzing magnetic devices, *J. Appl. Phys.*, **49**, 1791–1793.

Mehnen L., Pfützner H., Krismanic G., Leiss E., Krell C., 2000, 2D magnetization control by means of evolutionary algorithm, in *Proceedings of 6th Workshop on 2D Measurements*, Badgastein, Austria, pp. 122–130.

Michalski A., 2002, Magnetovision, *IEEE Instrum. Meas. Mag.*, **5**, 66–69.

Mikulec M., Havlicek V., Wiglasz V., Cech D., 1984, Comparison of loss measurements on sheets and strips, *J. Magn. Magn. Mater.*, **41**, 223–226.

Moghaddam A.J., Moses A.J., 1992, A nondestructive automated method of localised power loss measurement in grain oriented 3% silicon iron, *J. Magn. Magn. Mater.*, **112**, 132–134.

Moghaddam A.J., Moses A.J., 1993, Localised power loss measurement using remote sensors, *IEEE Trans. Magn.*, **29**, 2998–3000.

Mogi H., Yabumoto M., Mizokami M., Okazaki Y., 1996, Harmonic analysis of AC magnetostriction measurements under non-sinusoidal excitation, *IEEE Trans. Magn.*, **32**, 4911–4913.

Mohd Ali B.B., 1985, Computer mapping of grain structures in grain-oriented silicon iron, PhD dissertation, University of Cardiff, Cardiff, U.K.

Mohd Ali B.B., Moses A.J., 1989, A grain detection for grain-oriented electrical steels, *IEEE Trans. Magn.*, **25**, 4421–4426.

Mohri K., Fujimoto T., 1977, New grain detecton methods for grain oriented SiFe with coating, *Soft Masgnetic Materials Conference Proceedings*, Paper 19-1, Bratislawa, Slovakia.

Mohri K., Takeuchi S.I., Fujimoto T., 1979, Domain and grain observations using a colloid technique for grain-oriented SiFe with coating, *IEEE Trans. Magn.*, **15**, 1346–1349.

Morino H., Ishihara Y., Todaka T., 1992, Measuring method of magnetic characteristics in any direction for silicon steel, *J. Magn. Magn. Mater.*, **112**, 115–119.

Morino H., Ishihara Y., Todaka T., 1993, Measuring of magnetic characteristics of oriented silicon steel in an arbitrary direction, *Electr. Eng. Jpn.*, **113**(6), 1–9.

Moses A.J., 1998a, Measurement and analysis of magnetic fields on the surface of grain-oriented silicon steel sheet, *J. Phys. IV*, **8**, Pr2-561–Pr2-565.

Moses A.J., 1988b, Recent advances in experimental methods for the investigation of silicon iron, *Phys. Scr.*, **T24**, 49–53.

Moses A.J., 1992, Problems in modelling anisotropy of electrical steels, *Int. J. Appl. Electromagnet. Mater.*, **3**, 193–197.

Moses A.J., 1994, Rotational magnetization—Problems in experimental and theoretical studies of electrical steels and amorphous magnetic materials, *IEEE Trans. Magn.*, **30**, 902–906.

Moses A.J., Hamadeh S., 1983, Comparison of the Epstein square and a single strip tester for measuring the power loss of nonoriented electrical steels, *IEEE Trans. Magn.*, **19**, 2705–2710.

Moses A.J., Jones R.M., 1988, Application of ferromagnetic magnetoresistive sensors in mapping of steel structures, 49–56.

Moses A.J., Konadu S.N., 2001, Some effects of grain boundaries on the field distribution on the surface of grain oriented electrical steels, *Int. J. Appl. Electromagnet. Mech.*, **13**, 339–342.

Moses A.J., Leicht J., 2004, Measurement and prediction of iron loss in electrical steel under controlled magnetization conditions, *Przegl. Elektrotech.*, **80**(12), 1181–1187.

Moses A.J., Soinski M., 1996, Nonlinear dependencies between graoin orientation and anisotropy of magnetic properties of electrical steel sheets with Goss texture, in *Nonlinear Electromagnetic Systems*, IOS Press, Amsterdam, the Netherlands, pp. 446–449.

Moses A.J., Thomas B., 1973, The spatial variation of localized power loss in two practical transformer T-joints, *IEEE Trans. Magn.*, **9**, 655–659.

Moses A.J., Williams P.I., Hoshtanar O.A., 2005, A novel instrument for real time dynamic domain observation in bulk and micromagnetic materials, *IEEE Trans. Magn.*, **41**, 3736–3738.

Mthombeni T.L., Pillay P., Strnat R.M., 2007, New Epstein frame for lamination core loss measurements under high frequencies and high flux densities, *IEEE Trans. Energy Convers.*, **22**, 614–620.

Nafalski A., Moses A.J., 1990, Loss measurements of amorphous materials using single strip testers, *Phys. Scr.*, **40**, 532–535.

Nafalski A., Moses A.J., Meydan T., Abousetta M.M., 1989, Loss measurements on amorphous materials using a field compensated single strip tester, *IEEE Trans. Magn.*, **25**, 4287–4291.

Nakata T., Fujiwara K., Nakano M., Kayada T., 1990a, Effect of the construction of yokes on the accuracy of single sheet tester, *Ann. Fis.*, **B86**, 190–192.

Nakata T., Ishihara Y., Nakaji M., Todaka T., 2000, Comparison between the H-coil method and the magnetizing current method for the single sheet tester, *J. Magn. Magn. Mater.*, **215**, 607–610.

Nakata T., Ishihara Y., Takahashi N, Kawase Y., 1982, Analysis of magnetic fields in a single sheet tester using an H-coil, *J. Magn. Magn. Mater.*, **26**, 179–180.

Nakata T., Takahashi N., Fujiwara K., Nakano M., 1993, Measurement of magnetic characteristics along arbitrary directions of grain-oriented silicon steel up to high flux densities, *IEEE Trans. Magn.*, **29**, 3544–3546.

Nakata T., Takahashi N., Fujiwara K., Nakano M., 1994, Study of horizontal-type single sheet testers, *J. Magn. Magn. Mater.*, **133**, 416–418.

Nakata T., Takahashi N., Kawase Y., 1985, Factors affecting the accuracy of a single sheet testers using an H coil, in *Proceedings of the Soft Magnetic Materials Confernce*, pp. 49–51.

Nakata T., Takahashi N., Kawase Y., 1990b, Factors affecting the accuracy of a single sheet tester using a H-coil, *Ann. Fis.*, **B86**, 49–51.

Nakata T., Takahashi N., Kawase Y., Nakano M., 1984, Influence of lamination orientation and stacking on magnetic characteristics of grain-oriented silicon steel laminations, *IEEE Trans. Magn.*, **20**, 1774–1776.

Nakata T., Takahashi N., Nakano M., Muramatsu K., Miyake M., 1994b, Magnetostriction measurements with a laser Doppler velocimeter, *IEEE Trans. Magn.*, **30**, 4563–4565.

Nakata T., Yoshihiro K., Nakano M., 1987, Improvement of measuring accuracy of magnetic field strength in single sheet testers by using two H-coils, *IEEE Trans. Magn.*, **23**, 2596–2598.

Nakazaki O., Todaka T., Enokizono M., 2009, Iron loss evaluation in stator cores of rotating machines during manufacturing process, *Przegl. Elektrotech.*, 85(1), 74–78.

Nam Quoc Ngo, 2006, A new approach for design of wideband digital integrator and differentiator, *IEEE Trans. Circuits Syst. II*, **53**, 936–940.

Narita K., Yamaguchi T., 1974, Rotational hysteresis loss in silicon-iron single crystal with (001) surfaces, *IEEE Trans. Magn.*, **10**, 165–167.

Nencib N., Kedous-Lebouc A., Cornut B., 1994, 3D analysis of a rotational loss tester with vertical yokes, *J. Magn. Magn. Mater.*, **133**, 553–556.

Nencib N., Kedous-Lebouc A., Cornut B., 1995, 2D analysis of rotational loss tester, *IEEE Trans. Magn.*, **31**, 3388–3390.

Neumann H., 1939, Messung der Koerzitivkraft, ATM, 957–961.

Nicholson P.I., 1999, Development of a magnetic domain imaging system for electrical steel using a magnetoresistive sensor based stray field scanning technique, PhD dissertation, Cardiff School of Engineering, Cardiff, U.K.

Normann N., Mende H.H., 1980, Stray field measurements on grain oriented SiFe sheets, *J. Magn. Magn. Mater.*, **19**, 386–390.

O'Barr R., Lederman M., Schultz S., 1996, A scanning microscope using a magnetoresistive head as the sensing element, *J. Appl. Phys.*, **79**, 8067–8069.

Oral A., Bending S.J., Henini M., 1996, Real time scanning Hall probe microscopy, *Appl. Phys. Lett.*, **69**, 1324–1326.

Ouchi K., Iwasaki S., 1988, Analysis of perpendicular recording media using a biaxial vibrating sample magnetometer, *IEEE Trans. Magn.*, **24**, 3009–3011.

Overshott K.J., Blundell M.G., 1984, Power loss, domain wall motion and flux density of neighbouriong grains in grain-oriented 3% silicon iron, *IEEE Trans. Magn.*, **20**, 1551–1553.

Pacyna A.W., Ruebenbauer K., 1984, General theory of a vibrating magnetometer with extended coils, *J. Phys. E.*, **17**, 141–143.

Papamarkos N., Chamzas C., 1996, A new approach for the design of digital integrators, *IEEE Trans. Circuits Syst. I*, **43**, 785–791.

Pera T., Ossart F., Waeckerle T., 1993a, Numerical representation for anisotropic materials based on coenergy modeling, *J. Appl. Phys.*, **73**, 6784–6786.

Pera T., Ossart F., Waeckerle T., 1993b, Field computation in non-linear anisotropic sheets using the coenergy model, *IEEE Trans. Magn.*, **29**, 2425–2427.

Perevertev O., 2005, Measurement of the surface field on open magnetic samples by the extrapolation method, *Rev. Sci. Instrum.*, **76**, 104701.

Perevertev O., 2009, Increase of precision of surface magnetic field measurement by magnetic shielding, *Meas. Sci. Technol.*, **20**, 055107.

Permiakov V., Pulnikov A., Dupre L., De Wulf M., Melkebeek J., 2003, Magnetic properties of Fe-Si steel depending on compressive and tensile stresses under sinusoidal and distorted excitations, *J. Appl. Phys.*, **93**, 6689–6691.

Pfützner H., 1980a, Andwendungen of Hallgeneratoren im Vergleich zu anderen Methoden der Feldstärkeerfassung bei der Prüfung von Elektroblechen, *Z. Elektr. Inform. Energietechnik*, **10**, 534–546.

Pfützner H., 1980b, Computer mapping of grain structure in coated silicon iron, *J. Magn. Magn. Mater.*, **19**, 27–30.

Pfützner H., 1981, Neuartige zerstörungsfreie Kristallstruktur Analyse an isolierten Transformatorblechen, *Microchim. Acta., Suppl.*, **9**, 193–204.

Pfützner H., 1994, Rotational magnetization and rotational losses of grain oriented silicon steel sheets—Fundamental aspects and theory, *IEEE Trans. Magn.*, **30**, 2802–2807.

Pfützner H., 2000, Different designs of rotational single sheet testers—Suggestions for procedures of calibration, in *Proceedings of 6th Workshop 2D-Magnetic Measurements*, Badgastein, Austria, pp. 154–162.

Pfützner H., 2002, Present status of research on two-dimensional magnetization, in *Proceeedings of the 7th Workshop 2D-Magnetic Measurements*, Lüdenscheid, Germany, pp. 81–87.

Pfützner H., Bengtsson C., Leeb A., 1985, Domain investigation on coated unpolished SiFe sheets, *IEEE Trans. Magn.*, **21**, 2620–2625.

Pfützner H., Futschik K., Luo Y., 1982, Effect of bending on GO silicon iron sheets, *IEEE Trans. Magn.*, **18**, 1499–1501.

Pfützner H., Krismanic G., 2004, The needle method for induction tests: Sources of errors, *IEEE Trans. Magn.*, **40**, pp. pp. 1610–1616.

Pfützner H., Krimsanic G., Yamaguchi H., Leiss E., Chinag W.C., 2007, A study on possible sources of errors of loss measurement under rotational magnetization, *Przegl. Elektrotech.*, **83**(4), 9–13.

Pfützner H., Mulasalihovic E., 2009, Thin film technique for interior magnetic analysis of laminated machine cores, *Przegl. Elektrotech.*, **85**, 39–42.

Pfützner H., Schwarz G., Fidler J., 1983, Computer controlled domain detector, *Jpn. J. Appl. Phys.*, **22**, 361–364.

Pfützner H., Schönhuber P., 1991, On the problem of the field detection for single sheet testers, *IEEE Trans. Magn.*, **27**, 778–785.

Pfützner H., Schönhuber P., Futschik K., 1992, Novel nondestructive methods for analysis of crystalline and amorphous soft magnetic materials, in *Proceedings of 2nd International Symposium on Physics of Magnetic Materials*, Beijing, Japan, pp. 427–434.

Piermiakov V., Pulnikov A., Dupre L., Melkebeek J., 2005, 2D magnetic measurements under 1D stress, *Przegl. Elektrotech.*, **81**(5), 68–72.

Piermiakov V., Pulnikov A., Makaveev D., De Wulf M., Dupre L., Melkebeek J., 2003, Magnetic measurements under compressive and tensile stresses for nonoriented electrical steel, *PTB Bericht*, **E-81**, 15–21.

Pluta W., Kitz E., Krell C., Rygal R., Soinski M, Pfützner H., 2003, Practical relevance of rotational loss measurement of laminated machine cores, *Przegl. Elektrotech.*, 79, 151–154.

Pulnikov A., Piermikov V., De Wulf M., Melkebeek J., 2003, Measuring setup for the investigation of the influence of mechanical stresses on the magnetic properties of electrical steel, *J. Magn. Magn. Mater.*, **254–255**, 47–49.

Qiu Z.Q., Bader S.D., 2000, Surface magneto optic Kerr effect, *Rev. Sci. Instrum.*, **71**, 1243–1255.

Radley G.S., Moses A.J., 1981, Apparatus for experimental simulation of magnetic flux and power loss distribution in a turbogenerator stator core, *IEEE Trans. Magn.*, **17**, 1311–1316.

Richter H.J., Hempel K.A., Pfeiffer J., 1988, Improvement of sensitivity of the vibrating reed magnetometer, *Rev. Sci. Instrum.*, **59**, 1388–1393.

Ross G., 2006, Soft and hard magnetic materials: Standard measurement techniques for properties related to B(H) loop, in *Springer Handbook of Materials Measurement Methods*, Springer, Berlin, the Netherlands, Chapter 10.2.

Ross W., Hempel K.A., Voigt C., Dederichs H., Schippan R., 1980, High sensitivity vibrating reed magnetometer, *Rev. Sci. Instrum.*, **51**, 612–613.

Rygal R., Moses A.J., Derebasi N., Schneider J., Schoppa A., 2000, Influence of cutting stress on magnetic field and flux density distribution in non-oriented electrical steels, *J. Magn. Magn. Mater.*, **215–216**, 687–689.

Salz W., 1994, A two dimensional measuring equipment for electrical steel, *IEEE Trans. Magn.*, **30**, 1253–1257.

Samwell E.O., Bolhuis T., Lodder J.C., 1998, An alternative approach to vector vibrating sample magnetometer detection coil setup, *Rev. Sci. Instrum.*, **69**, 3204–3209.

Sandford R.L., Bennet E.G., 1933, An apparatus for magnetic testing of high magnetizing forces, *Bur. Stand. J. Res.*, **10**, 567.

Sasaki T., Imamura M., Takada S., Suzuki Y., 1985, Measurement of rotational power losses in silicon iron sheets using wattmeter method, *IEEE Trans. Magn.*, **21**, 1918–1920.

Schmidt N., Güldner H., 1996, A simple method to determine dynamic hysteresis loops of soft magnetic materials, *IEEE Trans. Magn.*, **32**, 489–496.

Scholes R., 1970, Application of operational amplifier to magnetic measurements, *IEEE Trans. Magn.*, **6**, 289–291.

Schönhuber P., Pfützner H., 1989, Hall sensor for automatic detection of domains ion coated silicon iron, *Phys. Scr.*, **40**, 558–560.

Schoppa A., Schneider J., Wuppermann C.D., 2000, Influence of the manufacturing process on the magnetic properties of non-oriented electrical steels, *J. Magn. Magn. Mater.*, **215–216**, 74–78.

Senda K., Ishida M., Sato K., Komatsubara M., Yamaguchit T., 1997, Localized magnetic properties in grain oriented silicon steel measured by stylus probe method, *Trans. IEE Jpn.*, **117A**, 941–949.

Senda K., Ishida M., Sato K., Komatsubara M., Yamaguchi T., 1999, Localized magnetic properties in grain oriented electrical steel measured by needle probe method, *Electr. Eng. Jpn.*, 126, 942–949.

Senda K., Kurosawa M., Ishida M., Komatsubara M., Yamaguchi T., 2000, Local magnetic properties in grain oriented steel measured by the modified needle probe method, *J. Magn. Magn. Mater.*, **215–216**, 136–139.

Shen D., Sabonnadiere J.C., Meunier G., Coulomb J.L., Sacotte M., 3D anisotropic magnetic field calculation in transformer joints, *IEEE Trans. Magn.*, **23**, 3783–3785.

Shilling J.W., Houze G.L., 1974, Magnetic properties and domain structure in grain oriented 3% SiFe, *IEEE Trans. Magn.*, **10**, 195–223.

Shimamura M., Okinori C., Tanaka H., Enokizono M., 2000, Approach to 2-dimensional high frequency magnetic characteristic measurement with high speed and accuracy, in *Proceedings of 6th Workshop on 2D Measurements*, Badgastein, Austria, pp. 138–146.

Shikoohi G.H., 1994, *Anisotropic* properties of low grade non-oriented electrical steel, in *Proceedings of International Conference ELMECO*, Lublin, Poland, September 8–9, 1994, pp. 141–146.

Shirkoohi G.H., Arikat M.A.M., 1994, Anisotropic properties of high permeability grain-orieneted 3.25% SiFe electrical steel, *IEEE Trans. Magn.*, **30**, 928–930.

Shirkoohi G.H., Kontopoulos A.S., 1994, Computation of magnetic field in Rogowski–Chattock potentiometer compensated magnetic testers, *J. Magn. Magn. Mater.*, **133**, 587–590.

Shirkoohi G.H., Liu J., 1994, A finite element method for modelling of anisotropic grain-oriented steels, *IEEE Trans. Magn.*, **30**, 1078–1080.

Shull R.D., McMichael R.D., Swartzendruber L.J., Leigh S.D., 2000, Absolute magnetic moment measurements of nickel spheres, *J. Appl. Phys.*, **87**, 5992–5994.

Sievert J.D., 1984, Determination of AC magnetic power loss of electrical steel sheet: present status and trends, *IEEE Trans. Magn.*, **20**, 1702–1707.

Sievert J., 1990, Recent advances in the one- and two-dimensional magnetic measurement technique for electrical sheet steel, *IEEE Trans. Magn.*, **26**, 2553–2558.

Sievert J., 1992, On measuring the magnetic properties of electrical sheet steel under rotational magnetization, *J. Magn. Magn. Mater.*, **112**, 50–57.

Sievert J., 1995, One- and two-dimensional magnetic phenomena in electrical steel and their measurement, in *Advanced Computational and Design Techniques in Applied Electromagnetic Systems*, Elsevier, Amsterdam, the Netherlands, pp. 639–642.

Sievert J., 2005, On the metrology of the magnetic properties of electrical sheet steel, *Przegl. Elektrotech.*, **81**, 1–5.

Sievert J., Ahlers H., 1990, Is the Epstein frame replaceable? *Ann. Fis.*, **B86**, 58–63.

Sievert J., Ahlers H., Brosien P., Cundeva M., Luedke J., 2000, Relationship of Epstein to SST results for grain oriented steel, in *Non-linear Electromagn. Systems*, Vol. 18, IOS Press, Amsterdam, the Netherlands, pp. 3–6.

Sievert J., Ahlers H., Enokizono M., Kauke S., Rahf L., Xu J., 1992, The measurement of rotational power loss in electrical sheet steel using a verical yoke system, *J. Magn. Magn. Mater.*, **112**, 91–94.

Sievert J., Ahlers H., Lüdke J., Siebert S., Pareti L., Solzi M., 1993, European intercomparison of measurements on permanent magnets, *IEEE Trans. Magn.*, **29**, 2887–2889.

Sievert J., Ahlers H., Siebert S., Enokizono M., 1990a, On the calibration of magnetometers having electromagnets with the help of cylindrical nickel reference samples, *IEEE Trans. Magn.*, **26**, 2052–2054.

Sievert J., Binder M., Rahf L., 1990b, On the reproducibility of a single sheet testers: Comparison of different measuring procedures and SST designs, *Ann. Fis.*, **B86**, 76–78.

Sievert J., Xu J., Enokizono M., Ahlers H., 1990c, Studies on the rotational power loss measurement problem, *Ann. Fis.*, **B86**, 35–37.

Sievert J. et al., 1996, European intercomparison of measurements of rotational power loss in electrical sheet steel, *J. Magn. Magn. Mater.*, **160**, 115–118.

Silvester P.P., Gupta R.P., 1991, Effective computational models for anisotropic soft B/H curves, *IEEE Trans. Magn.*, **27**, 3804–3807.

Smith C.H., Schneider R.W., Pohm A.V., 2003, High resolution giant magnetoresistance on chip arrays for magnetic imaging, *J. Appl. Phys.*, **93**, 6864–6866.

Snell D., 2008, Measurement of noise associated with model transformer cores, *J. Magn. Magn. Mater.*, **320**, e535–e538.

So M.H., Nicholson P.I., Meydan T., Moses A.J., 1995, Magnetic domain imaging in coated silicon iron using magnetoresistive sensors, *IEEE Trans. Magn.*, **31**, 3370–3372.

Soinski M., 1984, Application of the anisometric method for determining polar curves of induction, apparent core loss and core loss in cold-rolled electrical sheets of Goss texture, *IEEE Trans. Magn.*, **20**, 172–183.

Soinski M., Moses A.J., 1995, Anisotropy in iron based soft magnetic materials, in *Handbook of Magnetic Materials, v.8*, Elsevier, Amsterdam, the Netherlands, Chapter 4.

Somkun S., Moses A.J., Zurek S., Anderson P.I., 2009, Development of an induction motor core model for measuring rotational magnetostriction under PWM magnetisation, *Przegl. Elektr.*, **85**(1), 103–107.

Spuig P., Defrasne P., Martin G., Moreau M., Moreau Ph., Saint-Laurent F., 2003, An analog integrator for thousand second long pulses in Tore Supra, *Fusion Eng. Des.*, **66–68**, 953–957.

Stanbury H.J., 2010, International standards applied to magnetic alloys and steels, *IEEE Trans. Magn.*, **46**, 274–278.

Stata R., 1967, Operational integrators, Analog Devices Application Note No. AN-357.

Steongroever E., Ross G., 1997, *Magnetic Measuring Techniques*, Magnet-Physik, Cologne, Germany.

Street R., Day R.K., Dunlop J.B., 1987, Magnetic viscosity in BdFeB and SmCo5 alloys, *J. Magn. Magn. Mater.*, **69**, 106–112.

Stupakov O. Kikuchi H., Liu T., Takagi T., 2009, Applicability of local magnetic measurements, *Measurement*, **42**, 706–710.

Stupakov O., Wood R., Melikhov Y., Jiles D., 2010, Measurement of electrical steels with direct field determination, *IEEE Trans. Magn.*, **46**, 298–301.

Tamaki T., Fujisaki K., Yamada T., 2009, Hole effect on magnetic field uniformity in Single Sheet Tester, *Przegl. Elektr.*, **85**(1), 71–73.

Tejedor M., Garcia J.A., Carrizo J., 1990, Determination of anisotropy constants of amorphous magnetic alloy ribbons by torque magnetometer, *Ann. Fis.*, **B86**, 111–113.

Tejedor M., Rubio H., Gonzalez M., Elbaile L., Iglesias R., 1993, A system for the measurement of near surface magnetic properties, *Rev. Sci. Instrum.*, **64**, 2933–2937.

Tellez-Blanco J.C., Sato Turtelli R., Grössinger R., 1999, Giant magnetic viscosity in SmCo5Cu alloys, *J. Appl. Phys.*, **86**, 5157–5163.

Thouttuvelil V.J., Wilson T.G., Owen H.A., 1990, High frequency measurement techniques for magnetic cores, *IEEE Trans. Power Electron.*, **5**, 41–53.

Todaka T., Maeda Y., Enokizono M., 2009, Counterclockwise/clockwise rotational losses under high magnetic field, *Przegl. Elektrotech.*, **85**(1), 20–24.

Trout S.R., 1988, Use of Helmholtz coils for magnetic measurements, *IEEE Trans. Magn.*, **24**, 2108–2111.

Tseng C.C., 2006, Digital integrator design using Simpson rule and fractional delay filter, in *IEEE Proceedings of Vision Image Signal Processing*, 153, pp. 79–85.

Tumanski S., 1988, The application of permalloy magnetoresistive sensors for nondestructive testing of electrical steel sheets, *J. Magn. Magn. Mater.*, **75**, 266–272.

Tumanski S., 1998, The investigations of electrical steel nonuniformity, *J. Phys. IV*, **8**, Pr2-575–Pr2-578.

Tumanski S., 1999, Magnetovision, in *McGraw-Hill 2000 Yearbook of Science and Technology*, McGraw-Hill, New York, pp. 242–244.

Tumanski S., 2000a, Nondestructive testing of the stress effects in electrical steel by magnetovision method, in *Non-Linear Electromagnetic Systems*, Vol. 18, IOS Press, Amsterdam, the Netherlands, pp. 273–276.

Tumanski S., 2000b, The experimental verification of the condition of the magnetic material caused by different technological processes, *J. Magn. Magn. Mater.*, **215–216**, 749–752.

Tumanski S., 2002a, A multicoil sensor for tangential magnetic field investigations, *J. Magn. Magn. Mater.*, **242–245**, 1153–1156.

Tumanski S., 2002b, Which magnetizing circuit is suitable for two-dimensional measurements, in *Proceedings of 7th Workshop on 2D Measurements*, Lüdenscheid, Germany, pp. 151–157.

Tumanski S., 2002c, Investigations of two-dimensional properties of selected electrcial steel samples by means of the Epstein method, in *Proceedings of 7th Workshop on 2D Measurements*, Lüdenscheid, Germany, pp. 151–157.

Tumanski S., 2002d, A method of testing of the plane distribution of anisotropy, *IEEE Trans. Magn.*, **38**, 2808–2810.

Tumanski S., 2003, Investigations of the anisotropic behaviour of SiFe steel, *J. Magn. Magn. Mater.*, **254–255**, 50–53.

Tumanski S., 2005, New design of the magnetizing circuit for 2D testing of electrical steel, *Przegl. Elektrotech.*, **81**(5), 32–34.

Tumanski S., 2007, Scanning of magnetic field as a method of investigations of the structure of magnetic materials, *Przegl. Elektrotech.*, **83**(1), 108–112.

Tumanski S., Bakon T., 2001, Measuring system for two-dimensional testiong og electrical steel, *J. Magn. Magn. Mater.*, **223**, 315–325.

Tumanski S., Baranowski S., 2004, Single strip tester with direct measurement of magnetic field strength, *J. Electr. Eng.*, **55**, 41–44.

Tumanski S., Baranowski S., 2006, Magnetic sensor array for investigations of magnetic field distribution, *J. Electr. Eng.*, **57**, 185–188.

Tumanski S., Baranowski S., 2007, Single strip tester of magnetic materials with array of magnetoresistive sensors, *Przegl. Elektrotech.*, **83**, 46–49.

Tumanski S., Baranowski S., 2008, Analysis of heterogeneity of electrical steel, *J. Electr. Eng.*, **59**, 1–4.

Tumanski S., Fryskowski B., 1998, New method of texture and anisotropy analysis in Go SiFe steel, *J. Phys. IV*, **8**, Pr2-669– Pr2-672.

Tumanski S., Liszka A., 2002a, *On-Line Evaluation of Electrical Steel Structure and Quality*, INTRERMAG 2002 Paper No. DU09, Amsterdam, the Netherlands.

Tumanski S., Liszka A., 2002b, The methods and devices for scanning of magnetic fields, *J. Magn. Magn. Mater.*, **242–245**, 1253–1256.

Tumanski S., Liszka A., 2004, Analysis of magnetic field distribution in C-yoke system, *Int. J. Apl. Electromagn. Mech.*, **19**, 663–666.

Tumanski S., Pluta W., Soinski M., 2004, Analysis of magnetic field distribution in the sample of RSST device, in *Proceedings of Soft Magnetic Materials 16 Conference*, Düsseldorf, Germany, pp. 859–864.

Tumanski S., Stabrowski M., 1998, The magnetovision method as a tool to investigate the quality of electrical steel, *Meas. Sci. Technol.*, **9**, 488–495.

Tumanski S., Winek T., 1997, Measurements of the local values of electrical steel parameters, *J. Magn. Magn. Mater.*, **174**, 185–191.

Vandevelde L., Hilgert T.G.D., Melkebeek J.A.A., 2004, Magnetostriction and magnetic forces in electrical steel: finite element computations and measurements, *IEE Proc. Sci. Meas. Technol.*, **151**, 456459.

Waeckerle T., Rouve L.L., Talowski C., 1995, Study of anisotropic B-H models for transformer cores, *IEEE Trans. Magn.*, **31**, 3991–3993.

Wakabayashi D., Maeda Y., Shimoji H., Todaka T., Enokizono M., 2009, Measurement of vector magnetostriction in alternating and rotating magnetic field, *Przegl. Elektrotech.*, **85**(1), 34–38.

Wecker S.M., Morris P.R., 1978, Modern methods of texture analysis, *J. Appl. Cryst.*, **11**, 211–220.

Weiser B., Hasenzagl A., Booth T., Pfützner H., 1996, Mechanisms of noise generation of model transformer cores, *J. Magn. Magn. Mater.*, **160**, 207–209.

Weiss P., Planer V., 1908, Hysteresis in rotating field, *J. Phys.*, **4/7**, 5.

Wenk H.R., Van Houtte P., 2004, Texture and anisotropy, *Rep. Prog. Phys.*, **67**, 1367–1428.

Werner R., 1957, Einrichtung zur Messung magnetischer Eigenschaften von Blechen bei Wechselstrommagnetisierung, Austrian Patent 191015.

Wiglasz V., 1992, Comparison of magnetic measurements on sheets and Epstein samples, *J. Magn. Magn. Mater.*, **112**, 85–87.

Wilczynski W., Schoppa A., Schneider J., 2004, Influence of different fabrication steps of magnetic cores on their magnetic properties, *Przegl. Elektr.*, **80**(2), 118–122.

Wilkins F.J., Drake A.E., 1965, Instrument for measuring local power losses in uncut electrical steel sheet, *IEE Proc.*, **112**, 786–793.

Wilkins F.J., Drake A.E., 1970, Automatic measurement of local power losses in grain oriented silicon iron, *IEE Proc.*, **117**, 1048–1051.

Wood R., Anderson P., Moses A.J., Jenkins K., 2009, Divergence of flux in a grain-oriented electrical steel sheet locally magnetised by a single yoke system, *Przegl. Elektrotech.*, **85**, 31–33.

Wrona J., 2002, Magnetometers for thin film structures, PhD dissertation, AGH, Cracow, Poland.

Wrona J., Czapkiewicz M., Stobiecki T., 1999, Magnetometer for the measurements of the hysteresis loop of ultra-thin magnetic layers, *J. Magn. Magn. Mater.*, **196–197**, 935–936.

Wrona J., Stobiecki T., Czapkiewicz M., Rak R., Slezak T., Korecki J., Kim C.G., 2004, R-VSM and MOKE magnetometers for nanostructures, *J. Magn. Magn. Mater.*, **272–276**, 2294–2295.

Xu J., Sievert J., 1997, On the reproducibility, standardization aspects and error sources of the fieldmetric method for the determination of 2D magnetic properties of electrical sheet steel, in *Proceedings of 5th Workshop of 2D measurements*, Grenoble, France, pp. 43–54.

Yabumoto M., 2009, Review of teechniques for measurement of magnetostriction in electrical steels and progress towards standardisation, *Przegl. Elektrotech.*, **85**(1), 1–6.

Yamaguchi T., Narita K., 1976, Rotational power loss in commercial silicon iron laminations, *Electr. Eng. Jpn.*, 96, 15–21.

Yamaguchi H., Pfützner H., Hasenzagl A., 2008, Magnetostriction measurements on the multidirectional magnetization performance of SiFe steel, *J. Magn. Magn. Mater.*, **320**, e618–e622.

Yamaguchi T., Senda K., Ishida M., Sato K., Honda A., Yamamoto T., 1998, Theoretical analysis of localized magnetic flux measurement by needle method, *J. Phys. IV, France*, 8, Pr2717–Pr2720.

Yamamoto T., Takada S., Sasaki T., 1994, Estimation measuring error in digital DC fluxmeters, *J. Magn. Magn. Mater.*, **133**, 446–449.

Yanase S., Matsuno Y., Hashi S., Okazaki Y., 2007, 2D magnetic rotational loss of electrical steel at high magnetic flux density, *Przegl. Elektrotech.*, **83**(4), 31–34.

Young F.J., Schenk H.L., 1960, Method for measuring iron losses in elliptically polarized magnetic field, *J. Appl. Phys.*, **5**, S194–S195.

Zemanek I., 2006, Universal control and measuring system for modern classic and amorphous magnetic materials single/on-line testers, *J. Magn. Magn. Mater.*, **304**, e577–e579.

Zemanek I., Single sheet and on-line testing based on NMF compensation method, *Przegl. Elektr.*, **85**(1), 7983.

Zhu J.G., Lin Z.W., Guo Y.G., Huanf Y., 2009, 3D measurement and modelling of magnetic properties of soft magnetic composite, *Przegl. Elektrotech.* **85**(1), 11–15.

Zhu J.G., Ramsden V.S., 1993, Two dimensional measurement of magnetic field and core loss using a square specimen tester, *IEEE Trans. Magn.*, **29**, 2995–2997.

Zhu J.G., Ramsden V.S., 1997, Measurement and modelling of losses under two dimensional excitation in rotating electrical machines, in *Proceedings of 5th Workshop of 2D Measurements*, Grenoble, France, pp. 63–77.

Zhu J.G., Zhong J.J., Lin Z.W., Sievert J., 2002a, 3D magnetic property tester—Design, construction and calibration, in *Proceedings of 7th Workshop on 2D Measurements*, Lüdenscheid, Germany, 97102.

Zhu J.G., Zhong J.J., Lin Z.W., Sievert J., 2002b, Measurement of magnetic properties under 3D magnetic excitation, *IEEE Trans. Magn.*, **39**, 3429–3431.

Zieba A., Foner S., 1982, Detection coil, sensitivity function and sample geometry effect for vibrating sample magnetometers, *Rev. Sci. Instrum.*, **53**, 1344–1354.

Zijlstra H., 1967, *Experimental Methods in Magnetism*, North-Holland, Amsterdam, the Netherlands.

Zijlstra H., 1970, A vibrating reed magnetometer for microscopic particles, *Rev. Sci. Instrum.*, **41**, 1241–1243.

Zimmermann G., Hempel K.A., Dodel J., Schmitz M., 1996, A vectorial vibrating reed magnetometer with high sensitivity, *IEEE Trans. Magn.*, **32**, 416–420.

Zouzou S., Kedous-Lebouc A., Brissonneau P., 1992, Magnetic properties under unidirectional and rotational field, *J. Magn. Magn. Mater.*, **112**, 106–108.

Zurek S., 2009, Static and dynamic rotational losses in non-oriented electrical steel, *Przegl. Elektrotech.*, **85**(1), 89–92.

Zurek S., Meydan T., 2006a, A novel capacitive flux density sensor, *Sens. Act.*, **A129**, 121–125.

Zurek S., Meydan T., 2006b, Rotational power losses of magnetic steel sheets in circular rotational magnetic field in ccw/cw direction, *IEE Proc. Meas. Sci. Technol.*, **153**, 147–157

Zurek S., Meydan T., Moses A.J., 2008, Analysis of twisting of search coil leads as a method reducing the influence of stray fields on accuracy of magnetic measurements, *Sens. Actuators*, **A142**, 569–573.

Zurek S., Rygal R., Soinski M., 2009, Asymmetry of magnetic properties of conventional grain-oriented steel with relation to 2D measurement, *Przegl. Elektrotech.*, **85**(1), 16–19.

6

Magnetic Field Measurements and Their Applications

6.1 Environment Magnetic Fields

6.1.1 Earth's Magnetic Fields

In our life, the magnetic field is omnipresent. It is a field generated by Earth superimposed on field coming from outer space combined with magnetic pollution. Presence of this field causes also that magnetic materials act as a secondary magnetic field source due to the demagnetizing field. This field known as magnetic signature is often used for detection of magnetic objects, for example, sea ships. Moreover, magnetic history of the formation of the Earth's crust is available in rocks and stones and disciplines studying such effects are known as paleomagnetism and archeomagnetism.

Figure 6.1 presents the main elements of the geomagnetic field. The Earth's magnetic field can be simulated by an internal dipole—note that the symbolic magnet has its magnetic South Pole directed to the geographic North Pole. The real Earth's magnetic field is far from the idealized field of the dipole and therefore sometimes is divided into a dipole field and an additional field (see also Figure 6.13).

The location of geomagnetic pole (best fitting the dipole line pierce) is different than the geographic pole. Also the real magnetic pole (the place where inclination is 90° and horizontal component is zero) is in other place than the geographic or geomagnetic poles. Moreover, the geomagnetic and magnetic poles are wandering—even up to 40 km/year (Figure 6.2). The magnetic field weakens with the half-life of about 1400 years. Currently, it is about 15% weaker than it was 150 years ago.

The geomagnetic field is usually described by the horizontal component and total field. Direction of this field is described by declination (deviation from the geographic north direction) and inclination (deviation from the horizontal component). Because the magnetic pole is also moving, the components of the magnetic field are varying (Figure 6.2).

Moreover, the Earth's magnetic field in its long history (at least 3.5 billion years) has changed its polarity many times. The last reversal of the magnetic field occurred 780,000 years ago (Brunhes-Matuyama reversal) and 1,070,000 years ago (Jaramillo reversal).

Figure 6.3 presents the maps of the geomagnetic field. The total magnetic field is between 30 μT (24 A/m)

near the equator and 60 μT (48 A/m) near the poles.* For example, the magnetic field in Berlin is characterized by the horizontal component $B_H = 18.6$ μT, the total field $B = 48.8$ μT, the declination $D = +0.4°$, and inclination $I = 67.6°$. The Earth's magnetic field is usually described in nT or μT, but in geophysical organizations also often old unit $1\gamma = 1$ nT is commonly used.

The geomagnetic data have been investigated for the last hundred years by many national geophysical centers. All data are collected and available in International Association of Geomagnetism and Aeronomy. The Earth's magnetic field can be described by the spherical harmonic analysis developed by Carl Friedrich Gauss. This model is described as magnetic potential V by the following equation:

$$V(r,\phi,\theta) = a \sum_{l=1}^{L} \sum_{m=0}^{l} \left(\frac{a}{r}\right)^{l+1} (g_l^m \cos m\phi + h_l^m \sin m\phi) P_l^m \cos\theta$$

(6.1)

where

r is the radial distance from the Earth's center
L is the degree of expansion
ϕ is the East longitude
θ is the polar angle
g_l^m and h_l^m are the Gauss coefficients
$P_l^m \cos\theta$ are the Schmidt normalized associated Legendre functions

According to this model, the magnetic field potential can be calculated from Gauss coefficients determined every 5 years as *International Geomagnetic Reference Field* (IGRF) and then as definitive geomagnetic reference. There is a special software useful for calculation of the magnetic field from the IGRF data. At the Internet National Geophysical Data Center (address http://www.ngdc.noaa.gov/geomagmodels/IGRFWMM.jsp), a calculator that makes possible determination of the magnetic field for every point on Earth in chosen data between 1900 and present time is available. An example of calculated results is presented in Figure 6.4.

---

* The highest total magnetic field of about 67 μT is near the south magnetic pole and the smallest of about 23 μT near Sao Paulo (Brazil).

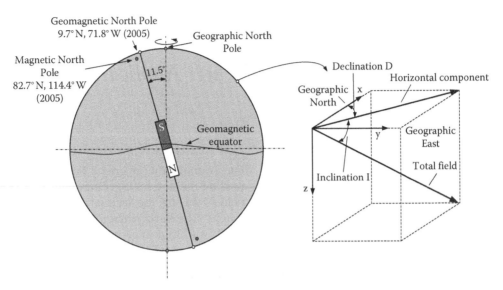

FIGURE 6.1
Earth's magnetic field—definitions.

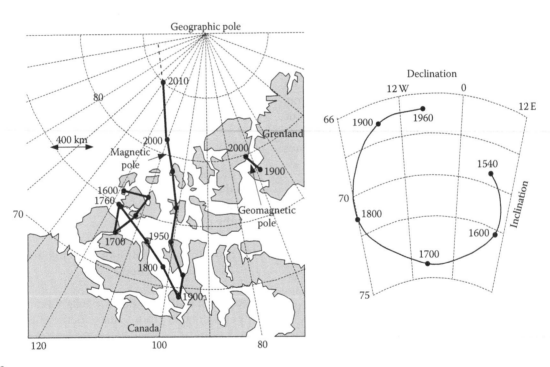

FIGURE 6.2
Movements of the magnetic and geomagnetic poles and variation of the declination and inclination of the magnetic field in London. (After *The Interior of the Earth*, Bott, M.H.P., Copyright (1971).)

The Earth's magnetic field (magnitude and direction) is varying of about 0.1% within a day under the influence of the Sun and other general space activity. Figure 6.5 presents an example of such daily variation—only the geophysical field is shown (thus without influence of human activity). During the magnetic storms resulting from activity of the Sun, the changes are more significant—even several percent during 1 day. Figure 6.6 presents an example of the changes of horizontal component during the magnetic storm. The storm starts with sudden changes of the magnetic field and it is necessary to wait more than 48 h to return to previous "normal" state. The number of storms increases significantly during the cyclically repeated increase of solar activity.

The mechanism of generation of the Earth's magnetic field is still not fully clear. There are several hypotheses

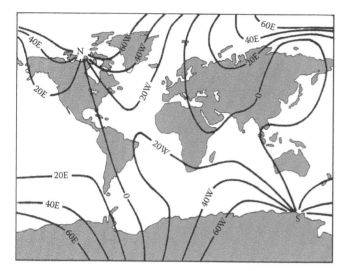

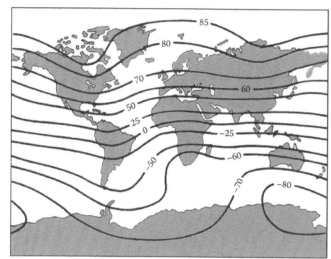

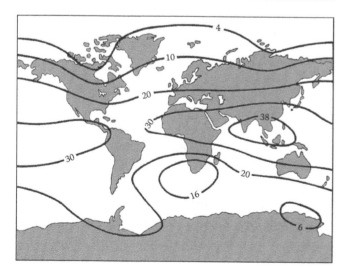

FIGURE 6.3
Maps of declination, inclination, and horizontal value of the Earth's magnetic field.

all assuming that the source (huge electromagnet) exists in the inner part of the globe. It cannot be permanent magnet because very high temperature of several thousand of Celsius grade is much larger than the Curie temperature and above which no permanent magnet can operate. For this reason, the field must be generated by electrical current in a "virtual coil." Such a coil can be created in the fluid part of the core composed mainly of good electrical conductors like iron and nickel.

The most commonly accepted model of the field generator is self-excited dynamo model proposed by Elsasser and Bullard (1956). In the molten part of the Earth core under the Coriolis force is forming a conducting material in the form of a disk (Figure 6.7). According to the Lorentz force in the magnetic field B, electric charges are moved to the edge of the disk. This current returns due to the thermal convection through a spiral way to the disk and this spiral operates as a coil generating magnetic field. This current acts as the positive feedback self-exciting dynamo. This model is only one of the hypotheses and does not explain all Earth's magnetic field phenomena.

Assuming that the Earth's magnetic field is generated only by one dipole, the external magnetic field B decreases as the cube of distance x (Figure 6.8):

$$B = B_0 \left(\frac{R}{x} \right)^3 \sqrt{1 + 3\cos^2 \theta} \qquad (6.2)$$

where
  B_0 is the magnetic field on the equator (about $30\,\mu T$)
  R is the radius of the Earth (about $6370\,km$)
  θ is the azimuth measured from the north magnetic pole

Thus, assuming that the interplanetary magnetic fields are at the level of $1\,nT$, we can believe that the Earth's magnetic field is extended to about 30 radii. But the real external magnetic field is strongly distorted by the solar wind. The border of the magnetosphere (*magnetopause*) on the solar side is compressed to the level of 10 Earth radii. On the other hand, the opposite side is stretched as *magnetotail* over a distance larger than 100 Earth radii.

The magnetosphere plays a crucial role for life on the Earth because it protects us from dangerous solar wind (like an umbrella). The ionized particles of the solar wind are trapped by the magnetic field. Part of them is collected in the Van Allen belts—internal at a distance of about one radius of Earth (containing mostly protons) and external at about 4 radii of Earth (containing mostly electrons).

Measurements of the magnetic field in the geophysical laboratories are commonly carried out by means

Start Date:			Step Size:	End Date:		
Day	Month	Year	Years	Day	Month	Year
16	July ▾	2010	0	16	July ▾	2010

Valid IGRF Range: 1900-2015

Compute Magnetic Field Values

Results:

	Declination	Inclination	Horizontal Intensity	North Component	East Component	Vertical Component	Total Field
Lat: 52° 14' 24" Lon: 21° 36" Elev: 0.00 m	+ East - West	+ Down - Up		+ North - South	+ East - West	+ Down - Up	
7/16/2010	4° 51'	67° 58'	18,690.6 nT	18,623.7 nT	1579.6 nT	46,196.4 nT	49,834.2 nT
Change per year	8' per year	1' per year	3.7 nT/year	0.1 nT/year	42.5 nT/year	32.5 nT/year	31.6 nT/year

FIGURE 6.4
Example of the geophysical data available by using NOAA calculator—determined for Warsaw at the day of writing of this book.

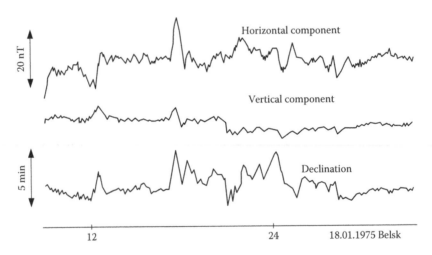

FIGURE 6.5
Example of the daily variations of the value and direction of magnetic field. (After Kadzilko-Hofmokl, M., Earth's magnetism, in *Encyclopedia for Modern Physic*, PWN, 1983.)

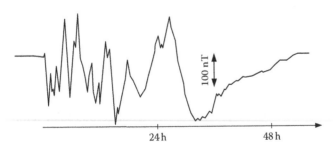

FIGURE 6.6
The example of variations of the horizontal component of the magnetic field over 48 h during the magnetic storm. (After Breiner, S., *Applications Manual for Portable Magnetometers*, Geometrics Application Note, 1999.)

of proton-free precession magnetometers and fluxgate sensors. For a magnetic exploration in outer space, better resolution is needed, and therefore cesium optical pumping and Overhauser magnetometers are also used.

6.1.2 The Space Research

Most of Earth's satellites (starting from Sputnik 3, Explorer 3 in 1958) were equipped with instruments for investigation of the magnetic field. Among them, there were highly specialized satellites designed for investigations of the magnetic field such as the Oersted European satellite launched in 1999 for mapping of the Earth's magnetic field, IMAGE satellite launched in

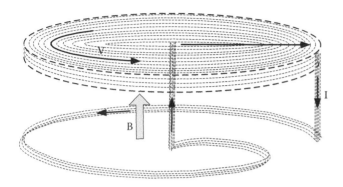

FIGURE 6.7
Self-excited dynamo model of the Earth's magnetic field source.

2000 for investigations of the magnetosphere (Imager for Magnetopause to Aurora Global Exploration), or THEMIS launched in 2007 for investigations of magnetic storms (Time History of Events and Macroscale Interactions during Substorms).

Also spacecrafts investigate the interplanetary magnetic field as magnetospheres of planets in our solar system (e.g., Pioneer 10 for investigation of Jupiter magnetosphere, MESSENGER for investigations of the Mercury magnetosphere, Mariner for investigations of Mars and Venus, and Voyager and Cassini for investigations of Saturn magnetosphere).

The first magnetometers used for testing of the space magnetic field were based on search coil magnetometers (Frandsen et al. 1969) or rubidium or helium optical pumping magnetometers (Slocum et al. 1963, Farthing and Foltz 1967, Smith et al. 1975). For testing of the vector value of the magnetic field, fluxgate magnetometers are commonly used. For example, Oersted satellite was equipped with Overhauser and fluxgate magnetometers. Statistics presented by Ness (1970) indicated that

of about 70 satellites launched until 1970, more than 40 were equipped with fluxgate magnetometers, 15 with search coil magnetometers, and 15 with optically pumped magnetometers. Recently, Overhauser magnetometers are most often used because of their small power consumption and high resolution.

The free proton precession magnetometers are used rather seldom because they require high power and optical magnetometers have better resolution. In space investigation, the resolution of 0.01 nT is recommended. The advantage of fluxgate magnetometers is that they are able to detect all three vector components of the magnetic field while resonance magnetometers detect mainly the scalar value. But it is possible to design the resonance magnetometers to measure also the vector components.

An example of the helium magnetometer used for the Pioneer Jupiter mission is presented in Figure 6.9. If we apply the rotating magnetic field H_s (rotating with angular velocity ω), the absorption of light is related to the change of energy:

$$\Delta E = \frac{K(H_x + H_s \sin \omega t)^2}{H_0^2 + (H_x + H_s \sin \omega t)^2 + (H_z + H_s \cos \omega t)^2} \quad (6.3)$$

where
H_x, H_z are components of the measured field
H_0 and K are constants

The amplitude of in-phase and quadrature components of the output signal are (Smith et al. 1975)

$$A_z = KH_z \left[\frac{H_s^3}{(H_0^2 + H_s^2)^2} \right] \cos \omega t \quad (6.4)$$

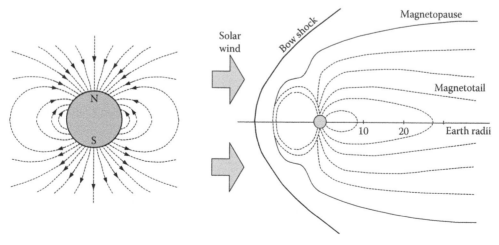

FIGURE 6.8
Idealized magnetic field generated by dipole and real magnetosphere.

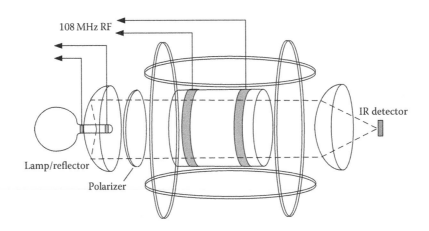

FIGURE 6.9
Vector helium magnetometer used in the Pioneer Jupiter mission (for simplicity only two pairs of coils are shown). (From Smith, E.J. et al., *IEEE Trans. Magn.*, 11, 962, 1975.)

$$A_x = KH_x \left[\frac{4H_s}{H_0^2 + H_s^2} - \frac{3H_s^3}{(H_0^2 + H_s^2)^2} \right] \sin \omega t \qquad (6.5)$$

Thus, by using phase coherent demodulation and rotating fields in two planes, it is possible to separate all three components of the measured field.

The measurement of the magnetic field in spacecrafts is a challenge for designers. If we need to measure the magnetic field with resolution of 0.01 nT, it is a significant problem to eliminate the stray internal field of electronic devices and spacecraft metal parts (especially ferromagnetic ones). Commonly, the sensors are placed away from the main body on the boom with length 8–10 m. Moreover, the whole spacecraft should have three-axis spin stabilization.

A two-magnetometer system was proposed to decrease the influence of the internal magnetic field (Ness et al. 1971). If we assume that the source of the internal magnetic field of the spacecraft can be simulated by the dipole source, the magnetic field decreases with distance x as $1/x^3$. Thus by comparison of the measurements of two magnetometers, we can calculate the value of the internal magnetic field.

The space magnetic field is strongly influenced by the solar wind, which is a plasma consisting of ions and protons (95%), helium nuclei, and other matter traveling at about 400 km/s. Although density of the ions is not very high (near the Earth it is about 6 ions/cm³), the magnetic field associated with such solar wind is about 2–5 nT.

Figure 6.10 presents the correlation between the measured interplanetary magnetic field and the velocity of the solar wind. Knowing the magnetic field related to the sun dipole, it can be calculated that the magnetic field in the distance between the Sun and the Earth should be smaller than 0.01 nT. But measurements proved that this field is of about several nT. This is the magnetic field generated by the solar wind (Figure 6.11) (Table 6.1).

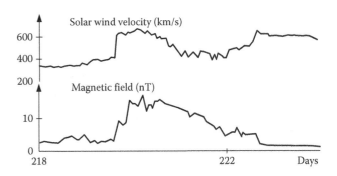

FIGURE 6.10
Example of measured interplanetary solar wind and associated magnetic field. (From Smith, E.J. et al., *IEEE Trans. Magn.*, 11, 962, 1975.)

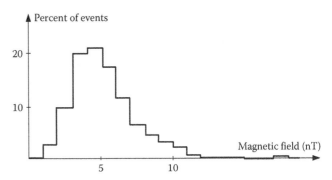

FIGURE 6.11
Statistics of measured interplanetary magnetic field. (From Smith, E.J. and Sonett, C.P., *IEEE Trans. Geosci. Electron.*, 14, 154, 1976.)

Almost all planets have the magnetosphere with exception of Venus and Mars. Mars exhibits only crustal magnetization but investigations of the meteorites revealed that many years ago Mars had a dynamo-like magnetic field at the level of about 1000 nT. It is not clear why this dynamo stopped; various theories try to explain this mystery, including a crash with a large asteroid.

It is also surprising why Venus with dimensions and structure similar to Earth does not have a dynamo-like

TABLE 6.1

Planetary Magnetic Field

Planet	B_{surface} (nT)	Tilt Angle (°)
Mercury	330	−11
Venus	<10 nT	
Earth	31,000	11.5
Mars	<10 nT	
Jupiter	428,000	9.6
Saturn	21,000	1
Uranus	23,000	58.6
Neptune	14,200	46.9
Earth's Moon	≈100	
Ganymede (Jupiter's Moon)	720	−4
Sun	100,000–400,000 in photosphere	

magnetic field. It is assumed that the reason for the lack of an internal magnetic field is the structure of the core with small thermal convection. On the other hand, Jupiter has exceptionally high-magnetic field (10 times higher than the Earth's magnetic field), which suggests a huge dynamo, especially if we take into account that Jupiter's volume is 1300 times larger than Earth's (Figure 6.12). A large magnetic field is supported by more (than Earth) rapid rotation and electrical currents flowing in the outer core of the planet, which is composed of metallic hydrogen.

The magnetic field of the Sun strongly influences life on our planet. It reverses direction every 11 years. In normal case, the magnetic storms caused by the Sun's activity appear once or twice per month. But during the reverse of a magnetic field, due to lack of magnetic braking of plasma, massive ejection of hot energy plasma results in increase of number of magnetic storms. The enlarged number of storms and substorms are dangerous not only for human life but also for electric power grids and communication systems.

6.1.3 Magnetic Exploration and Surveying

When we look at the maps of the Earth's magnetic field (Figure 6.3), we can see that they are very irregular. Indeed, the assumption that the magnetic field is generated by a single dipole is a large simplification. It is rather quadrupole or more complex spherical multipole harmonic function. Figure 6.13 presents the map of a non-dipole difference.

The magnetic anomalies are mainly a result of the complex tectonic structure of the Earth. On a smaller scale, local anomalies correlated with the crustal structure and minerals deposits are observed. A well-known example is the Kursk anomaly related to the one of the largest iron ore deposits. These local anomalies can be investigated by Earth's satellites and by marine and airborne gravity surveys. Figure 6.14 presents the map of lithosphere's* magnetic anomalies obtained by Magsat and POGO satellite measurements (Sabaka et al. 2000). In this map, the main local magnetic anomalies are indicated; letter B denotes the Kursk anomaly.

Recently, the maps of many parts of the Earth obtained by aeromagnetic surveying are being prepared. Figure 6.15 presents the magnetic map of Iowa state prepared by U.S. Geological Survey.

Geomagnetism (Campbell 2003, Lanza and Meloni 2006) and paleomagnetism (Butler 1998, Tauxe 2010) (discussed later in this chapter) are important disciplines of geophysics expanding our knowledge about our planet. But in a smaller scale, the magnetic surveying is an important method of gas, oil, and mineral exploration, buried object searching (magnetoarcheology), submarine or underground pipeline detection as well as rescue of skiers from avalanches. Most companies manufacturing low field equipment—Bartingthon (fluxgate instruments), Geometrics (proton and cesium magnetometers), GEM (proton, potassium, and Overhauser magnetometers), or Scintrex (proton and cesium magnetometres)—offer not only measuring instruments but also knowhow (Breiner 1999, Smekalova et al 2008) and tools of magnetic surveying, including software for preparation of the maps.

The interpretation of magnetic data is not easy; therefore, magnetic exploring is usually supported by other methods like gravity and electrical resistivity measurements, radar techniques, or electromagnetic inspection. There are two main sources of a magnetic field used for magnetic maps: induced magnetism and remanence. Induced magnetism means that the magnetic material (also paramagnetic and diamagnetic) is magnetized by an external (in our case mainly Earth's) magnetic field H:

$$M = \chi H \tag{6.6}$$

where χ is the magnetic susceptibility.

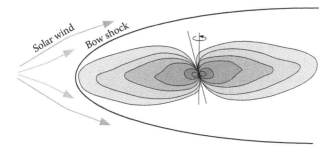

FIGURE 6.12
Magnetosphere of Jupiter.

---

* Lithosphere is the Earth's outer layer composed of crust and the upper mantle.

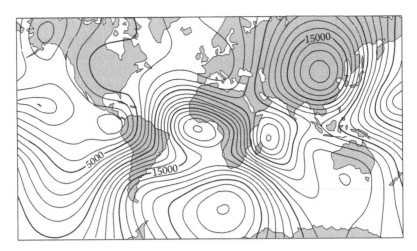

FIGURE 6.13
Map of non-dipole component of the Earth's magnetic field.

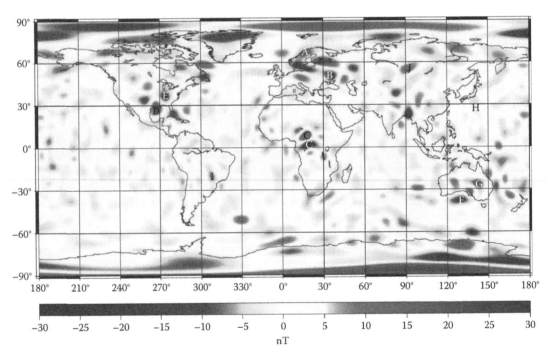

FIGURE 6.14
Map of magnetic field anomalies in the Earth's lithosphere (From Sabaka, J. et al., A comprehensive model of the near-Earth magnetic field, *Nasa Report*, NASA/TM-2000–209894, 2000. With permission.)

Susceptibility χ ($\chi = \mu_r - 1$) is one of the frequently tested material properties in geomagnetism. Figure 6.16 presents the susceptibility of main minerals. The largest possible magnetization is exhibited by iron oxides: not only hematite Fe_2O_3 and magnetite Fe_3O_4 but also other minerals have trace magnetic properties.

Besides the induced magnetization, the explored materials can exhibit magnetic remanence. Of course, the largest magnetic signatures are associated with ferromagnetic parts. Figure 6.17 presents the magnetic field generated by various ferromagnetic objects.

Figure 6.18 presents the principle of magnetic anomaly detection. The buried element is magnetized by the Earth's magnetic field. The moving sensor measures vector sum of the Earth's magnetic field and the magnetic field of the magnetized part. Therefore, the detected difference is rather small.

The main problem in detecting magnetic anomaly field is that the daily variation of Earth's magnetic field (see Figure 6.5) is often larger than the detected magnetic field difference. Therefore, often two measurements of magnetic field are necessary—magnetic anomaly field

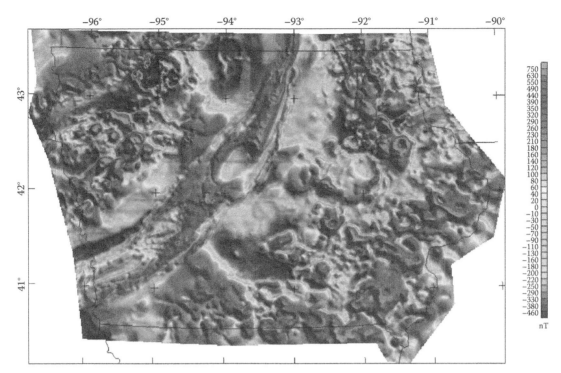

FIGURE 6.15
Magnetic map of Iowa state. (Courtesy of U.S. Geological Survey.)

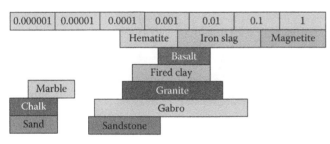

FIGURE 6.16
Susceptibility of the main minerals. (From Smekalova, T.N. et al., *Magnetic Surveying in Archeology*, House of Publishing of St. Petersburg State University, St. Petersburg, Russia, 2008.)

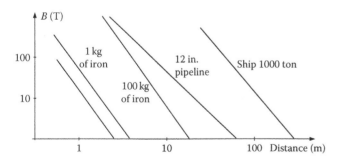

FIGURE 6.17
Magnetic signatures of ferromagnetic objects. (From Breiner, S., *Applications Manual for Portable Magnetometers*, Geometrics Application Note, 1999.)

and a reference. Usually, the measurement of magnetic field is substituted by a two-sensor gradient measurement (see Figure 6.19, left).

Figure 6.19 presents the typical instruments for magnetic anomaly mapping. The Overhauser magnetometer is equipped with geographical positioning system (GPS) as well as display for direct observation of the detected magnetic field.

The detected magnetic anomaly depends on many factors: direction of the Earth's magnetic field, depth of the object, its susceptibility, etc. Therefore, usually several components of magnetic field are measured. The example of two components response is presented in Figure 6.20.

Paleomagnetism is concentrated mainly on a study of magnetic properties of rocks (Butler 1998, Tauxe 2010). The iron oxide minerals record the magnetic field present during the cooling of the basalt lava below the Curie temperature. Also other diamagnetic and paramagnetic minerals memorized previous states of the magnetic field. Therefore by testing the natural remanent magnetization (NRM), it is possible to decipher the direction and value of the magnetic field from even billion years ago. The samples are usually obtained by drilling the holes in the rocks with careful documentation of the position of the sample. Another technique utilizes a column of sediment from a lake or sea bed. After determination of the value and direction of magnetization, it is possible

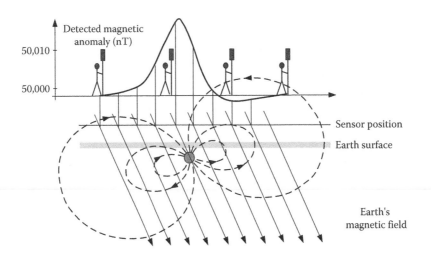

FIGURE 6.18
Principle of magnetic anomaly detection. (From Weymouth, J.W., Geological surveying of archeological sites, in *Archeological Geology*, G. Rapp (Ed.), Yale University Press, New Haven and London, 1985.)

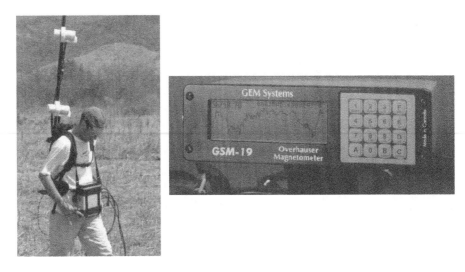

FIGURE 6.19
Equipment for magnetic anomaly mapping. (Courtesy of GEM Systems Inc., Ontario, Canada.)

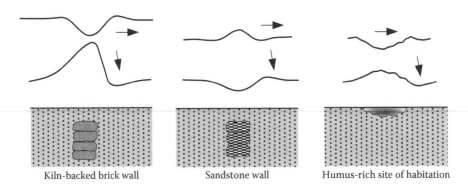

Kiln-backed brick wall   Sandstone wall   Humus-rich site of habitation

FIGURE 6.20
Examples of magnetic response of various detected objects. (From Breiner, S., *Applications Manual for Portable Magnetometers*, Geometrics Application Note, 1999.)

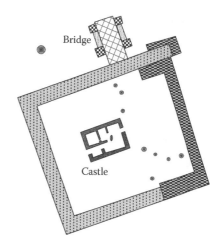

FIGURE 6.21
Examples magnetoarcheology results: magnetic anomaly map and revealed old castle surrounded by a wall. (From Smekalova, T.N. et al., *Magnetic Surveying in Archeology*, House of Publishing of St. Petersburg State University, St. Petersburg, Russia, 2008. With permission.)

to obtain data about history of Earth's magnetism and also about plate tectonics. It is also possible to perform magnetoarcheology investigations (Figure 6.21).

Usually following parameters are measured:

- Susceptibility, spontaneous magnetization, hysteresis, Curie temperature—thus all main intrinsic magnetic properties.
- Anisotropy.
- NRM: The sources of this magnetization are different: thermoremanent magnetization obtained by cooling the material, crystallization or chemical remanent magnetization obtained by growth of superparamagnetic grains, and detrital remanent magnetization when grains in sediment align with the magnetic field during or soon after deposition.

The remanent magnetization and susceptibility of the rocks can be extremely small. Therefore, the sample is usually sufficiently large; typically, a cylinder is 25 mm in diameter and 22 mm long. Also the measuring method should be exceptionally sensitive. For this reason, recently SQUID magnetometers are often used for detection of the remanence field.

Figure 6.22 presents the typical device for measurement of susceptibility. The sample under test is inserted in one of secondary coils (connected in opposite directions). Both coils are in uniform AC magnetic field generated by the primary coil. To detect the signal from the active coil, a lock-in amplifier is used. It is recommended that the operation be performed in a magnetically shielded room.

Remanent magnetization (and other parameters like anisotropy) is commonly tested with various magnetometers, including Foner magnetometer or Gouy

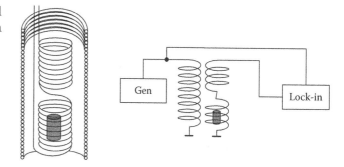

FIGURE 6.22
Examples of the of measurement methods of magnetic susceptibility. (Documentation of Advanced Physics Laboratory, University of Florida.)

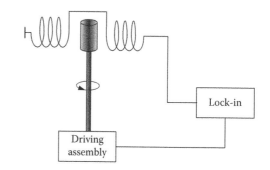

FIGURE 6.23
Principle of spinner magnetometer.

magnetic scale described in Section 5.4.1. Figure 6.23 presents another frequently used magnetometer—*spinner magnetometer* (with spinning sample). The sample is rotating at high rpm and the magnetic field is detected by a pickup coils, fluxgate sensor, or SQUID (De Sa 1963, Foster 1966, Jelinek 1966, Morris 1970, Flanders 1973, Hummervoll and Totland 1980). Also a portable version is reported (Chiron et al. 1981).

It should be noted that one of the first applications of the Zeeman effect (splitting of the spectral lines under the influence of the magnetic field) was investigation and measurement of the magnetic field of the Sun. Moreover, it was proved that increase of the Sun activity was correlated with changes of the Sun's magnetic field.

6.1.4 Compasses and Navigation

Magnetic needle compasses earlier commonly used by travelers, scouts, and soldiers are recently overshadowed by the GPS system. Indeed, most of GPS navigation devices are capable of calculating the azimuth, taking into account previous position; thus, the change of position is necessary. On the other hand, there are commercial small, simple, and cheap (available even in hand watch) magnetic sensor–based electronic compasses. As a sensor, the best candidates are fluxgate, magnetoimpedance, and magnetoresistive sensors. In the Internet shops, a great choice of electronic compasses, most often of unknown quality, is available.

But the Earth's magnetic field is a vector value directed to the magnetic North (not geographic one). Therefore even if there are simple electronic compasses, the main problems related to compass usefulness and reliability remain. One of them is the geographic changes of the magnetic field declination causing that on the U.S. territory the direction of the magnetic field changes ±20° with respect to the geographic North (Figure 6.24).

Other errors of azimuth determination are caused by nonexact positioning of the compass in the Earth's horizontal plane (Figure 6.25) (Caruso 1996). Therefore, it is necessary to add to the magnetic sensors also tilt measurement, for example, fluid sensors, accelerometers, or gyroscopes.

Yet another source of errors of the compass is the interference of the ferromagnetic objects in the neighborhood of the compass sensors. The presence of interference is usually tested by rotating the compass. Correct indication is when we obtain the circle as it is presented in Figure 6.26.

To detect the compass heading (azimuth), at least two sensors are necessary. Figure 6.27 presents an example of the sensor's arrangement. From the results of the measurement, we can detect the direction of the magnetic field according to the following relations:

$$
\begin{aligned}
\alpha &= 0 && \text{when } H_y = 0, H_x > 0 \\
\alpha &= 90 - \operatorname{arctg} \frac{H_x}{H_y} && \text{when } H_y > 0 \\
\alpha &= 180 && \text{when } H_y = 0, H_x < 0 \\
\alpha &= 270 - \operatorname{arctg} \frac{H_x}{H_y} && \text{when } H_y > 0
\end{aligned}
\tag{6.7}
$$

It is not always necessary to perform calculations according to the expressions (6.7). Figure 6.28 presents a simplified algorithm enabling to distinguish eight principal directions and Figure 6.29 presents a three-sensor compass with 30° resolution.

Among the possible magnetic field sensors, AMR sensors fulfill almost all requirements: they are simple in manufacturing, have sufficient sensitivity, and detect vector value of magnetic field. No wonder that one of the main manufacturers of the AMR sensors Honeywell company is also the market leader in digital compasses. Figure 6.30 presents one of the digital compasses of Honeywell. The sensors with dimensions $4 \times 4 \times 1.3\,\text{mm}$ comprises three AMR sensors, coils for offset, and set/reset operation, amplifier, analog-to-digital converter, and control unit with I^2C serial bus interface. The field

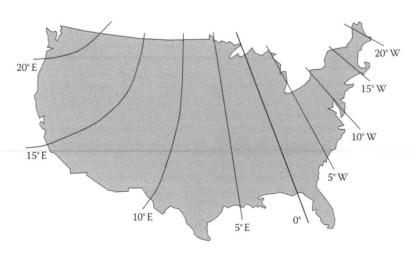

FIGURE 6.24
Declination of the magnetic field in the United States.

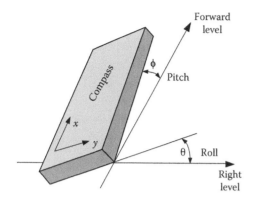

FIGURE 6.25
Compass tilt angles with respect to the Earth's horizontal plane. (From Caruso, M.J., Applications of magnetoresistive sensors in navigation systems, Honeywell Application Note, 1995.)

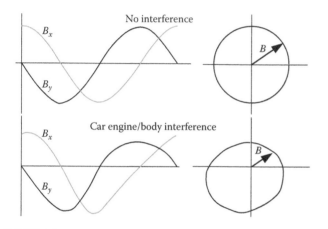

FIGURE 6.26
Interference of the ferromagnetic object on the magnetic field detected by the compass sensors. (From Caruso, M.J., Applications of magnetoresistive sensors in navigation systems, Honeywell Application Note, 1995.)

range of the compass is ±400 μT and resolution 0.7 μT. It enables 1°–2° compassing heading accuracy.

The digital compass HMC5384 is one of the ten various models offered by Honeywell (Figure 6.31). Others include microprocessor, three-axis MEMS accelerator

sensors, or MEMS gyroscopes. The model HMR35600 enables the azimuth detection with accuracy of 0.5° and resolution 0.1°. The smallest module HMC1052 two-axis AMR sensors has the dimension of $3 \times 3 \times 1$ mm.

In almost all compasses, it is possible to perform correction of the influence of ferromagnetic objects. In some of them, it is possible to select detection of magnetic North direction or geographic North direction. It is feasible because the internal memory comprises World Magnetic Model feature providing an automatic correction of the declination angle. Of course, the best solution is the digital compass with GPS module to determine the correction of the declination.

Such possibility is included in the most advanced Honeywell system—DRM 4000: Dead Reckoning Module. In this system (Figure 6.32), the compass module collaborates with a GPS system obtaining the actual position to introduce correction of the declination. The direction of movement is additionally analyzed by a three-axis MEMS accelerometer system.

In the navigation system, GPS collaborates with the compass device. But also reverse relationship is possible: during loss of the GPS data receiving (e.g., in a tunnel), the navigation system is able to forecast the navigation. Such system is known as dead reckoning system. Such navigation system can assist GPS in location-based services' applications and other telematic systems (Amundson 2003, 2006). The whole navigation system flowchart is presented in Figure 6.33.

Small dimensions and robustness of modern electronic compass systems allow that such systems can be embedded into cellular phone devices as well as hand watches. Three examples are presented in Figure 6.34.

6.1.5 Electromagnetic Pollution

The electromagnetic field (EMF) generated due to human activity is omnipresent. In many countries, finding a place for geophysical laboratory free of such fields is a nontrivial problem. Figure 6.35 presents the typical EMFs present in our environment.

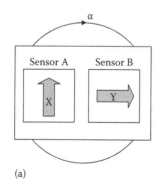

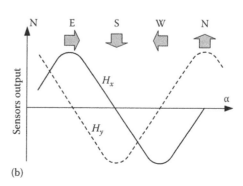

FIGURE 6.27
Principle of operation of electronic compass: two sensors arrangement (a) and the output signal of these sensors (b).

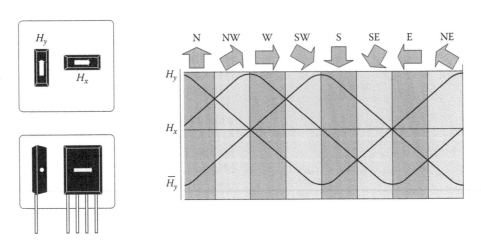

FIGURE 6.28
Simplified two-sensor compass with indication of the eight directions of magnetic field.

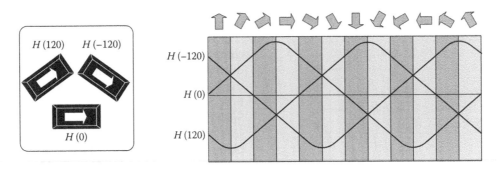

FIGURE 6.29
Three-sensor compass with 30° resolution. (From Wellhausen, H., *Elektronik*, 8, 85, 1989.)

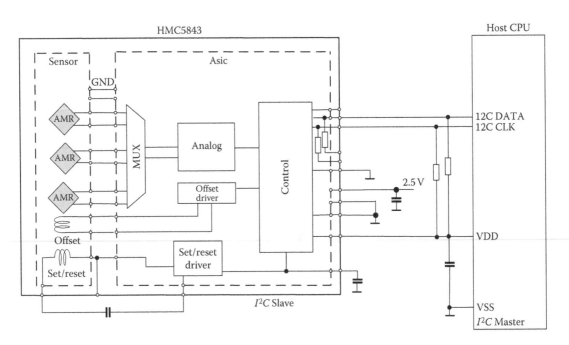

FIGURE 6.30
Three-axis digital compass model HMC5834 of Honeywell.

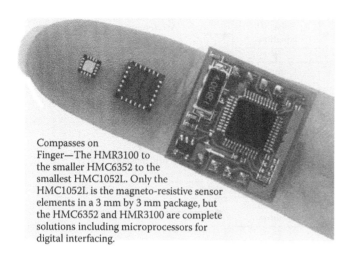

Compasses on Finger—The HMR3100 to the smaller HMC6352 to the smallest HMC1052L. Only the HMC1052L is the magneto-resistive sensor elements in a 3 mm by 3 mm package, but the HMC6352 and HMR3100 are complete solutions including microprocessors for digital interfacing.

FIGURE 6.31
Compasses of Honeywell. (From Amundson, M., The role of compassing in telematics, Analysis USA Focus, 2003.)

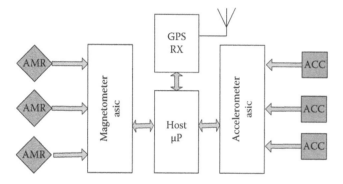

FIGURE 6.32
Principle of advanced navigation system for pedestrians. (From Honeywell Application Note No. AN219.)

Thus in our environment there exist EMFs from very low frequency 7.6 Hz (Schumann resonance; standing waves formed between Earth's surface and ionosphere) to hundreds of GHz (radars).* All these fields can influence the operation of various devices and can be dangerous if they exceed a certain level. Sometimes, these fields are known as electromagnetic pollution or electromagnetic smog.

The influence of EMF on the human health is continuously discussed and stirs up many emotions. No wonder that in Internet shops there is a large supply of measuring instrument for both electric and magnetic fields. International Commission on Non-Ionizing Radiation Protection (ICNIRP) published a comprehensive review

on the hundreds investigations of this risk (ICNIRP 1999). But the conclusion is ambiguous: "There is insufficient information on the biological and health effect of EMF exposure of human population and experimental animals to provide a rigorous basis for establishing safety factors over the whole frequency range and for all frequency modulations." Nevertheless the same commission published reference levels for general public exposure to time-varying magnetic fields (Table 6.2).†

Very wide range of frequency of EMF involves various measurement methods. Generally, all measuring instruments should be capable of measuring three components of magnetic field (to calculate the modulus value). Moreover, because for various frequency ranges there are various limitations, the measuring device should be equipped with frequency spectrum analysis. A different solution is to use filters selecting suitable bandwidth. Modern instruments can calculate the resultant field exposure according to standards indicating "percent of standard" value.

Figure 6.36 presents two most frequently measurement of magnetic exposure field: power frequency field (50 or 60 Hz) near the power delivery devices and high-frequency field near the antennas. Other frequently performed investigations are presented in Figure 4.43 where the household appliances are tested according to electromagnetic compatibility requirements. It means that these household electric devices are tested if they exceed safety level determined in standard EN50366.

The leader in the field of the EMF measurements used to test electromagnetic pollution is Narda Safety Test Solutions company. Figure 6.37 presents two main measuring instruments—ELT 400 for measurement of the field in the range 1 Hz–400 kHz and SRM 3006 for measurements in the range 9 kHz–6 GHz.

Figure 6.38 presents the block diagram of the ELT 400 device. Three sensors detect simultaneously the magnetic field in three spatial directions. The filters shape the measured values according to frequency bandwidth corresponding to standard requirements. Next the sum is calculated, rms (or peak) value. After square root operation, the average value of the magnetic field is displayed with information "percent of standard."

As the sensor, a three-axis coil sensor is used similar to that presented in Figure 4.13. Figure 6.39 presents the design of Narda sensor with three 100 cm² coils.

For low-frequency measurements of magnetic and electric fields, separate sensors are used. In the case of high-frequency EMF, direct relations between electric

---

* In Figure 6.35 following frequency bands are indicated: ELF—extremely low frequency, LF—low frequency, MF—medium frequency, HF—high frequency, VHF—very high frequency, UHF—ultra high frequency, SHF—super high frequency, EHF—extremely high frequency. The frequency bandwidth are also divided into: low frequency: to 30 kHz, radiofrequency: 30 kHz–300 GHz, microwave: 0.3–300 GHz.

† ICNIRP published also less stringent recommendations for occupational exposure and recommendations taking into account direct influence on human body: restrictions for current density for head and trunk (mA/m²) or body SAR (Specific Energy Absorption Rate) (W/kg) for RF range. There are also other recommendations as IEEE C95.1, IEC/EN 62233, 50366, 60335, or 50499 standards.

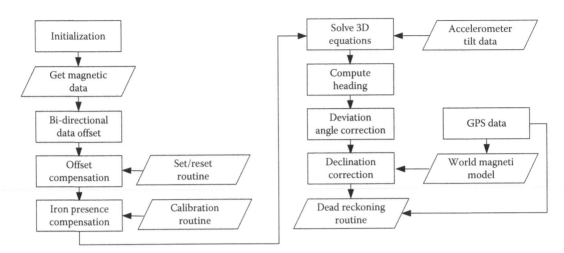

FIGURE 6.33
Flowchart of advanced navigation system for pedestrians. (From Honeywell Application Note No. AN219.)

FIGURE 6.34
Compass capabilities embedded into a cellular phone (iPhone model) or hand watches (Cassio and Timex models, respectively).

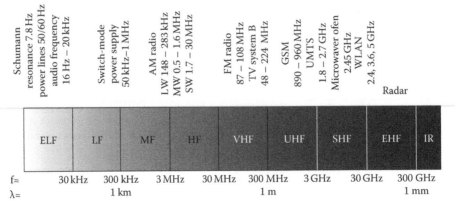

FIGURE 6.35
The frequency range of various EMFs present in our environment.

TABLE 6.2

Reference Levels for General Public Exposure to Time-Varying Magnetic Fields according to ICNIRP Recommendations

Frequency	Field Strength (A/m)	Flux Density (μT)	Wave Power Density (W/m²)
Up to 1 Hz	32,000	40,000	
1–8 Hz	$32,000/f^2$	$40,000/f^2$	
8–25 Hz	$4,000/f$	$5,000/f$	
25–800 Hz	$4/f$	$5/f$	
0.8–150 kHz	5	6.25	
0.15–10 MHz	$0.73/f$	$0.92/f$	
10–400 MHz	0.073	0.092	2
0.4–2 GHz	$0.0037f^{1/2}$	$0.0046f^{1/2}$	$f/200$
2–300 GHz	0.16	0.2	10

field E, magnetic field B, and power density S are as follows:

$$H = \frac{E}{Z_0} \quad \text{and} \quad S = \frac{E^2}{Z_0} \qquad (6.8)$$

where Z_0 is the free space impedance and $Z_0 = 377\ \Omega$.

Thus for high frequency, it is sufficient to use as the sensor three-axis dipole antenna as it is presented in Figure 6.39, right. But this relation is valid only for so-called *far field*. This far field is at the distance larger than 3λ, where λ is length of the wave and $\lambda = c/f$. As presented in Figure 6.35 for frequency of 300 MHz, this distance is equal to 3 m. If we are interested in measurement in near field at the distance smaller than 3λ, it is necessary to perform measurements of electric and magnetic fields separately. There are coil sensors designed for a high-frequency magnetic field. An example is presented in Figure 4.47.

Instead of magnetic field pollution, there is also magnetic pollution. The pollution of air is generated by power plants containing magnetic components (Evans and Heller 2003). This pollution is detected as the change of soil susceptibility.

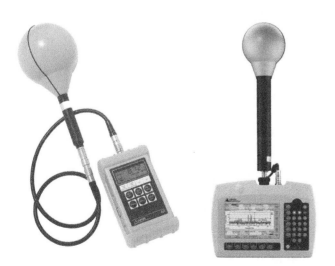

FIGURE 6.36
Two most frequently investigations of the electromagnetic pollution. (Courtesy of Narda Safety Test Solutions GmbH.)

6.2 Applications of Magnetic Field Measurements

6.2.1 Detection of Magnetic Objects: Magnetic Signature

As discussed earlier, the magnetic objects cause anomaly of the Earth's magnetic field and this effect is used in magnetoarcheology. The same effect can be used for detection of vehicles, ships, mines, or coins because all ferromagnetic objects have a magnetic signature, a characteristic magnetic field generated in the presence of the Earth's field. Especially important is the magnetic signature of the submarines because in water radar does not work and magnetic detection is the only method to investigate the underwater traffic.

The magnetic signature signals have many sources: permanent magnetic parts, magnetic field induced in soft magnetic parts, moving ferromagnetic elements, or moving current conducting parts (electric motors). The resultant field is a superposition of uniform Earth's field and magnetic disturbance field (Figure 6.40). Table 6.3 presents the detectable magnetic field of various objects. Although this field decreases quickly with the distance by applying conventional methods, it is possible to investigate the presence even of quite small parts.

FIGURE 6.37
Two instruments for measurement of low-frequency and high-frequency EMFs. (Courtesy of Narda Safety Test Solutions GmbH.)

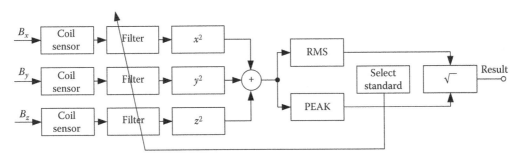

FIGURE 6.38
Block diagram of the ELT low-frequency meter. (From Braach, B., *Przegl. Elektrotech.*, 81, 35, 2005.)

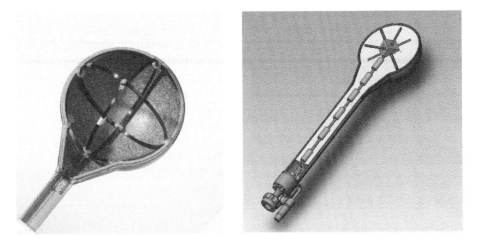

FIGURE 6.39
EMF sensors used for low-frequency and high-frequency range. (Courtesy of Narda Safety Test Solutions GmbH.)

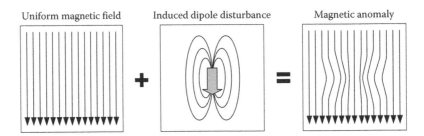

FIGURE 6.40
Magnetic field anomaly as a result of superposition of a uniform magnetic field and dipole disturbance. (From Lenz, 1990.)

Figure 6.41 presents a magnetic signature of a car. Such signature can be used for vehicle classification on the road, control the traffic, or control which places are occupied in a car park (Wolff et al. 2006). Figure 6.42 presents the dependence of the magnetic signature on the direction of moving. The sensor can be placed near the road or directly in the road.

Figures 6.41 and 6.42 present the magnetic signatures detected with AMR sensors. Similar results and method of estimation of vehicle speed by using GMR sensors or magnetoimpedance sensors were reported (Uchiyama et al. 2000, Sebastia et al. 2007).

Especially important is the magnetic signature of naval vessels. The most dangerous weapon for such ships is intelligent magnetically triggered mine or torpedo that is able to recognize the ship from its magnetic signature. This magnetic field is generated mainly due to induced magnetization in the ferromagnetic hull, but also by the parts of the ship magnetized by the Earth's magnetic field. Figure 6.43 presents a typical magnetic signature of a vessel.

To diminish the magnetic signature, certain routine procedures are performed. The first one is *de-perming*. By encircling the hull with a cable, the ship is demagnetized

TABLE 6.3

Detectable Magnetic Field of Various Objects

Object	Distance (m)	Signature Field (nT)
Ship (1000 ton)	300	0.5–1
DC train	300	1–50
Light aircraft	150	0.5–2
Automobile (1 ton)	30	1
Pipeline ($\phi = 30\,cm$)	15	12–50
Screwdriver	3	0.5–1
Revolver	3	1–2

Source: From Breiner, S., *Applications Manual for Portable Magnetometers*, Geometrics Application Note, 1999.

by huge AC current. But this technique removes only the permanent magnet component of the signature. Therefore, another method known as *degaussing* is also used. After the analysis of the magnetic field, the most magnetically active part is surrounded by the cable with a DC current. Figure 6.44 presents the magnetic signature after degaussing.

Because the radar devices do not work in water, a net of fluxgate sensors is commonly used to protect harbors and wharfs. The traffic of submarines is also investigated in water from air by Overhauser or proton magnetometers.

Magnetic methods are commonly used to detect buried metal object including mines. An example of such

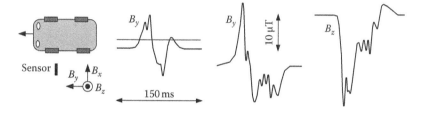

FIGURE 6.41

Magnetic field detected near the moving van (Silhouette). AMR sensor 30 cm above the ground and 30 cm from the car. (From Caruso, M.J. and Withanawasam, L.S., Vehicle detection and compass applications using AMR magnetic sensors, Honeywell Application Note, 1999.)

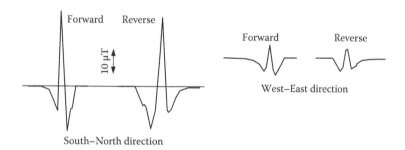

FIGURE 6.42

Magnetic signature of the car moving in different direction. (From Caruso, M.J. and Withanawasam, L.S., Vehicle detection and compass applications using AMR magnetic sensors, Honeywell Application Note, 1999.)

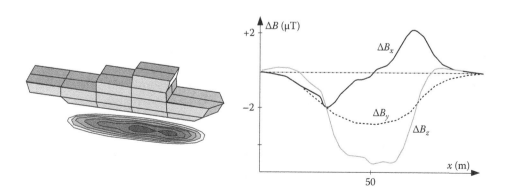

FIGURE 6.43

Calculated magnetic signature of the naval vessel. Magnetic field 5 m below keel. Stern 220 m eastward. (From Aird, G.J.C., Modeling the induced magnetic signature of naval vessels, PhD dissertation, University of Glasgow, Scotland, U.K., 2000.)

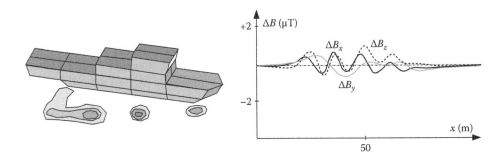

FIGURE 6.44
Calculated magnetic signature of the naval vessel from the Figure 6.43 after degaussing. (From Aird, G.J.C., Modeling the induced magnetic signature of naval vessels, PhD dissertation, University of Glasgow, Scotland, U.K., 2000.)

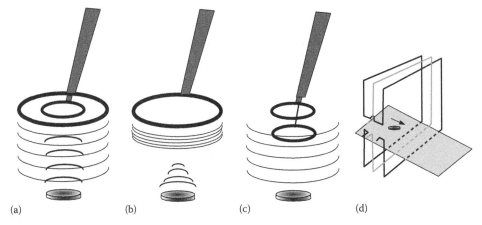

FIGURE 6.45
Various techniques of metal detectors: very low frequency VLF detector (a), pulse induction (b), DC gradient field detector (c), and magnetic baggage inspection.

device is presented in Figure 4.48. Figure 6.45 presents the principle of operation of the metal detectors.

The VLF detector (Figure 6.45a) consists of two inductive loops: transmitter and receiver. These coils are arranged in such a way that the signal of transmitter does not induce the signal in the receiver pickup coil. But if in the detected volume a conducting metal object appears, then the field generated by eddy currents is detected by the receiver coil. From the value and the phase shift between transmitted and received signal, it is possible to use the discriminator function to distinguish volume of the detected object, depth of position, and kind of material (Figure 6.46).

The pulse induction technique (Figure 6.45b) only one coil is used for detection of decay of pulse signal or echo signal during the break time. Pulse technique is less sensitive to soil mineralization magnetism. Figure 6.47 presents the example of the output signals in PI detecting system.

Both AC techniques are the active methods based on the response to the transmitted signal. Such methods can be effective for treasure hunters but in the case of mine detectors the transmitting signal can stimulate the mine to explosion. In such cases, passive DC gradient methods must be used (Figure 6.45c). Figure 6.48 presents the output signal from a three-dimensional fluxgate gradiometer system.

Of course, mine detection is the most complex problem because there is the war between mine and mine-detector designers (Siegel 2002). The nonmagnetic mines have been introduced (without metal parts) as well as magnetic mines equipped with sensitive magnetic field sensor triggered by magnetic anomaly.

Magnetic gates in airports operate similarly to the pulsed field detectors. Figure 6.45d presents the system for magnetic inspection of the baggage.

Magnetic systems are also used in the magnetic surveillance systems as described in Section 4.2.11 (Figure 6.49).

6.2.2 Mechanical Transducers

Many of the magnetic field sensors are most often used for measurements of other nonmagnetic values. Especially the Hall sensors and magnetic resistance (MR) sensors are frequently used as mechanical transducers.

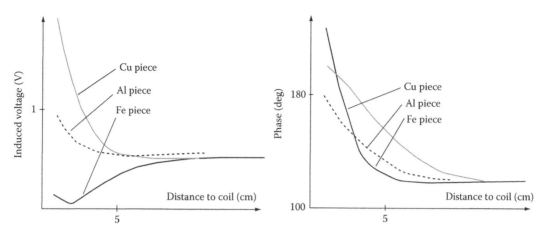

FIGURE 6.46
Voltage induced in the receiver coil in VLF detector. (From Sharawi, M.S. and Sharawi, M.I., Design and implementation of a low cost VLF metal detector with metal type discrimination capabilities, *IEEE International Conference on Signal Processing and Communications*, Dubai, pp. 480–483, 2007.)

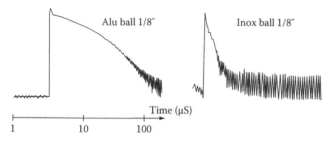

FIGURE 6.47
Example of the voltage response in the pulsed field metal detector. (From Ripka, P. et al., Bomb detection in magnetic soils: AC versus DC methods, *EXCO IEEE Sensors Conference*, Daegu, Korea, 2006.)

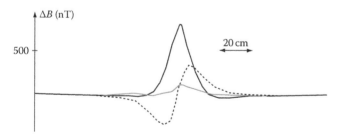

FIGURE 6.48
Three components of DC fluxgate gradiometer for the sample buried at the depth of 8 cm. (From Ripka, P. et al., Bomb detection in magnetic soils: AC versus DC methods, *EXCO IEEE Sensors Conference*, Daegu, Korea, 2006.)

Figure 6.49 presents several possibilities of using of the magnetic sensors as the displacement transducers. Usually, the magnet is moved with respect to the sensor but reverse operation is also possible. Instead of the magnet, the source of magnetic field can be realized in form of a recorded pattern (Figure 6.49d). The Hall and MR sensors are especially useful for such applications because they are relatively cheap, small, and with sufficient sensitivity to cooperate with simple magnets.

We can consider various displacements sensors:

- Sensors of displacement (Figures 6.49 and 6.50) where the relation $V_{out} = f(x)$ is realized. These sensors are used to convert the displacement into electric signal analog or digital.
- Proximity sensors (Figure 6.51) where the presence of moving object is detected. As an example, of such a sensor, the transducer where change of the distance between a magnet and a Hall sensor is starting the airbag system can be considered.
- One-point detector (Figure 6.52).
- Multidimensional position sensor (joystick) (Figure 6.56).
- Encoders for precise determination of position (Figures 6.53, 6.54 and 6.55). In the encoder system, usually the magnetic sensor cooperates with recorded magnetic tracks (as in Figure 6.49d). Encoders are commonly used as the feedback element for stepper motors.

Of course, we can measure linear displacement, angle position, or rotation.

Figure 6.51 presents various design possibilities of proximity transducer. Of course, instead of magnet, any ferromagnetic material can be used (magnetized by the magnetic field generated by the sensor) and also only conducting nonmagnetic part (magnetized by eddy currents).

Figure 6.53 presents two examples of the linear encoders where the magnetic sensor is moved along the scale. To obtain high resolution, the tracks should be recorded with high density. The resolution can be improved also by using two sensors. Figure 6.54 presents two methods of improving the resolution. In the first one, the incremental displacement is detected by using two scales

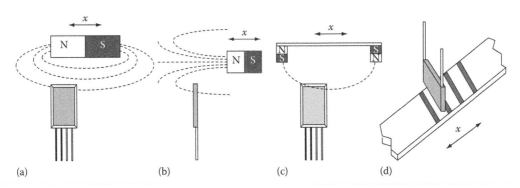

FIGURE 6.49
Methods of the displacement detection by using magnetic field sensors.

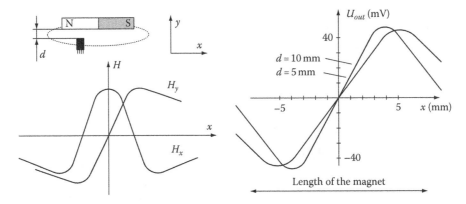

FIGURE 6.50
Example of the displacement transducer with the MR sensor KMZ10B of Philips.

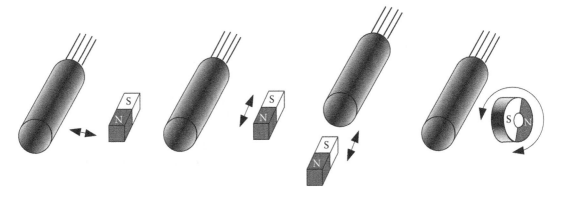

FIGURE 6.51
Various arrangements of the magnet-sensor proximity transducers. (After Cherry Electrical Products Application Note.)

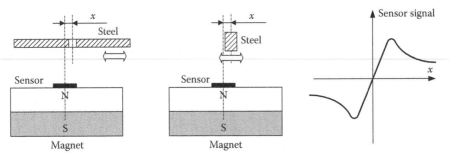

FIGURE 6.52
One-point position detectors.

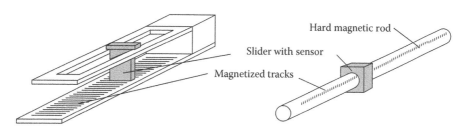

FIGURE 6.53
Two examples of the linear encoder systems.

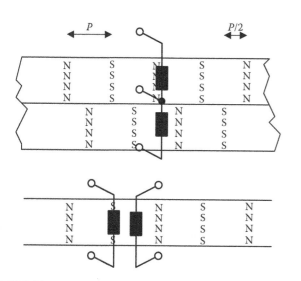

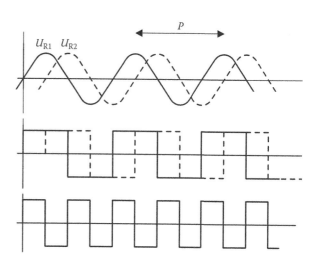

FIGURE 6.54
Incremental encoders with improved resolution.

shifted by a half of the pitch. Then in the output signal, the recorded period is represented by four pulses. In the second method, two sensors are shifted by a half of the pitch.

Three types of encoders: linear, ring, or rotary encoder most frequently used for control of the motors. Figure 6.55 presents the encoders of RLS company. The linear encoder detects position of the distance up to 100 m with resolution 1 μm. The ring encoder can operate up to 25,000 revolutions per minute (rpm) with the resolution of 37,680 counts per revolution. The rotary encoders can operate with speed up to 30,000 rpm with a resolution of 13 bit (it means 8,192 positions per revolution).

Figure 6.56 presents the multidimensional position manipulator known as joystick. Such type of manipulators is commonly used for computer games but the most important application is the operation as a robotic tactile sensor.

The AMR change of resistance depends on the angle ϑ between direction of the current and direction of magnetization (see Equation 2.56):

$$\frac{\Delta R_x}{R_x} = -\frac{\Delta \rho}{\rho}\sin^2\vartheta = -\frac{1}{2}\frac{\Delta\rho}{\rho} + \frac{1}{2}\frac{\Delta\rho}{\rho}\cos 2\vartheta \quad (6.9)$$

If we use relatively weak magnet rotating above the AMR sensor, we can measure component of this field and $V_{out} = H\cos\alpha$. This method is sensitive to the changes of magnetic field, temperature, and so on, and operates only for small angle deviation around 0° (AMR sensor does not work correctly for the magnetic field perpendicular to the sensor axis). Another solution is to use a strong magnet and this way to force that $\alpha = \vartheta$. According to the expression (6.9), we obtain relation $V_{out} = K\cos2\alpha$ where K is the constant.

Often, a two-sensor mode of the angle detection is used. There are two sensors: one detects $K\cos2\alpha$ value and the second $K\sin2\alpha$ value and we can compute linear dependence $V_{out} = K \cdot \text{arctg}(V_y/V_x) = K\alpha$. In the case of AMR sensors, two sensors should be mutually inclined by 45°. Figure 6.57 illustrates the principle of operation of such sensor and Figure 6.58 presents the layout of KMZ41 double-sensor angle detector developed by Philips and recently manufactured by NXP Semiconductors.

A similar principle of operation is realized by using two-dimensional Hall sensor developed by Sentron AG (Figure 6.59). This sensor with embedded integrated circuit enables to obtain linear 2V signal for the rotation of the magnet between 0° and 360°. Typically, an angular

FIGURE 6.55
Examples of the linear, ring, and rotational magnetic encoders. (Courtesy of RLS company.)

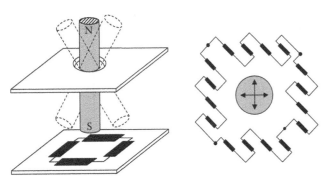

FIGURE 6.56
Position sensor used as manipulator (joystick). (From Nelson T.J., *IEEE Trans. Magn.*, 22, 394, 1986.)

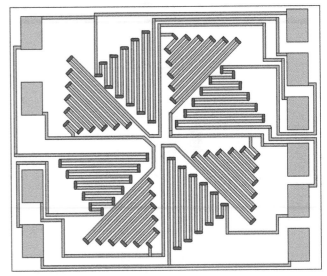

FIGURE 6.58
KMZ41 AMR sensor for the angle detection.

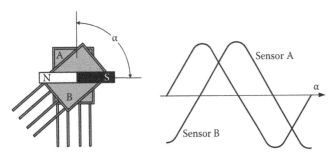

FIGURE 6.57
Angle transducer with two AMR sensors.

resolution of about 0.05° and overall accuracy of better than 1° is achieved. The same sensor can be used as the joystick device.

One of the most commonly used methods of rotational speed measurements (revolution) is application of magnetic sensors. Figure 6.60 illustrates the main methods: with rotating teeth of ferrous gear (a), with rotating magnets (b), or by the measurement of the magnetic field of eddy currents (c). The advantage of these methods is that they are noncontact and possible to realize in dirty and high-temperature environments.

Figure 6.61 illustrates the principle of operation of the rotational speed transducer. The output signal from the transducer depends on the distance d between the sensor and gear wheel and the wheel module $m = D/n$ (D = wheel diameter, n = number of the teeth). Figure 6.62 presents such dependence calculated for the KMZ10B AMR sensor of Philips. Recommended distance and wheel module are $m = 2$ and $d = 3$, respectively.

In the antilock braking system (ABS) system, it is necessary to measure the revolution of every wheel for which the magnetic sensors can be easily used. The inductive tachometers are disappointing at low speed and light sensors are not acceptable in dirty environment. Figure 6.63 presents the operating principle of an ABS system with magnetic sensors. Generally, the magnetic sensors are widely used in automobile industry (e.g., in the detection of steering wheel angle) (Treutler 2000, Kittel et al. 2003).

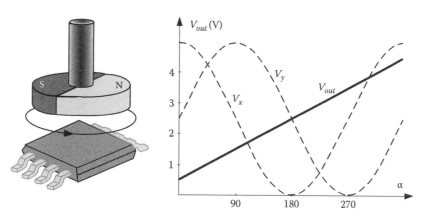

FIGURE 6.59
Two-dimensional Hall sensor 2SA-10 of Sentron as the angle transducer.

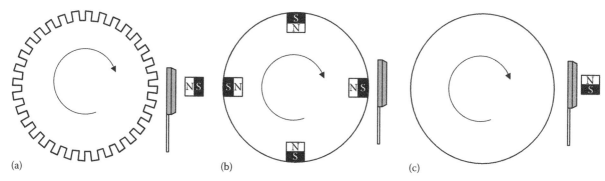

(a)   (b)   (c)

FIGURE 6.60
Main methods of rotation speed measurements.

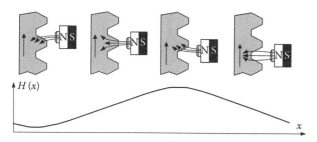

FIGURE 6.61
Principle of operation of the rotational speed transducer counting the magnetic field changes during the rotation of the ferrous gear teeth (Philip Technical Information).

If we are able to detect displacement, angle, speed etc. we can measure many other mechanical values. Figure 6.64 presents the transducers of pressure, acceleration, or torque. In the case of torque, we need to measure the difference angle $\Delta\alpha$ between the drive shaft and load shaft and then calculate the torque as (Miyashita et al. 1990):

$$T = \frac{\pi G D^4}{32L} \Delta\alpha \qquad (6.10)$$

where
G is the modulus of transverse elasticity
D is diameter of the torsion bar
L is the length of the torsion bar

6.2.3 Electrical Transducers

The main application of the magnetic sensors in electrical measurement is its application as current sensors. The very important advantage of such application is the noncontact measurement but even more important is the galvanic separation between input and output. Such magnetic separation is also possible in measuring transformers, but these devices are usually big and expensive and moreover operate only under AC currents, while magnetic sensors can "transform" both AC and DC currents.

Figure 6.65 presents the main methods of the current measurement by magnetic sensors. The magnetic field around the current conducting wire (Figure 6.65a) at the distance x can be measured according to the relationship

$$H = \frac{1}{2\pi x} I_x \qquad (6.11)$$

Also the current in the printed circuit board (Figure 6.65b) without cutting the current conducting track of the width w can be measured according to the approximated relation

$$H = \frac{1}{2w} I_x \qquad (6.12)$$

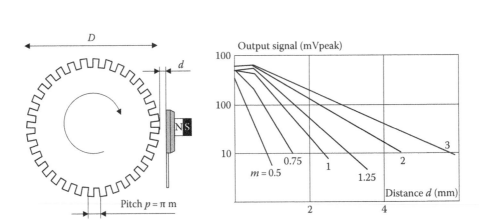

FIGURE 6.62
Dependence of the output signal of the rotational speed transducer based on the KMZ10B sensor (Philip Technical Information).

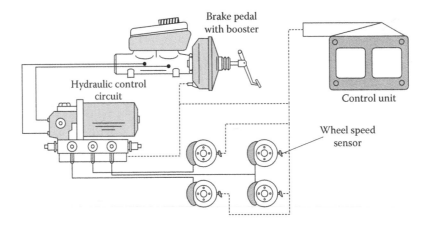

FIGURE 6.63
ABS system utilizing magnetic sensors for control of wheel revolutions. (From Graeger, V., *VDI Berichte*, 1009, 327, 1992.)

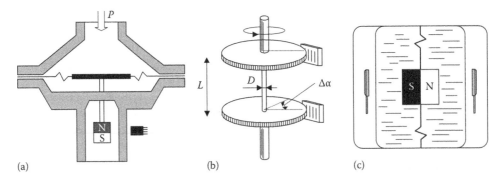

(a)   (b)   (c)

FIGURE 6.64
Transducers of pressure, torque, and acceleration basing on the magnetic sensors.

It is possible to measure the magnetic field generated by the coil (Figure 6.65c) of the length L, diameter $2a$ and number of turns n:

$$H = \frac{n}{2\sqrt{a^2 + L^2}} I_x \qquad (6.13)$$

And the currents are also recently commonly measured by means of the Rogowski coil (Figure 6.65d) where the output signal of the coli with length L, cross-section A and number of turn n is

$$V = \mu_0 \frac{nA}{L} \frac{dI_x}{dt} \qquad (6.14)$$

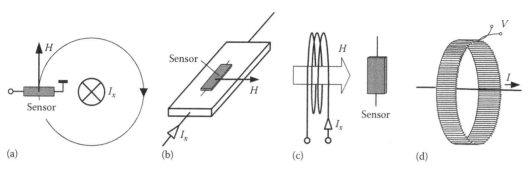

FIGURE 6.65
Main methods of current measurement by means of magnetic sensors.

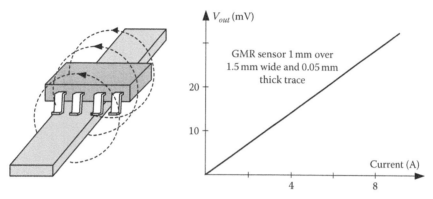

FIGURE 6.66
Direct measurement of the current using GMR magnetic field sensor.

Figure 6.66 presents the simple application of the GMR sensor for measurement of the current. Unfortunately, such method is sensitive to external magnetic field and therefore the method with two sensors is more recommended because it is significantly less sensitive to external magnetic field (Figure 6.67)

Recently as current transducers the Rogowski coil in form of closed ring is very often used (Figure 6.65d). Its main advantages are very high speed of operation (lack of ferromagnetic parts) and practically unlimited range of measured current (including plasma current). The drawback is that it is necessary to use integrating circuit at the output.

The best results are possible to obtain by using closed yoke system with a feedback. Two examples of such device are presented in Figure 6.68. There are many closed loop Hall sensor transducers on the market. Figure 6.69 presents typical transducers based on Hall sensors.[*] These transducers convert DC and AC (up to 100 kHz) current of the range from 25 to 500 A to the output current typically 100 mA. The accuracy of conversion is 0.5% and linearity 0.1%.

In many devices, as for example, interfaces of A/D converters, it is required to ensure galvanic isolation between input and output. Nonvolatile Electronics (NVE) developed GMR sensor–based microchip known as magnetic isolator, magnetic coupler, or IsoLoop. The principle of operation of such device is presented in Figure 6.70.

The magnetic coupler of NVE enables transmission of digital signal with speed of about 100 Mbps with delay smaller than 8 ns. It is possible to obtain more than five channels coupling. Figure 6.71 presents the comparison of a GMR and an optical couplers.

Figure 6.72 presents another possible application of the Hall or MR sensors as multipliers used for power conversion. Because the output signal of MR sensor depends on the magnetic field H_x (generated by the current I_x) and supply voltage V_x

$$V_{out} = KV_xI_x \qquad (6.15)$$

we can use such sensor as a power transducer. Figure 6.69 presents an example of the design of electrical power transducer that exhibited an accuracy of 0.5%. An important advantage of this transducer is possibility of short-time overload to even 10 times the nominal current.

[*] There are also open loop versions (without feedback) with voltage output.

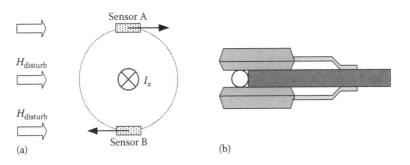

FIGURE 6.67
Dual sensor current transducer (Philips Technical Information).

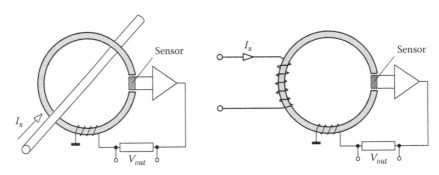

FIGURE 6.68
Current transducers with closed yoke and feedback.

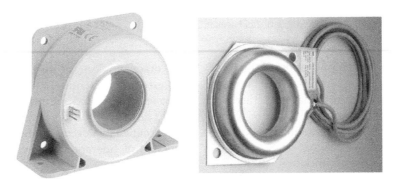

FIGURE 6.69
Example of the Hall sensor-based current transducers (left) and Rogowski coil transducer (right). (Courtesy of ABB.)

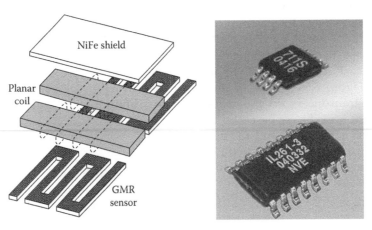

FIGURE 6.70
Principle of operation of the GMR galvanic isolation device.

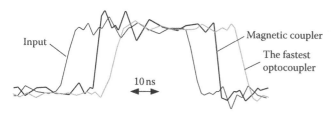

FIGURE 6.71
Comparison of the performances of magnetic and optical isolators. (From Mayers, J., *Sens. Technol. Des.*, 2002.)

6.3 Magnetic Diagnostics

6.3.1 Magnetic Imaging

In Section 5.4.2, the method of scanning of magnetic field distribution by means of an AMR sensor is presented.

In Section 6.1.3 methods of map creation of magnetic field are shown. These techniques can be classified as magnetic imaging techniques. Through magnetic imaging, we can analyze the magnetic properties of materials but also we can test other phenomena, for example, brain activity or quality of welded joints.

A widely known technique is *magnetic resonance imaging* (MRI) (described later). In testing of magnetic materials, magnetic domain imaging techniques are frequently used. Practically all sensors, SQUID, micro Hall sensors, and Kerr effect devices enable to obtain the magnetic image by scanning of magnetic field. Images can also be obtained by another technique: magnetic camera with an array of sensors.

Figure 6.73 presents the image of the bits recorded on the hard disk, a DVD, and Blu-ray disks. Such small areas need special microscope sensors. Figure 6.74 presents two extremely small Hall sensors developed

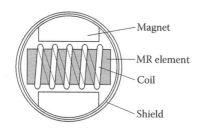

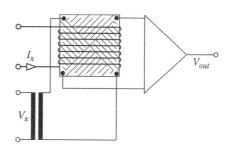

FIGURE 6.72
Design of AMR sensor-based electrical power transducer. (From Kwiatkowski, W. and Tumanski, S., *IEEE Trans. Magn.*, 20, 966, 1984.)

FIGURE 6.73
Microscopic images of hard disk, DVD, and Blu-ray disk surface obtained by the AFM and MFM (blu-ray image $4 \times 4\,\mu m$). (Courtesy of Nanomagnetics Instruments.)

FIGURE 6.74
Hall sensors for microscope purposes: 50 nm sensor, 1 μm sensor, and a part of the matrix composed of 16 384 sensors on the area 13×13 mm. (Courtesy of Nanomagnetics Instruments and MagCam NV.)

by Nanomagnetics Instruments. In the same figure, a matrix of 16,384 Hall sensors developed by MagCam NV for magnetic field camera is also presented.

Various Hall sensor–based microscopes have been developed (Chang et al. 1992, Oral et al. 1996a,b, Howells et al. 1999, Sandhu et al. 2001, 2002, 2004, Shimizu et al. 2004, Dinner et al. 2005, Kejik et al. 2006). Extremely small Hall sensor with dimensions $50 \times 50\,nm$ was prepared from $60\,nm$ thick thin film bismuth by focused ion beam milling. This sensor exhibited resolution $80\,\mu T\ Hz^{1/2}$ (Sandhu et al. 2004). A typical microsensor is prepared as a GaAs/AlGaAs heterostructure with dimensions $1 \times 1\,\mu m$ and resolution $4\,\mu T/Hz^{1/2}$ (Sandhu et al. 2001). The sensitivity and resolution can be significantly enhanced if the scanning is performed at low temperature.

Figure 6.75 presents an example of the Hall probe microscopy. The Hall probe is inserted with the tilt angle of about 1.5° for better detection of the close proximity

to the investigated surface (for instance, by detection of the tunneling current). Coarse positioning is performed by precise stepper motors but fine positioning is realized using piezoelectric scanning tube (PZT).

Figure 6.76 presents the application of the Hall sensors camera for testing of quality of welds. The main advantage of such a system is real-time imaging with the speed of 1–50 frames/s with 0.1 mm pixel resolution. Because the microsensors are very small, the sensitivity is modest—the smallest range of magnetic field is 0.1 mT. However, this is still sufficient to test the joints and detect the cracks (Jun and Lee 2008, Hwang et al. 2009).

Thin-film magnetoresistive sensors are commonly used for high-density reading of hard disk information and it is obvious that they should be capable of detecting the magnetic image with submicron resolution. Indeed application of AMR, GMR, and spin valve sensors for microscopy imaging has been reported (O'Barr et al. 1996, Yamamoto et al. 1996, 1997, Phillips et al. 2002,

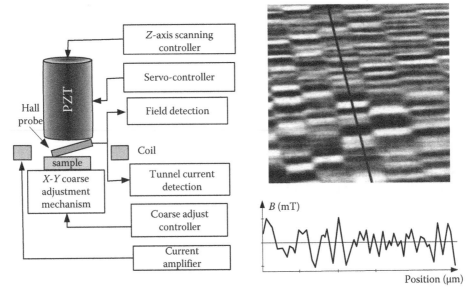

FIGURE 6.75
Hall sensors microscopy and the image of recorded tracks on the disk. (From Sandhu, A. et al., *Jpn. J. Appl. Phys.*, 40, 4321, 2001.)

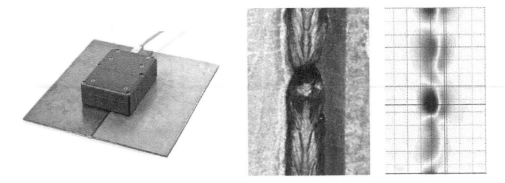

FIGURE 6.76
Hall sensors camera testing the welded joint. (Courtesy of MagCam NV.)

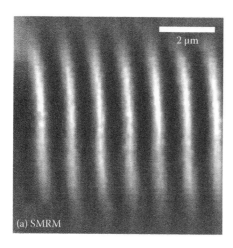 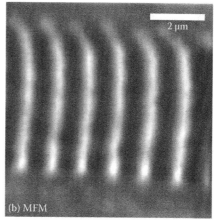

FIGURE 6.77

Comparison of the recorded track images obtained by scanning magnetoresistance microscopy (SMRM) (a) and MFM (b). (From Yamamoto, S.Y. and Schultz, S., *J. Appl. Phys.*, 81, 4696, 1997.)

Mazumdar and Xiao 2005, Takezaki et al. 2006, Sahoo et al. 2009). Figure 6.77 presents the comparison of images of recorded bits performed by scanning magnetoresistive microscopy. The quality of the picture is comparable with that obtained by magnetic force microscopy (MFM).

Conventionally for DNA analysis, a laser scanning microarray of fluorescently labeled dots is used. It is also possible to label the DNA sequence magnetically according to scheme presented in Figure 6.78. Biotinylated array of DNA dots is labeled by streptavidin covered

by magnetic nanoparticles. As a result, the microarray composed of thousands of beads is obtained.

To detect high density, very small magnetic field above the DNA array the sensors with μm size are required. The good candidate are GMR sensors and especially MTJ sensors (due high magnetoresistivity they can be exceptionally small) (Miller et al. 2001, Graham et al. 2003, 2004, Brzeska et al. 2004, Shen et al. 2008). Figure 6.79 presents the result of scanned magnetic field above magnetically labeled DNA array by MTJ sensor with

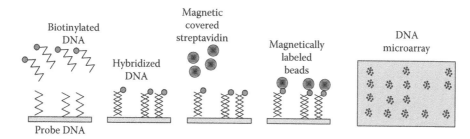

FIGURE 6.78

Principle of creating of magnetic labeled DNA microarray.

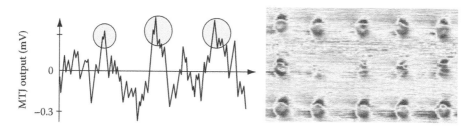

FIGURE 6.79

Results of scanning of the magnetic field above the DNA microarray (left, distance 1000 μm) and the detected image (right)—dimensions 700 × 1500 μm. (From Chan, M.L et al., Scanning magnetic tunnel junction sensor for the detection of magnetically labeled microarray, *14th International Conference on Solid State Sensors, Actuators and Microsystems*, Lyon, France, 2007; Chan M.L. et al., *IEEE Trans. Magn.*, 45, 4816, 2009.)

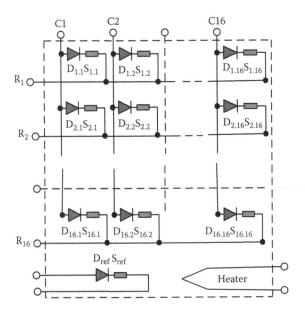

FIGURE 6.80
The MTJ sensors array as part of biochip for DNA analysis. (From Almeida, T.M. et al., Characterization and modelling of a magnetic biosensor, *Instrumentation and Measurement Technology Conference*, Sorrento, Italy, 2006.)

dimensions $2 \times 6\,\mu m$ and sensitivity 0.6%/mT. The signal and image are covered by noise but beads are detectable.

The best solution for fast analysis of expression of gene is the biochip microsystem containing the array of MTJ sensors (Baselt et al. 1998, Edelstein et al. 2000, Almeida et al. 2006, Cardoso et al. 2006, Lopes et al. 2010). Figure 6.80 presents typical sensor array. It is a $8 \times 8\,mm$ matrix of 16×16 sensors connected with diodes (256 sensors). The diodes are added to switch the device and to control the temperature. Each MTJ sensor with

dimensions $2 \times 10\,\mu m$ (composed of the following layers: 9 Ta/5 NiFe/25 MnIr/5 CoFeB/1.2 Al_2O_3/1.5 CoFeB/4.5 NiFe/3 Ta) exhibits a sensitivity of 25%/5 mT. Further improvement can be expected after substitution of Al_2O_3 barrier by MgO, which has much higher magnetoresistivity. The whole microsystem with dimensions of a credit card is equipped with electronic circuitry for addressing, reading the data, temperature controlling, fluid sample handling, and digital signal processing. The possibility of detection of 250 nm diameter magnetic nanoparticles was demonstrated.

Figure 6.81 presents field sensitivity and spatial resolution attainable by several magnetic imaging techniques. From this figure, it is visible that SQUID-based microscopy covers large area applications, especially the low-field area.

Indeed the SQUID sensor with its exceptional sensitivity is a good alternative to commonly used microscopy methods. Thin film SQUID sensors are small but the tested area can be even smaller because the pickup coil can be as small as several micrometers (Figure 6.82).

There are several review papers comprehensively describing scanning SQUID microscopy (Wikswo 1995, Kirtley and Wikswo 1999, Fagaly 2006). There are also hundreds papers describing more detailed SQUID microscopes (Staton et al. 1991, Mathai et al. 1992, 1993, Vu et al. 1993, Wellstood et al. 1997, Hasselbach et al. 2000, Kirtley 2002). Similarly as in other methods alternatively to scanning methods sensor array methods of magnetic imaging are used (Wikswo 1991, Espy et al. 2001, Krause et al. 2001, Gärtner et al. 2002, Clark et al. 2003, Matsuda et al. 2005) (Figure 6.83).

Of course, the best solution is when the sample and the pickup coil are at low temperature (Figure 6.80a,b).

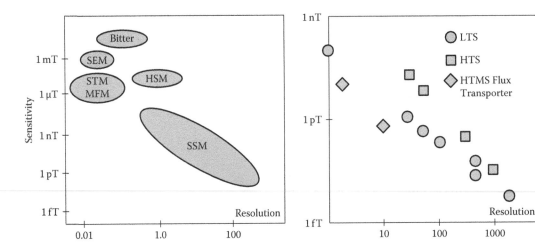

FIGURE 6.81
Field sensitivity and spatial resolution attainable by several magnetic structure imaging techniques: SEM, scanning electron microscope; STM, scanning tunneling microscope; MFM, magnetic force microscopy; HSM, Hall scanning microscopy; SSM, SQUID scanning microscopy (right side; data reported for SQUID microscopes). (From Vu, L.N. and Van Harlingen D.J., *IEEE Trans. Appl. Supercond.*, 1993, 3, 1918, 1993; Bending, 1999; Fagaly, R.L., *Rev. Sci. Instrum.*, 77, 101101, 2006.)

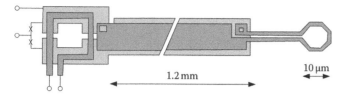

FIGURE 6.82
Micro-SQUID sensor for microscopy application. (From Kirtley, J.R. and Wikswo, J.P., *Ann. Rev. Mater. Sci.*, 29, 117, 1999.)

Not always it is possible to realize such strategy. If the sample is outside the Dewar, the main problem of resolution is the distance between sample and pickup coil. Nevertheless recently, the special design enables to scan the magnetic field with resolution of 50 μm and space distance between coil and sample as small as 15 μm (Kirtley and Wikswo 1999).

As presented in Figure 6.81, right side, there is a competition between resolution and sensitivity; higher sensitivity can be obtained only at the expense of resolution. However, as shown in this figure, a low-temperature SQUID LTS can be substituted by high-temperature SQUID HTS, which can be important in nondestructive industrial testing.

Figure 6.84 presents the magnetic image of the Martian meteorite determined by ultrahigh resolution

scanning microscope with a 500 μm diameter NbTi pickup coil separated by a 250 μm from the room temperature sample (scan step 200 μm).

The process of scanning is time-consuming and cannot be acceptable in many applications. Therefore various sensor array systems consisting of 3, 4, or 10 sensors are reported (Wikswo 1991, Espy et al. 2001, Krause et al. 2001, Gärtner et al. 2002, Clark et al. 2000, Matsuda et al. 2005). Especially, in the case of system for observation of the brain activity, magnetoencephalography (MEG), the real-time data analysis is required. Recently, MEG systems consisting of more than 200 pickup coils have been developed (Figure 6.85).

Kerr microscopy is often used for imaging of the domain structure. The principle of operation of typical Kerr microscope is presented in Figure 6.86. The polarized light is reflected from polished surface of the magnetized sample. The reflected beam is rotated and the angle of rotation depends on magnetization of the sample (the principle of the Kerr effect is presented in Section 2.6.3). As the light source, the mercury arc light is commonly used (but also laser beam can be applied).

The important advantage of the Kerr microscopy is that there is no need for scanning and the picture is obtained directly, for instance from the CCD camera.

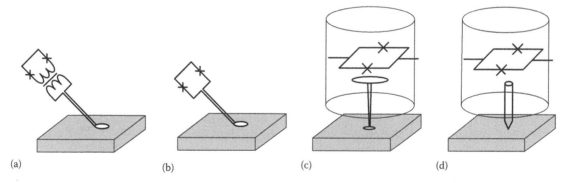

(a) (b) (c) (d)

FIGURE 6.83
Various strategies of microscope scanning system. (From Kirtley, J.R. and Wikswo, J.P., *Ann. Rev. Mater. Sci.*, 29, 117, 1999.)

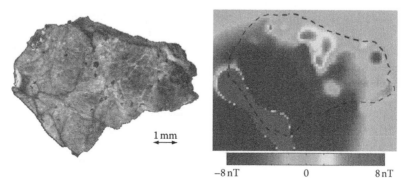

−8 nT   0   8 nT

FIGURE 6.84
Martian meteorite and its magnetic imaging. (From Weiss, B.P. et al., *Science*, 290, 791, 2000. With permission.)

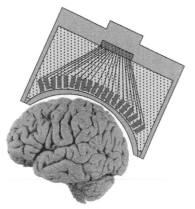
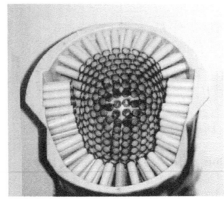

FIGURE 6.85
Sensor array for measurement of the magnetic field around the human head—principle and example of the coil array of the Magnes 3600 WH system showing the 248 independent magnetometer/gradiometer detector coils. (Courtesy of 4-D Neuroimaging.)

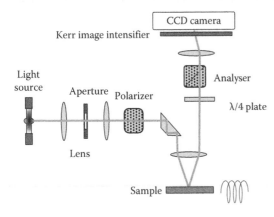

FIGURE 6.86
Principle of operation of the Kerr microscope.

This enables observation of the dynamics of the domain formation. The disadvantages are that the surface should be carefully prepared, and sometimes coating should be removed because light penetrates only to about 20 nm deep. To obtain satisfying image, the complex digital image processing is necessary (Hubert and Schäfer 1998).

Kerr microscopes manufactured by Neoarc Corporation are available on the market, for example, BH-786 V model with resolution 1 μm and the tested area 100×80 μm for 50× lens or 250×200 μm for 20× lens.

Kerr microscopes enable observation of the domain dynamics. Figure 6.87 presents the results of observation of the domain wall movement in real-time system (Moses et al. 2005, 2006). The fast camera was synchronized with alternating magnetizing source. From the repeated observations, it is visible that the process of domain formation is not fully repeatable; the locations of domain walls differ from cycle to cycle.

Specially designed Kerr microscopes with pulsed laser beam known as *time-resolved microscopes* enable making a snapshot of domain images with picosecond

time resolution and submicron spatial resolution (Kryder et al. 1990, Freeman and Smyth 1996, Hiebert et al. 1997, Back et al. 1999).

It is possible to perform magnetic imaging (including domain observation [Hubert and Schäfer 1998, Szmaja 2006]) by using classical microscopy methods such as scanning electron microscopy (SEM), transmission electron microscopy (TEM), scanning tunneling microscopy (STM), scanning atomic force microscopy (AFM), or magnetic force microscopy (MFM) (Petford-Long 2001). In the SEM, the electron beam reflected from the sample surface or transmitted through the sample (e.g., in the case of thin film samples) interacts with the sample. Various detected signals (secondary electrons, back scattered electrons, x-rays, light, etc.) can be affected by the topography of the surface as well as distribution of stray magnetic field.

Other principles represent microscopes that employ the tip detector (Figure 6.88):

- Tunneling microscopy (STM).
- Atomic force microscopy (AFM).
- Magnetic force microscopy (MFM).

The sharp tip with tip radius curvature at the atomic level (sub-nm size) is moved above the sample at the constant distance. In tunneling microscopy, the tunneling current is used as the output signal (thus the tip and the sample should be good electric conductors). In AFM, the tip is at the end of the silicon cantilever. Under the influence of atomic forces existing for atomic distance between tip and surface, the cantilever is deflected. This deflection is detected by laser beam.

The constant distance is assured due to the feedback. If the cantilever is oscillating, the frequency of oscillation depends on the distance between the sample and the tip. The feedback controls the oscillation frequency.

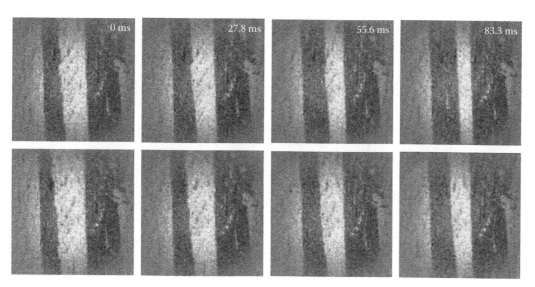

FIGURE 6.87
Comparison of images captured over the same area (1.2 × 1.7 mm) on two consecutive magnetizing 1 Hz cycles for GO steel. (From Moses, A.J. et al., *IEEE Trans. Magn.*, 41, 3736, 2005. With permission.)

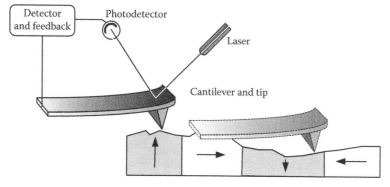

FIGURE 6.88
Principle of operation of atomic force/MFM.

In MFM, the tip is covered by magnetic material and the cantilever is deflected under the influence of magnetic force.

Figure 6.89 presents the image of the domain structure determined by MFM. The same area tested by AFM can be used to remove the influence of topography. The combination of various microscopy methods is often used to obtain satisfying results; for example, combination of Kerr microscope and MFM in one device is reported (Peterka et al. 2003). Also NanoMagnetic Instruments offers PPMS model with possibility to use scanning Hall probe microscopy, AFM, and MFM.

The traditional Bitter colloidal technique of observation of the domain structure is still preferred especially in industrial environment due its simplicity (compare Figure 5.134). Modern digital image processing enables significant improvement of this method, as presented in Figure 6.90.

6.3.2 Magnetic Nondestructive Testing

Electromagnetic techniques of nondestructive testing (NDT) are widely used because they are relatively simple (e.g., in comparison with x-ray methods). There are two main methods of electromagnetic NDT (Figure 6.91):

- Magnetic flux leakage (MFL)
- Eddy-current testing (ECT)

In the first method, the magnetic sample is magnetized with steady or low-frequency magnetic field and the defects are revealed in a form of magnetic anomaly (similarly as it was in the case of magnetoarcheology). Alternatively, the leakage flux can be detected around previously magnetized sample as remanence magnetization. Coil sensor, MR sensor, or Hall sensor can be used as the magnetic flux detectors. New perspectives on magnetic NDT are offered by SQUID sensors. The

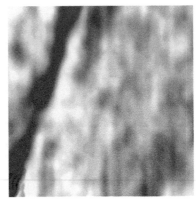

FIGURE 6.89

Domain structure of neodymium magnet determined by MFM microscopy (left). Scanned area $0.6 \times 0.6\,\mu m$, distance of the tip to the surface 100 nm. On the right side the same area (topography) determined by the AFM microscopy. (From Szmaja, W., Recent developments in the imaging of magnetic domains, in *Advances in Imaging and Electron Physics*, P.W. Hawkes (Ed.), Vol. 141, pp. 175–256, 2006. With permission.)

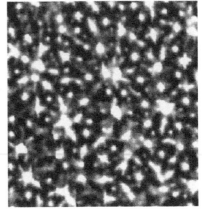

FIGURE 6.90

Domain structure of cobalt monocrystal determined by Bitter technique (left). Tested area $80 \times 80\,\mu m$. On the right side the same image after digital postprocessing. (From Szmaja, W., Recent developments in the imaging of magnetic domains, in *Advances in Imaging and Electron Physics*, P.W. Hawkes (Ed.), Vol. 141, pp. 175–256, 2006. With permission.)

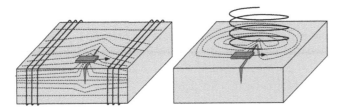

FIGURE 6.91

Main methods of magnetic NDT: flux leakage (left) and eddy currents (right).

most commonly used technique for the anomaly imaging is a *magnetic particle inspection* (MPI).

In the eddy-current method (ECT), the sample is magnetized by alternating magnetic field and magnetic field generated by eddy currents is detected. In comparison with the flux leakage method, the eddy-current technique can be used for all conducting materials, not only magnetic ones. Moreover, an important advantage of ECT is that by selecting frequency we can penetrate various depths of the sample.

Both these techniques are limited to volume just under surface or the surface itself. Recently developed NDT techniques are not limited only to detection of defects. More interesting and valuable is the possibility of prediction of appearance of the future defects. For example, tendency to corrosion can be detected by a SQUID sensor (Wikswo 1995, Jenks et al. 1997, Krause and Kreutzbruck 2002).

Besides the above presented two methods, other magnetic techniques can be used:

- Nuclear magnetic resonance (NMR) (mainly in medicine)

- Barkhausen noise and magnetoacoustic emission (MAE)
- Villari and other magnetostrictive effects (Kaleta et al. 1996, Baudouin et al. 2003)

The two main methods are affected by three important problems: technique of the magnetization of the sample, technique of the field detection, and interpretation of the data. Figure 6.92 presents the typical leakage magnetic field above various cracks. The experienced researcher can determine the shape of the crack from its magnetic signature.

But in the case of the MFL, the most convenient and simplest method is magnetic particle inspection (MPI) (Lovejoy 1993). It is sufficient to cover the magnetized sample with magnetic powder or colloid and simply to visually observe the investigated surface. Figure 6.93 shows two examples of images of the defected samples obtained with the MPI experiment.

Figure 6.94 presents various methods of the magnetization of the sample under test: by magnet, by

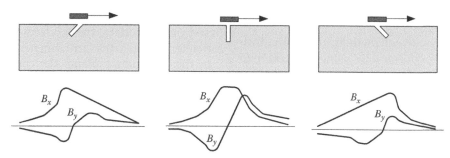

FIGURE 6.92
Examples of the leakage flux above the various cracks.

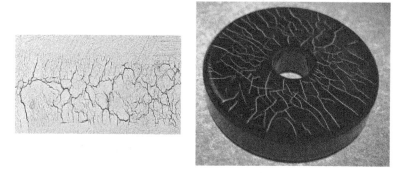

FIGURE 6.93
Two examples of the crack visualization by the MPI technique (right side in fluorescent light). (Courtesy of MR Chemie GmbH.)

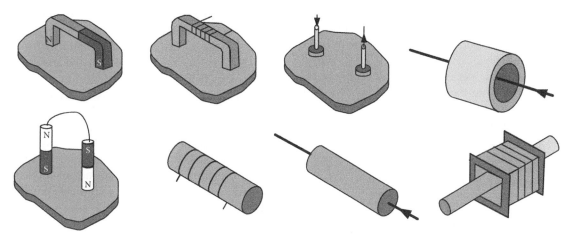

FIGURE 6.94
Various methods of magnetization of the investigated sample.

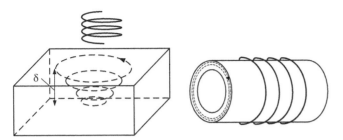

FIGURE 6.95
Generation of the eddy currents in the tested samples.

electromagnet, or directly by connecting the current source to the sample.

In the case of eddy-current method, the magnetization of the sample is performed by using of a coil with AC current (Figure 6.95). The depth of penetration of the eddy current depends on the frequency of the magnetic field and on the electric conductivity γ of the material (Table 6.4):

$$\delta = \frac{1}{\sqrt{\pi f \gamma \mu}} \quad (6.16)$$

Figure 6.96 presents typical techniques used in ECT. The tested specimen can be assumed as a core of an inductor (Figure 6.96a). In such case, the output signal is the impedance of the coil depending on the tested surface. Usually both components of the impedance $Z = R + j\omega L$ are analyzed because the phase of the impedance brings the information about the position of the defect. Often both components of permeability are also tested. Instead of a single-coil system, a two-coil system can be used where one coil is the reference coil (Figure 6.96b).

It is possible to detect the anomaly directly by using two coils: one primary coil for magnetizing and the second pickup coil for detecting induced voltage (Figure 6.96c). The sample plays a role of the core of the transformer. Alternatively, two differential secondary coils can be used to remove the common component (Figure 6.96d). Also in this case both components of the induced voltage can be analyzed.

The ECT is usually performed by moving of the probe and the analyzing the detected voltage. Modern

TABLE 6.4

Typical Depth of Eddy-Current Penetration in Millimeter

Frequency (Hz)	Copper with $\mu_r = 1$	Steel with $\mu_r = 100$
0.1	6.6	1.6
10	0.7	0.2
100	0.2	0.05
1000	0.07	0.02

instruments enable direct observation of the impedance patterns shown in the example in Figure 6.97.

Figure 6.98 presents the modern testing instrument developed by Rohmann GmbH. On the screen it is possible to observe the impedance pattern as well as output signal.

Förster (1952), Dodd (1977), Libby (1979), Cecco et al. (1981), and other authors calculated the so-called *impedance plane diagrams* for selected specimens and coil configurations (e.g., Figure 6.99, left). These diagrams are very useful because they allow determination of the value of impedance as well as its the change due to the crack presence, as demonstrated in Figure 6.99, right. Based on these data, the modern eddy-current testers can determine location and dimensions of the defect.

In the impedance plane, the component ωL_0 is the inductance of the search coil without specimen and the limiting frequency f_g is the frequency at which the argument of the Bessel function equals unity.* For example, for a solid cylinder or thick-walled tube with an encircling coil, the limiting frequency is

$$f_g = \frac{2}{\pi \mu_r \mu_0 \gamma D^2} = \frac{506.606}{\mu_r \gamma D^2} \quad (6.17)$$

Magnetic evaluation is useful not only for testing of the defects but also for determination of other properties of the material. For example, mechanical hardness is strongly correlated with coercivity. It is possible to test the hardening surface by eddy the current method (Kirel et al. 2000, Mercier et al. 2006).

Both magnetic Barkhausen effect (MBE) and magnetoacoustic emission effect (MAE) techniques are commonly used for testing of grain size, orientation, and presence of impurities or defects. The Barkhausen effect is mainly related to the sudden movements of domain walls while the MAE is also related to discontinuous irreversible non-180° domain wall motion. Thus, it is obvious that inhomogeneity of magnetic material influences both effects. As illustrated in Figure 6.100, both of these effects exhibit a different relationship to the magnetic field strength. It was proved that the magnetoacoustic effect is practically not present in nonmagnetostrictive materials (e.g., 6.5% SiFe steel) (Augustyniak et al. 2008).

The advantages of MBE and MAE techniques is simplicity of measurements: practically, all shapes of the samples are acceptable and signal is easy to detect. In the case of Barkhausen noise, it is sufficient to simply place the pickup coil on the surface of the sample. For the MAE, a piezoelectric transducer is commonly used as a sensor.

---

* The f_g does not have any physical meaning but it helps in impedance analysis when Bessel functions are employed.

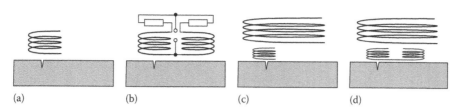

FIGURE 6.96
Various methods of ECT.

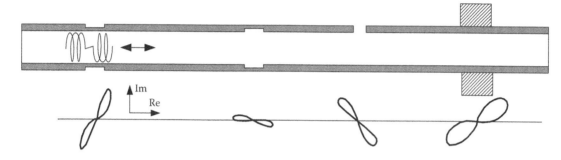

FIGURE 6.97
Patterns of impedance vector detected during the movement of differential probe inside the tube with various defects. (After Cecco, V.S. et al., *Eddy Current Manual*, Report: AECL-7523, Chalk River Nuclear Laboratories, Chalk River, Ontario, Canada,1981; Blitz 1991.)

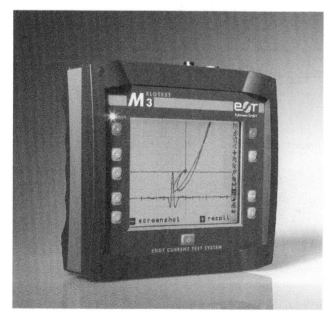

FIGURE 6.98
ECT instrument Elotest M3 (Courtesy of Rohmann GmbH.)

Unfortunately MBE and MAE methods are mainly qualitative in nature although correlation between increase of the stray field and Barkhausen noise is reported (Zurek et al. 2009). Figure 6.101 presents the comparison of the Barkhausen noise signatures of the bearing races after certain service time. Also other examples of the application of the MBE and MAE methods to NDT are reported (Mandal et al. 1999, Augustyniak et al. 2007, Piotrowski et al. 2010).

6.3.3 Magnetism in Medicine

Magnetism is used in many medical applications, especially as diagnostic tools (e.g., magnetic resonance, magnetocardiogram (MCG), MEG, and liver susceptometry). A comprehensive review on this subject is presented in the book of Andrä and Nowak (2007).

The most spectacular success in medicine diagnostic is related to MRI method. This method (Nobel Prize in 2003) is not as dangerous as a x-ray tomography and also gives often better image quality. MRI utilizes mainly the NMR (described in Section 2.7.2) because the human body is largely composed of water molecules therefore the resonance of ^{1}H nuclei is easy to detect.

In NMR, the nuclei are polarized by a strong DC magnetic field. Then the RF pulses of the magnetic field are applied at an angle 90° to the DC magnetic field. This causes that the magnetic moments of nuclei change the direction of magnetization with precession of frequency f_0. To obtain the resonance, the frequency of the RF signal should correspond with the resonance conditions described by the relationship

$$f_0 = \frac{1}{2\pi}\gamma B \tag{6.18}$$

where $\gamma/2\pi$ is a gyromagnetic ratio with a value 42.6 MHz/T for protons.

The relationship (6.18) can be used for positioning of the investigated part of the body (Figure 6.102). If to the constant magnetic field B_z we add magnetic field

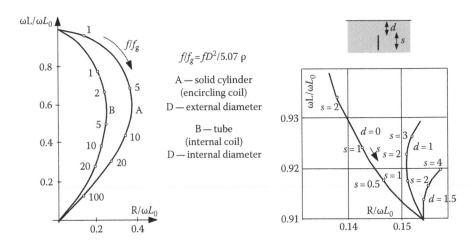

FIGURE 6.99
Normalized impedance plane and the change of impedance caused by the presence of the crack. (From Shull, P.J., *Nondestructive Evaluation*, Marcvel Dekker, New York, 2002.)

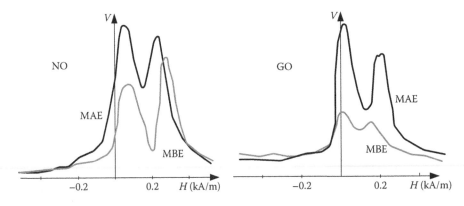

FIGURE 6.100
Examples of the envelope of Barkhausen noise and MAE envelopes (determined for NO and GO steel). (From Augustyniak, B. et al., *J. Appl. Phys.*, 93, 7465, 2003.)

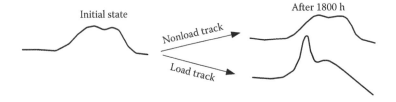

FIGURE 6.101
Barkhausen noise signal versus magnetic field of the bearing race. (From Bray, D.E. and McBride D., *Nondestructive Testing Techniques*, John Wiley & Sons, New York, 1992.)

increasing in space (gradient component $\Delta B_z/\Delta z$), we have just a narrow part of the body (slice) that corresponds with resonance conditions of the RF pulse. By changing the RF frequency, we can select different investigated slices. More often this selection is realized by the change of the gradient pulse. For a typical gradient, $\Delta B_z/\Delta z$ equal to 1 mT/m the change of frequency of about 40 Hz corresponds with change of the slice of about 1 mm.

Figure 6.103 presents a typical device for MRI diagnostic. It consists of a magnet or electromagnet, transmitter

RF coil that can also be used as receiver coil and gradient coils. Usually there are used also other coils like shim coils for improvement of the field uniformity.

If we use the rectangular pulse as the RF exciting signal, as response we obtain signal containing wide spectrum of frequencies (Figure 6.104a). Hence usually as the exciting RF signal, it is used sinc pulse (sin x/x) that is represented by almost rectangular pulse in the frequency domain (Figure 6.104b).

The NMR scanning process starts with gradient signal G_z and next RF excitation (Figure 6.105). As the response

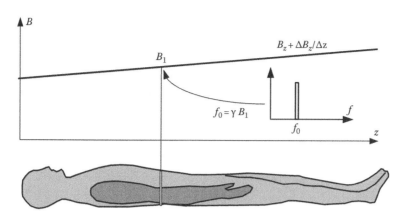

FIGURE 6.102
Positioning of the investigated slice by applying of the gradient field component.

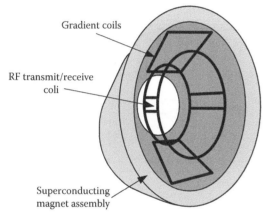

FIGURE 6.103
Tunnel used for MRI diagnostics. (From Guy et al. 1976.)

in the receiver coil is induced FID signal. Information about properties of the investigated substance usually brings the length of magnetization vector and the relaxation time T_2 (see Table 2.10). Because the frequency range of a sinc signal is extended to infinity, it is limited in time by reverse change of the gradient signal.

Unfortunately, due to magnetic field nonuniformity and other reasons (Kuperman 2000, Vlaardingerbroek and den Boer 2004), the response signal is decreasing faster according to T_2^* relaxation time (instead of T_2

time). To detect real T_2 relaxation time, commonly the spin echo method, presented in Figure 6.106, is used.

After 90° pulse, the decaying response signal is detected. The speed of decay is increasing due to increasing of the dispersion of the magnetic moments (Figure 6.106, right). Therefore, after time $TE/2$, the second pulse reverses the directions of magnetization by 180° (this pulse is stronger than 90° pulse because two times larger change of direction is forced). After this "trick," the directions of magnetic moments are again refocusing and the echo signal appears. The echo signal is decreasing according to $\exp(t/T_2)$ and therefore it is possible to determine this relaxation time.

The detected signal contains information about selected slice. To obtain the image, it is necessary to recover spatial distribution of the signal in the xy plane (the slice). This is realized by applying special technique known as frequency and phase encoding. During analysis of the signal, the special signals representing gradient in x direction G_{xz} and y direction G_{yz} (various values) are added (Figure 6.107). The two-dimensional Fourier analysis helps in the determination of the signal distribution in space; it can be assumed that the central harmonics (and phase) in the detected spectrum represent central line of the investigated plane.

The one-pulse sequence presented in Figure 6.107 takes about 10–30 ms. Thus the whole analysis is

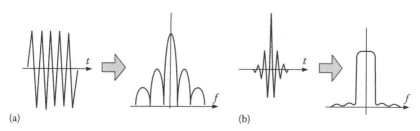

FIGURE 6.104
RF pulses and their frequency transform.

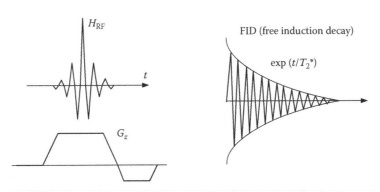

FIGURE 6.105
Pulsed NMR signals: excitation and response.

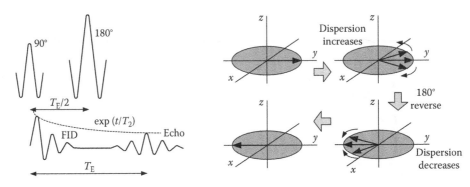

FIGURE 6.106
Spin echo method: signal and defocusing principle.

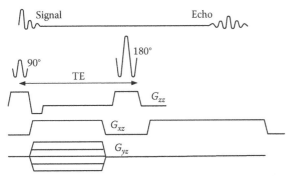

FIGURE 6.107
Pulse sequence in the spin echo method.

relatively slow. There are methods developed for significantly improvement of the speed of investigation: multiple slice acquisition in the same time, turbo spin echo (by inducting of several echoes for analysis), gradient echoes (Kuperman 2000, Vlaardingerbroek and den Boer 2004).

The image intensity I depends on the following factors:

$$I \propto N(H)e^{-TE/T_2}\left(1 - e^{-TR/T_1}\right) \qquad (6.19)$$

where
  $N(H)$ is the proton density
  TR is the repetition time
  TE is the echo time

The most important contribution is brought by the T_2 component (see Table 2.10). There were many methods developed for improvement of the contrast (Kuperman 2000, Vlaardingerbroek and den Boer 2004). Figure 6.108 presents the example of MRI picture of the brain with indicated activated areas.

Our nervous system is based on the transmission of electric currents; therefore the brain activity can be tested by magnetic methods: magnetoencephalography (MEG), heart activity by using magnetocardiography (MCG), stomach activity by using magnetogastrography, liver state by using liver susceptometry, etc. (Sternickel et al. 2006). Figure 6.109 presents the typical magnetic field levels generated by the human body.

It is obvious that the best method useful for such investigations is the SQUID method due to its exceptional sensitivity. Recently almost 90% of all operating SQUIDs are used in medical research, mostly of the human brain (Sternickel et al. 2006).

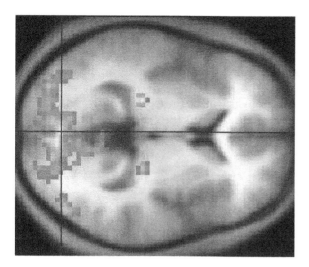

FIGURE 6.108
Typical magnetic resonance image of the brain slice showing the contrast between gray and white tissue and the activated areas. (From Wikimedia Commons.)

Table 6.5 presents the main areas of medical application of SQUIDs. In many cases, there is electrical alternative, more user-friendly techniques (electroencephalography, electrocardiography [ECG], electrogastrogram). Therefore although SQUID methods are commonly used in research centers they are rather seldom available in ordinary hospitals.

The main problem is not in low temperature and costs because nowadays these problems have been overcome. The main obstacle is that the detected magnetic fields are much smaller than the environmental magnetic fields. There are only several shielded rooms in the world with external magnetic field suppression to the range of the brain magnetic field (level of the fT). Therefore, other methods are used to enable the measurements of such small fields in unshielded environment.

The most commonly used method is the gradiometer mode, hardware (spatially distributed sensors) and software (called also as synthetic gradiometers). Modern digital signal-processing methods such as principal component analysis, blind signal separation, wavelet transform, neural networks, are very helpful in SQUID signal analysis. Relatively easy is detection of the signals of magnetocardiography in unshielded environment because the magnetic field is not so small. But recently it is also possible to analyze brain activity without shielded room by sophisticated methods of signal processing (Wikswo 1995, Koch 2001, Pizzella et al. 2001, Fagaly 2006, Andrä and Nowak 2007).

The magnetic methods are especially used for such diagnostic where there are no alternative. This is, for example,

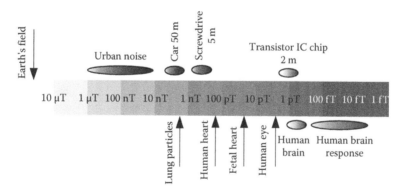

FIGURE 6.109
Biomagnetic and environmental fields. (After Nowak, H., Biomagnetic instrumentation, in *Magnetism in Medicine*, Andrä W. and Nowak H. (Eds.), Wiley-VCH Verlag, Weinheim, Germany, Chapter 2.2, 2007.)

TABLE 6.5

Selected Medical Methods Using SQUID

Organ Studied	Method	Range (pT)	Bandwidth (Hz)	Alternative
Brain	Magnetoencephalography	0.01–10	0.1–100	Electroencephalography
Heart	Magnetocardiography	1–100	0.01–100	ECG
Liver	Magnetic susceptometry	0.1–10	10	Biopsy
Lungs	Magnetic impurity detection	100–1000	0.1–10	—
Stomach	Magnetogastrography	1–20	0.05	Electrocardiography

Source: After Sternickel, K. and Braginski, A.I., *Supercond. Sci. Technol.*, 19, S160, 2006.

- Spatial analysis of the brain activity in real time
- Epilepsy diagnostic and medicine
- Fetal magnetocardiography
- Liver iron overload

Figure 6.110 presents the examples of the MEG signals detected in various parts of the head. The differences are significant and help in determination of the epilepsy center.

Figure 6.111 presents the spectacular example of the research of human brain activity by analyzing of an MEG signal. It is possible to detect the magnitude and frequency spectrogram as well as the location of the activation.

Figure 6.112 presents the example of the fetus magnetocardiography signal. Such noninvasive investigations

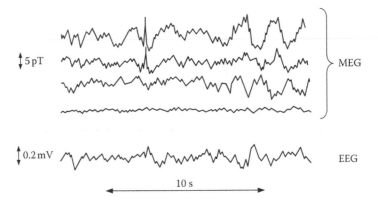

FIGURE 6.110
Examples of MEG signals detected for epileptic patient in unshielded environment in four selected segments (among 800 segments investigated). For comparison signal of electroencephalography. (After Sternickel, K. and Braginski, A.I., *Supercond. Sci. Technol.*, 19, S160, 2006.)

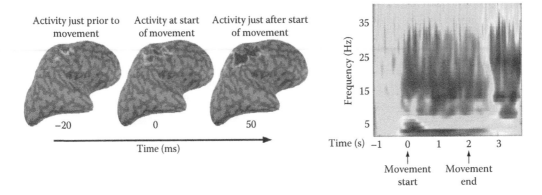

FIGURE 6.111
Examples of the analysis of an MEG signal. (Courtesy of Judith Schaechter, PhD, Athinoula A. Martinos Center for Biomedical Imaging, Massachusetts General Hospital, Boston, Massachusetts.)

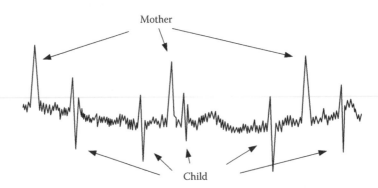

FIGURE 6.112
Example of the analysis of fetus MCG signal. (Courtesy of Low Temperature Division, University of Twente.)

are very difficult to perform by classic ECG method. The signal was determined in magnetically shielded room with noise level $10\,fT/Hz^{1/2}$ where the fetal heart signal was about $3\,pT$.

Similarly only magnetic method can be used for detection of the iron overload in the liver (the only alternative is biopsy). The liver of a healthy patient contains about $50\text{--}500\,\mu g$ of iron in 1 g of liver. To detect the liver pathology, the sensitivity of about $100\,fT/Hz^{1/2}$ is required.

References

Aird G.J.C., 2000, Modeling the induced magnetic signature of naval vessels, PhD dissertation, University of Glasgow, Scotland, U.K.

Almeida T.M., Piedale M.S., Cardoso F., Ferreira H.A., Freitas P.P., 2006, Characterization and modelling of a magnetic biosensor, *Instrumentation and Measurement Technology Conference*, Sorrento, Italy.

Amundson M., 2003, The role of compassing in telematics, *Analysis USA Focus*, **24**, 16–17.

Amundson M., 2006, Compass assisted GPS for LBS applications, Honeywell Application Note.

Andrä W., Nowak H. (Eds.), 2007, *Magnetism in Medicine*, Wiley-VCH Verlag, Berlin, Germany.

Augustyniak B., Piotrowski L., Chmielewski M.J., 2007, Creep damage zone detection in exploited power plant tubes with magnetoacoustic emission, *Przegl. Elektrotech.*, **83**(4), 93–98.

Augustyniak B., Piotrowski L., Radczuk M., Chmielewski M., Hauser H., 2003, Investigation of magnetic and magnetomechanical hysteresis properties of FeSi alloys with classical and mechanical Barkhausen effects and magnetoacousstic emission, *J. Appl. Phys.*, **93**, 7465–7467.

Augustyniak B., Sablik M.J., Landgraf F.J.G., Jiles D.C., Chmielewski M., Piotrowski L., Moses A.J., 2008, Lack of magnetoacoustic emission in iron with 6.5% silicon, *J. Magn. Magn. Mater.*, **320**, 2530–2533.

Back C.H., Heidmann J., McCord J., 1999, Time resolved Kerr microscopy: Magnetization dynamics in thin film write heads, *IEEE Trans. Magn.*, **35**, 637–642.

Baselt D.R. et al., 1998, A biosensor based magnetoresistance technology, *Biosens. Bioelectron.*, **13**, 731–739.

Baudouin P., Houbaert Y., Tumanski S., 2003, Magnetic local investigations of nonoriented electrical steels after tensile deformation, *J. Magn. Magn. Mater.*, **254–255**, 32–35.

Blitz J., 1997, *Electrical and Magnetic Methods of Nondestructive Testing*, Chapman and Hall, London, U.K.

Bott M.H.P., 1971, *The Interior of the Earth*, Elsevier Science, New York.

Braach B., 2005, New standards of exposure of magnetic fields, *Przegl. Elektrotech.*, **81**(9), 35–36.

Bray D.E., McBride D., 1992, *Nondestructive Testing Techniques*, John Wiley & Sons, New York.

Breiner S., 1999, *Applications Manual for Portable Magnetometers*, Geometrics Application Note.

Brzeska M. et al., 2004, Detection and manipulation of biomolecules by magnetic carriers, *J. Biotechnol.*, **112**, 25–33.

Butler R.F., 1998, *Paleomagnetism: Magnetic Domains to Geologic Terranes*, University of Arizona, Tucson, AZ, Electronic Edition.

Campbell W.H., 2003, *Introduction to Geomagnetic Fields*, Cambridge University Press, Cambridge, U.K.

Cardoso F.A. et al., 2006, Diode/magnetic tunnel junction cell for fully scalable matrix-based biochip, *J. Appl. Phys.*, **99**, 08B307.

Caruso M.J., 1995, Applications of magnetoresistive sensors in navigation systems, Honeywell Application Note.

Caruso M.J., 1996, Application of magnetic sensors for low cost compass systems, Honeywell Application Note.

Caruso M.J., Withanawasam L.S., 1999, Vehicle detection and compass applications using AMR magnetic sensors, Honeywell Application Note.

Cecco V.S., Van Drunen G., Sharp F.L., 1981, *Eddy Current Manual*, Report: AECL-7523, Chalk River Nuclear Laboratories, Chalk River, Ontario, Canada.

Chan M.L., Jaramillo G., Ahjeong S., Hristova K.R., Horsley D.A., 2009, Scanning magnetoresistance microscopy for imaging magnetically labeled DNA microarray, *IEEE Trans. Magn.*, **45**, 4816–4820.

Chan M.L., Jaramillo G.M., Horsley D.A., 2007, Scanning magnetic tunnel junction sensor for the detection of magnetically labeled microarray, *14th International Conference on Solid State Sensors, Actuators and Microsystems*, Lyon, France.

Chang A.M. et al., 1992, Scanning Hall probe microscopy, *Appl. Phys. Lett.*, **61**(16), 1974–1796.

Chiron G., Laj C., Pocachard J., 1981, A high sensitivity portable spinner magnetometer, *J. Phys. E*, **14**, 977–980.

Clark D.D., Espy M.A., Kraus R.H., Matlachov A., Lamb J.S., 2000, Weld quality evaluation using a high temperature SQUID array, *IEEE Trans. Appl. Supercond.*, **13**, 235–238.

De Sa A., 1963, A spinner magnetometer, *J. Sci. Instrum.*, **40**, 162–165.

Dinner R.B., Beasley M.R., Moler K.A., 2005, Cryogenic scanning Hall-probe microscope with centimeter scan range and submicron resolution, *Rev. Sci. Instrum.*, **76**, 103702.

Dodd C.V., 1977, Computer modeling for eddy current testing, in *Research Techniques in Nondestructive Testing*, Academic Press, New York.

Edelstein R.L. et al., 2000, The BARC biosensor applied to the detection of biological warfare agents, *Biosens. Bioelectron.*, **14**, 805–813.

Elsasser W.M., 1956, Hydromagnetic dynamo theory, *Rev. Modern Phys.*, **28**, 135–163.

Espy M.A., Matlashov A., Mosher J.C., Kraus R.H., 2001, Nondestructive evaluation with linear array of 11 HTS SQUIDS, *IEEE Trans. Appl. Supercond.*, **11**, 1303–1306.

Evans M.E., Heller F., 2003, *Environmental Magnetism*, Academic Press, San Diego, CA.

Fagaly R.L., 2006, Superconducting quantum interference device instruments and applications, *Rev. Sci. Instrum.*, **77**, 101101.

Farthing W.H., Foltz W.C., 1967, Rubidium vapor magnetometer for near earth orbiting spacecraft, *Rev. Sci. Instrum.*, **38**, 1023–1030.

Flanders P.J., 1973, Magnetic measurements with the rotating sample magnetometer, *IEEE Trans. Magn.*, **9**, 94–109.

Förster F., 1952, Theoretische und experimentalle Grundlagen der Zerstörungsfreine Werkstoffprüfung mit Wirbelstromverfahren, *Z. Metallkd.*, **43**, 163–171.

Foster J.H., 1966, A paleomagnetic spinner magnetometer using a fluxgate gradiometer, *Earth Planet. Sci. Lett.*, **1**, 463–466.

Frandsen A.M.A., Holzer R.E., Smith E.J., 1969, OGO search coil magnetometer experiment, *IEEE Trans. Geosci. Electron.*, **7**, 61–74.

Freeman M.R., Smyth J.F., 1996, Picosecond time resolved magnetization dynamics of thin film heads, *J. Appl. Phys.*, **79**, 5898–5900.

Gärtner S., et al., 2002, Nondestructive evaluation of aircraft structures with a multiplexed HTS rf SQUID magnetometer array, *Phys. C*, **372–376**, 287–290.

Graeger V., 1992, Aktiver Drehzahlksensor für ABS/ASR Systeme, *VDI Berichte*, 1009, 327–337.

Graham D.L., Ferreira H.A., Freitas P.P., 2004, Magnetoresistive based biosensors and biochips, *Trends Biotechnol.*, **22**, 455–462.

Graham D.L., Ferreira H.A., Freitas P.P., Cabral J.M.S., 2003, High sensitivity detection of molecular recognition using magnetically labeled biomolecules and magnetoresistive sensor, *Biosens. Bioelectron.*, **18**, 483–488.

Guy C., Ffytche D., 2005, *An Introduction to the Principles of Medical Imaging*, Imperial College Press, London, U.K.

Hasselbach K., Veauvy C., Mailly D., 2000, MicroSQUID magnetometry and magnetic imaging, *Phys. C*, **332**, 140–147.

Hiebert W.K., Stankiewicz A., Freeman M.R., 1997, Direct observation of magnetic relaxation in a small permalloy disc by time resolved scanning Kerr microscopy, *Phys. Rev. Lett.*, **79**, 1134–1137.

Howells G.D. et al., 1999, Scanning Hall probe microscopy of ferromagnetic structures, *J. Magn. Magn. Mater.*, **196–197**, 917–919.

Hubert A., Schäfer R., 1998, *Magnetic Domains*, Springer, Berlin, Germany

Hummervoll R., Totland O., 1980, An automatic spinner magnetometer for rock specimens, *J. Phys. E*, **13**, 931–935.

Hwang J., Lee J., Kwon S., 2009, The application of a differential type Hall sensor array to the nondestructive testing of express train wheels, *NDT E Int.*, **42**, 34–41.

ICNIRP, 1999, *Guidelines for Limiting Exposure to Time Varying Electric, Magnetic and Electromagnetic Fields*.

Jelinek V., 1966, A high sensitivity spinner magnetometer, *Studia Geophys. Geod.*, **10**, 58–78.

Jenks W.G., Sadeghi S.S.H., Wikswo J.P., 1997, SQUIDs for nondestructive evaluation, *J. Phys. D*, **30**, 293–323.

Jun J., Lee J., 2008, Nondestructive evaluation of a crack on austenitic steinless steel using a sheet type induced current and a Hall sensor array, *J. Mech. Sci. Technol.*, **22**, 1684–1691.

Kadzilko-Hofmokl M., 1983, Earth's magnetism, in *Encyclopedia for Modern Physic*, PWN.

Kaleta J., Tumanski S., Zebracki J., 1996, Magnetoresistors as a tool for investigations of the mechanical properties of ferromagnetic materials, *J. Magn. Magn. Mater.*, **160**, 199–200.

Kejik P., Boero G., Demierre M., Popovic R.S., 2006, An integrated micro-Hall probe for scanning magnetic microscopy, *Sens. Actuators*, **129**, 212–215.

Kirel L.A., Tsys O.T., Nagovilsyn V.S., Michailova O.M., Kamardin V.M., Michailova N.A., 2000, Magnetic tests of hardness and structure of rail steel, *Russ. J. Nondestr. Test.*, **36**, 675–680.

Kirtley J., 1996, Imaging magnetic fields, *IEEE Spectrum*, **33**, December, 41–48.

Kirtley J.R., 2002, SQUID microscopy for fundamental studies, *Phys. C*, **368**, 55–65.

Kirtley J.R., Wikswo J.P., 1999, Scanning SQUID microscopy, *Ann. Rev. Mater. Sci.*, **29**, 117–148.

Kittel H., Siegle H., May U., Farber P., 2003, Magnetoresistive Kfz-Sensoren zur Erfassung von Weg, Winkel and Drehzahl, *Proceedings and Symposium on Magnetoresistive Sensoren*, Lahnau, Germany, pp. 36–45.

Koch H., 2001, SQUID Magnetocardiography: Status and perspectives, *IEEE Trans. Appl. Supercond.*, **11**, 49–59.

Krause H.J., Gärtner S., Wolters N., Hohmann R., Wolf W., Schubert J., Zander W., Zhnag Y., 2001, Multiplexed SQUID array for nondestructive evaluation of aircraft structures, *IEEE Trans. Appl. Supercond.*, **11**, 1168–1171.

Krause H.J., Kreutzbruck M., 2002, Recent developments in SQUID NDE, *Phys. C*, **368**, 70–79.

Kryder M.H., Koeppe P.V., Liu F.H., 1990, Kerr effect imaging of dynamic processes in magnetic recording heads, *IEEE Trans. Magn.*, **26**, 2995–3000.

Kuperman V., 2000, *Magnetic Resonance Imaging*, Academic Press, San Diego, CA

Kwiatkowski W., Tumanski S., 1984, Application of the thin film permalloy magnetoresistive sensors in electrical measurements, *IEEE Trans. Magn.*, **20**, 966–968.

Lanza R., Meloni A., 2006, *Geomagnetism. Principles and Applications—An Introduction to Geologists*, Springer, Berlin, Germany.

Lovejoy D., 1993, *Magnetic Particle Inspection*, Kluwer Academic Publishers, Dordrecht, the Netherlands

Libby H.L., 1979, *Basic Principles and Techniques of Eddy Current Testing*, Robert E. Krieger Publishing Company, Huntington, NY.

Lopes P.A.C. et al., 2010, Measuring and extraction of biological information on new handheld biochip-base microsystem, *IEEE Trans. Instrum. Meas.*, **59**, 56–62.

Mandal K., Dufour D., Atherton D.L., 1999, Use of magnetic Barkhausen noise and magnetic flux leakage signals for analysis of defects in pipeline steels, *IEEE Trans. Magn.*, **35**, 2007–2017.

Mathai A., Song D., Gim Y., Wellstood F.C., 1992, One-dimensional magnetic flux microscope based on the DC superconducting quantum interference device, *Appl. Phys. Lett.*, **61**, 598–600.

Mathai A., Song D., Gim Y., Wellstood F.C., 1993, High resolution magnetic microscopy using a DC SQUID, *IEEE Trans. Appl. Supercond.*, **3**, 2609–2612.

Matsuda M., Nakamura K., Mikami H., Kuriki S., 2005, Fabrication of magnetometers with multiple SQUID arrays, *IEEE Trans. Appl. Supercond.*, **15**, 817–820.

Mayers J., 2002, Magnetic couplers in industrial systems, *Sens. Technol. Des.*

Mazumdar D., Xiao G., 2005, Scanning magnetoresistive microscopy study of quasi-static magnetic switching in mesoscopic square dots: Observation of field-driven transition between flux-closure states, *IEEE Trans. Magn.*, **41**, 2226–2229.

Mercier D., Lesage J., Decoopman X., Chicot D., 2006, Eddy current and hardness testing for evaluation of steel decarburizing, *NDT E Int.*, 39, 6520660.

Miller M.M. et al., 2001, A DNA array sensor utilizing magnetic microbeads and magnetoelectronic detection, *J. Magn. Magn. Mater.*, **225**, 138–144.

Miyashita K., 1990, Non-contact magnetic torque sensor, *IEEE Trans. Magn.*, **26**, 1560–1562.

Morris P., 1970, Three new high sensitivity spinner magnetometers for paleomagnetic research, *J. Phys. E*, **3**, 819–821.

Moses A.J., Williams P.I., Hoshtanar O.A., 2005, A novel instrument for real time dynamic domain observation in bulk and micromagnetic materials, *IEEE Trans. Magn.*, **41**, 3736–3738.

Moses A.J., Williams P.I., Hoshtanar O.A., 2006, Real time dynamic domain observation in bulk materials, *J. Magn. Magn. Mater.*, **304**, 150–154.

Nelson T.J., 1986, Shear-sensitive magnetoresistive robotic tactile sensor, *IEEE Trans. Magn.*, **22**, 394–396.

Ness N.F., 1970, Magnetometers for space research, *Space Sci. Rev.*, **11**, 459–554.

Ness N.F., Behannon K.W., Lepping R.P., Schatten K.H., 1971, Use of two magnetometers for magnetic measurements on the spacecraft, *J. Geophys. Res.*, **76**, 3564–3573.

Nowak H., 2007, Biomagnetic instrumentation, in *Magnetism in Medicine*, Andrä W., Nowak H. (Eds.), Wiley-VCH Verlag, Weinheim, Germany, Chapter 2.2.

O'Barr R., Lederman M., Schultz S., 1996, A scanning microscopy using a magnetoresistive head as the sensing element, *J. Appl. Phys.*, **79** (8), 6067–6069.

Oral A., Bending S.J., Henini M., 1996a, Scanning Hall probe microscope of a superconductors and magnetic materials, *J. Vac. Sci. Technol.*, **B14**(2), 1202–1205.

Oral A., Bending S.J., Henini M., 1996b, Real time scanning Hall probe microscopy, *Appl. Phys. Lett.*, **69**, 1324–1326.

Peterka D., Enders A., Haas G., Kern K., 2003, Combined Kerr microscope and magnetic force microscope for variable temperature ultrahigh vacuum investigations, *Rev. Sci. Instrum.*, **74**, 2744–2748.

Petford-Long A.K., 2001, Magnetic imaging, *Lect. Notes Phys.*, **569**, 316–331.

Phillips G.N., Eisenberg M., Draaisma E.A., Abelmann L., Lodder J.C., 2002, Performance of focused beam trimmed yoke-type magnetoresistive heads for magnetic microscopy, *IEEE Trans. Magn.*, **38**, 3528–3535.

Piotrowski L., Augustyniak B., Chmielewski M., 2010, On the possibility of the application of magnetoacoustic emission intensity measurements for the diagnosis of thick-walled objects in the industrial environment, *Meas. Sci. Technol.*, **21**, 035702.

Pizzella V., Della Pena S., Del Gratta C., Romani G.L., 2001, SQUID systems for biomagnetic imaging, *Supercond. Sci. Technol.*, **14**, R79–R114.

Ripka P., Wčelak J., Kašpar P., Lewis A.M., 2006, Bomb detection in magnetic soils: AC versus DC methods, *EXCO IEEE Sensors Conference*, Daegu, Korea.

Sabaka J., Olsen N., Langel L.A., 2000, A comprehensive model of the near-Earth magnetic field, *Nasa Report*, NASA/TM-2000-209894.

Sahoo D.R., Sebastian A., Häberle W., Pozidis H., Elefheriou E., 2009, Magnetoresitive sensor based scanning probe microscopy, *9th Nanotechnology Conference*, Genova, Italy, pp. 862–865.

Sandhu A., Kurosawa K., Dede M., Oral A., 2004, 50 nm Hall sensor for room temperature scanning Hall probe microscopy, *Jap. J. Appl. Phys.*, **43**(2), 777–778.

Sandhu A., Masuda H., Oral A., Bending S.J., 2001, Room temperature sub-micron magnetic imaging by scanning Hall probe microscopy, *Jpn. J. Appl. Phys.*, **40**, 4321–4324.

Sandhu A., Masuda H., Oral A., Bending S.J., Yamada A., Konagai M., 2002, Room temperature scanning Hall probe microscopy using GaAs/AlGaAs and Bi micro-Hall probes, *Ultramicroscopy*, **91**, 97–101.

Sebastia J.P., Lluch J.A., Vizcaino J.R.L., 2007, Signal conditioning for GMR sensors applied to traffic speed monitoring sensors, *Sens. Act.*, **A137**, 230–235.

Siegel R., 2002, Land mine detection, *IEEE Instr. Meas.*, **5**, 22–28.

Sharawi M.S., Sharawi M.I., 2007, Design and implementation of a low cost VLF metal detector with metal type discrimination capabilities, *IEEE International Conference on Signal Processing and Communications*, Dubai, pp. 480–483.

Shen W., Schrag B.D., Carter M.J., Xiao G., 2008, Quantitative detection of DNA labeled with magnetic nanoparticles using arrays of MgO-based magnetic tunnel junction sensors, *Appl. Phys. Lett.*, **93**, 033903.

Shimizu M., Saitoh E., Miyajima H., Masuda H., 2004, Scanning Hall probe microscopy with high resolution of magnetic field image, *J. Magn. Magn. Mater.*, **282**, 369–372.

Shull P.J., 2002, *Nondestructive Evaluation*, Marcvel Dekker, New York.

Slocum R.E., Reilly F.N., 1963, Low field helium magnetometer for space applications, *IEEE Trans. Nucler Sci.*, **10**, 165–171.

Smekalova T.N., Voss O., Smekalov S.L., 2008, *Magnetic Surveying in Archeology*, House of Publishing of St. Petersburg State University, St. Petersburg, Russia

Smith E.J., Connor B.V., Foster G.T., 1975, Measuring of the magnetic field of Jupiter and the outer solar system, *IEEE Trans. Magn.*, **11**, 962–980.

Smith E.J., Sonett C.P., 1976, Extraterrestrial magnetic fields: Achievements and opportunities, *IEEE Trans. Geosci. Electron.*, **14**, 154–171.

Staton D.J., Ma Y.P., Sepulveda N.G., Wikswo J.P., 1991, High resolution magnetic mapping using a squid magnetometer array, *IEEE Trans. Magn.*, **27**, 3237–3240.

Sternickel K., Braginski A.I., 2006, Biomagnetism using SQUIDs: Status and perspectives, *Supercond. Sci. Technol.*, **19**, S160–S171.

Szmaja W., 2006, Recent developments in the imaging of magnetic domains, in *Advances in Imaging and Electron Physics*, P.W. Hawkes (Ed.), Vol. 141, pp. 175–256.

Takezaki T., Yasigawa D., Sueoka K., 2006, Magnetic field measurement using scanning magnetoresistance microscope with spin-valve sensor, *Jap. J. Appl. Phys.*, **45**(3B), 2251–2254.

Tauxe L., 2010, *Essentials of Paleomagnetism*, University of California Press, Berkeley.

Treutler C.P.O., 2000, Magnetic sensors for automotive applications, *Proceedings of 3rd European Conference on Magnetic Sensors and Actuators*, Dresden, Germany.

Uchiyama T., Mohri K., Itho H., Nakashima K., Ohuchi J., Sudo Y., 2000, Car traffic monitoring system using MI sensor built-in disk set on the road, *IEEE Trans. Magn.*, **36**, 3670–3672.

Vlaardingerbroek M.T., den Boer J.A., 2003, *Magnetic Resonance Imaging*, Springer, Berlin, Germany

Vu L.N., Van Harlingen D.J., 1993, Design and implementation of a scanning microscope, *IEEE Trans. Appl. Supercond.*, **3**, 1918–1921.

Weiss B.P., Kirschvink J.L., Baudenbacher F.J., Vali H., Peters N.T., Macdonald F.A., Wikswo J.P., 2000, A low temperature transfer of ALH8400 from Mars to Earth, *Science*, **290**, 791–795.

Wellhausen H., 1989, Elektronischer Kompass, *Elektronik*, **8**, 85–89.

Wellstood F.C., Gim Y., Amar A., Black R.C., Mathai A., 1997, Magnetic microscopy using SQUID, *IEEE Trans. Appl. Supercond.*, **7**, 3134–3138.

Weymouth J.W., 1985, Geological surveying of archeological sites, in *Archeological Geology*, G. Rapp (Ed.), Yale University Press, New Haven and London.

Wikswo H., 1991, A review of SQUID magnetometry applied to nondestructive evaluation, *IEEE Trans. Magn.*, **27**, 3231–3234.

Wikswo J.P., 1995, SQUID magnetometers for biomagnetism and nondestructive testing: Important questions and initial answers, *IEEE Trans. Appl. Supercond.*, **5**, 74–120

Wolff J., Heuer T., Gao H., Weinmann M., Voit S., Hartmann U., 2006, Parking monitor system based on magnetic field sensors, *IEEE Intelligent Transportation Systems Conference*, Toronto, pp. 1275–1279.

Yamamoto S.Y., Schultz S., 1997, Scanning magnetoresistance microscopy: Imaging with a MR head, *J. Appl. Phys.*, **81**(8), 4696–4698.

Yamamoto S.Y., Vier D.C., Schultz S., 1996, High resolution contact recording and diagnostics with a raster-scanned MR head, *IEEE Trans. Magn.*, **32**, 3410–3412.

Zurek S., Marketos P., Tumanski S., Patel H.V., Moses A.J., 2009, Correlation between surface magnetic field and Barkhausen noise in grain oriented electrical steel, *Przegl. Elektrotech.*, **85**(1), 111–114.

Index

Printed and bound by CPI Group (UK) Ltd, Croydon, CR0 4YY

01/11/2024

01782605-0019